From the Quantum to the Multiverse

Probing the Universe

2nd edition

Don Hainesworth, M.Sc., M.Eng.

© 2025 Don Hainesworth, M.Sc., M.Eng.. All rights reserved.

No part of this book may be reproduced, stored in a retrieval system, or transmitted by any means without the written permission of the author.

AuthorHouse™
1663 Liberty Drive
Bloomington, IN 47403
www.authorhouse.com
Phone: 833-262-8899

Because of the dynamic nature of the Internet, any web addresses or links contained in this book may have changed since publication and may no longer be valid. The views expressed in this work are solely those of the author and do not necessarily reflect the views of the publisher, and the publisher hereby disclaims any responsibility for them.

Any people depicted in stock imagery provided by Getty Images are models, and such images are being used for illustrative purposes only.
Certain stock imagery © Getty Images.

This book is printed on acid-free paper.

ISBN: 979-8-8230-4120-1 (sc)
ISBN: 979-8-8230-4158-4 (e)

Library of Congress Control Number: 2025900712

Print information available on the last page.

Published by AuthorHouse 01/30/2025

authorHOUSE®

FROM THE QUANTUM TO THE MULTIVERSE

Probing the Universe
2ND Edition

Don Hainesworth, M.Sc., M.Eng.

Preface

Quantum theory is among the great intellectual achievements of the 20th century, and how this came about is interesting in itself. Quantum theory was once widely held to resist any realist interpretation and to mark the advent of a postmodern science characterized by paradox, uncertainty, and the limits of precise measurement. It seems that there is a realm of reality in the subatomic or micro-physical domain.

The success of the *Aspect Experiment* in Paris in 1982 marked the end of the contemplation period, with the first direct experimental proof that even the most unusual aspects of QM are a literal description of the way things really are in the real world.

Further analyses of QM have led scientists to ponder the possibility of multiple dimensions.
Extra dimensions have changed the way physicists think about the Universe. And because the connections of extra dimensions in the Cosmos could connect to many more well-established physics ideas; extra dimensions are a way to approach older, already verified facts about the Universe. As a consequence of this, physicists have postulated the real possibility of parallel universes.

Introduction

The story is largely told in chronological order, but where appropriate it gives way to a more logical account in which complete topics are presented in largely self-contained units. The main focus in the writing of this book has been to communicate the central ideas and scientific concepts of the nature of the Quantum and of the Multiverse. I have attempted to present a comprehensive overview of the subject at a level which carries the reader beyond the generalizations necessary in philosophy and science books. The text focuses on the logical development of the individual topics and gives only the main historical interconnections.

The chapters in the book contains no advanced scientific or mathematical concepts, and no complicated scientific formulas are written down for analysis. However, there are some scientific related concepts in various chapters that overlap into subsequent chapters. This is stressed in order to reinforce a particular idea and maintain historical continuity.

It is not assumed that the reader has an understanding of Quantum Physics theory. Therefore, the text provides the reader with enough historical and philosophical information to secure his or her confidence in understanding the behavior and makeup of the Quantum and of the Multiverse.

Contents

Ch 1—From Cosmic Designers to Natural Philosophers ... 1

Ch 2—Ancient Greek philosophies concerning Nature ... 3

Ch 3—Astrology & Astronomy ... 16

Ch 4—Classical Physics ... 44

Ch 5—Optics and Newtonian Mechanics ... 59

Ch 6—The Nature of Light ... 75

Ch 7—Chemistry of the Elements ... 93

Ch 8—Electricity and Magnetism ... 130

Ch 9— Radioactivity and the Atom ... 146

Ch 10—Light Quanta ... 158

Ch 11—Special Relativity (Space and Time) ... 165

Ch 12—Early Quantum Physics ... 203

Ch 13—Quantum Physics ... 225

Ch 14— Many Worlds Theory & Philosophies and Paradoxes in Physics ... 279

Ch 15—Extra Dimensions ... 332

Ch 16—Parallel Worlds ... 338

Ch 17—The Fabric of Space-Time ... 342

Ch 18—In Search of the Multiverse ... 363

Ch 19— The latest discoveries of the Cosmos ... 377

Ch 20— Recent Advancements in Observational Technology ... 389

Ch 21—Development of advanced telescopes - James Webb Space Telescope (JWST) Extremely Large Telescope (ELT) B. Enhanced imaging and data analysis techniques ... 394

Ch 22— Discovery of primordial galaxies ... 398

Ch 23—Dark matter and dark energy ... 402

Ch 24—Evidence supporting the existence of dark matter & Implications of dark energy on cosmic expansion ... 406

Ch 25—Black holes and gravitational waves recent black hole observations ... 409

Ch 26—LIGO's ongoing discoveries of gravitational waves and Exoplanet research ... 413

Epilogue ... 417

Glossary ... 456

References ... 468

Ch 1—From Cosmic Designers to Natural Philosophers

1.1 A great cosmic Designer

On Earth, fossil evidence might support the idea of a Great Designer. Perhaps some species are eliminated when the Designer is dissatisfied, and new, improved versions are created. However, this concept raises some concerns. Wouldn't an all-knowing Designer be capable of creating the perfect species from the start? The fossil record suggests a process of trial and error in the development of organisms, reflecting a lack of foresight—qualities that seem at odds with the notion of a skilled cosmic Designer.

1.2 The observation of natural patterns

Around the world, various cultures have created myths to explain the origins and evolution of the Universe. While these creation stories differ in many ways, some share common themes in their descriptions of how the Universe began. These belief systems emerged, in part, to help humans make sense of their surroundings and their place in the vast Universe. Similarly, science has developed as a tool for understanding nature and its processes. Although both the mystical believer and the scientist seek to understand our world, the methods they use to achieve this shared goal are very different.

It wasn't until the 6th century BC, during a surge of intellectual tolerance, that philosophers were, for the first time, free to set aside traditional mythological explanations of the macrocosm and develop their own rational theories. In prehistoric times, natural patterns were identified through constant observation of the Cosmos. For instance, it was noted that the Sun consistently rises in the east and sets in the west. It was also observed that different constellations appear in different seasons, with the same constellations always rising at the beginning of fall. A new constellation never suddenly emerged from the east. There was a clear order in the movements of the heavens, with predictable patterns in the motions of the planets and stars.

1.3 The beginning of the Universe

The origin of the macrocosm had been debated for millennia. In various early cosmologies within Christian, Jewish, and Muslim traditions, the Universe was believed to have begun relatively recently. St. Augustine, in his book *The City of God*, presented an argument about the creation of the Universe. He observed that civilization is advancing and suggested that humanity, and possibly the Universe as well, could not have existed forever. If it had, he argued, we would have made far more progress by now.

St. Augustine accepted 5000 B.C. as the date of the Universe's creation, aligning with the account in the book of Genesis from the Old Testament. This date is not far from the last Ice Age, around 10,000 B.C., when civilization as we know it began. In contrast, Greek philosophers like Aristotle were uncomfortable with the concept of a creation or first cause, as it implied too much divine intervention. They believed that both humanity and the Universe were eternal, having existed forever. To address the argument of progress, they proposed that periodic floods and other natural disasters repeatedly reset human civilization to its early stages. However, this explanation fails to consider how or why humanity has never faced extinction during these numerous setbacks.

1.4 Precision of measurement

For centuries, humans hunted wild animals that migrated with the changing seasons. Food was abundant at certain times but scarce at others. The invention of agriculture ensured that crops could be planted and harvested at the right times, and annual gatherings of nomadic tribes were scheduled accordingly. The ability to interpret the "cosmic calendar" became a matter of life and death.

Over time, we have gained much knowledge from our ancestors. They discovered that the more accurately one understood the positions and movements of the Sun, Moon, and stars, the better one could predict when to hunt, sow, and harvest. As measurement precision improved, record-keeping became essential, leading to the development of astronomy from the need for accurate observation of the heavens. This, in turn, spurred the creation of mathematics and writing. For our ancestors, understanding cosmic and earthly phenomena was crucial for survival. They also built devices to track the passing of seasons. Our nomadic ancestors likely felt a deep connection to the heavens, venerating the Sun, Moon, and planets. However, many centuries later, the rise of scientific inquiry began to challenge mysticism and superstition, transforming what had been largely an empirical science.

Apart from the Sun and the Moon, only five planets were visible to the naked eye. To the ancients, these planets, along with the stars, comprised the entire known universe. Over time, they came to believe that the Sun, Moon, and stars directly influenced the Earth and all its inhabitants. This belief eventually developed into what we now know as astrology.

Astrology claims that the position of the planets in certain constellations at the time of your birth significantly shapes your future. Astrologers studied the movements of the planets, wondering if events would repeat when, for instance, Venus rose in the constellation of Capricorn, as it had in the past. This deterministic view of the Cosmos developed across many cultures. However, astrology's accuracy can be challenged. Take, for example, the lives of twins. There are numerous instances where one twin dies in childhood while the other lives to old age. These twins were born in the same place, perhaps within seconds of each other, with the same planets rising at their birth. If astrology were valid, how could such twins have such drastically different destinies?

Ch 2—Ancient Greek philosophies concerning Nature

2.1 Cosmology - the study of the Universe

For millennia, theories about the nature of the heavens have existed. The word "Universe" means "all that is." Humans have long pondered the structure of the Universe, its origins, and its future. Today, cosmologists and theologians find little common ground on these questions. The relationship between the Church and scientists on matters of the Universe has not always been harmonious.

Cosmos is a Greek term referring to the order and harmony of the Universe, contrasting with *Chaos*. It suggests a profound interconnectedness among all natural elements. Cosmology is the study of the Universe's nature, while cosmogony focuses on its origin and evolution.

Primitive humans must have pondered deeply as they gazed at the night sky. Ancient philosophers commonly believed that the stars were fixed on a single, massive sphere surrounding the central planet Earth. Early societies observed that the Sun moved across the heavens with regularity. Similarly, the stars appeared to revolve around the Earth at roughly the same rate as the Sun. Given that there was no sensation of motion on Earth, it was assumed that only the Sun and the stars were in motion, while the Earth remained stationary. Consequently, the earliest cosmologists placed the Earth at the center of a constantly revolving Universe.

One of the earliest and most rudimentary models of the Universe portrayed the stars as being fixed to a distant, solid sphere that completed a full rotation in 23 hours and 56 minutes. The Sun and Moon were associated with two inner spheres closer to Earth. The Sun's sphere was believed to complete a rotation every 24 hours, while the Moon's sphere had a rotation period of approximately 24 hours and 50 minutes.

For centuries, many cultures, including the Greeks, meticulously recorded the movements of the celestial lights in the night sky. They observed that while nearly all of the thousands of visible lights appeared to move together across the sky, five of them, excluding the Moon, did not follow this pattern. The Greeks identified only these five distinct lights because they were the only ones visible to the naked eye.

These unusual objects moved unpredictably across the night sky, sometimes straying from a regular path and then returning. Through careful observation, their true nature was eventually revealed. They were identified as *"wandering stars"* or *"planets,"* with the term "planet" derived from the Greek word for *"wanderer."* These planets were named after various pagan gods: Mercury, the messenger; Venus, the goddess of love; and Mars, the god of war. Jupiter was regarded as the chief deity, while Saturn was considered Jupiter's father.

Aristotle

These planets, along with the Sun and the stars, were thought to be part of geocentric spheres. As these spheres rotated independently at different rates in the Cosmos, some ancient philosophers began to believe that they might collide and produce sounds. A few even claimed to hear a faint, musical sound emanating from the celestial spheres. They speculated that this could only be the music of angels and other heavenly beings, creating a harmonious symphony throughout the Universe.

2.2 The Alexandrian library

In Alexandria, a renowned city in Egypt founded by Alexander the Great around 300 B.C., the intellectual pursuit that eventually led to space exploration began. Alexander, a great Greek military leader, and his former bodyguard established the city, promoting respect for diverse cultures and the open-minded pursuit of knowledge. He honored the gods of other nations and even collected exotic creatures, including an elephant for his teacher, Aristotle. The city was known for its grand architecture and style.

The greatest marvel of Alexandria was its library and its associated scholarly community. Today, only the damp and forgotten cellar of the Serapeum, once a hub for scholars, remains of that legendary library. The scholars there explored the entire Cosmos.

The library was a hub for scholars and researchers delving into philosophy, physics, astronomy, mathematics, biology, medicine, and literature. It was here that humans first began to systematically compile and organize world knowledge. The Greek kings of Egypt who succeeded Alexander were deeply committed to learning, supporting research and fostering an environment conducive to intellectual excellence. The library featured ten large research halls, each dedicated to a different subject, and included botanical gardens, a zoo, dissecting rooms, an observatory, and a grand dining hall for critical discussions.

At the heart of the library was its vast collection of books. Organizers searched through all the cultures and languages of the world, acquiring libraries from various countries. Commercial ships arriving in Alexandria were inspected by officials not for contraband but for books. Scrolls were borrowed, copied, and then returned. It is estimated that the library housed over half a million volumes, each a handwritten papyrus scroll. However, as classical civilization declined, the library and its immense collection were destroyed, leaving only a small fraction of its works and a few scattered fragments.

2.3 The origins of modern astronomy

The foundations of modern astronomy date back to the ancient Greeks around 340 B.C. At that time, the prevailing view was that the Earth was a flat disk encircled by the Sun, stars, and other celestial objects. However, some thinkers disagreed with this view. One such dissenter was the Greek philosopher Aristotle, who argued in his book *On the Heavens* that the Earth was a sphere rather than a flat plate.

In his book, Aristotle used lunar eclipses as a key argument. He reasoned that these eclipses occurred when the Earth moved between the Moon and the Sun, casting its shadow on the Moon. Aristotle observed that the Earth's shadow was always round, which would be expected if the Earth were a sphere. In contrast, if the Earth were a flat disk, its shadow would only appear round if the eclipse occurred when the Sun was directly beneath the center of the disk. At other times, the shadow would be elongated into an elliptical shape (an elongated circle or oval).

In addition to Aristotle's arguments, the Greeks had another reason to believe the Earth was round. If the Earth were flat, a ship approaching from the horizon would first appear as a tiny, featureless dot. As it came closer, details like its sails and hull would gradually become visible. However, what is actually observed is that the sails of the ship appear first, followed by the hull. The fact that the ship's mast, which is higher than the hull, is the first part to become visible over the horizon suggests that the surface of the water is curved, indicating that the Earth is round.

Aristotle also believed that the Earth was stationary, with the Sun, Moon, planets, and stars moving in circular orbits around it. He held this view because he felt, for mystical reasons, that the Earth was the center of the Universe, and that circular motion was the most perfect. In the 2nd century A.D., the Greek scholar Ptolemy, deeply invested in his studies, developed this concept into a comprehensive model of the heavens.

Other notable Greeks included *Hipparchus*, who estimated star brightness and mapped constellations; *Euclid*, who systematized geometry; Herophilus, who identified the brain, not the heart, as the seat of intelligence; *Heron of Alexandria*, who invented gear trains and steam engines and wrote *Automata*, the first book on robotics; *Apollonius of Perga*, who described conic sections like ellipses, parabolas, and hyperbolas, which shape the orbits of celestial bodies; *Archimedes*, the greatest mechanical genius before *Leonardo da Vinci*; and *Ptolemy*, who compiled much of what is now considered pseudo-science in astrology. Among these great figures was Hypatia, a mathematician and astronomer, the last luminary of the library, whose death was tied to the library's destruction seven centuries after its founding.

2.4 Aristarchus of Samos

Aristarchus (310–230 BC) was a Greek astronomer and mathematician from the island of Samos, Greece. He was the first to propose a heliocentric model of the solar system, placing the Sun at the center instead of the Earth. Despite this revolutionary idea, his theories were largely overshadowed by the geocentric views of Aristotle and Ptolemy until nearly 1,800 years later, when Copernicus revived and expanded upon them, with further contributions from Johannes Kepler and Isaac Newton. Today, we have greatly advanced beyond the science of the ancient world. Thanks to our continued exploration, we now understand that the Universe is far older than Aristarchus could have imagined—approximately 15 to 20 billion years old, dating back to the explosive event known as the Big Bang.

In the early Universe, there were no galaxies, stars, planets, or any form of life—just a uniform, radiant fireball filling all of space. The transition from the chaotic aftermath of the Big Bang to the organized Cosmos we are now beginning to understand is the most incredible transformation of matter and energy we've ever glimpsed. As descendants of the Big Bang, we are driven to comprehend the Cosmos from which we emerged.

2.5 The Ionian influence

Over time, the Ionian influence and experimental methods spread to mainland Greece, Italy, and Sicily. In an era when the concept of air was barely understood—people knew about breathing, but the idea of an invisible substance like wind was almost unimaginable—the first recorded experiment on air was conducted by a physician named Empedocles, who lived around 450 B.C. Some accounts suggest he claimed to be a god, though it may have been that his intelligence led others to think so. Empedocles believed that light travels very fast, though not infinitely so. He also taught that there was once a much greater variety of living beings on Earth, but many species must have failed to survive. In his view, only those species with traits like craftiness, courage, or speed managed to persist. Like Anaximander and Democritus, Empedocles anticipated some aspects of Darwin's theory of evolution by natural selection.

Empedocles, through an experiment, discovered the existence of an invisible substance—air—which he believed to be matter so finely divided that it could not be seen. This early notion of atoms was later expanded upon by Democritus, a philosopher from the Ionian colony of Abdera in northern Greece. Democritus proposed that numerous worlds had spontaneously formed from diffused matter in space, evolved, and eventually decayed. Remarkably, at a time when impact craters were unknown, he theorized that worlds could sometimes collide. He envisioned some worlds drifting alone through the darkness of space, while others were accompanied by multiple suns and moons. He also speculated that some worlds were inhabited, while others lacked plants, animals, or even water. Democritus believed that the simplest forms of life emerged from a kind of primordial ooze. He taught that perception—such as the belief that one is holding a pen—was purely a physical and mechanistic process, and that thinking and feeling were attributes of matter arranged in a sufficiently fine and complex way, rather than the result of some spirit infused into matter by the gods.

Democritus

2.6 Atoms - Greek atomic physics

Many popular accounts of the history of science credit the ancient Greeks with the early conception of atoms, often celebrating them for their insight into the true nature of matter. However, this view is somewhat exaggerated. It is accurate that Democritus of Abdera, who died around 370 B.C., proposed that the complex nature of the world could be understood if everything was composed of different kinds of unchangeable atoms, each with its own unique shape and size.

Democritus, who lived in the 5th century B.C., introduced the concept that everything in nature is composed of small, indivisible particles called atoms. He believed that these atoms combine to form all the substances we observe in the world, famously stating, "The only existing things are atoms and empty space; all else is mere opinion." Although Democritus was more of a philosopher than a scientist—lacking the mathematical tools to calculate how atoms combine—he credited his teacher, Leucippus, with the idea, though it is uncertain whether Leucippus existed. Contrary to popular belief, the concept of the atom did not originate in the modern Western world. Democritus developed thought experiments to demonstrate the validity of his ideas. For instance, he argued that when cutting an apple, the knife must pass through empty spaces between atoms; otherwise, the knife would encounter impenetrable atoms, making the apple impossible to cut. In another exercise, Democritus conceptualized calculating the volume of a cone or pyramid by stacking an extremely large number of thin plates, tapering in size from the base to the apex. This idea laid the groundwork for the "theory of limits," a precursor to differential and integral calculus. Unfortunately, most of Democritus' work was destroyed, delaying the development of calculus until the time of Isaac Newton. Had his writings survived, the foundations of calculus might have been established much earlier, perhaps even by the time of Christ.

In contrast to Democritus' theory of indivisible atoms, the more influential philosopher Aristotle proposed that everything in the Universe was composed of four fundamental "elements": fire, earth, air, and water. Aristotle's theory was far more popular and enduring, leading to the near abandonment of the atomic theory by the time of Christ. Since there was no mathematical model or empirical way to measure atoms in ancient Greece, Democritus' idea remained purely speculative. Laboratory experimentation, which could have provided evidence for atoms, didn't exist at the time. It wasn't until the early 17th century, when Galileo developed the scientific method, and later in the 20th century, when Albert Einstein provided evidence for atoms, that Democritus' concept was validated.

Democritus led a somewhat unconventional life. He was uncomfortable around women, children, and romantic pursuits, largely because he felt they distracted from his intellectual endeavors. However, he deeply valued friendship and believed that cheerfulness was the goal in life, even dedicating significant philosophical work to understanding the nature and origins of enthusiasm. Despite traveling to Athens to meet Socrates, he was too shy to approach him.

Democritus was a close friend of Hippocrates and held strong views on governance, preferring poverty in a democracy over wealth in a tyranny. He was critical of the prevailing religions of his time, considering them harmful. He also rejected the belief in immortal souls and gods, maintaining that such ideas did not exist.

The ancient natural philosophers, from Thales to Democritus, are often referred to as "Pre-Socratic" in historical and philosophical accounts. They maintained a dominant philosophical approach until the arrival of Socrates, Plato, and Aristotle, possibly influencing them to some degree. However, the Ionians followed a distinct and often opposing tradition, one that aligns more closely with modern scientific thought. Their influence, though powerful, lasted only two or three centuries, and the loss of their insights between the Ionian Awakening and the Renaissance is an irreplaceable void in human intellectual history.

Pythagoras

Before the rise of great philosophers like Socrates, Plato, and Aristotle, there was Pythagoras, arguably one of the most influential figures associated with the island of Samos. A contemporary of Polycrates in the 6th century B.C., local tradition claims that Pythagoras spent time living in a cave on Mount Kerkis. He is also credited with being the first person to deduce that the Earth is a sphere. His reasoning may have stemmed from observing the curved shadow of the Earth during a lunar eclipse or noticing how ship masts were the last to disappear over the horizon, suggesting the Earth's curvature.

Pythagoras, or perhaps his disciples, is credited with discovering the Pythagorean Theorem, which states that the sum of the squares of the two shorter sides of a right triangle equals the square of the longer side. He developed a method of mathematical deduction to prove this theorem in general terms. Science owes him the honor of establishing the modern tradition of mathematical argument, which is foundational to all scientific inquiry. Additionally, Pythagoras was the first to use the term "Cosmos" to describe the Universe as a well-ordered and harmonious system.

Many Ionians believed that the Universe's harmony could be uncovered through observation and experimentation, a method central to modern science. Pythagoras, however, took a different approach. He argued that the laws of nature could be understood through pure reasoning rather than experiments. He and his followers were mathematicians, not experimentalists, and their views were shaped by mysticism, as Pythagoras founded a religious order. This order eventually gained control of the state, establishing a governance based on the rule of spiritual leaders.

The Pythagoreans reveled in the precision of mathematical proofs, seeing in the simple relationships of right triangles a glimpse of cosmic order. This abstract, perfect world of mathematics stood in stark contrast to the chaotic everyday reality. They believed their mathematical discoveries revealed a divine realm, a perfect reality of the gods, with our world being just an imperfect reflection. Their ideas would deeply influence Plato and, later on, Christianity.

The Pythagoreans were captivated by regular solids—symmetrical three-dimensional shapes with faces that are all the same regular polygon. The cube, with its six square faces, is the simplest example. While there are countless regular polygons, only five regular solids exist. Among these, the dodecahedron, with its twelve pentagonal faces, was considered particularly enigmatic and potentially dangerous due to its mystical association with the Cosmos. The Pythagoreans linked the other four regular solids to the four classical elements—earth, fire, air, and water. They speculated that the fifth solid represented a "fifth" element, which they believed to be the substance of the heavenly bodies.

The Pythagoreans, enamored with whole numbers, believed that everything, including all other numbers, could be derived from them. However, they faced a doctrinal crisis when they discovered that the square root of two ($\sqrt{2}$)—the ratio of the diagonal to the side of a square—was irrational. This meant that $\sqrt{2}$ could not be expressed as the ratio of any two whole numbers, regardless of their size. In mathematical terms, "irrational" originally referred only to numbers that cannot be expressed as ratios. For the Pythagoreans, however, it signified a troubling challenge to their worldview, echoing the modern sense of "irrational" as something fundamentally unsettling. Rather than sharing these groundbreaking discoveries, the Pythagoreans chose to suppress knowledge of √2 and the dodecahedron, keeping these revelations hidden from the outside world.

The Pythagoreans considered the sphere to be the epitome of perfection, as all points on its surface are equidistant from its center. They also regarded circles as perfect shapes. Accordingly, they believed that planets moved in circular paths at constant speeds, insisting that any variation in speed during an orbit would be improper for the Cosmos. To them, non-circular motion seemed flawed and unsuitable for the celestial bodies orbiting the Earth.

A disdain for practical matters prevailed in the ancient world. Influenced by the Pythagoreans, Plato argued that while astronomers should contemplate the heavens, they should not spend time observing them. However, there were notable exceptions to this Pythagorean philosophy. One significant exception was their fascination with whole-number ratios in musical harmonies, which was clearly based on observation and experimentation with the sounds produced by plucked strings.

Pythagoras and Plato made significant strides in science by recognizing that the Cosmos is knowable and has a mathematical foundation. However, their disdain for experimentation and preference for mysticism hindered scientific progress. After a prolonged period during which scientific inquiry languished, the Ionian approach—sometimes preserved through scholars at the Alexandrian library—was eventually rediscovered and revitalized.

Pythagoras and Plato made significant contributions to science by recognizing that the Cosmos is knowable and has a mathematical foundation. However, their aversion to experimentation and reliance on mysticism hindered scientific progress. After a long period during which scientific methods lay dormant, the Ionian approach—partially preserved by scholars at the Alexandrian library—was eventually rediscovered and revitalized.

The Western world experienced a reawakening, with experiment and open inquiry regaining their respectability. Forgotten texts and fragments were rediscovered and studied. Figures like Leonardo, Columbus, and Copernicus were inspired by or independently revisited aspects of ancient Greek thought. While modern science is deeply rooted in Ionian traditions, we also contend with enduring superstitions and ethical dilemmas arising from ancient contradictions.

Platonists and their Christian successors held the peculiar belief that the Earth was flawed and corrupt, while the heavens were seen as perfect and divine. For example, Anaximander of Miletus posited that the Sun was a hole in a fire-filled ring surrounding the Earth, and that the Moon and stars were merely holes in the firmament revealing hidden fires.

The term "cosmology" is derived from the Greek word *kosmeo*, meaning "to order" or "to organize," reflecting the belief that the Universe is understandable and worthy of study. The Greeks were keen to identify and analyze patterns within the Cosmos, engaging in debates to determine the most compelling theories.

Pythagoras of Samos played a crucial role in this rationalist movement starting around 540 B.C. His passion for mathematics led him to demonstrate how numbers and equations could explain scientific phenomena. One of his early breakthroughs was explaining musical harmony through numerical relationships. His work laid the groundwork for future generations, who applied mathematical principles to various scientific fields, from cannonball trajectories to chaotic weather patterns. Pythagoras famously declared that "everything is number."

2.7 Theory and observation

Simplicity and elegance are secondary to the most crucial aspect of any scientific theory: it must accurately reflect reality and be open to testing. This is where the theory of celestial music falls short.

A valid scientific theory must make predictions about the Universe that can be measured or observed. If experimental or observational results align with these predictions, the theory may be accepted and integrated into a broader scientific framework. Conversely, if the theory's predictions are incorrect or conflict with observations, it must be discarded, regardless of its simplicity or elegance. Therefore, the ultimate test for any scientific theory is its ability to be tested and its compatibility with reality.

Cosmopolitan Alexandria had surpassed Athens as the intellectual center of the Mediterranean, with its library standing as the most esteemed institution of learning in the world. The groundbreaking work of Eratosthenes, Aristarchus, and Anaxagoras advanced scientific thought in ancient Greece, thanks to their reliance on logic, mathematics, observation, and experimentation. However, the Greeks were not the sole contributors to the foundations of science.

Empedocles, a philosopher who predated Socrates, proposed that everything was composed of four elements—Earth, water, air, and fire—each corresponding to one of the classical elements. Some ancients felt that this was insufficient, suggesting that beyond the "sub-lunar" region (within the Moon's orbit), there must be a more perfect fifth element, known as quintessence, which comprised the substance of the heavenly bodies.

Before ancient Greeks could calculate celestial distances or sizes, they first had to establish that the Earth is a sphere. This idea gained acceptance as philosophers noted that ships gradually disappear over the horizon, with only their masts visible last, which could only be explained if the sea's surface were curved. Observations of lunar eclipses, where the Earth casts a round shadow on the Moon, further supported this view. Additionally, the round shape of the Moon suggested that a spherical shape was natural, bolstering the hypothesis of a spherical Earth.

Yet, a spherical Earth raised a common question: what prevents people in the southern hemisphere from falling off? The Greek solution posited that the Universe had a central point to which everything was drawn. The Earth's center was believed to align with this universal center, making the Earth itself static and causing everything on its surface to be attracted toward the center.

Thus, this force would keep inhabitants grounded, regardless of their location on the globe, even if they were in the southern hemisphere. However, the ancient Greeks didn't engage in science as we understand it today. Apart from Archimedes, Greek thinkers primarily focused on theoretical ideas rather than experimental methods. Their concepts were far advanced compared to their technological capabilities. For example, proving the existence of atoms requires precise instruments, which the Greeks lacked.

Archimedes

Aristotle explained that the four natural elements each had a designated natural place, and that motion represents an attempt to return to these positions. However, the Greeks were not the only culture to contribute to early scientific thought; the Babylonians also played a role. Although it is widely agreed among philosophers and historians of science that the Babylonians were not "true" scientists—primarily because they attributed the workings of the Universe to deities and myths—their extensive collection of measurements and records of celestial and planetary positions was a significant endeavor. Nonetheless, this empirical approach fell short of true science, which aspires to explain such observations by uncovering the fundamental nature of the Universe.

2.8 The beginnings of physics

In addition to their achievements in astronomy, the ancient Greeks also made significant strides in what we now refer to as physics. Among the most notable contributors was Archimedes, often regarded as one of the greatest Greek scientists. Born into a royal family in the 3rd century B.C., Archimedes gained renown as an inventor, scientist, and mathematician.

One of Archimedes' most significant contributions, and the one for which he is most renowned, is the discovery of the principle of buoyancy, also known as Archimedes' Principle. This principle states that a body submerged in water is buoyed up by a force equal to the weight of the water displaced by the object. Alexandria, as previously mentioned, was the greatest city the Western world had ever seen, attracting people from all nations who came to live, trade, and learn. Its harbors were bustling with merchants, scholars, and tourists daily.

Hypatia

In Alexandria, a bustling hub where Greeks, Egyptians, Nubians, Arabs, Syrians, Hebrews, Italians, Persians, Gauls, and other cultures mingled, the last prominent scientist associated with its renowned library was Hypatia. Born around 370 A.D., Hypatia was a mathematician, physicist, astronomer, and head of the Neoplatonic School of philosophy. She adeptly navigated male-dominated disciplines and was celebrated for her beauty.

By Hypatia's era, Alexandria, under Roman rule, was in decline. The city's classical vibrancy had waned because of slavery, and the rising Christian Church was consolidating power and seeking to suppress pagan traditions. Hypatia became a focal point of these intense social and political forces.

The Archbishop of Alexandria, Cyril, held a grudge against Hypatia because of her close relationship with the Roman governor and her embodiment of knowledge and science, which the early Church equated with paganism. Despite the growing danger, Hypatia persisted in teaching and writing. In 415 A.D., she was violently attacked by a fanatical mob of Cyril's followers, who dragged her from her chariot, stripped her, killed her, and burned her remains. Her works were soon destroyed, and her name fell into obscurity. Cyril was later canonized as a saint, while the splendor of the Alexandrian library became a mere memory, its remnants lost shortly after Hypatia's death. It must be noted that some historians suggest that Hypatia's death may not have been caused by Christians.

None of the physical contents of that magnificent library survive intact today. In modern Alexandria, few people have a genuine appreciation or detailed knowledge of the Alexandrian library or the great Egyptian civilization that preceded it for thousands of years.

2.9 The Birth of Science

As mentioned earlier, the ancient Greeks did not invent science; this distinction belongs to the Babylonians, who began their study of the heavens about 5,000 years ago. The Babylonians, residing in what is now Iraq, were driven to understand celestial phenomena to determine the best times for harvesting. They worshipped the Sun, Moon, and the five visible planets—Mercury, Venus, Mars, Jupiter, and Saturn—and meticulously tracked their movements across the sky. This astronomical knowledge helped them develop a calendar.

The Greeks inherited some of this Babylonian knowledge and made significant strides in understanding the natural world. Through their observations of the heavens, they began to develop the foundations of physics. For instance, Pythagoras proposed that the Earth was spherical and centered in the universe. Aristotle expanded on Pythagoras' ideas, positing that the Earth was immovable and fixed at the center of a rotating cosmos.

During this period, the Greeks had not yet observed or measured the apparent shift in the position of stars relative to the Earth, leading to the abandonment of the heliocentric model proposed by Aristarchus. Despite Aristarchus' belief in his correct theory, concrete proof had to wait until Western civilization emerged from the Dark and Middle Ages. Today, we know that stars are so distant that their parallax (apparent shift in position) is not visible to the naked eye. It wasn't until the Renaissance, with the invention of telescopes, that observations of stellar parallax became possible. In 1838, Friedrich Bessel used a telescope to make the first successful measurements of a star's parallax.

If the Babylonians were not the first proto scientists, what about the Egyptians? The Egyptians were undoubtedly more advanced than the Greeks in terms of technological inventions. However, these achievements are examples of technology, not science. Technology focuses on practical applications—making life (and death) more comfortable and efficient—while science is driven by the pursuit of understanding the world. Scientists are motivated by curiosity rather than by practical comfort or utility.

Although scientists and technologists have different objectives, science and technology are often conflated, likely because scientific discoveries frequently lead to technological advancements. For instance, scientists spent decades exploring electricity, which technologists then applied to create light bulbs and other devices. In ancient times, technology developed without the support of science, allowing the Egyptians to be skilled technologists without an understanding of scientific principles. The goals of Greek scientists were similar to those described two millennia later by *Henri Poincaré*:

The scientist does not study nature because it is useful; he studies it because he delights in it, and he delights in it because it is beautiful. If nature were not beautiful, it would not be worth knowing, and if nature was not worth knowing, life would not be worth living.

Thus, while the Egyptians were technologists, not scientists, Eratosthenes and his peers were scientists, not technologists. The Greeks demonstrated that determining the Sun's diameter relies on knowing its distance, which in turn depends on measuring the distance to the Moon, and that requires knowing the Earth's diameter. Eratosthenes' significant breakthrough was his realization that the Moon aligns perfectly with the Sun during a solar eclipse.

Ch 3—Astrology & Astronomy

3.1 Claudius Ptolemaeus

Modern popular astrology traces its origins to Claudius Ptolemaeus, commonly known as Ptolemy. Working as a philosopher and astronomer in the Library of Alexandria during the 2nd century A.D., Ptolemy systematized the ancient Babylonian astrological tradition. Many of today's terms related to planetary movements, such as the "solar" and "lunar" houses or the "Age of Aquarius," are derived from Ptolemy's work.

Ptolemy

The Old Testament primarily survives through Greek translations produced at the Library of Alexandria. The Ptolemies invested their vast wealth in acquiring texts from Greece, Africa, Persia, India, Israel, and other regions. Their efforts went beyond mere collection; they actively supported and funded scientific research, leading to new discoveries. For instance, Euclid's textbook on geometry, developed under their patronage, was studied for twenty-three centuries and played a crucial role in inspiring scientists like Kepler, Newton, and Einstein.

Faced with the challenge of explaining increasingly complex astronomical observations and the apparent immobility of the Earth at the center of the Universe, the ancient Greeks developed more intricate models to account for their findings. Their geocentric theory was refined to accommodate new data, but it seldom questioned the underlying assumption that the Earth must be the central focus of the cosmos. This adherence to the belief that humanity, with its grand achievements and divine deities, deserved a central position in the Universe led them to add complexities rather than reassess their fundamental cosmological framework. Even today, science can sometimes be swayed by emotion rather than reason, as seen in the Greeks' preference for elaborate explanations over acknowledging a potential foundational error.

Despite the increasing complexity of the ancient Greek models, which predate Ptolemy, they were eventually found to be inadequate for accurately explaining planetary motions. According to contemporary cosmology, planets were expected to move at a constant and uniform speed relative to the distant stars. However, observations showed that planets did not follow a steady path; instead, they would occasionally slow down, stop, and then reverse direction temporarily. Ptolemy was the first to address this phenomenon of retrograde motion, developing what is now known as the Ptolemaic system to explain these irregularities.

3.2 Ptolemy's Universe

In the second century A.D., Ptolemy developed a comprehensive geocentric model, where the planets, Moon, and Sun orbited the Earth in circular paths. To account for the complex planetary motions, Ptolemy introduced a sophisticated model involving two types of circular motion: "deferents" and "epicycles."

In this model, each planet moved along a small circle, or epicycle, which itself orbited a larger circle, or deferent, centered around the Earth. The combination of these circular motions aimed to replicate the observed movements of the planets, including their occasional retrograde motion.

Ptolemy's model also included eight rotating spheres surrounding the Earth, each larger than the previous one. The outermost sphere, beyond which the observable Universe did not extend, was fixed, ensuring that the stars appeared to rotate together across the sky. The inner spheres carried the planets, which were not fixed like the stars but moved on their respective spheres along their epicycles.

Despite the complexity of this model, it was able to approximate planetary paths relative to the Earth quite closely. However, even this intricate system had some small errors.

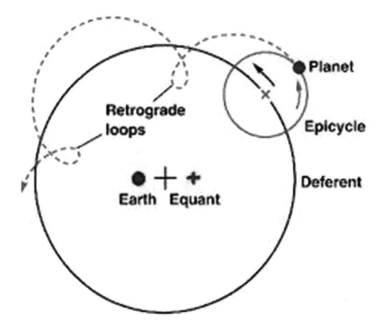

Ptolemaic system

Figure 3-1 Deferents and epicycles model

To align his model with observed facts, Ptolemy introduced additional epicycles, allowing him to account for the complex paths of the planets, which could not be explained by simple circular orbits alone. As new and more precise observations came in, Ptolemy further refined his model by adding even more epicycles, leading to a system of 40 interconnected circles. This intricate network accurately reflected the astronomical observations of his time.

Despite its complexity, Ptolemy's model was effective for predicting the positions of celestial bodies. However, it required the assumption that the Moon's orbit brought it twice as close to Earth at certain times, causing it to appear twice as large. Ptolemy recognized this flaw, but his model was still widely accepted. It was embraced by the Christian Church as it fit with Scriptural views of the universe, leaving ample space beyond the fixed stars for heaven and hell.

Ptolemy's celestial spheres, once imagined in medieval times as being made of crystal, are the reason we still speak of the "music of the spheres" and the "seventh heaven." With Earth at the center of the universe and everything revolving around terrestrial events, there was little incentive for advancing astronomical observations.

Endorsed by the Church during the Dark Ages, Ptolemy's model hindered the progress of astronomy for nearly a thousand years. It wasn't until 1543 that a radically different hypothesis emerged to explain planetary motion, introduced by Nicholas Copernicus, a Polish Catholic cleric

Figure 3-2 Ptolemaic Worldview

Copernicus audaciously proposed a heliocentric model, placing the Sun, rather than Earth, at the center of the universe. In this model, Earth was relegated to being just one of the planets, orbiting the Sun in a perfect circular path. Ptolemy had briefly considered a similar heliocentric idea but dismissed it because it conflicted with Aristotelian physics and seemed inconsistent with observed evidence of Earth's rotation.

Ptolemy's model was published in "The Mathematical Collection," a comprehensive work spanning 13 volumes. After the destruction of the Library of Alexandria, the Arabs preserved Ptolemy's writings, referring to them as "al Magiste," or "The Greatest" in Arabic. During their long occupation of Spain, the Arabs introduced the book to Europe, where it was known as the Almagest and studied for over a millennium.

3.3 A New Perspective on the Cosmos

Not long after Ptolemy's death in 170 A.D., much of Greek knowledge was lost to Western civilization, which diverged significantly during the Dark and Middle Ages. The sciences and arts did not reach the Greek level of achievement again until the 17th century Renaissance. However, in 1514, Nicholas Copernicus proposed a simpler model of the cosmos. To avoid accusations of heresy, he initially published his work anonymously. His model placed the Sun at the center of the universe, with the Earth and other planets orbiting it in circular paths.

Copernicus had encountered newly discovered classical texts while studying medicine in Italian universities, which reignited interest in earlier astronomical ideas. Although he acknowledged Aristarchus' precedence in his manuscript, he omitted this reference before publication. In a letter to Pope Paul III, Copernicus wrote, "According to Cicero, he had thought the Earth was moved... I myself also began to meditate upon the mobility of the Earth." This suggests that he may have drawn inspiration from Aristarchus. Unfortunately, it would be nearly a century before Copernicus' heliocentric model gained serious consideration.

Enters Copernicus

Born in 1473 in Torun, Poland, into a prosperous family, Copernicus was appointed a canon at the Freudenberg cathedral chapter, thanks in part to the influence of his uncle, Lucas, the Bishop of Ermland. Although he studied law and medicine in Italy and primarily served as a physician and secretary to his uncle, Copernicus's true passion lay in astronomy. His interest began when he acquired a copy of the Alphonsine Tables as a student. This amateur astronomer's growing obsession with planetary motion would eventually establish him as a pivotal figure in the history of science.

3.4 The Charge of Heresy

The journey toward a rational understanding of the universe began in the 15th century with Nicholas Copernicus. Like many of his contemporaries, Copernicus grew frustrated with Ptolemy's model of the cosmos, which failed to account for new astronomical observations. He proposed a groundbreaking idea: not all celestial bodies needed to orbit the Earth.

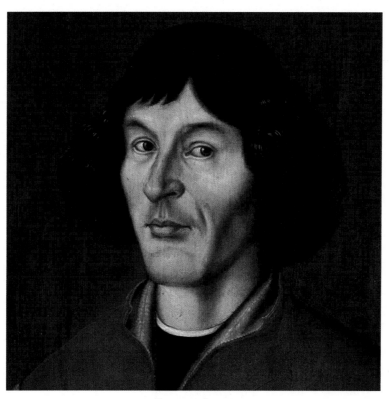

Copernicus

Copernicus proposed that the Sun, not the Earth, was the center of the solar system, with Earth and the other planets orbiting it in circular paths. This idea mirrored Aristarchus's 17 centuries earlier and suggested that placing the Sun at the center could simplify astronomical explanations. In 1514, Copernicus, a Polish priest, introduced this heliocentric model.

Initially, fearing accusations of heresy, Copernicus shared his model anonymously. While his system was simpler and more elegant than Ptolemy's, it still did not perfectly match observations. It took several decades for Copernicus's ideas to gain traction, but eventually, astronomers Johannes Kepler and Galileo Galilei publicly supported his theory.

3.5 Copernicus Challenging Ptolemy's World

The Copernican model explained the apparent motion of the planets as well as Ptolemy's system but faced significant opposition. In 1616, the Catholic Church placed Copernicus' work on its list of forbidden books, where it remained until 1835, pending correction by ecclesiastical authorities. Some of Copernicus' supporters even argued that he may not have genuinely believed in a Sun-centered universe but proposed it merely as a practical tool for calculating planetary motions.

In Copernicus' heliocentric model, the Earth, along with Venus, Mars, Jupiter, and Saturn, orbited the Sun in circular paths. This model was met with outrage by Church authorities, who adhered to the doctrine that the Earth must be the center of creation. Proposing otherwise was deemed heretical.

Martin Luther, who recorded his dinner-table discussions about Copernicus, famously said: "There is talk of a new astronomer who claims that the Earth moves, and the sky, Sun, and Moon stay still, as if someone were stationary while the ground and trees moved around him. This fool wants to turn the entire field of astronomy upside down." Luther dismissed Copernicus as an "upstart astrologer" and argued, "This fool wishes to overturn the whole science of astronomy. But Sacred Scripture tells us that Joshua commanded the Sun to stand still, not the Earth." Luther condemned Copernicus as "a fool who contradicted Holy Writ."

Copernicus' assistant, Rheticus, a staunch believer in scientific inquiry over religious doctrine, supported Copernicus unwaveringly. Rheticus spent three years at Frauenburg, reviewing and encouraging Copernicus on his manuscript. Despite this support, Copernicus feared the repercussions from the Church and hesitated to publish his theory.

Eventually, in his late 60s, with encouragement from friends, Copernicus decided to publish his work. In the spring of 1543, his groundbreaking book, *"De Revolutionibus Orbium Coelestium"* ("On the Revolutions of the Heavenly Spheres"), was finally released. Unfortunately, Copernicus, who had suffered a stroke in late 1542, was bedridden and fighting to stay alive. His copies arrived just in time, and though his memory and mental vigor had declined, he managed to see his work in print. His publisher delivered an advanced copy to him shortly before his death at age 70.

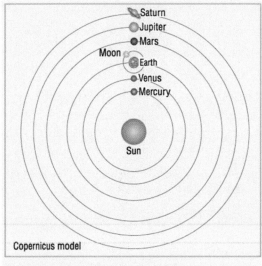

Figure 3-3: The Copernican Model of the Solar System

Copernicus had fulfilled his mission by presenting a compelling argument for the Sun-centered model, echoing Aristarchus' earlier ideas. Despite this, many scientists and theologians dismissed Copernicus's propositions as heretical. The established belief in Ptolemy's cosmology, which had persisted for over a millennium, remained dominant. The "upstart" theory was not taken seriously, and more evidence was needed to validate the heliocentric model, despite its simplicity and greater accuracy in explaining planetary retrograde motions. Even after Copernicus's death, his followers continued to champion the Sun-centered model, arguing that its predictive success was due to the Sun's true central position in the Universe.

This inevitably led to a strong reaction from the Church. In February 1616, a committee of the Inquisition officially declared that the Sun-centered model of the Universe was heretical. Consequently, in March 1616, over sixty years after its publication, Copernicus's *De Revolutionibus* was banned.

3.6 Summary of the Copernican system vs the Ptolemaic system

To the eye, the sky seemed like a dome with stars affixed to it. Throughout the day, this dome appeared to rotate around an axis close to the North Star. At that time, the Earth's surface was thought to be stationary and the eternal foundation of the Universe. Terms like "above" and "below" were considered absolute, and poetic or philosophical speculations about the heavens or the underworld required no further explanation.

The Ptolemaic system, formulated by Ptolemy around 150 A.D., encapsulated this worldview. Although Ptolemy had gathered extensive observations about the Sun, Moon, and planets and developed methods to predict their movements, he maintained the belief that the Earth was stationary, and the stars revolved around it at vast distances. According to his model, the stars' orbits were composed of circles and epicycles based on terrestrial geometry, yet astronomical space itself was not viewed as a geometric object. Instead, the orbits were thought to be attached to crystal spheres, which were arranged in concentric shells to form the sky.

Greek thinkers had long recognized the Earth's spherical shape and began exploring beyond the geocentric worldview, with figures like Aristarchus in the 3rd century B.C. making early advances. However, it took many centuries after the decline of Greek civilization for other cultures to accept the Earth's spherical shape as a physical reality. This acceptance marked a significant shift away from mere observational evidence and was a crucial step toward a deeper understanding of the Universe.

This shift did not initially lead to conflict between the objective and subjective views of the world or between scientific inquiry and the Church. Such strife emerged only later, with Copernicus' heliocentric model in 1543, which moved the Earth from the center of the Universe and placed the Sun at its core. This revolutionary model was significant because it displaced Earth, humanity, and individual self-importance, positioning the Earth as a mere satellite of the Sun, orbiting in the vast expanse of space.

In Copernicus' model, planets of equal significance orbit the Sun, reducing humanity's central role in the Universe to that of self-importance alone. In 1543, Nicholas Copernicus introduced his cosmology, positioning Earth as one of seven planets revolving around the Sun. While other astronomers had suspected this arrangement, they feared challenging the established Ptolemaic system and did not openly promote the theory. Copernicus, unable to definitively prove his model, faced skepticism over how Earth could maintain its orbit without continuous external force, as Aristotle had posited. Without visible cosmic forces pushing Earth, it was assumed to be stationary.

Copernicus, however, extended his reasoning to the other planets: if invisible forces accounted for their motion, they could also apply to Earth. His model applied scientific principles and geometric laws directly to astronomical space, replacing Ptolemaic crystal spheres with real orbital paths. In this system, the Sun is at the center, and planets, including Earth, orbit it. Earth rotates on its axis, while the Moon orbits Earth. The fixed stars, at vast distances, are suns like ours, stationary in space.

Copernicus' revolutionary model simplified the explanation of astronomical phenomena that the Ptolemaic system had addressed with complex and artificial theories. It provided clear, intelligible explanations for the alternation of day and night, the seasons, the phases of the Moon, planetary orbits, and other celestial events, all amenable to straightforward calculations.

3.7 The Copernicus model finding its place as part of reality

Soon, Copernicus' circular orbits proved inadequate for explaining new observations, revealing that the actual orbits were far more complex. The challenge was whether to introduce artificial constructs like the epicycles of the Ptolemaic system or to refine the existing model without such complications.

Despite being simpler and more elegant than the Ptolemaic system, Copernicus' model still left the Earth seemingly motionless. Like the ancient Greeks, his contemporaries could not detect Earth's motion against the distant stars, which led to limited acceptance of the Copernican model.

Kepler

The debate between the Earth-centered and Sun-centered models reached a dramatic climax in the 16th and 17th centuries through the efforts of a man who, like Ptolemy, was both an astrologer and an astronomer. Living in an era where intellectual freedom was severely restricted, and religious persecution was rampant, this individual challenged the prevailing view of the heavens populated by angels and divine forces. His courageous and solitary work laid the foundation for the modern scientific revolution.

Johannes Kepler's groundbreaking work in 1618 revolutionized our understanding of planetary motion, rescuing the Copernican system during a crucial period. He discovered that planetary orbits are not perfect circles around the Sun, but ellipses with the Sun at one focus. Kepler's three laws of planetary motion simplified the description of these orbits, detailing their shapes, the speeds at which planets travel, and the relationship between orbital periods and the dimensions of the ellipses.

Kepler, along with Galileo Galilei, began to publicly advocate for the Copernican theory, despite its predictions not perfectly aligning with observations. The end of the Aristotelian-Ptolemaic system came in 1609 when Galileo began using the newly invented telescope to observe the night sky.

3.8 Enters Johannes Kepler

Johannes Kepler was born in Germany in 1571 into a family plagued by hardship. His father, a troubled man with a criminal record, and his mother, who faced exile after being accused of witchcraft, contributed to his tumultuous upbringing. This background left Kepler feeling insecure and neurotic, and he described himself as a "little dog" in his own writings.

As a boy, Kepler was sent to a Protestant seminary in Maulbronn, which served as a rigorous training ground for future clergy, aimed at preparing them to challenge Roman Catholicism with theological arguments. Despite his troubled personal life, Kepler displayed a stubborn intelligence and fierce independence.

Though classical sciences had been dormant for over a millennium, the late Middle Ages saw a revival of ancient Greek knowledge. Kepler, exposed to Euclidean geometry, saw it as a reflection of cosmic perfection and divine glory. He would later write, "Geometry existed before Creation. It is co-eternal with the mind of God... Geometry provided God with a model for Creation... Geometry is God Himself."

For Kepler, God was not just a figure of divine retribution but the creative force behind the universe. His fascination with the natural world and his desire to understand the divine mind became an all-consuming obsession, shaping his character despite the imperfections of his external environment.

Superstition was a common refuge for people struggling against the hardships of famine, disease, and religious conflict. In an era marked by fear and uncertainty, many found solace in the predictability of the stars, and astrological beliefs thrived in the courtyards and taverns of Europe. Kepler, who remained ambivalent about astrology throughout his life, pondered whether there might be hidden patterns beneath the apparent chaos of daily existence. If the world was created by God, shouldn't it be scrutinized? Wasn't all of creation a reflection of the divine harmonies in God's mind? Yet, Kepler's exploration of this "book of Nature" would remain unread for a long time.

In 1589, Kepler left Maulbronn to study for the clergy at the esteemed University of Tübingen. His experience there was only marginally satisfactory. However, amidst the vibrant intellectual environment, his exceptional talent was quickly recognized by his professors, one of whom introduced him to the provocative ideas of the Copernican model.

Kepler's belief in a heliocentric universe aligned deeply with his religious views, where the Sun symbolized God, around whom everything revolved. Though he was on the path to becoming a clergyman, he accepted a secular job offer, possibly because he felt unsuited for the ecclesiastical life. He moved to Graz, Austria, to teach mathematics at a secondary school and later prepared astronomical almanacs and horoscopes. "God provides every creature with its sustenance," he wrote, "For the astronomer, He has provided astrology."

Despite being a brilliant thinker and writer, Kepler struggled as a teacher. His lectures were often muddled and hard to follow, resulting in very few students attending. By his second year, no one came.

In 1598, the tensions that would lead to the Thirty Years' War erupted. The local Catholic archduke, zealous in his faith, declared he'd rather "make a desert of the country than rule over heretics." Protestants like Kepler lost their rights, and his school was closed. Religious practices deemed heretical were banned, and the townspeople were forced to publicly declare their religious allegiance. Those refusing to convert to Catholicism faced fines, torture, or exile. Kepler chose exile, later stating, "Hypocrisy I have never learned. I am serious about faith. I do not play with it."

Leaving Graz, Kepler, his wife, and stepdaughter embarked on the challenging journey to Prague. Their marriage was already strained. His wife, chronically ill and grieving the recent loss of two children, was described as "stupid, sulking, lonely, and melancholy." She neither understood nor appreciated Kepler's work, and, having grown up among the rural gentry, she looked down on his financially unstable profession. Kepler, in turn, either scolded or ignored her, later reflecting, "My studies sometimes made me thoughtless; but I learned my lesson, I learned to have patience with her."

Pythagoras in the 6th century B.C., along with Plato, Ptolemy, and all Christian astronomers before Kepler, believed that planets moved in perfect circular paths. The circle was considered a "perfect" shape, and the planets, far from earthly imperfections, were thought to share this mystical perfection. Even Galileo, Tycho, and Copernicus were committed to circular planetary motion, with Copernicus stating that "the mind shutters" at any alternative, as it seemed unworthy of a divinely perfect creation.

Initially, Kepler also tried to explain planetary motion by imagining the Earth and Mars moving in circular orbits around the Sun. The influence of the Pythagorean tradition is clear in Kepler's work. He embraced the idea of a mathematically harmonious universe, believing that simple numerical relationships governed planetary motion. As he wrote, "the Universe was stamped with the adornment of harmonic proportions."

Yet, like the Pythagoreans, Kepler initially held that only uniform circular motion was possible. However, his observations repeatedly contradicted this belief. Unlike many of his predecessors, Kepler valued empirical evidence and real-world observations. Ultimately, it was this commitment to observation that led him to abandon circular orbits and recognize that planets move in elliptical paths.

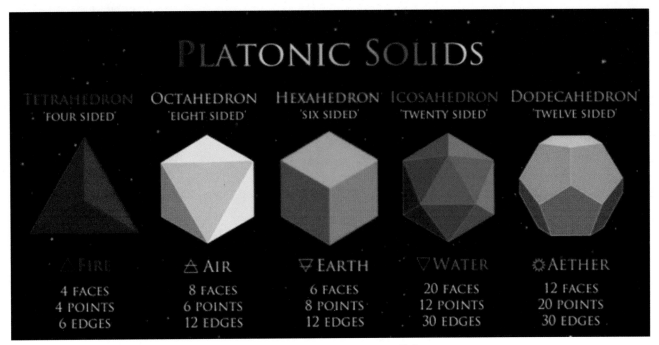

Figure 3-4 Platonic Solids

Kepler's pursuit of planetary harmony was both driven and delayed for over a decade by his fascination with Pythagorean doctrine. At the time, only six planets were known—Mercury, Venus, Earth, Mars, Jupiter, and Saturn. Kepler wondered why there were only six planets and why they had the spacing that Copernicus had calculated. These were questions no one had posed before. Ancient Greek mathematicians had identified five regulars, or "Platonic," solids—geometrical shapes with identical polygonal faces. Kepler believed this was linked to the number of planets, theorizing that the five solids, nested within each other, determined the distances of the planets from the Sun. In these perfect forms, he thought he had discovered the hidden structure of the universe, calling his insight "The Cosmic Mystery." To Kepler, the alignment between these solids and the planets could only be explained by "The Hand of God, Geometer."

Overwhelmed by his geometrical findings, Kepler marveled at being divinely chosen for such a revelation, despite seeing himself as sinful. He submitted a research proposal to the Duke of Württemberg, offering to oversee the construction of a three-dimensional model of his nested solids to showcase the beauty of the holy geometry. The duke rejected the proposal, suggesting instead that Kepler make a less costly version from paper, which he eagerly attempted. Kepler later wrote, "The intense pleasure I have received from this discovery can never be told in words... I shunned no calculation, no matter how difficult. Days and nights I spent in mathematical labors, until I could see whether my hypothesis would agree with the orbits of Copernicus or whether my joy was to vanish into thin air."

Despite his efforts, Kepler could not make the solids align accurately with the planetary orbits. Yet the theory's elegance convinced him that the observational data, not his hypothesis, must be wrong—a conclusion many theorists have drawn throughout scientific history. Only one man had access to more precise planetary data: the Danish nobleman and astronomer, Tycho Brahe, who was then working under the patronage of the Roman Emperor Rudolf II.

By chance, at the suggestion of Emperor Rudolf II, Tycho Brahe invited Kepler—whose reputation as a mathematician was rising—to join him in Prague. Before long, the promising young astronomer became part of Tycho's prestigious observatory. Known simply as "Tycho" by most scientists, his observatory was the finest in the world, home to the most accurate astronomical data of the time.

Kepler saw Tycho's observatory as a sanctuary from the turmoil of the time, a place where his "Cosmic Mystery" would finally be validated. He hoped to become a close collaborator of Tycho Brahe, who had spent thirty-five years meticulously measuring the universe, long before the invention of the telescope. However, Kepler's hopes were soon dashed. Tycho, a flamboyant and eccentric figure, did not live up to Kepler's expectations. Reflecting on Tycho's wealth, Kepler once remarked, "Tycho is superlatively rich but knows not how to make use of it. Any single instrument of his costs more than my whole family's fortunes put together."

3.9 Enters Tycho Brahe

Though Tycho was raised in the tradition of Ptolemaic astronomy, his meticulous observations led him to question the ancient view of the universe. He owned a copy of Copernicus' *De revolutionibus* and was sympathetic to its ideas, but rather than fully embracing them, Tycho developed his own hybrid model—a compromise between Ptolemy and Copernicus.

Tycho once wrote, "I have been occupied in alchemy as much as in celestial studies since my 23rd year." However, he believed both alchemy and astronomy held dangerous secrets, safe only in the hands of princes and kings, from whom he sought support. He followed a long tradition of scientists who felt only the elite should have access to such knowledge, saying, "It serves no useful purpose and is unreasonable to make such things generally known."

Tycho Brahe

In contrast, Kepler was far more open. He lectured on astronomy in schools and published extensively, often at his own expense. The difference between Tycho and Kepler, separated by a single generation, is striking. In 1588, nearly fifty years after Copernicus' death, Tycho published *De mundi aetherei recentioribus phoenomenis* ("Concerning the New Phenomena in the Ethereal World"), in which he proposed that while the planets orbited the Sun, the Sun itself orbited the Earth.

Tycho's model allowed the Sun to serve as the hub for the planets but kept the Earth at the center of the universe. He was hesitant to move the Earth from its central position, believing this was the only way to explain why objects fall toward the center of the Earth.

Over many years, Tycho Brahe meticulously gathered observations of Mars and other planets as they moved through the constellations. His data, the most accurate of its time, were collected before the invention of the telescope. Kepler, eager to make sense of these observations, focused intensely on Mars, particularly its puzzling retrograde loops in the night sky.

Tycho tasked Kepler with analyzing Mars because its motion was the most challenging to reconcile with the traditional notion of circular orbits. While Tycho held onto the Ptolemaic system, Kepler had already embraced the Copernican model, having been introduced to it by a professor during his studies at the University of Tübingen, where he graduated in 1588.

At the time, the most significant challenge in astronomy was accurately predicting Mars' path—a problem that had confounded Copernicus' assistant, Rheticus. Kepler recounted how

Rheticus, frustrated by the complexity of Mars' motion, reportedly "appealed to his guardian angel as an Oracle" for guidance.

However, Tycho only gave Kepler access to his prized data in fragments, withholding much of his observations. Kepler lamented, "Tycho gave me no opportunity to share in his experiences. He would only, in the course of a meal and in between other matters, mention, as if in passing, today the figure of the apogee of one planet, tomorrow the nodes of another... Tycho possesses the best observations... He also has collaborators. He lacks only the architect who would put all this to use."

Tycho, the era's foremost observational genius, and Kepler, its greatest theoretician, both knew that neither could fully succeed without the other. Yet, Tycho, protective of his life's work, was unwilling to share credit with the younger Kepler. The collaboration between them was tense, teetering on mutual mistrust, and they quarreled frequently during the remaining 18 months of Tycho's life.

Tragically, Tycho's death came after a dinner hosted by the Baron of Rosenberg, where he "placed civility ahead of health" and, despite consuming much wine, resisted leaving the table. This led to a fatal urinary infection, worsened by his refusal to heed medical advice. On his deathbed, Tycho entrusted his invaluable observations to Kepler, repeating words like a man composing a poem in his final hours.

On his deathbed, Tycho Brahe repeated the words, "May I not have lived in vain." But his fears were unfounded—Kepler would ensure that Tycho's meticulous observations found their purpose. Ironically, Tycho's death may have been necessary for his work to truly flourish. While alive, he guarded his notebooks jealously, intent on publishing a solo masterpiece, never seeing Kepler as an equal. As a Danish nobleman, Tycho could not reconcile working closely with Kepler, a man of humble peasant origins. Yet, the full significance of Tycho's observations could only be realized through the mathematical brilliance of Kepler.

After Tycho's death, Kepler, now the imperial mathematician, managed to pry Tycho's valuable observations from his reluctant family. His belief that the planets' orbits were governed by the five Platonic solids fared no better with Tycho's data than it had with Copernicus'. Later discoveries of Uranus, Neptune, and Pluto further disproved Kepler's "Cosmic Mystery," as no additional Platonic solids could explain their distances from the Sun.

Kepler's theory also overlooked the existence of Earth's moon, and Galileo's discovery of Jupiter's four large moons posed another challenge. Yet, rather than becoming discouraged, Kepler embraced the search for more satellites and speculated on how many moons each planet should have.

In a letter to Galileo, he wrote: "I immediately began to think how there could be any addition to the number of planets without overturning my *'Mysterium Cosmographicum'*... Euclid's five regular solids do not allow more than six planets around the Sun... I am so far from disbelieving the existence of the four *circumjovial* planets that I long for a telescope, to discover two around Mars... six or eight around Saturn, and perhaps one each around Mercury and Venus."

While Mars does have two small moons, and Kepler Ridge is named after him in honor of this guess, he was wrong about Saturn, Mercury, Venus, and the number of moons around Jupiter. Despite these inaccuracies, the reasons behind the number of planets and their relative distances from the Sun remain largely a mystery.

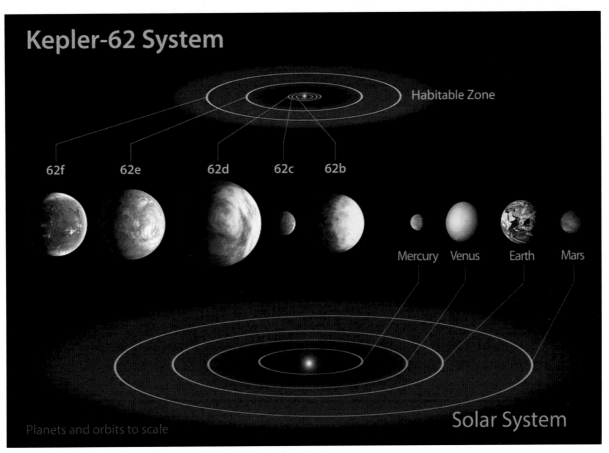

Figure 3-5 Kepler's Solar System

3.10 Uncovering the Laws of Planetary Motion

With access to Tycho's observations, Kepler initially believed he could solve the problem of Mars and correct the inaccuracies in the Sun-centered model within eight days. However, it took him eight years of rigorous and painstaking work, involving calculations that filled nine hundred folio pages and seventy trials on Mars's motion.

Kepler used the precise measurements provided by Tycho and his assistants to demonstrate that Mars's orbit was not a circle, but an elongated shape known as an ellipse. By adapting Copernicus's theory, Kepler showed that planets move in elliptical orbits rather than perfect circles. This adjustment allowed the predictions to finally align with the observed data.

During his "battle with Mars," Kepler discovered that Mars moved faster when closer to the Sun and slower when farther from it. He also found that a line connecting the planet to the Sun swept out equal areas in equal times, leading to his first law of planetary motion: a planet orbits the Sun in an ellipse with the Sun at one focus.

Additionally, Kepler observed that as Mars traversed its elliptical orbit, it swept out a wedge-shaped area within the ellipse. When close to the Sun, Mars covered a large arc in a given time, but this arc represented a smaller area because it was nearer to the Sun. When farther from the Sun, Mars covered a shorter arc in the same time, but this arc represented a larger area. This led to his second law of planetary motion: planets sweep out equal areas in equal times.

Ultimately, Kepler realized that his findings extended beyond Mars to all known planets. Through years of detailed calculations, he identified a mathematical relationship between each planet's orbital period and its distance from the Sun.

Kepler knew he had uncovered the fundamental laws of planetary motion and was particularly proud of his final law, the harmonic law, which defines the relationship between a planet's orbital period and its distance from the Sun. His pride in these discoveries led him to write a book titled *Harmony of the World*.

Despite the advances that elliptical orbits brought to Copernicus's model, Kepler considered them a provisional solution. He held a preconceived notion that ellipses were less perfect than circles, a belief he shared with Aristotle. The idea of planets moving in such imperfect paths seemed too unappealing to be the ultimate truth. Additionally, Kepler struggled to reconcile elliptical orbits with his belief in magnetic forces driving planetary motion. Though incorrect about magnetic forces, he was insightful in recognizing that some force must be responsible for the planets' orbits. The true explanation came later in 1687 when Sir Isaac Newton published *Philosophiæ Naturalis Principia Mathematica*, a landmark work in the physical sciences.

Copernicus's circles and Kepler's ellipses represented what modern science calls a "kinematic" or *"phoronomic"* description of orbits—mathematical formulations of motion without explaining their causes. The causal theory of motion, or dynamics, was established by Galileo. Newton applied this concept to celestial motions and, through a novel interpretation of Kepler's laws, introduced the concept of mechanical force into astronomy. Newton's law of gravitation not only explained deviations from Kepler's laws but also addressed perturbations in orbits revealed by more advanced observational methods.

This dynamic perspective necessitated a more precise formulation of space and time, which Newton provided in his work. His theory remained the prevailing view until Einstein's General Theory of Relativity, which redefined our understanding of gravitation.

Kepler's model of the Solar System was both simple and elegant, and it accurately predicted planetary paths. However, it was largely dismissed as representing reality. Most philosophers, astronomers, and Church leaders viewed it as a useful tool for calculations but remained convinced of an Earth-centered Universe. Their skepticism stemmed partly from Kepler's failure to address key issues like gravity.

They questioned how the Earth and other planets could orbit the Sun when everything around us seemed to be attracted to the Earth. Additionally, Kepler's use of ellipses, in contrast to the traditional doctrine of circles, was met with ridicule. Since ellipses could not be derived from circles and epicycles, many saw a compromise as unattainable.

3.11 The lifelong quest of Johannes Kepler to understand the Heavens

After three years of calculations, Kepler believed he had determined the correct values for a Martian circular orbit, aligning with ten of Tycho's observations within two minutes of arc. However, his excitement soon turned to dismay when two additional observations from Tycho deviated by as much as eight minutes of arc. This discrepancy led Kepler to realize that his fixation on circular orbits had been misguided. Acknowledging that the Earth, plagued by wars, famine, and suffering, was far from perfect, Kepler began to consider that planets might also be made of imperfect material, like Earth itself. He proposed that if planets were imperfect, their orbits might be too. After experimenting with various oval-shaped curves and encountering some errors that initially led him to reject the correct solution, Kepler eventually turned to the elliptical formula first described by Apollonius of Perga in the Alexandrian library. This ellipse perfectly matched Tycho's observations.

3.12 Kepler's Law Diagrams

Upon discovering the solution to the mystery of planetary orbits, Kepler exclaimed, "O, Almighty God, I am thinking Thy thoughts after Thee." His breakthrough hinged on the first point of his new model: that planetary orbits are elliptical. The second and third points of his model logically followed from this revelation.

The distinction between circular and elliptical orbits could only be made through precise measurements and a willingness to accept the new facts. Despite this, Kepler was unsettled by the need to abandon the concept of circular orbits and question his belief in the Divine Geometer. Having cleared traditional models of circles and spirals, he introduced a geometry based on stretched-out circles or ellipses.

Kepler demonstrated that as a planet moves along its elliptical orbit around the Sun, an imaginary line connecting the planet to the Sun sweeps out equal areas in equal times. This important principle reveals how a planet's speed varies throughout its orbit, contradicting Copernicus's notion of constant planetary speeds.

Though the geometry of ellipses had been studied since ancient Greek times, no one had previously proposed ellipses as the shape of planetary orbits. One reason was the prevailing belief in the circle's sacred perfection, which blinded astronomers to other possibilities. Another reason was that most planetary ellipses are only slightly elliptical, making them appear nearly circular under normal observation.

Kepler achieved his breakthrough by rejecting the ancient belief that planets move in perfect circles or combinations of circles. Even Copernicus had adhered to this circular dogma, which Kepler identified as a fundamental flaw in Copernicus's model. Specifically, Kepler challenged the following assumptions made by his predecessor:

1. The planets move in perfect circles.
2. The planets move at constant speeds.
3. The Sun is at the center of these orbits.

Kepler's new equations for planetary orbits matched observational data, finally revealing the true structure of the Solar System. He corrected Copernicus's errors by demonstrating that:

1. Planets move in ellipses, not perfect circles.
2. Planets vary their speed continuously.
3. The Sun is not precisely at the center of these orbits.

Kepler's first law states: A planet moves in an ellipse with the Sun at one of its two foci.

Kepler's second law states that a planet sweeps out equal areas in equal times. This means that the time it takes for a planet to travel from point A to point B is the same as from point C to point D, and the shaded areas swept out during these intervals are equal.

Kepler's third law, also known as the harmonic law, describes the precise relationship between the size of a planet's orbit and its orbital period around the Sun. This law applies to all planets, including those discovered after Kepler's death, such as Uranus, Neptune, and Pluto.

These laws are applicable to every planet, asteroid, comet, and other celestial objects in the Universe.

3.13 General overview of Kepler's laws

Law of Orbits: Each planet orbits the Sun in an elliptical path, with the Sun located at one of the two foci.

Law of Areas: A planet moves around the Sun such that a line connecting the Sun to the planet sweeps out equal areas in equal time intervals.

Harmonic Law: The squares of the orbital periods (T) of any two planets are proportional to the cubes of their average distance (d) from the Sun. Mathematically, $T = k \sqrt{d^3}$, where k is a constant depending on the units used for time and distance.

Kepler developed his theory by combining Copernicus's postulates with Tycho's observations, leading to what are now known as Kepler's laws, published in the early 17th century. He demonstrated that planets orbit the Sun in elliptical paths rather than perfect circles, with varying degrees of eccentricity and the Sun positioned at one focus of the ellipse. Kepler found that planets move faster when they are closer to the Sun (perihelion) and slower when they are farther from the Sun (aphelion). He also discovered that a planet's orbital period is related to its average distance from the Sun by the formula $T = k \sqrt{d^3}$. Kepler's observations revealed that Mars orbits the Sun in an ellipse, and though other planets' orbits are less elliptical, his work with Mars led to these discoveries. If Tycho had directed Kepler to study a planet like Venus, Kepler might not have uncovered the true nature of planetary orbits. The elliptical orbits explain why planets are described as constantly falling toward the Sun but never actually reaching it. Kepler's first law of planetary motion states: A planet moves in an ellipse with the Sun at one focus.

Years later, Kepler formulated his third and final law of planetary motion, which defines the relationship between the orbits of different planets and accurately describes the solar system's dynamics. He detailed this law in his book *The Harmonies of the World*. For Kepler, "harmony" encompassed the orderly and beautiful nature of planetary motion, the existence of mathematical laws governing that motion, an idea rooted in Pythagoras, and even the musical notion of the "harmony of the spheres."

While the orbits of Mercury and Mars are noticeably elliptical, the orbits of other planets are so close to circular that their true shapes are indistinguishable even in highly accurate diagrams. From Earth, our vantage point for observing planetary motion, the inner planets orbit quickly, Mercury, named after the swift messenger of the gods, exemplifies this. Venus, Earth, and Mars orbit more slowly, while the outer planets, like Jupiter and Saturn, move majestically and slowly, fitting their status as the kings of the gods.

Kepler's third law, or harmonic law, states that the squares of the planets' orbital periods (the time required for one complete orbit) are proportional to the cubes of their average distance from the Sun. This means that more distant planets move more slowly, following a precise mathematical relationship:

$$P^2 = a^3,$$

where P represents the planet's orbital period around the Sun, measured in years, and a denotes the planet's average distance from the Sun, measured in astronomical units (AU).

An astronomical unit is the average distance between Earth and the Sun. For instance, Jupiter is located 5 astronomical units from the Sun and

$$a^3 = 5 \times 5 \times 5 = 125.$$

With the establishment of Kepler's three laws, the motion of the Moon and planets became predictable, revealing a universe that adhered to these principles. People could now understand why Venus has a year of 225 days while Mars has a year of 687 days, information that was known but previously unexplained. Kepler demonstrated that planets orbit the Sun in precise mathematical paths and times, consistently following his laws. These same laws govern our own motion as we, anchored by gravity to Earth, travel through interplanetary space. Kepler's laws extend beyond our solar system, as they are observed to apply universally, even to the motion of distant galaxies.

Having extracted secrets from Nature, the laws of planetary motion, Kepler endeavored to find some still more fundamental underlying cause, some influence of the Sun on the kinematics of worlds. The planets sped up on approaching the Sun and slowed down on retreating from it. Somehow the distant planets sensed the Sun's presence.

Magnetism also was an influence felt at the distance, and in a stunning anticipation of the idea of universal gravitation, Kepler suggested that the underlying cause was akin to magnetism. Kepler once wrote 'My aim in this is to show that the celestial machine is to be likened not to a divine organism but rather to a clockwork.

Magnetism is, of course, not the same as gravity, but Kepler's fundamental innovation here is nothing less than astonishing. He proposed that quantitative physical laws that apply to the Earth are also the underpinnings of quantitative physical laws that govern the heavens. It was the first *nonmythical* explanation of motion in the heavens; it made the Earth a province of the Cosmos. Kepler became the last scientific astrologer and the first astrophysicist. Exactly eight days after Kepler's discovery of his third law, the incident that unleashed the "Thirty Years" War transpired in Prague. The war's convulsions consumed an entire generation and shattered the lives of millions, Kepler among them. He lost his wife and son to an epidemic carried out by the soldiery, his royal patron was deposed, and he was excommunicated by the Lutheran Church for his uncompromising individualism on matters of doctrine. Kepler was a refugee once again. The conflict, portrayed by both the Catholics and the Protestants as a holy war, was an exploitation of religious extremism by those hungry for land and power. The savaged population of Europe stood helpless as plowshares and pruning hooks were literally beaten into swords and spears. Waves of rumor and paranoia swept through the countryside, enveloping especially the powerless. Among the many scapegoats chosen were elderly women living alone, who were charged with witchcraft. In Kepler's little hometown of Weil der Stadt, roughly three women were tortured and killed as witches every year between 1615 and 1629. Katharina Kepler was a belligerent old woman. She engaged in disputes that annoyed the local nobility, and she sold hallucinogenic drugs to the local townspeople. Kepler believed that he himself had contributed to her arrest.

Amidst his personal turmoil, Kepler rushed to Wurttemberg to find his seventy-four-year-old mother imprisoned in a Protestant secular dungeon, facing threats of torture similar to those experienced by Galileo in a Catholic prison. Determined to defend her, Kepler approached the situation as a scientist, seeking natural explanations for the events that led to her witchcraft accusations. He investigated various claims, including minor ailments attributed to her supposed spells. His research proved successful, showcasing the triumph of reason over superstition.

Though his mother was ultimately sentenced to exile with a death penalty should she return to Wurttemberg, Kepler's vigorous defense resulted in a decree from the duke prohibiting further witchcraft trials based on such flimsy evidence.

3.14 Invention of the telescope

In 1608, with the advent of the telescope, precise measurements of the heavens became possible, a field Kepler referred to as "lunar geography." He described the Moon as being covered with mountains and valleys, reflecting the lunar craters Galileo had recently observed with the first astronomical telescope.

Kepler noted the extreme temperature variations on the Moon due to the long lunar day and night, accurately describing it as experiencing an "intemperate climate" with severe heat and cold. He also proposed an intriguing view of the Moon's craters, likening them to the face of a boy scarred by smallpox. He correctly argued that these craters were depressions rather than mounds. However, he did not yet realize that most lunar craters were formed by local explosions caused by falling meteoroids, which created large circular cavities. This understanding of crater formation came later.

It is remarkable that the search for extraterrestrial life began in the same generation as the telescope's invention and with Kepler, the era's foremost theoretician. In addition to his scientific discoveries, Kepler also authored one of the earliest works of scientific fiction. He imagined a journey to the Moon, where travelers would stand on its surface and watch Earth slowly rotate in the sky above. He believed that changing our perspective could reveal the workings of the universe.

Kepler speculated about the future of space exploration, envisioning celestial ships with sails adapted to the heavenly winds, carrying explorers—both human and robotic—through the vastness of space. These voyages would be guided by the three laws of planetary motion that Kepler uncovered through his lifetime of dedication and discovery.

The turmoil of the war stripped Kepler of much of his financial support, leading him to spend his final years struggling to secure funds and sponsors. He continued casting horoscopes, this time for Duke Wallenstein, just as he had for Rudolf II. Kepler spent his last years in the town of Sagan, under Wallenstein's control. Tragically, the Thirty Years' War ultimately destroyed his grave. Despite these hardships, Kepler remained committed to the pursuit of hard truth over cherished illusions.

3.15 Developers of the methods of modern science

Kepler's approach to science closely resembles that of modern scientists. He began with observational data from Tycho Brahe's observatory and used it to develop a model explaining the motions of celestial bodies. Kepler's quest to find harmony in the heavens eventually laid the groundwork for Isaac Newton's work, completed thirty-six years after Kepler's death.

However, it was Galileo Galilei, Kepler's contemporary, who significantly advanced the methods used by modern scientists. Although Galileo was not the first to develop scientific theories, conduct experiments, or demonstrate the power of invention, he excelled in all these areas. He was a brilliant theorist, a master experimentalist, and a meticulous observer.

In 1610, Galileo used the newly invented telescope to observe the night sky and discovered Jupiter's moons. This discovery provided a smaller-scale model of the planetary system and supported Copernican ideas with optical evidence. Galileo's greatest contribution, however, was the development of the principles of mechanics. Newton applied these principles to planetary orbits in 1687, thus completing the Copernican world system.

3.16 The nature of a Scientific Theory

A theory is considered robust if it meets two key criteria: it must accurately explain a broad range of observations using a model with minimal arbitrary elements, and it must make clear predictions about future observations. For example, Aristotle's belief in Empedocles' theory of four basic elements—earth, air, fire, and water—was simple but failed to make specific predictions. In contrast, Newton's theory of gravity, which posits that bodies attract each other with a force proportional to their masses and inversely proportional to the square of the distance between them, accurately predicts the movements of celestial bodies like the Sun, Moon, and planets.

Modern scientists develop theories or models to explain natural phenomena. They derive conclusions about the behavior of what they are studying and design experiments to test these conclusions. Physical theories are always provisional: they cannot be definitively proven, only tested. While a theory may remain consistent with experimental results over time, it can be disproven by a single contradictory observation. If new experiments align with a theory's predictions, our confidence in it increases. However, any observation that contradicts the theory requires us to either revise or discard it, though the competence of the observer can always be questioned.

Developing a comprehensive theory to describe the entire Universe is extremely challenging. Instead, scientists often break the problem into smaller parts, creating partial theories that address specific classes of observations while simplifying or ignoring other factors. While this approach has led to significant progress, it may be limited if the Universe is fundamentally interconnected. The ultimate goal of science remains to formulate a unified theory that comprehensively describes the entire Universe.

3.17 Enters Galileo

Born in Pisa on February 15, 1564, into an esteemed Florentine family, Galileo is often hailed as the father of science due to his remarkable achievements. At 17, he enrolled at the University of Pisa to study medicine, a choice made to satisfy his father's desire for a secure profession. However, Galileo shifted his focus to mathematics and science after being inspired by Archimedes' work in one of his courses.

Galileo

His father was not pleased with Galileo's decision. Besides his undeniable intellect, Galileo's success as a scientist was fueled by his immense curiosity about the world. Aware of his inquisitiveness, he once asked, "When shall I cease from wondering?" This curiosity was paired with a rebellious spirit; he did not accept ideas simply because they came from teachers, theologians, or ancient authorities. For instance, while Aristotle had philosophized that heavy objects fall faster than lighter ones, Galileo challenged this notion with experiments that disproved Aristotle's claim.

3.18 Using the tools at his disposal

A few years after graduating, Galileo began researching math and physics, presenting his findings at the Florentine Academy. His remarkable talent quickly gained attention, and by age 26, he was appointed as a professor of mathematics at the University of Pisa. There, Galileo embarked on studying mechanics and challenged Aristotle's straightforward views on motion. According to Aristotle, objects fall to their "natural place" based on their composition—heavy objects, being more "earthy," fall faster than lighter ones.

Galileo conducted his research in Pisa and is famously associated with the idea of dropping objects from the Leaning Tower of Pisa, though he did not actually perform this experiment. The high speed of falling stones made it difficult for him to measure accurately with the rudimentary clocks of his time. Instead, Galileo used a makeshift timing device: wine bottles filled with water, each with a hole at the bottom and markings to indicate the water level as it drained. To address the challenge of measuring motion accurately, he devised a long, smooth inclined plane to slow down the descent of wooden balls, allowing him to time their movement more precisely.

3.19 Creating the modern scientific method

Galileo conducted hundreds of experiments using his inclined plane, varying its angle and the mass of the balls. He observed that all balls, regardless of mass, rolled down the plane simultaneously. By using increasingly steeper angles, he demonstrated that, in the extreme case of vertical fall—which he couldn't measure directly—all balls would fall with the same acceleration and hit the ground at the same time. Galileo was the first to design experiments specifically to test his theories of motion. He mathematically proved that all objects, irrespective of their mass, accelerate towards the ground at the same rate and should fall together if dropped at the same time. His meticulous approach established the foundation of the modern scientific method, which is still in use today.

One of Galileo's thought experiments involved his inclined plane arranged in the shape of a letter V. A ball rolling down one incline would speed up, then roll up the second incline, slowing down until it stopped at a height slightly lower than its starting point. To study this, Galileo adjusted the second incline, making it longer and eventually horizontal. Theoretically, if the second incline were infinitely long, the ball would roll forever, trying to reach the original height.

Galileo recognized that friction would eventually slow the ball and bring it to a stop in real life. However, in his thought experiment, he considered a frictionless scenario where a perfectly smooth ball on a perfectly smooth track would continue to roll indefinitely. This concept was groundbreaking, as traditional beliefs, influenced by Aristotle, suggested that objects require a continuous force to remain in motion. Galileo's insight into this idea was profound, as he removed real-world complications to focus on the underlying principles of motion.

Galileo also anticipated the principle of relativity, which Albert Einstein would later expand upon in his general theory of relativity. Galileo's principle of relativity states that motion is relative; one cannot determine their speed or motion without an external reference frame. In essence, Galileo's relativity theory asserts that without an external reference, you cannot detect whether you are moving and, if so, at what speed.

3.20 Galileo's observation of the night sky using the telescope

In 1609, Galileo began using the newly invented telescope to observe the night sky. When he turned it toward Jupiter, he discovered several small moons orbiting the planet. This finding contradicted the Aristotelian and Ptolemaic view that all celestial bodies must orbit the Earth, suggesting instead that not everything was centered on Earth. At the same time, Kepler refined Copernicus's heliocentric theory by proposing that planets move in elliptical orbits rather than circular ones. This adjustment made the theory's predictions align with observational data, dealing a significant blow to Ptolemy's model.

When Kepler learned about Galileo's telescope observations, he likely assumed Galileo had invented the device. However, the telescope was actually invented by Hans Lippershey, a Flemish spectacle-maker, who patented it in October 1608. Upon hearing of Lippershey's invention, Galileo quickly set out to build his own telescopes.

Galileo improved Lippershey's basic design, transforming it into a powerful instrument. While he benefited commercially from his advancements, he also recognized the telescope's immense scientific value. His observations allowed him to see farther and more clearly into space than anyone before him. When Herr Wackher informed Kepler about Galileo's telescope, Kepler immediately saw its potential.

In the spirit of intellectual freedom promoted by the Dutch, the University of Leiden offered Galileo a professorship. At that time, Galileo had been forced by the Catholic Church to recant his support for the heliocentric model under threat of torture. Galileo's first astronomical telescope was an enhancement of the Dutch design, and with it, he discovered sunspots, the phases of Venus, the Moon's craters, and the four large moons of Jupiter, now known as the Galilean satellites.

Galileo detailed his ecclesiastical struggles in a 1615 letter to the Grand Duchess Christina. His observations revealed the Moon's rugged surface, contradicting the Ptolemaic belief in celestial perfection. He also observed sunspots—cooler patches on the Sun's surface. In January 1610, Galileo made a groundbreaking discovery: what he initially thought were stars near Jupiter were actually moons orbiting the planet. This provided clear evidence that not all celestial bodies orbited the Earth, challenging Ptolemy's Earth-centered model.

Galileo, who corresponded with Kepler, recognized that his discovery supported the heliocentric model proposed by Copernicus and refined by Kepler. Despite his conviction in the heliocentric theory, he sought further evidence to challenge the prevailing Earth-centered view. To resolve the debate, clear predictions were needed to test and differentiate between the competing models. Galileo's observations strengthened the case for the heliocentric model, though the final validation awaited a deeper understanding of gravity and the reasons we do not sense the Earth's motion. Despite contradicting common sense, the heliocentric model demonstrated that science often transcends intuitive beliefs.

At this point in history, every astronomer should have embraced the Sun-centered model, but a major shift did not occur. Most astronomers, having devoted their lives to the belief that the Universe revolved around a static Earth, found it difficult to make the intellectual or emotional leap to a heliocentric view. When Francesco Sizi learned about Galileo's discovery of Jupiter's moons, which suggested that the Earth was not the center of everything, he proposed a bizarre counterargument: "The moons are invisible to the naked eye and thus have no influence on the Earth, making them useless and therefore non-existent." Similarly, the philosopher Giulio Libri refused to look through a telescope on principle. When Galileo died, he humorously suggested that perhaps he might finally see the sunspots, Jupiter's moons, and the phases of Venus on his way to heaven.

The Catholic Church was equally resistant to abandoning its doctrine of a geocentric Universe. Despite Jesuit mathematicians confirming the superior accuracy of the heliocentric model, theologians conceded its predictive power but still refused to accept it as a valid representation of reality. Galileo, though a devout Catholic, was also a fervent rationalist who reconciled his faith with his scientific beliefs. He argued that scientists were best equipped to comment on the material world, while theologians should address spiritual matters. As he put it: *"Holy Writ was intended to teach men how to go to Heaven, not how the heavens go."*

Had the Church criticized the heliocentric model based on poor data or weak arguments, Galileo and his peers would have been open to discussion. Instead, the Church's criticisms were purely ideological. Galileo chose to disregard the cardinals' objections and continued to advocate for a new vision of the Universe. In 1623, he saw a chance to challenge the establishment when his friend, Cardinal Maffeo Barberini, was elected as Pope Urban VIII. However, despite his hopes of convincing the Church to accept the heliocentric model, Galileo was ultimately condemned for heresy.

Figure 3-6 Galileo Roman Inquisition

Galileo's trial and punishment marked one of the darkest moments in the history of science, a victory of irrationality over reason. At the end of the trial, Galileo was compelled to recant and deny the validity of his arguments. However, despite the Church's claims, the Universe continued to follow its own scientific laws, and the Earth indeed orbited the Sun.

3.21 The Ultimate Question

Over the next century, the Sun-centered model gradually gained acceptance among astronomers, thanks to improved observational evidence from advanced telescopes and theoretical breakthroughs that clarified the model's underlying physics. Additionally, the passing of a generation of conservative scientists who clung to outdated theories facilitated this shift, as new, more accurate ideas emerged.

Galileo and Kepler's bravery in advocating for the heliocentric model was not universally shared. Even in more progressive regions like Holland, their ideas faced resistance. For instance, in an April 1634 letter, René Descartes wrote from Holland: "You doubtless know that Galileo was recently censured by the Inquisitors of the Faith and that his views on Earth's movement were condemned as heretical. I must confess that the arguments in my own treatise, which included the doctrine of Earth's motion, were so interconnected that disproving one would invalidate all my arguments. Despite believing them to be based on solid evidence, I would not wish to defend them against the Church's authority. I prefer to live in peace and continue my life quietly, adhering to the motto to live where one must live unseen."

Ch 4—Classical Physics

4.1 Early History of Celestial Observations

Throughout history, it has been observed that a few celestial objects move differently from the stars, which follow periodic orbits. Apart from the Sun and Moon, the planets Mercury, Venus, Mars, Jupiter, and Saturn were known to traverse individual paths. These movements intrigued navigators and calendar makers, and the planets were often associated with religious and mystical significance. Various civilizations developed methods to predict their motions.

The 2nd-century Hellenistic astronomer Claudius Ptolemy posited that the Earth was the center of the Universe. According to the Ptolemaic model, planetary orbits were thought to align with their apparent movements across the sky. To explain the observed bends and loops, Ptolemy's model used complex mechanisms of spheres and gears.

During medieval times, Ptolemy's Earth-centered view was widely accepted, serving as both a practical and scientific reference. However, when the Ptolemaic model was eventually challenged, the question of the frame of reference became open for debate. Nicolaus Copernicus proposed replacing the Earth with the Sun as the central reference point, a shift that greatly improved the description of planetary motions within the solar system. Today, we understand that the Sun is just one of millions of stars in our galaxy, which itself is part of countless galaxies observable through powerful telescopes.

In the 17th century, the German astronomer Johannes Kepler (1571-1630) used the meticulous observations of his Danish predecessor, Tycho Brahe (1546-1601), to refine the Copernican model. Kepler demonstrated that planets move along elliptical paths rather than circular ones, with the Sun positioned at one focus of each ellipse. He also discovered that a line connecting a planet to the Sun sweeps out equal areas in equal times, and that the orbital periods of planets are proportional to the 3/2 power of the major axis of their orbits.

Johannes Kepler's lifelong quest to understand planetary motions and uncover a celestial harmony ultimately bore fruit thirty-six years after his death through the work of Isaac Newton (1642-1727). Newton, a British mathematician and physicist, built upon Kepler's discoveries to formulate the general laws of mechanics and specifically those of gravitation. To fully appreciate Newton's genius, one must recognize that he developed Kepler's laws into a comprehensive physical theory by creating essential mathematical tools. Newton invented differential and *integral calculus*; techniques crucial for analyzing variable quantities like planetary movements over time. Using these calculus methods, Newton derived the fundamental principles of motion from Kepler's empirical laws, thus explaining planetary trajectories in complete orbits.

4.2 Enters Sir Isaac Newton

Isaac Newton was born under tragic circumstances on Christmas Day 1642, though some records date his birth to January 4, 1643. His father had passed away just three months before, and he was so tiny that, as his mother later recounted, he could have fit into a quart mug. Newton's early life was marked by illness and a sense of abandonment. He was known for being quarrelsome, unsociable, and remained a virgin until his death, choosing celibacy to avoid distractions from his scientific work.

As an infant, Newton's mother remarried a sixty-three-year-old rector named Barnabas Smith, who refused to accept Newton into his home. Consequently, Newton was raised by his grandparents. Over the years, he developed a deep resentment towards his mother and stepfather. As a college student, he even included in a list of childhood grievances his thoughts of "threatening my father and Mother Smith to burn them and the house over them."

At age eleven, Newton's mother, Hannah, hoped he would help manage her properties. However, Newton struggled with the work and had no interest in agriculture. Fortunately, one of Hannah's brothers intervened, persuading her to send Isaac to school instead. Initially, Newton struggled academically, but he eventually excelled and became the top student in his classes. As graduation approached, the principal encouraged him to apply to college. However, Hannah was determined for her son to take over the Woolsthorpe estate. It was again with the help of her brother and the school principal that Hannah reluctantly agreed to let Newton pursue higher education.

Even as a young man, Isaac Newton showed impatience with questions he deemed trivial, such as whether light was a "substance" or how gravity could act through a vacuum. He entered Trinity College, Cambridge, in 1660 and graduated with a B.A. in 1665. Eager to continue his studies, he faced an unexpected setback when the Great Plague forced the university to close. Newton returned home in June 1665, and the university remained closed until April 1667.

Newton was not initially a standout student; he was more captivated by the works of Copernicus, Kepler, and Galileo than by the standard curriculum focused on Aristotle and Ptolemy. Like Kepler, Newton was not immune to the superstitions of his time and engaged with mysticism. At the Stourbridge Fair in 1663, at the age of twenty, he bought a book on astrology out of curiosity. Unable to understand part of it due to his lack of *trigonometric* knowledge, he subsequently purchased a trigonometry book, only to struggle with its geometrical arguments. This led him to read Euclid's *Elements of Geometry*, which sparked his invention of *differential calculus* within two years.

As a student, Newton was deeply intrigued by light and the Sun. He dangerously stared at the Sun's reflection in a looking glass, which damaged his eyes and caused him to see the Sun's image even in darkness. He wrote, "In a few hours I had brought my eyes to such a pass that I could look upon no bright object with either eye but I saw the Sun before me, so that I dared neither write nor read but to recover the use of my eyes, shut myself up in my chamber, made it dark three days together, and used all means to divert my imagination from the Sun. For if I thought upon him, I presently saw his picture though I was in the dark."

In 1666, at the age of twenty-three, during his time away from Cambridge due to the plague, Newton spent nearly two years in isolation at Woolsthorpe, his birthplace. This period of solitude allowed his mind to fully unleash its power, leading him to conduct experiments on light and develop his theory of colors.

Isaac Newton

Newton began his astronomical observations, dedicating countless nights to tracking comets. During this period, he also developed the foundational concepts of his law of universal gravitation, which became central to his celestial mechanics. Realizing he needed a new mathematical tool to complete his calculations for gravity—one that did not yet exist—he set about creating differential and integral calculus. This innovation, along with his discoveries in light and the groundwork for the theory of universal gravitation, came from a 22-year-old recent college graduate during his remarkable year of 1666.

A similar leap in physics occurred nearly 240 years later with Albert Einstein's "Miracle Year" of 1905. When asked about his remarkable achievements, Newton's response was simply, "By thinking upon them." His work was so groundbreaking that Isaac Barrow, his teacher at Cambridge, resigned his chair of mathematics in favor of Newton five years after he returned to college.

In his mid-forties, Newton was described by his servant as someone who never engaged in leisure activities like riding, walking, or bowling. He viewed any time not spent on his studies as wasted and seldom left his room except for lectures, where he often spoke to an empty audience, sometimes reading to the walls due to the lack of listeners.

Newton's engagement with the Holy Scriptures was as intense as his work in physics. He early on rejected the conventional Christian belief in the Trinity, which he saw as a misinterpretation of Scripture, according to biographer John Maynard Keynes. Later in life, Newton concluded that the Bible was not inerrant, believing that the doctrine of the Trinity resulted from later falsifications. He believed in a singular revealed God but went to great lengths to keep this belief a secret throughout his life.

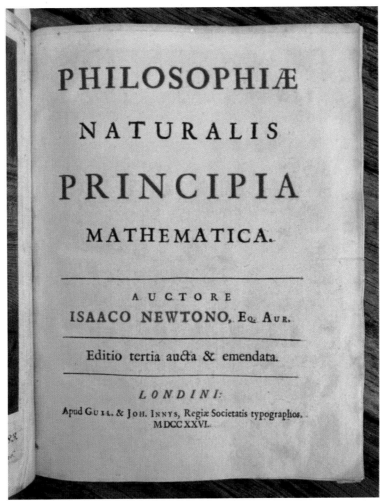

Figure 4-1 The Principia (Calculus)

During his time at college, Newton's peers were unaware of the groundbreaking work he was doing. He discovered the law of inertia—the principle that a moving object continues in a straight line unless acted upon by an external force. He realized that the Moon would fly off in a straight line if not continually redirected by a force, which he identified as gravity. This force, acting at a distance without any physical connection between the Earth and the Moon, was pivotal in shaping Newton's laws of universal gravitation. Using Kepler's third law, Newton mathematically defined the nature of this force, demonstrating that it governed the motion of the Moon, the recently discovered moons of Jupiter, and any falling object on Earth.

Newton's insight into the Moon's orbit around the Earth, which had been a longstanding belief, marked the first time someone connected two different phenomena under a single force. This universality is central to Newtonian gravitation: the same law applies everywhere in the Universe. Although Newton did not acknowledge his debt to Kepler directly in his work *"Principia,"* he later admitted in a 1686 letter to Edmund Halley that his law of gravitation stemmed from Kepler's work.

Newton's laws could derive all three of Kepler's empirical laws of planetary motion from theoretical principles. While Kepler's laws were based on Tycho Brahe's meticulous observations, Newton's theoretical laws provided a simpler mathematical framework from which all observations could be explained. Newton's work in *"Principia"* was revolutionary, laying out a comprehensive framework for understanding the cosmos.

Later in life, Newton became the president of the Royal Society and Master of the Mint, where he focused on combating counterfeiting. Despite his growing reclusiveness and rumored emotional struggles, possibly due to psychogenic ailments from heavy metal poisoning, Newton's intellectual prowess remained exceptional. In 1697, he solved the *"brachistochrone"* problem—a challenge posed by Bernoulli—by inventing a new branch of mathematics called the *"calculus of variations."* His solution, submitted anonymously, was so distinct that Bernoulli recognized Newton's work by its brilliance.

Newton's final posthumous work, *"The Chronology of Ancient Kingdoms Amended,"* explored various subjects including astronomical calibrations of historical events, an architectural reconstruction of Solomon's Temple, and provocative claims about constellations and ancient deities.

Kepler and Newton signify a pivotal shift in human understanding, revealing that simple mathematical laws govern all aspects of nature. Their work demonstrated that the same principles apply both on Earth and in the heavens, showing a deep resonance between human thought and the workings of the universe. Our current global civilization and our exploration of the Cosmos owe much to their groundbreaking insights.

Galileo's work laid the foundation for physics, but it was Newton who fully developed the field. He built a comprehensive system of the world that operated with remarkable precision. Despite being competitive and sometimes secretive about his discoveries, Newton's humility before the grandeur of nature was evident. As he reflected shortly before his death, he saw himself as a child playing on the seashore, discovering small treasures while the vast ocean of truth remained largely unexplored. To the world, Isaac Newton is celebrated as one of the greatest scientific geniuses in history.

4.3 Developing Newtonian Physics - Newton's Universe

Our modern understanding of motion traces back to Galileo and Newton. Before them, people followed Aristotle's belief that the natural state of an object was to be at rest and that it only moved when driven by a force. Aristotle also taught that heavier objects fell faster than lighter ones because they had a stronger pull toward the Earth. He believed that the laws governing the universe could be deduced through reasoning alone, without the need for observation. Consequently, no one tested whether objects of different weights fell at different speeds—until Galileo.

Legend claims Galileo disproved Aristotle's theory by dropping weights from the Leaning Tower of Pisa, though this story is likely untrue. Instead, Galileo rolled balls of varying weights down a smooth slope. This was like heavy objects falling but easier to measure. His experiments showed that objects, regardless of their weight, increased in speed at the same rate. For example, a ball rolling down a slope that drops by one meter for every ten meters traveled will accelerate by one meter per second in the first second, two meters per second after two seconds, and so on. The same principle applies to falling objects, though air resistance can affect lighter objects like feathers. In a famous experiment on the Moon, astronaut David R. Scott confirmed that a feather and a hammer fall at the same rate without air resistance.

Galileo's findings laid the foundation for Newton's laws of motion. Through Galileo's experiments, Newton realized that the effect of a force is to change an object's speed, not just set it in motion. This led to Newton's first law, published in 1687, which states that an object will continue moving in a straight line at a constant speed unless acted upon by a force.

Newton's second law explains how objects react to forces. It states that an object will accelerate at a rate proportional to the force applied. The more force, the greater the acceleration. However, the greater the object's mass, the less it will accelerate for the same force. A powerful engine accelerates a car quickly, but a heavier car will accelerate more slowly with the same engine.

In addition to his laws of motion, Newton developed the theory of gravity, explaining how to calculate the gravitational force between two bodies. His theory states that every object attracts every other object with a force proportional to their masses. If one object's mass doubles, the force between them also doubles. If one object's mass triples and the other's doubles, the force is six times stronger.

Newton's laws explain why all objects fall at the same rate. A heavier object has more gravitational force pulling it down, but it also has more mass, which reduces the rate of acceleration. These two effects cancel out, so objects of different weights fall at the same speed.

Newton's law of gravity also shows that the farther apart two objects are, the weaker the force between them. The gravitational force decreases by the square of the distance between them, meaning that if the distance doubles, the force becomes one-quarter as strong. This principle allows precise predictions of planetary orbits.

The key difference between Aristotle's and Newton's views is that Aristotle believed in a state of rest, particularly that the Earth was stationary. Newton's laws, however, show that there is no absolute standard of rest in the universe.

Newton built on Galileo's groundbreaking discoveries about motion to develop a complete theory of the Universe. Galileo had shown that an object moving at a constant speed in a straight line will continue to do so indefinitely unless acted upon by a force. Newton took this idea further by considering how applying a force changes an object's motion. He realized that an object's motion can only change if an unbalanced force (the net force after accounting for all other forces) acts on it.

For example, to keep a planet moving in orbit around the Sun, a force must be acting on it. This aligned with Kepler's observations of planetary motion, where he suggested that the Sun exerted a force on the planets to maintain their orbits. Galileo had discovered that no force is required to keep an object moving, as long as it maintains a constant speed in a straight line. Newton synthesized these insights in 1666—his "miracle year"—creating the foundation of what we now call classical physics.

4.4 Newton's universal law of gravitation

Newton's analysis revealed that applying a force to an object changes its state of motion. If the object was stationary, the force made it move; if it was already moving, the force could speed it up, slow it down, or alter its direction, depending on how the force was applied.

Many have heard the famous story of Newton discovering gravity when an apple fell from a tree, perhaps hitting him on the head. According to his friend William Stuckey, during a warm evening while drinking tea in Newton's garden under an apple tree, Newton recalled how the fall of an apple had once triggered his thoughts on gravity. Though the apple didn't strike his head, this moment did inspire Newton to consider the force of gravity, leading to his first laws of motion. In essence, Newton's first law says that without any forces acting on it, an object remains in its current state of motion.

To understand Newton's second law, consider driving a car. Pressing the gas pedal causes the engine to supply the force that accelerates the car, while pressing the brake applies a force that slows it down. Friction with the road allows you to steer, and without it—like on icy or wet roads—control is lost, often leading to accidents.

Newton's second law also states that moving a more massive object requires a larger force. This concept is captured in the famous equation $F = ma$, where F represents force, m is mass, and a is acceleration. While not as widely recognized as Einstein's $E = mc^2$, this equation is just as crucial, forming the foundation for analyzing any object's motion.

Newton's third law of motion—the law of action and reaction—adds that forces always come in pairs. You can't apply a force without an equal and opposite reaction force. For example, when you step on the gas pedal, the car's engine turns the wheels. On slick ice, the wheels may spin without moving the car, but on dry pavement, the ground provides the friction that allows the car to accelerate. The wheels exert a reaction force on the car, causing it to move forward.

Newton's three laws of motion can be summarized as follows:

1. **Law of Inertia**: An object will remain at rest or continue moving in a straight line at a constant speed unless acted upon by an external force.
2. **Law of Acceleration**: When a force is applied to an object, it changes the object's motion. This change depends on the object's mass—heavier objects require more force to accelerate.
3. **Law of Action and Reaction**: For every action, there is an equal and opposite reaction. If you apply a force to an object, the object applies an equal force back on you.

4.5 Newton's masterpiece

Newton's second law of motion led him to realize that a planet's orbit around the Sun requires a force pulling it toward the Sun. He deduced that this force must come from the Sun itself. Similarly, he reasoned that the force keeping the Moon in orbit around the Earth must come from the Earth.

Newton then wondered if the same force that pulls an apple to the ground also keeps the Moon in orbit. Could gravity be responsible for both? Recalling Kepler's third law, which relates a planet's orbital period to its distance from the Sun, Newton calculated the force exerted by the Sun on the planets. He derived an equation showing that this gravitational force is inversely proportional to the square of the distance between the planet and the Sun.

4.6 The Earth and Moon attraction

If the Moon is held in its orbit around the Earth by the gravitational pull between the two, the Moon should technically be falling toward the Earth. In reality, both the Earth and the Moon fall toward each other, but because the Earth is much more massive, its motion is negligible. If the gravitational force between the Earth and Moon were suddenly removed, the Moon would shoot off into space in a straight line. This is similar to twirling a ball on a string—if the string snaps, the ball flies off in a straight line. The gravitational pull of the Earth, like the string, constantly pulls the Moon from its straight path into a circular orbit.

Proving that this force is the same one that pulls an apple to the ground required mathematical tools that didn't yet exist. Newton, a brilliant mathematician, invented those tools—what we now call calculus. With this new method, Newton demonstrated that the same force pulling an apple from a tree keeps the Moon in its orbit around Earth. Even more remarkable, he extended this discovery to apply to all objects, formulating it as a universal law.

Newton's universal law of gravitation states that every object in the Universe attracts every other object with a force proportional to their masses and the distance between them. This law explains the movements of the planets around the Sun, the orbits of Jupiter's moons, and even the behavior of the particles in Saturn's rings.

Newton's law of gravitation revealed that the Universe operates like clockwork, with predictable motions. Today, we rely on Newtonian physics to calculate the orbits of planets and launch spacecraft that travel vast distances, using the gravitational pulls of planets like Venus and Earth to reach distant destinations, such as Saturn, precisely when expected.

4.7 Two views of the Universe

In Newton's vision of the Universe, it operates like clockwork. His equations of motion could, in theory, reveal everything about the cosmos. If you had enough time and computational power, you could take the current state of the Universe, input all the variables into his equations, and rewind the clock to reveal the entire history of the Universe, all the way to its beginning. You could also fast forward to predict the future, discovering what would happen in the next second, year, or even century, and eventually uncover the ultimate fate of the Universe.

Newton laid out this comprehensive view of the Universe in his work *Principia Mathematica*. In it, he not only proposed how objects move through space and time but also created the mathematics—calculus—necessary to analyze these motions. One of his key revelations was the law of universal gravitation, which states that every object attracts every other object in the Universe. The strength of this attraction depends on the masses of the objects and the distance between them. Newton showed that the same force that causes an apple to fall to the ground also governs the orbits of the Moon and planets.

While the story of an apple falling on Newton's head is likely a myth, Newton himself mentioned that the idea of gravity came to him as he sat under a tree and observed an apple fall. From this simple observation, Newton concluded that the same gravitational force causes the Moon to orbit the Earth and the planets to follow elliptical paths around the Sun. By discarding the ancient idea of celestial spheres, Newton and his contemporaries also opened up the possibility that the stars were like our Sun, but much farther away.

However, Newton realized that his theory posed a problem: if gravity was universal, why didn't all the stars eventually pull each other together and collapse into a single point? In a letter to Richard Bentley in 1691, Newton suggested that if there were an infinite number of stars evenly spread throughout infinite space, there would be no central point for them to collapse into. However, this was a flawed idea, as we now know that gravity in an infinite, static Universe would eventually lead to instability. The stars would either collapse together or drift apart, with no balance possible.

Philosophers like Heinrich Olbers also pointed out another issue with an infinite static Universe—why isn't the night sky as bright as the Sun? If the Universe were filled with an infinite number of stars, every line of sight should end at a star, making the night sky blindingly bright. Olbers suggested that intervening matter might absorb the light, but this matter would eventually heat up and glow just as brightly. The only explanation was that the Universe hadn't existed forever. There had to be a point in time when the stars "turned on," and their light hadn't reached us yet.

Newton's heliocentric model gradually gained widespread acceptance because it provided the first logical explanation for the invisible forces that govern planetary motion. He condensed the work of his predecessors into his three laws of motion: (1) An object at rest will remain at rest, and an object in motion will stay in motion unless acted upon by an external force; (2) A force applied to an object causes acceleration in proportion to the force and inversely to the object's mass; and (3) Every action has an equal and opposite reaction.

These laws became the foundation of classical mechanics and still hold today, except in extreme conditions like high speeds or strong gravitational fields. Newton's principle of gravitation, which explains the acceleration of falling objects and the orbits of celestial bodies, revealed that every object in the Universe exerts an attractive force on every other object, proportional to their masses and inversely proportional to the square of the distance between them.

To explain the acceleration of falling objects and the orbits of the Moon and planets, Newton introduced his law of universal gravitation. He stated that every object in the Universe attracts every other object. The force between two objects, x and y, is proportional to the product of their masses ($m_x * m_y$) and inversely proportional to the square of the distance (d) between them.

Finally, Newton's work made clear how the planets orbit the Sun—they are constantly falling toward it due to gravity, but their inertia keeps them in a stable orbit. The same is true for the Moon orbiting Earth and Jupiter's moons orbiting their planet. Although Newton couldn't explain the origin of gravity, his laws proved its existence and described its effects, revolutionizing our understanding of the Cosmos.

4.8 Orbital geometry according to Newton

Johannes Kepler demonstrated that planetary orbits are not perfect circles. In fact, it's rare for any object's orbit to be circular; there is almost always some variation. For example, the Moon's distance from Earth changes as it orbits, and Earth is about two million miles closer to the Sun in January than in July. Some asteroids follow highly eccentric, elongated paths, while other objects passing through the solar system are pulled by the Sun's gravity, make a single pass, and then never return.

Newton explained that all orbital paths, whether closed or open, follow a "conic section" shape. These include circles, ellipses, parabolas, and hyperbolas. You can visualize these by shining a flashlight on a flat surface in a dark room. When held straight down, the beam forms a circle; tilt the flashlight slightly, and it becomes an ellipse. Tilt it further to form a parabola or a hyperbola.

Newton realized that if an object in a circular orbit around the Sun received a slight nudge, its path would distort into an ellipse. A stronger push could turn the orbit into a parabola, sending the object out of the solar system forever. An even larger force would create a hyperbolic path. Such orbital shapes are common in space, with parabolas and hyperbolas being frequent for objects passing through gravitational fields.

The gravitational pull between all celestial bodies causes slight shifts. For instance, as the Moon orbits Earth, it displaces Earth slightly, just as Earth causes a small displacement in the Sun's position as it orbits. Even the smallest satellite in orbit slightly affects Earth's position. Though tiny objects like Sputnik didn't move Earth much, the Moon's gravitational pull is strong enough to shift Earth's center of gravity by about 2,900 miles.

4.9 Newton's Absolute Space and Time

Newton's theory of gravitation, which clarified the mechanics behind planetary and satellite orbits, also raised questions about the broader implications for the universe. If every object attracts every other object, even at great distances, this could have significant effects on the overall structure of the cosmos.

Early attempts to measure stellar distances used the "parallax" method, which involved observations from two widely separated points. With Earth's orbit spanning about 186 million miles, this baseline allowed astronomers to triangulate star positions. Stars appeared displaced against the background of fainter stars when observed from Earth in April compared to October, confirming their vast distances. However, the cumulative gravitational effects of countless stars, potentially infinite in number, could be substantial even across cosmic distances.

Newton did not fully connect the pervasive gravitational field of the universe with his laws of inertia. Instead, he attributed inertia to "absolute space," delving into theoretical rather than empirical realms. This led him to question why the Moon doesn't fall into the Earth or the Earth into the Sun. According to his view, both the Moon and Earth were in motion relative to this absolute space.

An experiment designed to prove Earth's rotation involved a heavy weight suspended by a long wire. The weight was set in motion and, after accounting for air currents and disturbances, it was observed that the plane of the pendulum's swing rotated over time. This was taken as evidence that Earth rotates relative to absolute space.

Newton also proposed the concept of "absolute time," suggesting that time flows uniformly and is unaffected by external factors. This assumption, though intuitively appealing, lacked empirical support. Modern relativistic physics has since shown that both absolute time and absolute space are illusions.

As scientific understanding advanced and measurement techniques improved, astronomers tested Newton's theories more rigorously. While Newton's laws accurately described many celestial phenomena, one persistent discrepancy remained. In the realm of logic and science, even a single exception can challenge the validity of a theory.

4.10 The perihelion of Mercury

The perihelion of Mercury shifts by 43 seconds of arc per century, a phenomenon first observed by astronomer Leverrier in 1845. This precession was undeniable but remained unexplained by Newton's laws, highlighting a significant anomaly in the otherwise flawless Newtonian model of the Universe. Despite various attempts, a satisfactory explanation only emerged with Einstein's General Theory of Relativity. For a long time, the success of Galilean-Newtonian mechanics—explaining both terrestrial and celestial movements—dominated scientific thought. By the mid-19th century, interpreting physical phenomena through Newtonian mechanics was seen as the pinnacle of research. However, this focus led to a neglect of whether the foundational assumptions were sufficient to support the entire framework. Einstein was the first to emphasize the critical role of the equivalence of inertial and gravitational mass in the foundation of physical sciences.

4.11 Analytical Mechanics

The challenge in analytical mechanics is to determine the motion of an object given the forces acting on it, as described by the equation *mb=K*. This equation provides the acceleration, or the change in velocity. To find the velocity from acceleration, and then the object's position from velocity, involves solving complex integral calculus problems, especially when forces vary intricately with position and time. For instance, deriving the position change in uniformly accelerated motion along a straight line illustrates this issue. The problem becomes even more complex in a plane where the motion is influenced by a constant force with a specific direction, such as in the case of a falling or thrown object. Here, one can approximate the continuous motion by breaking it into a series of uniform motions, with each transition accomplished through impulses.

In cases where forces act continuously, the approximate method can be replaced by a rigorous approach using integral calculus. Mechanical problems often involve multiple bodies interacting with each other, with forces depending on their unknown motions, making the calculations complex.

When the velocity of a planet is known at all times, differential calculus can determine the rate of change of velocity, or acceleration, at any moment. Conversely, integral calculus allows one to compute the distance traveled over a given period from the velocity. Newton realized that Kepler's law, which states that the line between a planet and the Sun sweeps out equal areas in equal times, implies that any change in a planet's velocity must be directed toward the Sun, not perpendicular to it.

Newton derived three key conclusions, building on Galileo's earlier work:

1. The Sun's influence causes continuous changes in a planet's velocity, meaning that these changes, or accelerations, follow relatively simple laws. Newton dismissed the medieval idea that each body had a fixed position in the Universe, asserting instead that a body not influenced by others would maintain its velocity, with external forces causing changes or accelerations.
2. From Kepler's equal-areas law, Newton inferred that the Sun's force on each planet is attractive and directed toward the Sun.
3. Newton also deduced that gravitational attraction between two bodies depends on their distance from each other and is inversely proportional to the square of that distance. For example, if the distance between two bodies is doubled, the gravitational force becomes one-fourth of its original strength.

During Newton's time, only gravity and the forces of push and pull were quantitatively studied. Other forces like electromagnetism and the weak and strong nuclear forces had not yet been discovered, though magnetism was known since ancient times.

Newton aimed to develop a general theory encompassing all known and future forces. He intended his theory of gravitation to be a comprehensive example of this vision. To describe forces broadly, Newton formulated his three fundamental laws:

1. A body remains in uniform rectilinear motion unless acted upon by an external force.

2. When a force acts on a body, it accelerates in the direction of the force, with the force being equal to the product of the body's mass and its acceleration (f = ma).
3. For every action, there is an equal and opposite reaction. Thus, each planet exerts an attractive force on the Sun, causing a small acceleration due to the Sun's much greater mass compared to the planets.

Among all known forces, gravity is notable for its universal presence. All objects have mass, defined as the ratio of applied force to resulting acceleration. The gravitational force between two bodies depends on their masses and follows the inverse-square law, meaning the force diminishes with the square of the distance between them. Other forces may vary with distance in different ways.

A summary of Newtonian Mechanics

In Newtonian mechanics, mass plays a dual role: it affects both the relationship between force and acceleration and the magnitude of gravitational attraction. This dual function leads to the principle of equivalence. According to this principle, in a given gravitational field, the acceleration due to gravity is the same for all objects, regardless of their mass. For two bodies with masses *m* and *M* falling freely toward the Earth, the ratio of the gravitational forces they experience is proportional to their masses, *m/M*. However, their accelerations remain equal. Thus, gravitational acceleration is independent of the properties of the falling bodies and depends solely on the masses and distributions of the objects creating the gravitational field. The principle of equivalence states that mass equally measures a body's resistance to acceleration and its gravitational attraction.

On Earth's surface, gravitational acceleration is approximately 32 feet per second squared (ft/s^2) or 10 meters per second squared (m/s^2) in metric units. This means that, in the absence of air resistance, the speed of a freely falling object increases by 32 feet per second every second. For example, an object starting from rest will reach a speed of 96 feet per second by the end of the third second of free fall. This rate of acceleration is commonly represented by the symbol ggg. However, because the Earth is not a perfect sphere, ggg varies slightly depending on location: it is stronger near the poles and weaker at higher altitudes or near the equator.

This uniform acceleration due to gravity is unique, as other forces like electric fields cause particles with different charge-to-mass ratios to accelerate differently. Instruments like mass spectrographs exploit this principle to separate particles based on their properties. In contrast, a gravitational mass spectrograph would be impractical because gravitational forces would cause all particles to accelerate identically, regardless of their properties.

Newton recognized the universality and uniformity of gravitational acceleration but did not fully understand its implications. It was left to the German physicist Albert Einstein (1879-1955) to interpret this concept, known as the principle of equivalence, and use it to revolutionize our understanding of space and time.

In his work "Principia," Newton introduced a law stating that objects remain at rest or continue in uniform motion unless acted upon by a force. He explained that gravity, the same force causing objects to fall to Earth, was responsible for the elliptical orbits of planets around the Sun. Newton named this force *"gravity"* and developed the mathematical framework to describe how objects react to gravitational forces. His work demonstrated that the gravitational pull of the Sun causes planets, including Earth, to move in elliptical orbits, aligning with Kepler's predictions. Newton's laws applied universally, from falling apples to celestial bodies, marking the dawn of modern physics and astronomy. Consequently, the concept of rest became understood as relative to its surroundings.

Choice of a proper frame of reference

Newton understood the deep implications of choosing an appropriate frame of reference. His fundamental laws of mechanics dealt with accelerations—changes in the velocities of physical bodies—rather than their absolute velocities. These accelerations were related to distances between bodies, such as Earth's distance from the Sun or the Moon's distance from Earth. While the choice of a reference frame did not affect the measurement of distances between objects, it did influence the calculation of accelerations resulting from mutual attractions and repulsions, which were to be measured relative to a universal norm tied to the chosen frame of reference.

Anyone who has traveled in a car or airplane knows that when the motion is constant and straightforward, one instinctively uses the vehicle as the frame of reference. For example, a passenger can walk down the aisle of a plane at one mile per hour while traveling at several hundred miles per hour, with the ground speed barely affected by their movements. However, during takeoff or landing, the passenger feels pressed against the seat or pulled toward the back of the plane, demonstrating that the plane is an unsuitable frame of reference during these times. In such cases, the accelerations experienced are not solely due to external forces but also due to the changing frame of reference.

To determine which frames of reference are suitable, consider two scientists: one on Earth and the other in a vehicle moving at constant speed in a straight line. If they both observed the flight of the same bird, they would disagree on its speed at any moment, each using their own frame of reference. However, they would agree on the bird's change in velocity (speed and direction). This indicates that for experimental purposes, observers who are in straight-line, unaccelerated motion relative to each other can be considered equivalent.

Newton posited that to truly measure dynamics without distortion, there must be a class of observers for whom a body not acted upon by external forces would experience no acceleration. Such observers, called inertial observers, view the motion of a force-free body as unaccelerated. The concept of inertial frames—those at rest or in uniform straight-line motion relative to each other—became a core aspect of Newtonian mechanics. This idea, known as the "principle of relativity," asserts that no frame of reference is privileged; all inertial frames are equally valid, and there is no absolute rest.

For about two hundred years, the principle of relativity was accepted without question, with scientists believing that only acceleration was absolute, while the rest was relative. However, Newton himself harbored doubts about the completeness of his understanding of the Universe. Despite the success of his gravitational laws in explaining phenomena from the flight of an arrow to the trajectory of a comet, he suspected there was more to discover.

Ch 5—Optics and Newtonian Mechanics

This chapter explores the various components and concepts that paved the way for the development of optics theory and the framework of Newtonian mechanics established by Sir Isaac Newton. Newton's contributions to science were groundbreaking and continue to influence our understanding today. In this chapter, you will be introduced to his significant discoveries and those of his contemporaries.

5.1 Fundamental Laws of Classical Mechanics

Historically, mechanics began with the study of equilibrium, known as "statics," and this starting point is also the most logical progression of the field. The key concept in statics is force, which originates from the subjective experience of physical exertion when we perform tasks. The most obvious example of force is weight, which causes objects to fall downward, making it a natural unit of measurement. Originally, a piece of metal, chosen by decree of the Church or state, served as the unit of weight. Today, an international congress sets these units. In technical contexts, the standard unit of weight is based on a specific piece of platinum kept in Paris, known as the pond.

To measure and compare the weights of different objects, we use a balance. Two objects are considered to have the same weight if they do not disturb the balance when placed on its scales. If we place these two equal-weight objects on one side of the balance, and another object on the other side that maintains equilibrium, this new object has twice the weight of either original object. By continuing this process, starting with the unit of weight, we can create a system for determining the weight of any object.

5.2 Equilibrium and the concept of Force

The principle at work here is that, in a state of equilibrium, all forces cancel each other out. This aligns with Newton's principle of "action and reaction." When this equilibrium is disrupted by reducing or removing one of the forces, motion results. Force, therefore, tends to produce motion. This marks the beginning of dynamics, which aims to uncover the laws governing such motion.

As velocity changes become increasingly frequent and smaller in magnitude, the path traced out—originally a polygon—becomes indistinguishable from a curved line. This represents motion with continuously changing velocity, or non-uniform motion, which can either be accelerated or decelerated. To accurately measure velocity and its rate of change (acceleration), we must employ the methods of differential calculus.

The concept of acceleration can also be applied to non-uniform motion by observing the motion over such a short interval of time that it can be treated as uniformly accelerated within that period. In this way, acceleration becomes continuously variable.

All these definitions gain precision and practicality when carefully studying the subdivision of time into small intervals, during which the quantities involved can be considered constant. This approach leads to the concept of limiting values, which are the foundation of differential calculus and its inverse, integral calculus.

The study of motion, or kinematics, precedes the full development of dynamics, the mechanics of forces. Kinematics can be viewed as a kind of geometry of motion, where each motion is graphically represented in a plane with coordinates of space (x) and time (t). The principle of relativity highlights the importance of treating time as a coordinate alongside spatial dimensions. Velocity and acceleration are thus vector quantities, having both direction and magnitude.

5.3 Motion in Space

Our method of graphical representation originates from motion in space, where we use spatial coordinates (x, y, z) and must include time as a fourth coordinate. However, our visual abilities are limited to three-dimensional space. This is where the symbolic language of mathematics steps in to assist us. Through analytical geometry, we can handle the properties and relationships of spatial configurations purely through calculation, without relying on visual interpretation or drawing diagrams.

This approach is much more powerful than graphical construction. Most importantly, it isn't limited to three dimensions but can be extended to spaces with four or more dimensions. In mathematics, the concept of a space with more than three dimensions is not mystical; it simply means we are working with objects determined by more than three numerical values. For instance, the position of a point at a specific moment in time is described by four values: the three spatial coordinates (*x, y, z*) and the time (*t*).

Once we grasp the use of *xyt-space* to represent motion in a plane, it becomes easier to handle three-dimensional motion by thinking in terms of four-dimensional *xyzt-space*. This framework allows us to apply familiar geometric laws to the study of motion, offering a more comprehensive and efficient way to analyze it.

5.4 Circular Motion

As previously mentioned, this is an example of accelerated motion, since the direction of the velocity is constantly changing. If the motion were not accelerated, the point would continue moving in a straight line with a constant velocity, *v*.

However, in this case, the point must remain on the circle, which requires an additional velocity or acceleration directed toward the center, known as "centripetal acceleration" (sometimes considered a fictitious force). While the magnitude of the velocity remains constant, v, because the motion is uniform, the direction continuously changes. The centripetal acceleration is equal to the square of the velocity divided by the radius of the circle. This principle forms the foundation of one of the earliest and most significant empirical validations of Newton's theory of gravitation.

5.5 Dynamics – The Law of Inertia

The age-old question of how forces generate motion will be examined here. The simplest scenario is one where no forces are present. In this case, a body at rest will remain at rest. The ancients recognized this but also believed the reverse—that motion always required a force to sustain it. This assumption leads to difficulties when considering why a thrown stone or spear continues to move even after leaving the hand that set it in motion.

Clearly, the hand imparts the initial force, but once released, its influence ends. Ancient thinkers struggled to explain what forces maintained the motion of the thrown object. Galileo was the first to offer the correct perspective. He realized it was a mistake to assume that force is always needed to maintain motion. Instead, the right question is: what property of motion is related to force—whether it's the body's position, velocity, acceleration, or a combination of these?

Philosophical reflection cannot answer this question; we must turn to nature for answers. Nature reveals that force affects changes in velocity, but no force is required to maintain motion if the velocity's magnitude and direction remain constant. Conversely, without force, both the magnitude and direction of velocity remain unchanged, meaning an object at rest stays at rest, and an object moving uniformly in a straight line continues to do so.

This "law of inertia" is not as obvious as its simple expression might suggest. In our experience, we rarely encounter objects free from external influences. Imagining bodies traveling in straight paths through space with constant velocity brings up the question of an absolutely straight path in space that is absolutely at rest—an issue we will address later. For now, we will interpret the law of inertia as Galileo intended, in a more limited sense.

5.6 Impulses

We have described the acceleration of nonuniform motion as a limiting case of rapid changes in velocity during brief periods of uniform motion. To explore this further, we must first understand how a single, sudden change in velocity is caused by the application of force. In this case, the force must act over a very short period, creating what is known as an "instantaneous" or "impulsive" force. The effect of this impulsive force depends not only on its magnitude but also on the duration of its action, however brief. Thus, we introduce a new quantity known as *"impulse"* (J).

5.7 Mass and Momentum

Weight acts downward, creating pressure on the spheres (or objects) resting on a table, but it does not generate any horizontal force. Through experiment or reasoning, we can identify a variation in the resistance of spheres to impacts, which we refer to as "inertial resistance." This resistance is defined as the ratio of *impulse* (J) to *velocity* (v) when measured from rest. The term "mass" has been assigned to represent this concept, and it is denoted by the symbol m, with the equation $m = J/v$, where *m* represents mass.

thus, $m = \mathbf{J/v}$, where m = mass.

This formula indicates that for a given body, increasing the impulse (*J*) leads to a proportional increase in velocity (*v*), with their ratio always remaining constant, represented by the value mass (*m*). Once mass is defined this way, its unit is no longer arbitrary, as the units for both velocity and impulse are already established. Mass thus has dimensionality.

It is known that in the absence of forces, velocity remains constant in both direction and magnitude. Therefore, forces are linked to changes in velocity and acceleration. Inertia reflects an object's resistance to motion, while momentum represents the resistance to stopping, slowing, or changing the direction of motion (i.e., the need for a continuous force or drive).

Mathematical verbal notation of changes in speed is given as follows:

Motion velocity = speed = 1ˢᵗ prime.

Motion acceleration = change (increase) in velocity = 2ⁿᵈ prime.

Motion impulse = short durational increased change (jerk) of acceleration = 3ʳᵈ prime

In everyday language, "mass" is often understood as the amount of substance or matter, and the concept of substance is taken as self-evident. However, in physics, mass has no meaning beyond what is defined by its formula: it is the measure of a body's resistance to changes in velocity.

The law of impulses can be expressed more generally as *mw=J*, which describes how a change in velocity *www* occurs when a body experiences an impulse *J*. The product of mass *mmm* and velocity (*v*) is known as momentum (*p*). The quantity *mv*, where *w* is the change in velocity, represents the change in momentum resulting from the impulse *J*.

5.8 Force and Acceleration

Before exploring the notable similarity between mass and weight mentioned earlier, we will apply the established laws to situations where forces act continuously. Although a rigorous formulation of these theorems requires the methods of infinitesimal calculus, the following explanation provides a basic understanding of the relevant concepts.

A continuously acting force generates motion with a continuously changing velocity. Each instance of force application results in a change in velocity *w*. This is the law of motion in dynamics for continuous forces, which states that a force produces an acceleration proportional to its magnitude. If a force *K* acts on a body, the momentum *p=mv* of the body changes such that the rate of change of momentum per unit time equals the force *K*. This formulation of the law is valid only for linear motion where the force acts in the same direction as the motion.

5.9 Weight and Mass

We've observed a notable parallel between mass and weight. Heavier bodies resist acceleration more than lighter ones. To illustrate this, consider the experiment of setting spheres into motion on a smooth horizontal table using impacts. Let's take two spheres, *A* and *B*, where *B* is twice as heavy as *A* (i.e., *B* balances exactly with two spheres like *A* on a scale). When equal blows are applied to both spheres, *A* rolls away at twice the velocity of *B*. This shows that sphere *B*, which is twice as heavy as *A*, resists changes in velocity twice as strongly.

In other words, bodies with twice the mass also have twice the weight. More generally, the ratio of masses *m* is proportional to the ratio of their weights *G*. The ratio of weight to mass is a constant, denoted by *g* and we write,

$$G/m = g \text{ or } G = mg.$$

The experiment used to demonstrate the law is admittedly quite basic, but numerous other phenomena support the same principle. Notably, all bodies fall at the same rate regardless of their weight. For instance, the experiment simplifies the situation by ignoring the resistance encountered during the rolling of a sphere, which depends on its mass distribution (known as the moment of inertia).

In experiments where two spheres roll simultaneously down an incline, it is assumed that only gravity affects their motion. To eliminate air resistance, these experiments are typically conducted in a vacuum. When two spheres of different weights but similar appearances roll down an inclined plane, they reach the bottom at the same time.

The weight of an object is the force driving its motion, while its mass determines its resistance to this force. If weight and mass are proportional, a heavier body will experience a stronger driving force but will also resist acceleration more. Consequently, both heavy and light bodies will fall or slide down at the same rate. Replacing force with weight *G* confirms this observation and we get,

$$mb = G = mg.$$

Thus, all bodies experience the same vertical acceleration due to gravity, regardless of whether they fall from rest or are thrown. This acceleration, denoted by *g*, has the value

$$g = 981 \text{ cm./sec.}^2 \text{ or } (32 \text{ ft./sec.}^2).$$

In another perspective, *gravitational* and *inertial* mass are equivalent. Gravitational mass is defined as weight divided by *g*, while the term "inertial mass" is used to refer to proper mass. Newton was already aware that this equivalence holds precisely.

5.10 Celestial Mechanics

The general form of Newton's law represented a significant advancement in describing planetary orbits. Originally, Newton's law was derived from Kepler's laws and merely served as a concise summary of them.

However, a key development was proving that, under Newton's law, the orbit of a body around a central attracting body must be a Keplerian ellipse. This result extends to closed periodic orbits known as ellipses, but not to *hyperbolic* orbits, which are not closed.

The complexity increases when considering both bodies in motion and adding additional bodies into the problem. This leads to the *"n-body problem,"* which mirrors the true conditions in our solar system, where each celestial body exerts gravitational influence on every other body. Thus, Kepler's ellipses are approximate due to mutual gravitational interactions. These interactions, while small over long periods, result in deviations from the strict Keplerian paths, known as *"perturbations."*

In Newton's time, perturbations were recognized, and subsequent centuries of observational refinements have amassed a vast amount of data that Newton's theory successfully accounted for, marking one of the greatest triumphs of science.

Mathematicians from various countries contributed to the development of the *"theory of perturbations."* While a rigorous solution for the three-body problem remains elusive, calculations can precisely predict celestial motions over extensive time spans. Newton's theory has consistently aligned with new observations, except for one notable case.

This exception involves Mercury, the planet closest to the Sun. While Newton's laws predict that planetary orbits, *perturbed (disturbed)* by gravitational interactions, should show gradual changes, Mercury's orbit displays a small but persistent deviation. This deviation, *43 seconds of arc per century*, was first calculated by the astronomer Leverrier in 1845. Despite efforts to explain it with hypothetical additional masses, the deviation remains unexplained by Newtonian mechanics.

The fact that this deviation occurs with Mercury suggests there may be a fundamental flaw in Newton's theory. Any refinement to Newton's laws must stem from a more general and fundamentally sound principle. Albert Einstein achieved this with his theory of relativity, addressing the limitations of Newtonian mechanics.

5.11 The Relativity Principle of Classical Mechanics

In exploring the grand problems of the Cosmos, we often overlook our starting point—Earth. The laws of dynamics, initially discovered through terrestrial experiments, were extended to astronomical space, despite Earth's rapid orbit around the Sun. This raises a question: How did Galileo uncover laws on a moving Earth that, according to Newton, should only apply in a space absolutely at rest?

We touched on this when discussing Newton's ideas about space and time. We noted that the seemingly straight path of a sphere or ball rolling on a table is slightly curved due to Earth's rotation. This curvature is imperceptible because the path and observation time are so short relative to Earth's rotation. However, Earth is also moving in its orbit around the Sun at an enormous speed of 30 *km/s*.

This orbital motion is a form of rotation, and one might expect it to influence terrestrial events similarly to Earth's axial rotation, albeit to a lesser degree due to the tiny curvature of its orbit. Nonetheless, the forward motion of Earth is nearly uniform and straight over short periods, making the effects of this vast speed seem negligible.

In practice, all mechanical events on Earth behave as if this immense forward motion did not exist. This observation aligns with the relativity principle of classical mechanics, which states that the laws of mechanics are the same in any coordinate system moving uniformly and rectilinearly through Newton's absolute space, as they are in a coordinate system at rest.

5.12 Revisiting Absolute Space and Absolute Time

The principles of mechanics, as developed here, were inspired in part by Galileo's work and in part formulated by Newton himself. For Newton, these generalized definitions and laws seemed disconnected from terrestrial experiments and were initially applicable only to astronomical phenomena.

In developing these laws, Newton had to make specific assumptions about space and time. Without such definitions, even the fundamental law of inertia—stating that a body not acted upon by a force will move uniformly in a straight line—would be meaningless. Consider the table where we first experimented with the rolling sphere. To an observer on another planet, the sphere's path, which appears straight to someone on Earth, would seem curved because of Earth's rotation. While the sphere travels in a straight line relative to an Earth-bound observer, this path would trace a curve for an observer who is not rotating with the Earth.

5.13 Rectilinear or straight motion of objects

In experiments conducted on Earth—such as observing a sphere rolling on a table—the path of the freely moving body is not perfectly straight but slightly curved. This subtle curvature goes unnoticed due to the short distances involved in these experiments compared to the vast size of the Earth. Often in science, limited observation leads to the discovery of significant facts. If Galileo had access to the precise observational tools of later centuries, the sheer complexity of the data might have obscured the discovery of fundamental laws. For instance, Kepler might not have deciphered the planetary orbits if he had known their true accuracy from the start, as Kepler's ellipses are only approximations that deviate from actual orbits over long periods.

A similar situation occurred in modern physics with the analysis of spectral lines; the abundance of observational data initially complicated and delayed the discovery of simple relationships.

Newton faced the challenge of finding a reference system where the law of inertia, along with other mechanical laws, would apply consistently. Choosing the Sun as a reference would not have solved the problem, as it too might be discovered to be in motion. This led Newton to conclude that a reference system based on material bodies could never serve as a solid foundation for the law of inertia. Instead, he saw the law of inertia—and its inherent connection to Euclidean space, particularly the straight line—as the natural starting point for understanding dynamics in astronomical space.

Similarly, the concept of time, expressed through uniform motion due to inertia, would not hold if one used the Earth's rotation period as a time unit because of Earth's irregular motion. Thus, Newton concluded the existence of absolute space and absolute time.

5.14 Newton's Law of Attraction

Newton achieved a major breakthrough by developing a dynamical theory of planetary orbits, which we now refer to as celestial mechanics. He extended Galileo's concept of force to explain the motions of the stars. Rather than relying on speculative hypotheses, Newton arrived at his laws by systematically and rigorously analyzing the known facts of planetary motion. These facts were encapsulated in Kepler's three laws, which succinctly summarized all the observational data of the time.

We must here state Kepler's laws in full. They are the following:

1. The planets move in ellipses with the Sun at one of the foci.
2. The radius vector drawn from the Sun to the planet describes equal areas in equal times.
3. The cubes of the major axes of the ellipses are proportional to the squares of the periods of revolution

The fundamental law of mechanics relates the acceleration b of a motion to the force k that produces it. Knowing the acceleration b, which is determined by the motion's trajectory, allows for the calculation of k. Newton realized that Kepler's laws provided a sufficient basis to determine b, and thus, the force k could be calculated using the relation $k=mb$.

At the time, ordinary mathematics was inadequate for this task, but Newton had access to differential and integral calculus, which were developed almost simultaneously by Leibniz in 1684 and form a cornerstone of modern mathematics. Despite this, Newton chose not to use these methods in his seminal work, *Philosophiæ Naturalis Principia Mathematica*. Instead, he used classical geometric methods.

The planetary orbits are ellipses with minor eccentricities, meaning they are nearly circular. It is reasonable to approximate planetary orbits as circles around the Sun, an assumption that aligns with Kepler's first law, since a circle is a special case of an ellipse.

The second law now implies that each planet moves in its orbit with a constant speed. We know that in such circular motion, the acceleration is directed towards the center of the circle. The formula for this acceleration is given by:

$$b = v^2/r,$$

where *v* is the orbital speed and *r* is the radius of the circle. Next, we examine Kepler's third law, which, for a circular orbit, asserts that the ratio of the cube of the orbit's radius (r^3) to the square of the orbital period (T^2) has the same value constant (*C*) for all planets:

$$r^3/T^2 = C \text{ or } r/T^2 = C/r^2.$$

If we insert this in the value for b above, we get,

$$b = 4(pi)^2 C/r^2.$$

Thus, the value of the "centripetal acceleration" depends solely on the distance of the planet from the Sun, being inversely proportional to the square of this distance. This acceleration is independent of the planet's properties, such as its mass. Since the quantity C in Kepler's third law is constant for all planets, it pertains only to the Sun's characteristics and not to those of the planets.

Remarkably, the same principle applies to elliptical orbits, though the calculation is more complex. The acceleration always points towards the Sun, located at one of the foci of the ellipse, and is given by the formula, where *r* is the length of the radius vector.

5.15 Limited Absolute Space

We initially justified the use of absolute space and absolute time by pointing out that without them, the law of inertia would lack meaning. However, we must now examine whether these concepts truly deserve to be termed *"real"* in the context of physics. A concept is considered to represent physical reality only if it corresponds to something that can be measured or observed in the world of phenomena. While a detailed discussion of the philosophical concept of reality is beyond scope here, it is clear that the criterion of reality described aligns with how the term "reality" is used in the physical sciences. Concepts that do not meet the standard of being measurable or observable have gradually been removed from the framework of physics.

In this sense, a *"fixed spot"* in Newton's absolute space lacks physical reality, which aligns with the principle of relativity. If we were to assume that a particular reference system is at rest in space, then any reference system moving uniformly and rectilinearly relative to it could equally be considered at rest. Mechanical events would occur identically in both systems, and neither system would be preferred over the other. A body appearing stationary in one system would exhibit uniform rectilinear motion from another perspective, and vice versa. Thus, Newton's concept of absolute space loses much of its concrete meaning. A space without any physically identifiable points is ultimately a highly abstract idea.

The principle of relativity, therefore, asserts that there are infinitely many equivalent inertial systems—each moving uniformly and rectilinearly with respect to one another—in which the laws of mechanics apply in their classical form. This clearly shows how deeply the concept of space is intertwined with mechanics. It is not space that imposes form on objects, but rather the objects and their physical laws that define space. This view evolves further and culminates in Einstein's general theory of relativity.

In Newton's mechanical universe, objects move predictably and can be precisely calculated using the three laws of motion and the universal law of gravitation. In Newtonian physics, space and time are absolute and fixed. For example, a sailor walking along the centerline of his boat at *4 mph*, while the boat sails at *20 mph* relative to land, calculates his speed relative to the land as *24 mph*. Conversely, walking towards the stern at the same speed would result in a speed of *16 mph* relative to the land.

Newton incorporated Galileo's principle of relativity in his development of mechanics and explicitly stated it in his work, *Principia Mathematica*. He recognized that uniform motion does not affect the laws of mechanics.

5.16 Galilean Transformations

The question is how to transition from one coordinate system to another that is moving relative to the first. At this point, we need to briefly address the concept of equivalent (equally valid) coordinate systems and the so-called *"transformation equations"* that allow us to switch between them through calculation.

In geometry, coordinate systems serve as tools to fix the relative positions of one object with respect to another. To do this, we assume that the coordinate system is rigidly attached to one of the objects. The coordinates of points on the other object then provide a complete description of its relative position. It doesn't matter whether the coordinate system is rectangular, oblique, polar, or even more general in shape. Similarly, the orientation of the system relative to the first object is unimportant—unless the orientation changes. In that case, the change in position of the coordinate system with respect to the object must be clearly specified.

5.17 Inertial Forces

Having acknowledged that the individual points in Newton's concept of absolute space lack physical significance, we must now ask what remains of the concept itself. What persists is the idea that the resistance of bodies to acceleration, which Newton interpreted as the influence of absolute space, is still relevant. For example, the locomotive that starts a train must overcome the resistance caused by inertia. Inertial effects emerge whenever accelerations occur, and these accelerations are simply changes in velocity relative to absolute space. Since changes in velocity have the same value in all inertial reference frames, we can still use this notion.

However, reference frames that are accelerated with respect to inertial frames are not equivalent to them, nor are they equivalent to each other. Only inertial reference frames are equivalent to other inertial frames. A body with no external forces acting on it is still subject to inertial forces, so its motion is typically neither uniform nor straight. A vehicle starting or stopping, for instance, is an example of an accelerated system. The jolt felt when a train starts or stops is a familiar example of the inertial force described here.

5.18 Centrifugal Forces and Absolute Space

In Newton's perspective, the presence of inertial forces in accelerated systems supports the existence of absolute space, or more specifically, the privileged status of inertial reference frames. Inertial forces are particularly evident in rotating reference frames, where they manifest as centrifugal forces. Newton relied heavily on these forces to reinforce his doctrine of absolute space. One illustrative example is the Moon's motion around the Earth. According to Newton, without an absolute rotational motion around the Earth, the Moon would fall toward it due to gravitational attraction. Imagine a coordinate system centered at the Earth's core, with the xy-plane containing the Moon's orbit and the s-axis always passing through the Moon. If this system were absolutely stationary, the Moon would be subject only to the gravitational force directed toward the Earth's center, given by the equation:

$$K = k\, Mm/r^2,$$

Thus, the Moon would fall toward the Earth along the x-axis. The fact that this does not happen seemingly confirms the absolute rotation of the coordinate system (x, y). This rotation generates a centrifugal force that balances the gravitational force (K), resulting in the equation:

$$mv^2 = k\, Mm/r^2.$$

This formula is, of course, nothing other than Kepler's third law.

5.19 Universal Gravitation

The law of acceleration discussed in the previous section shares a key characteristic with the gravitational force on Earth (weight): it is entirely independent of the nature of the moving object. When we calculate the force based on the acceleration, we find that it is similarly directed towards the Sun. This force represents an attraction, expressed as:

$$K = mb = m(4\pi^2 C/r^2).$$

It is proportional to the mass of the moving body, similar to how the weight of an object on Earth is expressed as:

$$G = mg.$$

This realization led Newton to the idea that both the force of gravity on Earth and the force acting on celestial bodies might have the same origin. Today, this concept seems so self-evident that we might not appreciate the boldness and creativity it required. Newton envisioned that the motion of planets around the Sun and the Moon around the Earth could be understood as a form of "falling," governed by the same laws and forces as a stone falling to the ground. The reason that planets and the Moon don't crash into their central bodies is due to inertia, which results in a balancing centrifugal force. Newton first applied this idea of universal gravitation to the Moon, whose distance from Earth was measured directly. Using the Earth as the central body and the

Moon as the "planet," with the distance between them denoted as *r* and the Moon's revolution period as *T*, Newton applied his law of gravitation to connect this to Earth's gravitational acceleration *g*. The significance of this result lies in how it redefined the force of weight. In ancient times, weight was thought of as a pull towards an absolute "below" experienced by all objects on Earth. With the discovery of the Earth's spherical shape, weight was reinterpreted as a pull toward Earth's center. Earth itself, as a planet, is attracted to the Sun and similarly attracts the Moon. This is an approximation of the true interactions where the Sun, Moon, and Earth mutually attract each other. Although the Sun's massive size prevents significant acceleration, and the Moon's small mass is relatively negligible, a precise theory must account for these interactions, known as "perturbations."

Newton's law of universal gravitation takes the symmetrical form:

$$K = k \, (mM/r^2)$$

This equation describes the gravitational force between two masses, m and M, where r is the distance between them, and k is the gravitational constant.

In simple terms, the law states: Two bodies attract each other with a force that is directly proportional to the product of their masses and inversely proportional to the square of the distance between them. This mathematical formulation is considered one of Newton's most profound achievements in explaining the dynamics of the Universe.

5.20 The Law of Energy

There is a law that simplifies motion problems and provides a broad understanding of motion: the law of "conservation of energy." This principle has become crucial in advancing the physical sciences. While we cannot fully formulate or prove it here, we can illustrate its meaning with simple examples. In celestial mechanics, this law holds perfectly, where ideal dynamics, as previously developed, are strictly valid. However, on Earth, this is not the case, as all motion is subject to friction, which transforms energy into heat. Machines, which generate motion, convert thermal, chemical, electrical, and magnetic forces into mechanical forces, meaning the law of energy in its strict mechanical form does not always apply. Yet, the law can be upheld in a broader sense. If we consider heat energy (Q), chemical energy (C), electromagnetic energy (W), and other forms, then the law holds that for closed systems, the sum of energies,

$$E = T + U + C + W...$$

is always constant.

5.21 The Optics of moving Bodies

We now turn to the theory of the elastic ether in relation to space, time, and relativity. In previous optical investigations, we ignored the positions or movements of bodies that emit, receive, or propagate light. Now, we must address these aspects. In "mechanics," space is considered empty wherever material bodies are absent. However, in "optics," space is thought to be filled with ether, a hypothetical medium with mass, density, and elasticity. According to this view, Newton's conception of space and time can be applied to a Universe filled with ether, no longer consisting of isolated masses separated by empty space but rather fully filled with this thin, nearly rigid substance. In this model, the coarse masses of matter float in the ether, and both ether and matter interact with mechanical forces, moving according to Newton's laws. The question that remains is whether this theory aligns with observational evidence.

Let us now revisit the principle of relativity in classical mechanics. According to this principle, absolute space exists only in a limited sense, as all inertial systems moving in a straight line and at a constant velocity relative to each other can be equally considered as being at rest. The first hypothesis regarding the luminiferous ether is as follows: In astronomical space, far from material bodies, the ether is at rest within an inertial system. If this were not the case, parts of the ether would experience acceleration, leading to the emergence of centrifugal forces, which would alter its density and elasticity. We would then expect the light from a star to provide evidence of such changes. This hypothesis formally adheres to the classical principle of relativity.

5.22 The Velocity of Light

The determination of the most important property of light—its velocity, which will be central to our following reflections—was made independently of the debate over light's nature. It was already evident from observations that the velocity of light was extremely high. Galileo attempted to measure it using lantern signals, but these efforts failed due to the extremely short time intervals light takes to travel over Earthly distances. Accurate measurements were only achieved by considering the vast distances between celestial bodies.

In 1676, Olaf Romer was the first to calculate the speed of light, ccc, using astronomical observations, specifically the eclipses of Jupiter's moons. Each eclipse occurs when a moon passes into Jupiter's shadow at intervals equal to its orbital period, τ. When viewed from Earth, these signals take additional time, L/c, where L is the distance between Jupiter and Earth. If L changes during the time of the moon's revolution, the observed intervals between eclipses, $τ+l/c$, are either lengthened or shortened depending on whether L increases or decreases.

James Bradley (1727) discovered another phenomenon caused by the finite speed of light: all fixed stars appear to trace an annual motion, which mirrors the Earth's orbit around the Sun.

5.23 The Fundamental Laws of Optics

Mechanics is both the historical and logical foundation of physics, but it represents only a small portion of the broader field. So far, to address the problem of space and time, we have relied solely on mechanical observations and theories. However, we must now consider what other branches of physics, such as optics, electricity, and magnetism, reveal about the nature of space. These fields are particularly relevant because light, along with electric and magnetic forces, propagates through empty space. The idea that physical phenomena could travel across vast cosmic distances led to the hypothesis that space is not truly empty, but rather filled with an extremely fine, imponderable substance called ether, which serves as the medium for these processes. While modern usage of the term "ether" refers only to certain physical states or fields in space, historically, ether was viewed as a real substance, capable of motion and possessing physical properties.

The science of optics can be traced back to Descartes, whose *Dioptrics* (1638) introduced the fundamental laws of light propagation: reflection and refraction. Reflection was known to the ancients, but refraction was only recently discovered by Snell (around 1618). Descartes proposed the idea of ether as the carrier of light, a precursor to the wave theory. Robert Hooke (1667) hinted at the wave theory, which was later clearly articulated by Christiaan Huygens (1678). Newton, their contemporary, is associated with the opposing corpuscular theory. Before discussing the conflict between these theories, it is essential to outline the core ideas of each.

The corpuscular theory posits that luminous bodies emit fine particles that travel in accordance with the laws of mechanics and create the sensation of light when they strike the eye. In contrast, the wave theory draws an analogy between light propagation and the movement of water or sound waves, requiring the existence of an elastic medium—luminiferous ether—that permeates all transparent bodies. In this theory, the ether's particles oscillate around their equilibrium positions, and the light wave is the result of this motion rather than the movement of the particles themselves.

There is, however, a notable objection to the wave theory. Waves are known to bend around obstacles, as seen with water waves or sound waves moving around corners. Yet light seems to travel in straight lines, and when an opaque object with sharp edges is placed in its path, it casts a shadow with a clear boundary. This observation led Newton to reject the wave theory. Though he did not commit to a definitive hypothesis, he suggested that light moves outward from a luminous body "like ejected particles." His successors interpreted this as support for the corpuscular theory, and Newton's authority lent this theory credence for over a century. Yet, by this time, Grimaldi had already discovered (though published posthumously in 1665) that light, too, could bend around corners.

At the edges of sharp shadows, a faint illumination appears in the form of alternating bright and dark fringes, a phenomenon known as the diffraction of light. This discovery played a crucial role in making Huygens a strong advocate of the wave theory. He viewed diffraction as the first and most significant argument supporting the idea that two light rays can cross paths without disturbing each other, much like how water waves intersect without interference. In contrast, if light consisted of emitted particles, collisions or disturbances would inevitably occur between them.

Huygens was also successful in explaining both the reflection and refraction of light using the wave theory. He employed what is now called the Huygens principle, which states that every point struck by light acts as the source of a new spherical light wave. This principle led to a fundamental difference between the wave theory and the corpuscular theory, a difference that ultimately led to experimental confirmation of the wave theory.

When a ray of light travels through air and strikes the boundary of a denser medium, such as glass or water, it bends or "refracts," becoming more steeply inclined toward the surface. The corpuscular theory explains this by suggesting that light particles are attracted by the denser medium, accelerating as they enter and being deflected toward the normal. This implies that light should travel faster in denser mediums. However, Huygens' wave theory assumes the opposite: when the light wave hits the boundary, it generates elementary waves at every point, and if these waves move more slowly in the denser medium, the refracted wave is correctly bent in accordance with Huygens' construction.

5.24 Elements of the Wave Theory – Interference

Newton's most significant achievement in optics was his discovery that white light could be separated into its component colors using a prism. His meticulous examination of the spectrum led him to conclude that these individual colors were the fundamental, indivisible constituents of light. As the founder of the theory of color, Newton's work remains valid today, despite the criticisms made by Goethe. His groundbreaking discoveries were so influential that they stifled independent thought in the generations that followed. Newton's rejection of the wave theory of light also delayed its acceptance for nearly a century. However, there were some proponents, such as the notable mathematician Leonhard Euler, who continued to support the wave theory in the 18th century.

The revival of the wave theory can largely be credited to Thomas Young's work in 1802. Young introduced the principle of interference to explain the colored rings and fringes Newton had observed in thin layers of transparent materials. Interference plays a crucial role in all finer optical measurements and forms the foundation of key discoveries in the theory of relativity.

To understand interference, it is important to review the nature of waves. Waves arise when individual particles of a medium oscillate periodically around their equilibrium positions, with neighboring particles differing in their phase of motion. The time required for one complete oscillation is called the period (T), and the number of vibrations per second is the frequency (v). These quantities are related by the formula:

$$vT = v = 1/T \text{ or } T = 1/v.$$

In the context of light waves, frequency correlates directly with the sensation of color, meaning that a light wave of a specific frequency generates a specific color in the eye. The complex relationship between physical light waves and the perception of color is not fully explored here.

Light waves originating from a small source are spherical, meaning that all particles on the same spherical wave surface (with the light source at the center) are in the same phase of vibration. This is referred to as being in the same "state of vibration" (e.g., all crests or troughs of the wave). Refraction or other factors may distort this spherical wave, altering the shape of its wave surfaces. Whether the individual particles vibrate parallel or perpendicular to the direction of propagation—whether the waves are longitudinal or transverse—will not be specified at this point.

The distance from one wave crest to the next is known as the wavelength and is denoted by the symbol λ. The distance between any two consecutive planes of equal phase also has this same value, λ. According to the wave theory, light is not composed of a stream of material particles but rather is a state of motion. However, two vibration impulses occurring simultaneously can cancel each other out, much like two people attempting to do opposite things end up obstructing each other and accomplishing nothing. This initial challenge to the wave theory was resolved when the theory was further refined, revealing that waves do indeed "bend around corners," but only in areas that are on the same scale as the wavelength itself.

Ch 6—The Nature of Light

6.1 Light Wave / Particle Phenomena

Light was a motif of the age: the central theme of the era, symbolizing both the intellectual awakening of free thought and religion and serving as a subject of scientific investigation. This was evident in Snell's study of refraction, Leeuwenhoek's invention of the microscope, and Huygens' development of the wave theory of light.

Isaac Newton admired Christiaan Huygens, calling him "the most elegant mathematician" of their time and a true successor to the mathematical traditions of ancient Greece—a high compliment then, as it is now. Newton believed that light acted like a stream of tiny particles, which he inferred from the sharp edges of shadows. He suggested that red light consisted of the largest particles and violet the smallest. In contrast, Huygens proposed that light behaved like a wave, traveling through a vacuum much as ocean waves move through water. This wave theory naturally explained many phenomena, such as diffraction, and in the years that followed, Huygens' view became the accepted theory.

Enters Christian Huygens

In Italy, Galileo had proclaimed the existence of other worlds, and Giordano Bruno had speculated about alien life forms, both of which led to their persecution. In contrast, in Holland, Christiaan Huygens, a physicist and astronomer who believed in both ideas, was celebrated and honored. His father, Constantijn Huygens, was a master diplomat, poet, composer, and close friend of the English poet John Donne. Constantijn admired the painter Rubens and "discovered" a young Rembrandt van Rijn, appearing in several of his works. After meeting Christiaan, Descartes remarked, "I could not believe that a single mind could occupy itself with so many things and equip itself so well in all of them." The Huygens household was rich with goods from around the world, frequently hosting distinguished thinkers. Immersed in this environment, young Christiaan excelled in languages, drawing, law, science, engineering, mathematics, and music, reflecting his broad and diverse interests.

"The world is my country," he declared, "science my religion." The microscope and telescope, both developed in early 17th-century Holland, symbolize the expansion of human vision into the realms of the incredibly small and the infinitely large. This era marked the beginning of our observations of atoms and galaxies. Christiaan Huygens had a passion for grinding and polishing lenses for astronomical telescopes, even constructing one that measured five meters in length. His discoveries made through this telescope alone would have secured his legacy in the history of human endeavors.

Huygens

His discoveries with the telescope alone would have secured his place in the annals of human achievement. Following in the footsteps of Eratosthenes, Huygens was the first to measure the size of another planet. He speculated that Venus is completely enveloped in clouds, drew the first surface features of Mars, and recognized that Saturn is surrounded by a ring system that does not make contact with the planet.

Huygens's contributions extended far beyond astronomy. A significant challenge for marine navigation during this time was determining longitude. While latitude could be easily measured by observing stars—where southern constellations become visible as one travels further south—longitude required precise timekeeping. An accurate shipboard clock could indicate the time at the home port, while the rising and setting of the Sun and stars would determine local shipboard time. The difference between these two times would provide the necessary longitude.

To address this issue, Huygens invented the pendulum clock, building on principles discovered earlier by Galileo. Although the clock was not entirely successful for calculating positions across the vast oceans, it represented a significant advancement. Huygens was pleased to see the Copernican View of the Earth as a planet in motion around the Sun gain widespread acceptance, even among ordinary people in Holland. He remarked that Copernicus was recognized by all astronomers, except for those who were "somewhat slow-witted or bound by the superstitions imposed by mere human authority."

During the Middle Ages, Christian philosophers often argued that because the heavens complete a full circle around the Earth every day, they could not be infinite in extent; thus, the existence of an infinite number of worlds—or even a large number—was deemed impossible. The revelation that the Earth rotates rather than the sky moving had significant implications for the uniqueness of our planet and the potential for life beyond it. Copernicus proposed that not only the solar system, but the entire Universe is heliocentric, while Kepler contended that the stars do not possess planetary systems.

Sir Isaac Newton

Giordano Bruno was likely the first to explicitly propose the concept of a vast—indeed, infinite—number of other worlds orbiting different suns. However, many believed that the idea of a plurality of worlds naturally stemmed from the theories of Copernicus and Kepler. As a product of his era, Huygens regarded science as his religion and argued that the planets must be inhabited; otherwise, he contended, God would have created these worlds without purpose.

6.2 Waves vs. Particles

Given his success with particle physics, it's no surprise that Newton approached the behavior of light through a particle lens. Observations showed that light rays travel in straight lines, and their reflection off a mirror resembles how a ball rebounds from a solid surface. Newton constructed the first reflecting telescope, explained "white light" as a combination of all the colors in the rainbow, and made significant advancements in optics—all based on the assumption that light consists of tiny particles, or corpuscles. He noted that light rays bend when transitioning from a less dense medium, like air, to a denser one, such as water or glass. However, even during Newton's time, alternative explanations existed. Huygens, a contemporary of Newton and born thirteen years earlier in 1629, proposed that light is not a stream of particles but rather a wave, similar to ripples on the surface of a body of water, propagating through an invisible medium known as "luminiferous ether."

Similar to ripples created by a pebble dropped into a pond, light waves in the ether were envisioned to radiate outward in all directions from a light source. The wave theory effectively explained both reflection and refraction, just as well as the corpuscular theory did. Consequently, the two theories diverged in their predictions based on observational evidence.

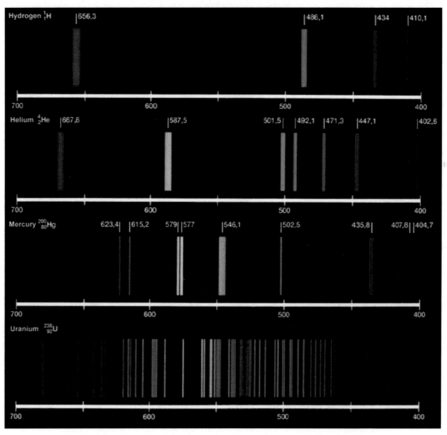

Figure 6.1 Two waves of different frequencies

Three hundred years ago, the evidence strongly supported the corpuscular theory, leading to the wave theory being largely dismissed. However, by the early 19th century, the standing of these two theories had nearly flipped. In the 18th century, very few took the wave theory seriously, with Swiss mathematician Leonhard Euler being one of the notable exceptions. Euler, a leading mathematician of his time, made significant contributions to geometry, calculus, and trigonometry. He developed many techniques that underpinned the modern mathematical descriptions used in physics today. Another supporter of the wave theory was Benjamin Franklin, but their ideas were largely overlooked until critical experiments were conducted by Thomas Young and Augustin Fresnel in the early 19th century.

Young leveraged his understanding of wave motion on water surfaces to design an experiment testing whether light propagates similarly. When considering water waves, it's important to visualize ripples rather than large breakers. A key characteristic of a wave is how it temporarily raises and then lowers the water level as it passes; the height of the wave's crest above the calm water is its amplitude, which corresponds to how much the water level dips as the wave moves through. A series of these ripples, such as those created by a pebble dropped into a pond, are spaced regularly, with the distance between crests defined as the wavelength.

As the pebble creates waves that radiate outward in circles, the waves on the surface of the pond may appear almost flat, though they are technically three-dimensional. In contrast, light travels in a spherical geometric shape from its source rather than dispersing flatly. The frequency of the wave—how many crests pass a fixed point, like a rock, each second—tells us about the wave's characteristics. The velocity of the wave, or the speed at which each crest moves, is determined by multiplying the wavelength by the frequency.

When two ripples spread across the water, they create a complex pattern of interference. Where both waves coincide to lift the water's surface, a more pronounced crest form. Conversely, when one wave creates a crest while the other forms a trough, they can cancel each other out, resulting in an undisturbed water level.

These phenomena are known as constructive and destructive interference, and they are quite observable. If light behaves as a wave, then a corresponding experiment should also demonstrate similar interference patterns among light waves. This is precisely what Young discovered.

Enters Thomas Young

Born in 1773, Young was the eldest of ten siblings and displayed remarkable talent from an early age. By the age of two, he was reading fluently and had read the entire Bible twice by the time he was six.

Young

A master of over a dozen languages, Young made significant strides in deciphering Egyptian hieroglyphics. He studied medicine at the universities of Edinburgh and Göttingen, graduating in 1796. Although he practiced medicine throughout his life, his poor bedside manner hindered his effectiveness as a doctor. Financially secure from a bequest by an uncle, Young was able to pursue various intellectual interests. Despite his medical training, he was more captivated by science and often neglected his patients.

While in medical school, Young discovered how the lens of the eye changes shape when focusing at different distances. He also identified that astigmatism results from imperfections in the curvature of the cornea. His exploration of the eye led him to investigate the nature of light, where he demonstrated that light creates an interference pattern, a hallmark of wave behavior.

From his calculations, he determined that the wavelength of light was much smaller than previously believed, with the longest wavelength in the visible spectrum belonging to red light—less than one-thousandth of a millimeter. This small wavelength explains why light casts sharp shadows and appears not to bend around corners; only tiny objects, like Young's pinholes, can reveal this bending.

Young's fascination with light drove him to compare its properties with those of sound, ultimately leading to a reevaluation of Newton's particle theory. He devised an experiment that would mark the beginning of the end for Newton's ideas. By shining monochromatic light onto a screen with a single slit, he allowed a beam of light to spread out and strike a second screen with two narrow, closely spaced slits. These slits acted as new sources of light, or as Young described them, "centers of divergence, from whence the light diffracted in every direction." What he observed on a third screen placed some distance behind the slits was a central bright band flanked by alternating dark and bright bands, demonstrating the wave nature of light.

6.3 Demonstration that light is a wave

A century after Newton, in 1802, Thomas Young refined Grimaldi's experiments. As previously mentioned, Young directed a beam of light through a pinhole he created in a screen. This light then spread out and passed through a pair of pinholes he had placed side by side on a second screen. Using a third screen, he observed the resulting pattern of dark and bright regions. Young understood that this pattern indicated that light behaves as a wave, similar to how water ripples interact with one another, creating comparable patterns.

6.4 Coherent beams of light

Young's interference experiment, as it is known today, was remarkably clever. You cannot produce these interference patterns with a typical light source. In waves, the height of the crest above the undisturbed water level is known as the amplitude, and in a perfect wave, this is equal to the depth the water is pushed down as the wave passes. A series of ripples, like those created by a stone dropped into a pond, follow each other with regular spacing called the wavelength, which is measured from one crest to the next. Around the point where the pebble hits the water, the waves spread out in circles. The ripples on the surface of the pond appear almost flat, although they are technically three-dimensional. However, light travels in a spherical wave from its source, not in a flat dispersion.

The number of wave crests passing a fixed point, like a rock, in one second, is the wave's frequency. The frequency is the number of wavelengths passing per second, so the velocity of the wave (the speed at which each crest advances) is the product of the wavelength and the frequency.

When two sets of ripples spread across the water, a more complex pattern form. Where both waves raise the water's surface, a more pronounced crest appears, but when one wave forms a crest while the other creates a trough, they cancel each other out, leaving the water undisturbed. These phenomena are known as **constructive and destructive interference** and are easy to observe. If light behaves as a wave, then a similar experiment should create interference patterns among light waves. This is precisely what Young discovered.

6.5 Resistance to Young's pinhole experiment

Young's experiment is considered a landmark in the history of science, providing definitive evidence that light behaves as a wave. The interference pattern he observed could only be explained by wave behavior. Particles bouncing off the edges of the pinholes, as some might have suggested, would not produce such a symmetric pattern unless their motion was somehow coordinated, which was implausible. One might expect this irrefutable proof to immediately establish the wave nature of light, but it didn't. Young's findings contradicted the long-held views of Isaac Newton, and many English physicists, loyal to Newton's authority, were unwilling to accept Young's conclusions.

To explain the appearance of the bright and dark "fringes" in his experiment, Young used a simple analogy. Imagine two stones dropped simultaneously and close together into a calm lake. Each stone generates waves that spread out across the water. As these ripples meet, where two wave crests or two wave troughs overlap, they combine to form a new, larger crest or trough, a phenomenon called constructive interference. Conversely, where a crest meets a trough, the two waves cancel each other out, leaving the water undisturbed—this is destructive interference.

In Young's experiment, light waves emerging from the two slits interfere similarly before reaching the screen. Bright fringes occur where constructive interference happens, and dark fringes are the result of destructive interference. Young recognized that only the wave theory of light could explain these patterns. Newton's particle theory would have produced two bright images of the slits with darkness in between, making an interference pattern of bright and dark fringes impossible.

When Young first proposed the concept of interference and presented his results in 1801, he faced harsh criticism, particularly from those defending Newton. He tried to defend his work by publishing a pamphlet where he expressed his admiration for Newton, but also noted, "Much as I venerate the name of Newton, I am not therefore obliged to believe that he was infallible. I see, not with exultation, but with regret, that he was liable to err, and that his authority has, perhaps, sometimes even retarded the progress of science." Despite his effort, only one copy of the pamphlet was sold.

In an effort to convince his doubters, Young tried to explain that Newton wasn't entirely opposed to the wave theory of light. In fact, Newton had acknowledged that different colors of light corresponded to different wavelengths. However, when people hold strong beliefs, logical arguments often aren't enough to change their minds. It wasn't until 1818, when two French physicists developed a complete wave theory of light grounded in mathematics, that the wave theory finally gained acceptance. Since then, it has remained unchallenged.

It was a French civil engineer, Augustin Fresnel, who followed Young's work and helped solidify the wave theory, stepping out of Newton's shadow. Fresnel, fifteen years younger than Young, independently rediscovered interference and many other findings that Young had already made. Fresnel's experiments, however, were far more extensive and rigorously designed. His mathematical analysis was so thorough that by the 1820s, the wave theory began to win over notable figures in the scientific community. Fresnel demonstrated that the wave theory offered better explanations for various optical phenomena than Newton's particle theory, and he addressed a major objection to the wave theory: that light couldn't bend around corners. He argued that it could, but because light waves are millions of times smaller than sound waves, the bending is minimal and difficult to observe. Waves only bend significantly around obstacles that are similar in size to their wavelength. Since sound waves are much longer, they can easily travel around most objects they encounter, unlike light.

One way to finally settle the debate between the two rival theories was to find predictions where they differed. In 1850, experiments in France showed that light travels slower through dense media like glass or water compared to air. This was exactly what the wave theory predicted, while Newton's corpuscular theory failed to explain the slower speed.

Despite the significance of Young's experiment, it didn't immediately cause a revolution in the scientific world, especially in Britain. There, opposition to any idea that contradicted Newton's was considered almost heretical and even unpatriotic. Newton had died in 1727, and less than a hundred years later, in 1805, Young made his discoveries. Newton had been the first scientist to be knighted for his achievements, and his reputation remained untouchable. In the midst of the Napoleonic Wars, it was perhaps fitting that a Frenchman, Augustin Fresnel, would take up this "unpatriotic" idea and cement the wave theory of light.

In time, the scientific community accepted that light was a form of wave motion traveling through what they called the ether. However, one question remained: what exactly was "waving" in a beam of light? This mystery was finally solved in the 1860s and 1870s when Scottish physicist James Clerk Maxwell developed the theory of electromagnetic radiation. Maxwell showed that light consists of waves involving oscillating electric and magnetic fields, much like how water waves consist of alternating crests and troughs. This discovery completed the theory of light and revealed its true nature as electromagnetic radiation.

In 1887, just 124 years ago, Heinrich Hertz successfully transmitted and received electromagnetic radiation in the form of radio waves, which, like light waves, are part of the same spectrum but have much longer wavelengths. With Hertz's discovery, the wave theory of light seemed complete—only for it to be challenged by one of the most profound revolutions in scientific thought since Newton and Galileo. By the late 19th century, the idea that light could be made up of particles seemed unthinkable—except to a genius named Albert Einstein.

In 1905, Einstein demonstrated that the particle theory of light could explain the photoelectric effect, where electrons are ejected from a metal when exposed to light. This led to the development of quantum mechanics, which merged the wave and particle theories. Today, we understand that light can behave as both a wave and a particle, depending on the circumstances. This wave-particle duality defies our everyday intuition, but it perfectly matches what experiments reveal about the nature of light.

There's something remarkable about this union of opposites, and it's poetic that our modern understanding of light emerged from the ideas of two very different thinkers: Newton, who championed particles, and Huygens, who advocated for waves—both lifelong bachelors, leaving behind a legacy that continues to inspire awe.

6.6 Historical overview of the nature of light

Scientists have been trying to understand the nature of light for centuries. The ancient Greeks believed light consisted of small particles traveling in straight lines that entered our eyes and allowed us to see. Isaac Newton later supported this particle theory of light. However, the

English scientist Thomas Young challenged this idea, proposing that light behaves as a wave and designed an ingenious experiment to prove it.

Newton's work on the theory of colors was groundbreaking and shaped much of what we know today. In 1666, during his "miracle year," Newton conducted experiments that would lay the foundation for our understanding of color. He observed that a beam of light passing through a prism split into a spectrum of colors—violet, blue, green, yellow, orange, and red. This phenomenon had been known since Aristotle's time, but no one understood why it happened until Newton's experiments.

In his famous experiment, Newton used a glass prism and a small hole in the shutter of a darkened room to create a narrow beam of light. When the light passed through the prism, it spread into a spectrum of colors on the opposite wall. While this wasn't a new observation, Newton noticed something unusual: although the hole in the shutter was circular, the pattern on the wall was oval. At the time, people believed that the prism itself altered the color of light, but this theory didn't explain the odd shape of the spectrum.

Determined to find the answer, Newton varied the size of the hole and the position of the prism. No matter what he did, the spectrum remained unchanged. He even added a second prism to test whether the light would change color again, but it did not—red remained red, blue remained blue. This led Newton to conclude that colors were an intrinsic property of light itself, not something the prism created.

Newton's detailed mathematical explanations for optical phenomena such as reflection and refraction were revolutionary. However, his particle theory of light struggled to explain why, when light hit a glass surface, some particles were transmitted while others were reflected. To address this, Newton suggested that light particles caused disturbances in the ether—an invisible medium thought to fill space. He called these disturbances "Fits of easy Reflection and easy Transmission" and linked them to the color of light, with red corresponding to the largest disturbances and violet to the smallest.

Meanwhile, Christiaan Huygens, a Dutch scientist thirteen years older than Newton, offered a different view. By 1678, Huygens had developed a wave theory of light that also explained reflection and refraction. His theory, published in *Traité de la Lumière* in 1690, proposed that light traveled through the ether in waves, much like ripples on a pond's surface. Huygens argued that if light were made of particles, as Newton claimed, there should be collisions when two beams of light crossed paths—but none were observed. Just as sound waves don't collide, Huygens reasoned that light must also behave as a wave.

Both Newton's particle theory and Huygens' wave theory explained some aspects of light's behavior, but they diverged in other areas. One significant difference was how they predicted light would behave when casting shadows. Newton's particles, traveling in straight lines, would create sharp-edged shadows, while Huygens' waves, bending around obstacles, would produce slightly blurred edges. This bending of light, known as diffraction, had been observed by the Italian Jesuit and mathematician Francesco Grimaldi. In 1665, Grimaldi described how an opaque object placed in a narrow shaft of sunlight cast a shadow larger than expected, with fringes of colored light around the edges—evidence supporting the wave theory of light.

These debates about the nature of light continued for centuries, and it wasn't until much later that a more complete understanding emerged.

Newton was well aware of Francesco Grimaldi's discovery of diffraction and later conducted his own experiments to explore the phenomenon, which seemed to support Huygens' wave theory. However, Newton argued that diffraction was the result of forces acting on light particles, a feature he saw as inherent to the nature of light. Due to Newton's towering reputation, his particle theory, a curious blend of particle and wave concepts, became the dominant view.

Newton's influence was so great that it shaped scientific thought for decades after his death in 1727. His particle theory of light remained largely unchallenged, bolstered by tributes like Alexander Pope's famous lines: "Nature and Nature's Laws lay hid in Night; God said, Let Newton be! And all was Light." It wasn't until the early 19th century that the English polymath Thomas Young questioned this orthodoxy, leading to a revival of the wave theory of light.

Young, using his knowledge of how waves move across the surface of a pond, designed an experiment to test whether light behaved similarly. Like ripples in water, Young reasoned, light waves might also interfere with one another. When ripples from two sources interact, they can either reinforce each other, creating a higher crest (constructive interference), or cancel each other out, leaving a flat surface (destructive interference). If light was a wave, it should display similar interference patterns.

Young's famous double-slit experiment demonstrated just that, showing interference patterns among light waves. Despite the groundbreaking nature of his work, the British scientific community was slow to accept Young's findings. With Newton regarded as a national hero, any challenge to his theories was seen as unpatriotic. Newton had been the first scientist to be knighted, and his influence still loomed large less than a century after his death. In fact, it took the work of a French physicist, Augustin Fresnel, during the Napoleonic Wars to further develop the wave theory of light.

Fresnel's meticulous experiments, backed by mathematical analysis, eventually convinced many scientists of the validity of the wave theory. By the 1820s, it was widely accepted that light behaved as a wave, though questions remained about the exact nature of what was "waving" in a light beam.

In the 1860s and 1870s, the Scottish physicist James Clerk Maxwell provided a more complete answer. He demonstrated that light waves were in fact oscillations of electric and magnetic fields—what we now call electromagnetic radiation. Maxwell's theory revealed that light waves, like water waves, have crests and troughs, but these correspond to varying strengths of electric and magnetic fields. This marked a significant advancement in understanding light, and in 1887, Heinrich Hertz successfully transmitted and received electromagnetic radiation in the form of radio waves, which have much longer wavelengths than visible light.

The wave theory of light seemed complete—until the end of the 19th century, when Albert Einstein entered the scene. In 1905, Einstein proposed that light could also behave as a stream of particles, or quanta, to explain the photoelectric effect, where electrons are ejected from a metal surface when exposed to light. His work reintroduced the idea of light as corpuscular, overturning centuries of thought.

Today, quantum mechanics reconciles both perspectives, and light is understood to behave as both a particle and a wave, depending on the situation. This wave-particle duality may challenge common sense, but it aligns perfectly with experimental evidence. There is something profound in this union of opposites, and it's fitting that Newton and Huygens, though never married, are symbolically the parents of modern light theory.

6.7 The true nature of light

We have established that light, along with all forms of electromagnetic energy, as well as electric and magnetic fields, travels at a constant speed through the Universe. This speed remains consistent when measured from all non-accelerating reference points. But what exactly is electromagnetic energy, and how does it propagate? Scientists have developed two primary models to explain this phenomenon: the "particle theory" and the "wave theory."

These terms are straightforward; light exhibits certain properties that suggest it is made up of particles, while in other contexts, it behaves like a wave. Both models offer valuable insights, yet they also complicate our understanding of what light truly is, much like how atomic theory challenges us to define the nature of matter.

6.8 The ultraviolet catastrophe of radiation

At the turn of the 20th century, evidence began to surface suggesting that describing light solely as a wave was inadequate to account for all its observed properties. Two key areas of study highlighted this limitation. The first involved the heat radiation emitted by hot objects. At sufficiently high temperatures, this radiation becomes visible, leading us to describe the object as "red hot" or even "white hot" at higher temperatures. Since red light corresponds to the longest wavelength in the optical spectrum, it appears that longer wavelengths (associated with lower temperatures) can be emitted more easily than shorter wavelengths; indeed, longer-wavelength heat radiation is referred to as "infrared."

Following the development of Maxwell's theory of electromagnetic radiation and advancements in the understanding of heat—an area to which Maxwell contributed significantly—physicists sought to grasp the properties of heat radiation. It was established that temperature is related to energy: the hotter an object, the more heat energy it contains. Maxwell's theory predicted that the energy of an electromagnetic wave should depend solely on its amplitude and be independent of its wavelength.

One might expect, therefore, that a hot body would radiate across all wavelengths, becoming brighter without changing color as its temperature rises. However, detailed calculations indicated that the number of possible waves of a given wavelength increases as the wavelength decreases, suggesting that shorter-wavelength heat radiation should be brighter than longer wavelengths, remaining consistent at all temperatures. If this were the case, all objects would appear violet in color, with brightness varying from low at lower temperatures to high at higher temperatures, which is not what we observe. This discrepancy between theory and observation became known as the "ultraviolet catastrophe."

6.9 The Particle theory of Light

The concept that light may consist of discrete particles, or corpuscles, originated with Isaac Newton. Today, we understand that light behaves as a stream of particles known as quanta or photons, which are the smallest possible packets of electromagnetic energy. As such, photons are indivisible; it is impossible to have a fraction of a photon. According to the corpuscular theory of light, if you flash a light beam on a surface for progressively shorter durations, there will come a point where only a few photons—perhaps four or even two—strike the surface, and ultimately just one photon. After that, you cannot have half a photon.

With each subsequent flash of light, the probability of detecting a photon decreases. For example, after one photon, the chance of receiving another on the next flash is 50%, then 25% for the next, 12.5% for the one after that, and so on. While there may be rare instances where two photons are emitted nearly simultaneously, the key point is that a photon is never divided into parts. Furthermore, experiments have shown that a beam of light exerts physical pressure on any object it illuminates.

The energy carried by a single photon can vary based on its color: a red photon carries less energy than a green photon, which, in turn, carries less energy than a violet photon. By knowing the energy of a photon, we can determine the type and amount of energy it represents. A single photon contains only an incredibly small amount of energy, typically in the range of quintillionths of a joule for visible light, septillionths or octillionths of a joule for radio signals, and a few trillionths of a joule for X-rays. However, for any given type of electromagnetic energy or frequency, all photons possess the same energy, meaning that any amount of energy must be a multiple of that value.

The relationship between the mass of a photon and the energy it carries is described by the formula

$$E=mc^2,$$

where E represents energy, m denotes mass in joules and grams, and c is the speed of light.

6.10 Electromagnetic energy as a Wave

Electromagnetic energy exhibits wave-like behavior under certain conditions. This can be clearly demonstrated by shining a beam of light through two narrow slits positioned closely together. The resulting interference pattern strongly supports the idea that light behaves as a wave. Much like ripples on a pond, light waves tend to bend around sharp edges and diffract through narrow openings.

The wave theory of light suggests the existence of some medium to transmit electromagnetic energy. However, this idea conflicts with other observations. If waves can travel through a complete vacuum, it raises the question of how they propagate without a medium. While the notion of a conductive medium aligns with our intuition, it is not necessarily a requirement for the transmission of electromagnetic forces.

All waves arise from oscillations, and electromagnetic waves are produced by the vibrations of electrically charged particles. Wave motion is characterized by two key properties: frequency (*f*) and wavelength (λ, represented by the Greek letter lambda). These two properties are interrelated through the equation from the constant factor c:

$$c = \lambda f c,$$

where λ is measured in meters, *f* in oscillations per second, and *c* in meters per second.

Radio signals have wavelengths that can be measured in meters, while visible light waves are measured in millionths of meters (microns) or in units' of 10^{10} meters, known as Angstrom units. Electromagnetic radiation can exist across a wide range of wavelengths. It's important to note that electromagnetic waves travel more slowly in substances like water compared to their speed in a perfect vacuum, meaning the speed of light ccc is reduced in such materials. Additionally, the speed of light varies depending on the substance through which it passes.

6.11 Particle / Wave duality

This intriguing phenomenon necessitates a refinement of the fundamental principle: we must measure the speed of light within the same medium and at the same wavelength—if the medium is not a vacuum—to ensure that ccc remains constant under all conditions. The wave theory and the particle theory of light present two distinct models for the same phenomenon, which may lead one to question whether scientists truly understand what light is. Ultimately, we can only say that light is a form of energy. It exhibits properties of both particles and waves; at times, the particle model offers a clearer explanation for observed data, while at other times, the wave model is more descriptive.

From this author's perspective, one key aspect of particle-wave duality is that the particle attribute of light represents a more concentrated and denser form of energy, whereas its wave attribute is a less concentrated form of energy propagating through space-time at a significantly higher velocity (the speed of light). Inertia may play a crucial role in the differing propagation speeds of particles and waves. However, both forms of energy are fundamentally interconnected, as suggested by Einstein's equation *E=mc²*.

6.12 Light as electromagnetic radiation

Another commonly observed wavelike phenomenon is **electromagnetic radiation**, which includes the radio waves that transmit signals to our radios and televisions, as well as light itself. These waves vary in frequency and wavelength: for instance, typical *FM* radio signals have a wavelength of about 3 meters, while the wavelength of light depends on its color—approximately 4×10^{-8} meters for blue light and 7×10^{-8} meters for red light, with other colors falling in between.

Unlike water waves and sound waves, light waves do not require a vibrating medium (such as water or air) to propagate. In fact, light waves can travel through a vacuum, which is evident from our ability to see the light emitted by the Sun and stars.

The wave-like nature of light posed a significant challenge for scientists in the 18th and 19th centuries. Some theorized that space was not truly empty but filled with an undetectable substance called **ether**, believed to facilitate the oscillation of light waves. However, this hypothesis encountered difficulties when it became clear that the properties necessary to support the high frequencies of light waves could not be reconciled with the idea that ether offered no resistance to the movement of objects (like Earth) through it.

It was James Clerk Maxwell who, around 1860, demonstrated that the ether concept was unnecessary. During this time, the field of electricity and magnetism was advancing, and Maxwell formulated a set of equations (now known as **Maxwell's equations**) that encompassed these principles. He showed that one type of solution to these equations corresponded to the existence of waves formed by oscillating electric and magnetic fields, which could propagate through empty space without needing a medium. The speed of these electromagnetic waves was determined by fundamental constants of electricity and magnetism, and when calculated, it matched the measured speed of light. This finding directly supported the idea that light is an electromagnetic wave, a model that also applies to various other phenomena, including radio waves, infrared radiation (heat), microwaves, and X-rays.

6.13 Interference

Direct evidence for light behaving as a wave can be observed through **interference**. This phenomenon occurs when two waves of the same wavelength interact. When these waves are **in phase**, or aligned perfectly, they combine to create a wave with twice the amplitude of either original wave. Conversely, when they are **out of phase** (in **anti-phase**), they cancel each other out entirely.

In situations where the waves are partially out of sync, they result in a combined amplitude that lies between the two extremes. Interference serves as crucial evidence for the wave nature of light, as no classical model can adequately explain this effect. For instance, if we considered two streams of classical particles, the total number of particles would simply be the sum of those in each beam, and they wouldn't be able to cancel each other out as waves do.

However, we will soon explore evidence that light also exhibits particle-like properties in certain situations. This more nuanced understanding of light leads us to the concept of **wave-particle duality**.

6.14 The Universal Speed of Light

To define absolute rest and absolute motion, it is necessary to identify one specific inertial frame of reference that not only remains unaccelerated but also possesses a unique quality that distinguishes it from other inertial frames. This special quality would make it the preferred frame for describing all processes in the Universe, including the speed of light. However, it seems that there was nothing in Newton's conception of the Universe that would provide a way to select this singular frame.

6.15 Measuring the Speed of Light

James Clerk Maxwell's theory of electromagnetism reveals that light is an electromagnetic wave traveling at a speed of 300,000 kilometers per second (*kps*) in a spherical geometric pattern. His equations demonstrate that changing electric and magnetic fields create and sustain each other, even in areas devoid of electric charges or moving magnets. Maxwell illustrated how these two interlocking fields propagate through space as light or other forms of electromagnetic radiation. However, prior to Maxwell, other scientists also endeavored to understand the nature of light and to measure its speed.

6.16 How light speed was measured during the Renaissance Era

With modern instruments, we can accurately measure the exceptionally high speed of light. In earlier times, attempts to gauge this speed were more rudimentary. Galileo was among the first to employ a scientific method for this purpose, using two people positioned on distant hills to flash lanterns at each other. However, his initial experiments were unsuccessful; he struggled to measure time accurately, let alone the minuscule fraction of a second it took for light to travel between the hills. Despite these challenges, Galileo's efforts demonstrated that the speed of light is finite. In contrast, his contemporary, the French philosopher René Descartes, claimed that light traveled instantaneously.

6.17 Measuring the velocity of light by means of astronomical objects

The determination of light's most significant property, its velocity, serves as the foundation for our subsequent discussions and was established independently of the ongoing debate regarding the nature of light. Observations consistently indicated that light travels at an extraordinarily high speed. Although Galileo attempted to measure this speed using lantern signals, he was unsuccessful, as light covers terrestrial distances in incredibly brief intervals. Accurate measurements were only possible by utilizing the vast distances between celestial bodies.

Olaf Römer was the first to calculate the speed of light (c) in 1676 through astronomical observations, specifically by studying the eclipses of Jupiter's moons. An eclipse occurs when a moon enters Jupiter's shadow, which happens at regular intervals equal to the moon's orbital period (τ). If L represents the distance from Jupiter to Earth, the light signals reach Earth after a time delay of L/c. During this period, if l represents the change in distance L while the moon completes its revolution, the observer on Earth perceives the eclipses at gradually varying intervals of τ + l/c. Consequently, the observed orbital periods on Earth are either longer or shorter than the true periods observed from Jupiter, depending on whether the distance L is increasing or decreasing.

In 1727, James Bradley discovered another consequence of the finite speed of light: all fixed stars appear to exhibit a common annual motion, which corresponds to the Earth's rotation around the Sun.

6.18 Roemer's accurate light measurement

About 70 years after Galileo's experiment, the young Danish astronomer Olaus Römer became the first to estimate the speed of light, utilizing the moons of Jupiter instead of distant hills. He faced challenges, including disagreements with his superior, the renowned astronomer Jean Dominique Cassini, after whom the rings of Saturn are named.

Light travels through the vacuum of space at a modern speed of approximately 300,000 kilometers per second (186,000 miles per second), though its precise speed is 299,792,458 kilometers per second. Given that the Earth's circumference is around 40,000 kilometers (25,000 miles), light would take just over a tenth of a second to circle the globe. Römer's work in 1676 demonstrated that light travels at a finite speed, contradicting the prevailing belief that light was instantaneous.

When observing the moons of Jupiter, Römer noticed that they periodically vanished as they passed behind the planet. While these eclipses should occur at regular intervals, he found that they were not evenly spaced. Rather than attributing this to variations in the moons' orbits, he proposed that if light traveled instantaneously, we would see the eclipses at the exact moments they occurred, as if they were synchronized to a cosmic clock. However, since light takes time to travel, we observe each eclipse after it has occurred.

This delay varies with Jupiter's distance from Earth. If Jupiter remained at a constant distance, the delay would be consistent. But as Jupiter occasionally moves closer, the light from successive eclipses has less distance to travel, arriving earlier. Conversely, when Jupiter moves away, the eclipses appear later. Römer noted that the timing of the eclipses of one of Jupiter's moons varied with Earth's position relative to Jupiter; they appeared earlier when Earth was approaching and later when moving away. By measuring this variation, he calculated the speed of light. Although his estimate of 140,000 miles per second fell short of the modern value of 186,000 miles per second due to inaccuracies in measuring Earth's distance to Jupiter, Römer's accomplishment in proving that light travels at a finite speed was groundbreaking, coming just eleven years before Newton published his *Principia Mathematica*.

6.19 Roemer and Jupiter's Moons

At just 21 years old, Olaus Römer was hired by one of Cassini's assistants to work at the Paris Observatory, led by the prominent astronomer Jean Dominique Cassini. However, Römer didn't merely assist; he took on one of the observatory's significant challenges. Cassini had observed irregularities in the motion of one of Jupiter's moons, Io (named after a lover of Jupiter, the Roman equivalent of Zeus). The timing of Io's emergence from behind the planet seemed inconsistent and unpredictable.

Cassini tasked his assistants with improving observations and recalibrating their calculations. Römer, however, suspected the issue lay elsewhere. He believed that the variations in Io's visibility were due to the changing distances between Earth and Jupiter as they orbited the Sun. Depending on their positions in their respective orbits, the two planets could be closer together or farther apart, affecting how long it took light to travel from Io to Earth.

Cassini dismissed Römer's theory, holding firmly to the belief that light traveled instantaneously, regardless of the distances involved. Undeterred, Römer meticulously reviewed years of observational data from Cassini's observatory. He calculated the variations in the timings of Io's eclipses based on the distances between the two planets and was convinced he had the correct explanation. Determined to share his findings, Römer decided to present his conclusions to the Academy of Sciences in Paris independently, as Cassini did not support his work.

After five years in the observatory, Römer confidently announced that Io would reemerge from behind Jupiter exactly ten minutes later than Cassini had predicted. Cassini calculated that Io would reappear on November 9, 1676, at 5:25:45 PM. That night, astronomers gathered to observe. As the clock struck 5:25:45, Io was nowhere to be seen. Even at 5:30, there were no signs of the moon. But at 5:30:45, Io finally reappeared. Römer's prediction was proven correct.

Römer's friend, Christian Huygens, used this data to calculate an initial value for the speed of light, estimating it at 227,000 kilometers (140,000 miles) per second—about 24 percent lower than today's accepted value. Cassini never acknowledged his error, and many European astronomers remained skeptical of the notion that light had a finite speed. It wasn't until about 50 years later, when other methods were employed to measure the speed of light, that Römer's conclusions were fully validated.

Ch 7—Chemistry of the Elements

7.1 The history and nature of Chemistry

Chemistry is the scientific study of the composition, structure, properties, and reactions of matter, particularly at the atomic and molecular levels. Along with physics, chemistry is a fundamental branch of science and is also closely linked to biology. This field examines not only the composition and its changes but also the energy transformations associated with matter.

Through the study of chemistry, we aim to uncover the general principles that govern the behavior of all matter. Empirical chemistry has been practiced since ancient times, with early civilizations engaging in various chemical processes such as winemaking, glassmaking, pottery, dyeing, and basic metallurgy. For instance, ancient Egyptians demonstrated considerable knowledge of chemical processes, as evidenced by excavations of tombs dating back to around 3000 B.C., which revealed artifacts like gold, silver, copper, and iron, alongside pottery, glass beads, vibrant dyes, and well-preserved royal mummies. While these developments were significant, they were primarily empirical, relying on trial and error without a coherent theoretical framework for understanding matter.

Philosophical ideas regarding the properties of matter lagged behind those related to astronomy, mathematics, and *pseudo* or early physics. Greek philosophers made notable advancements in the philosophical exploration of material concepts, laying the groundwork for chemistry as an intellectual and scientific discipline. They introduced key ideas such as elements, atoms, atomic shapes, and chemical combinations. Influenced by Aristotle, they believed that all matter was composed of four fundamental elements: earth, air, fire, and water. Although the Greek philosophers had insightful ideas that closely mirrored the foundations of modern chemistry, their primary limitation was a lack of systematic experimentation.

7.2 The Aristotelian quartet

Aristotle's Quartet: The Elements in Antiquity

In 1624, French chemist Étienne de Clave was arrested for heresy—not for challenging religious doctrine or political authority, but for his radical ideas about the elements. De Clave believed that all substances were composed of just two elements, water and earth, combined with three fundamental principles: mercury, sulfur, and salt. While not an entirely new concept—prominent pharmacist Jean Béguin had similarly argued for these five basic components in his influential textbook *Tyrocinium Chymicum* (1610)—de Clave's ideas were seen as heretical because they contradicted the ancient Greek elemental system endorsed by Aristotle.

Aristotle's view, inherited from his teacher Plato and ultimately traced back to the philosopher Empedocles, held that the universe was made up of four elements: earth, air, fire, and water. This model, dating back to the Golden Age of Athens in the fifth century BC, became deeply entrenched in the Western intellectual tradition. After the fall of Rome and the subsequent turmoil of the Dark Ages, medieval Europe emerged with a reverence for ancient scholars, including Aristotle, whose teachings were merged with Christian doctrine. Questioning Aristotle's elemental theory was akin to challenging religious authority.

By the late 17th century, thinkers like Galileo, Newton, and Descartes began to shake the intellectual foundations of the West, allowing for new scientific ideas. However, in 1624, such challenges were still considered dangerous. De Clave and other French intellectuals planned to debate a "non-Aristotelian" theory of the elements at the Paris home of nobleman François de Soucy. Their attempt was swiftly shut down by parliamentary order, and de Clave was arrested.

This conflict wasn't truly about science. Rather, the use of legal force to suppress de Clave's ideas reflected a desire to maintain the status quo. Like Galileo's trial before the Inquisition, the debate over the elements was less a matter of "truth" and more a struggle for power, driven by the dogmatic fervor of the Counter-Reformation.

Free from such constraints, the ancient Greeks themselves explored the concept of the elements with much greater flexibility. Aristotle's four-element theory was just one of several competing models. In fact, by the sixteenth century, Swiss scholar Conrad Gesner had identified at least eight different elemental systems proposed between the time of Thales in the sixth century BC and that of Empedocles. Despite the 1624 condemnation of non-Aristotelian theories, the variety of ancient views made it increasingly difficult to uphold Aristotle's quartet as the definitive model, eventually reigniting the debate over the true nature of matter.

7.3 What are things made from

The Periodic Table of Chemistry lists all known elements, and aside from the slowly growing row of human-made elements, it is comprehensive. The term "elements" refers to the foundational components of matter. Today, we know there are about 92 naturally occurring elements, not just one, four, or five. The Periodic Table is a triumph of science, but it doesn't fully capture the complexity of the question: "What are things made of?"

Even though the table lays out atomic building blocks, these atoms are more varied and intricate than the table implies. Atoms themselves are not fundamental or immutable; they are composed of subatomic particles. While the table may offer a modern answer, it doesn't tell the full story of how humanity has wrestled with the question of matter through history.

In seeking to understand the elements, we shouldn't just present the modern table as a rebuttal of Aristotle's or earlier ideas, like those held before the 18th century. Rather, we must explore how the question of what things are made of has been answered throughout time. This exploration becomes essential, especially in our age, where composition is crucial to everything—from the lead in Antarctic snow to mercury contamination in South American fish. Composition determines which elements sustain life and which are deadly, as seen in calcium supplements that prevent bone diseases and arsenic-tainted wells that poison populations.

Surprisingly, life itself is composed of relatively few elements. The bulk of living organisms consists of just four—carbon, nitrogen, oxygen, and hydrogen—along with essential elements like phosphorus, which forms DNA, and sulfur, which gives proteins their structure. Metals like iron, magnesium, sodium, and potassium play vital roles in transporting oxygen, capturing solar energy, and transmitting nerve impulses. Altogether, 11 elements form the basis of life, with another 15 essentials in trace amounts, including elements like arsenic and bromine, which can be toxic in excess but are still necessary for life.

The uneven distribution of elements across Earth has shaped human history, driving trade, exploration, exploitation, and conflict. Southern Africa's natural wealth in gold and diamonds is a prime example. Rare and valuable elements like tantalum and uranium, vital for modern technology, are often mined under hazardous conditions in impoverished regions.

By the mid-20th century, all naturally occurring stable elements were known, and experiments in nuclear energy introduced us to heavier, short-lived radioactive elements. Modern chemical analysis reveals the delicate balance of elements in nature, with minerals seasoning oceans, air, and even our drinking water. That's why a bottle of mineral water isn't just H^2O—it's a cocktail of sodium, potassium, chlorine, and more.

Elements are not static; they can be harmful, as seen in the transition away from lead pipes and lead-based paints. Aluminum cooking utensils, once popular, now face scrutiny for their potential role in causing dementia. Folklore also shapes our understanding of elements—copper bracelets are said to cure arthritis, while selenium supplements are touted to boost fertility, despite selenium contamination devastating California's ecosystems.

Ultimately, the story of the elements is more than a list of atoms. It's a narrative about how we interact with the very stuff of the universe—culturally, scientifically, and socially.

7.4 The various Greek schools of thought regarding the nature of matter

The concept of elements is closely linked to the idea of atoms, but the two aren't inseparable. Plato believed in the four classical elements—earth, air, fire, and water—but didn't fully embrace the notion of atoms. Other Greek philosophers, however, believed in atoms but didn't reduce all matter to a few basic components. Thales of Miletus (c. 620–c. 555 BC), one of the earliest thinkers to explore the nature of the physical world, proposed that everything was derived from a single substance: water.

This idea has roots in ancient myths, where many cultures, including the Hebrews, believed the world emerged from a primordial ocean. Yet, the Milesian school that Thales founded didn't agree on what the "first matter," or *prote hyle*, was. Anaximander (c. 611–547 BC), Thales' successor, avoided specifics by suggesting that the first substance was *apeiron*—the indefinite and unknowable. Anaximenes (d. c. 500 BC) argued instead that air was the primary element, while Heraclitus (d. 460 BC) considered fire the fundamental substance of creation.

Some historians suggest these early philosophers were driven by a desire for unity, trying to reduce the diversity of the natural world to a simpler framework. While this tendency is evident in Greek thought, there was also a practical reason for considering fundamental elements: the observable changes in nature. Water freezes or evaporates, wood burns and becomes ash, and food is consumed, much of it disappearing through digestion. If one material could transform into another, perhaps they were different forms of the same underlying substance. The idea of elements likely arose from the need to understand these daily transformations, not just from a theoretical quest for unity.

Anaximander believed that change occurred through opposing qualities like hot and cold, or dry and moist. When Empedocles (c. 490–c. 430 BC) proposed his four elements—earth, air, fire, and water—he argued that their transformations were driven by conflict. Empedocles himself wasn't a typical "sober" philosopher; legend portrays him as a magician who could raise the dead and who supposedly died by leaping into Mount Etna, convinced of his own divinity. In his cosmology, the elements were governed by two forces: Love, which caused mixing, and Strife, which led to separation. These forces were in constant flux—Love bringing things together and Strife pulling them apart—and this dynamic applied not only to the elements but to human life and society as well.

Empedocles' four elements didn't represent a multiplication of *prote hyle* but rather a way of interpreting its complexities. Aristotle, while agreeing that a single "primal substance" existed, found it too abstract and distant to be a useful basis for understanding matter. Instead, he adopted Empedocles' elements as intermediaries between the mysterious essence of matter and the tangible world. Aristotle's approach—reducing vast cosmic questions to more manageable concepts—helped cement his lasting influence on Western thought.

Aristotle shared Anaximander's view that the qualities of heat, cold, wetness, and dryness are central to both transformation and our perception of the elements. According to Aristotle's framework, each element possesses two of these qualities, allowing one element to transform into another by altering one of its properties. For example, wet and cold water could become dry and cold earth by converting wetness into dryness.

It's easy to picture these ancient philosophers as part of an intellectual circle where ideas were shared, debated, praised, or criticized, all while remaining largely theoretical "armchair" scientists who avoided hands-on experimentation. This same image applies to the discussions around the nature of atoms.

Leucippus of Miletus (5th century BC) is generally credited with introducing the idea of atoms, though little else is known about him. He believed that all atoms were composed of the same "primal" substance but took on different shapes in various materials. His student, Democritus (c. 460–370 BC), named these particles *atomos*, meaning uncuttable or indivisible. To reconcile atomic theory with the classical elements, Democritus proposed that the atoms of each element had shapes that explained their properties. Fire atoms, for instance, were thought to be immiscible with others, while the atoms of the other three elements entangled to form solid, tangible matter.

Aristotle, on the other hand, believed that the four elements of Empedocles were infused with two key qualities, allowing for their transformation. The real debate between the atomists and their opponents wasn't about whether matter was made of tiny particles, but about what separated them. Democritus claimed that atoms moved through a void, while other philosophers rejected the concept of "nothingness," arguing that the elements filled all of space. Anaxagoras (c. 500–428 BC), a teacher of Pericles and Euripides, countered that matter was infinitely divisible, with no limit to how small particles could become.

Aristotle argued that air would fill any void between atoms, but this idea becomes problematic when you consider that air itself is made of atoms. Plato, however, had a different perspective on the elements. While not an atomist like Democritus, he did propose that the four Empedoclean elements were composed of fundamental "atom-like" particles. Plato's affinity for geometry led him to suggest that these particles had regular, mathematical shapes: the *Platonic solids*. He assigned each element a corresponding shape—earth as a cube, air as an octahedron, fire as a tetrahedron, and water as an icosahedron. These shapes, in turn, were composed of two basic types of triangles, which Plato considered the true "fundamental particles" of nature. By rearranging these triangles, the elements could be transformed into one another.

Plato also introduced a fifth solid, the dodecahedron, which did not fit with the others because it couldn't be constructed from the same triangles. He associated this shape with the heavens, thus introducing the idea of a fifth element, which Aristotle later called *aether*. This celestial substance, according to Aristotle, had no role in the formation of earthly matter.

Greek philosophers often linked their four-element theory to other symbolic systems, including a set of four primary colors. Empedocles identified these colors as white, black, red, and a vaguely defined yellow (*ochron*), corresponding to water, earth, fire, and air, respectively. This tradition persisted through the ages. Renaissance artists like Leon Battista Alberti associated colors with elements in their work—fire with red, air with blue, water with green, and earth with an "ash color." Such associations likely influenced painters of the time in their approach to mixing and using colors. The "fourness" of these fundamental elements extended beyond art to medicine, where the Greek physician Galen (c. 130–201 AD) connected the elements to the balance of four bodily humors: red blood, white phlegm, black bile, and yellow bile.

While the classical elements no longer hold sway in modern chemistry, they remain significant due to their resonance with human experience. Earth, water, air, and fire aren't just substances but archetypes of different physical states. Earth represents solids, water stands for liquids, air symbolizes gases, and fire—a more elusive concept—is a glowing plasma formed by the heat-induced excitation of molecules. Fire, in particular, was seen as a symbolic representation of light and transformation.

In ancient philosophy, elements were not strictly tied to specific substances. When Plato referred to "water," he didn't just mean the water in rivers but anything that flowed, like molten metal. Likewise, "earth" represented not only soil but also flesh, wood, and metals. These elements could be converted into one another because their "atoms" shared geometric commonalities. For Anaxagoras, all substances were mixtures of the four elements, and transformations occurred when the proportions of one element increased while others diminished. This blending of elements contrasts sharply with today's view of elements as pure, indivisible substances.

Aristotle's endorsement of Empedocles' elements ensured their dominance until the seventeenth century, leading to the decline of atomism. While Epicurus (341–270 BC) and later the Roman poet Lucretius in *De rerum natura* (56 BC) preserved atomism, it was largely condemned in the Middle Ages. The atomistic worldview resurfaced in the seventeenth century, particularly through the work of Pierre Gassendi (1592–1655), who saw atoms as the building blocks of a mechanical universe. However, resistance to this new vision persisted. Even progressive thinkers like Marin Mersenne (1588–1648) supported the 1624 condemnation of atomistic ideas, believing they were linked to alchemical and heretical thought. Nonetheless, alchemy had much more to say about the elements as new ideas began to emerge.Top of Form

From a modern perspective, it may seem odd that many of the substances we now recognize as elements—such as gold, silver, iron, copper, lead, tin, and mercury—weren't considered elements in antiquity, despite their impressive purity. Metallurgy, one of the oldest technical arts, had surprisingly little impact on early theories of the elements until after the Renaissance. In the Aristotelian view, metals, aside from liquid mercury, were simply forms of "earth." Alchemy later expanded these ideas, introducing more sophisticated notions of matter and its transformation, which helped bridge the gap between ancient and modern conceptions of elements.

Initially, the idea of a single *prote hyle* (prime matter) didn't advance a practical theory of matter, and the Aristotelian four-element theory offered limited explanations for the significant differences between materials like lead and gold. Society clearly valued these differences, but the four-element model fell short. A more refined understanding was needed, particularly for metals. Gold and copper were the first metals discovered in their natural, elemental form, with evidence of gold mining in regions like Armenia and Anatolia as early as 5000 BC. Similarly, copper use was ancient, especially in Asia. Copper was often found in ore form—compounds like copper carbonate, known as malachite and azurite—rather than as pure metal. These ores were used as pigments and glazes, and it's believed that copper smelting, which dates back to 4300 BC, likely arose accidentally during the glazing of stone ornaments in the Middle East. Around the same time, the creation of bronze, an alloy of copper and tin, began. Lead was smelted from galena by 3500 BC, although it didn't become common until a millennium later. Tin appears to have originated in Persia around 1800–1600 BC, while iron was first smelted in Anatolia around 1400 BC.

The timeline of metal discoveries reflects the difficulty of extracting pure metals from their ores. For instance, iron binds tightly to oxygen in the ore haematite (ochre), requiring intense heat and charcoal to separate them. As the variety of known metals grew, a classification system became necessary. Early conventions relied on symbolic correspondences, with the seven known metals linked to the seven celestial bodies and the seven days of the week. Since all metals shared similar traits—shininess, density, and malleability—it was natural to assume they differed only by degree, not by kind.

This assumption became central to alchemy. Alchemists believed that if metals could transform into one another within the earth over time, they might be able to accelerate this process and turn base metals into gold. Efforts to transmute metals into gold likely date back to the Bronze Age, but by the 18th century AD, these attempts were no longer random. They were guided by the sulfur-mercury theory developed by the Arabic alchemist Jabir ibn Hayyan. However, "Jabir" is more accurately viewed as a school of thought, as many writings were attributed to him that he couldn't have written, leading to doubts about whether he existed at all. The Jabirian tradition intriguingly reworked the Aristotelian elements, implicitly accepting them but adding a new layer of complexity between these fundamental substances and the reality of metals.

According to Jabir, the "fundamental qualities" of metals stem from the Aristotelian categories of hot, cold, dry, and moist. However, the more immediate qualities are defined by two "principles": sulphur and mercury. Jabir proposed that all metals are mixtures of these two principles, with base metals being impure forms and silver and gold representing a higher state of purity. The ultimate goal of alchemy, the creation of the "Philosopher's Stone," was believed to come from the purest combination of sulphur and mercury, which could supposedly transform base metals into gold. Some scholars equate Jabir's sulphur and mercury with the Aristotelian elements of fire and water.

However, it is clear that these principles do not refer to the physical substances we recognize today—yellow sulphur and liquid mercury, which were known even to ancient alchemists. Instead, Jabir's sulphur and mercury were more akin to the four classical elements—idealized substances only imperfectly realized in the material world. In this sense, the "Jabirian system" retained the four classical elements but largely disregarded them, just as Aristotle's elements made room for but overshadowed the idea of a universal *prote hyle*. This reflects an evolving focus in alchemy, paying nominal respect to Aristotle while engaging with more practical matters about the nature of substances.

The next evolution in alchemical thought introduced a third principle to Jabir's system: salt. While sulphur and mercury were seen as components of metals, salt was considered essential for living bodies. This addition expanded alchemical theory beyond metallurgy to encompass the entire material world. The three-principle theory—sulphur, mercury, and salt—is generally attributed to the Swiss alchemist Paracelsus (1493–1541), although its origins likely predate him. Paracelsus claimed that these three principles "form everything that lies in the four elements." They were not meant to be elements in the strict sense but rather material manifestations of the ancient elements.

By the end of the seventeenth century, thinking had shifted further. There was no longer a need to reconcile these ideas with Aristotle's framework, and the alchemical principles came to be seen as elements in their own right. Jean Béguin, for example, proposed a five-element scheme consisting of mercury, sulphur, salt, phlegm, and earth. He believed that none of these elements existed in a pure form; each contained traces of the others.

7.5 The decline of alchemy

Greek civilization was followed by the rise of the Roman Empire, which excelled in military, political, and economic matters. The Romans engaged in various practical chemical arts, including metallurgy, enameling, glassmaking, and pottery, but contributed little to advancing theoretical knowledge. After the fall of Rome, Europe entered the Dark Ages, a period marked by the rapid decline of learning and culture.

In contrast, knowledge flourished in the Middle East and North Africa during this time. Arabic cultures made significant contributions that would later prove invaluable to the development of modern chemistry. Notably, the Arabic numeral system, which included the use of zero, gained widespread acceptance, algebra was developed, and "alchemy"—a precursor to modern chemistry—was practiced extensively.

One of the more fascinating chapters in the history of chemistry occurred during the era of the alchemists, between 500 and 1600 A.D. At the time, gold was seen as the most perfect of all metals, and the alchemists were driven by two primary goals: to discover a way to extend human life indefinitely and to transmute base metals like iron, zinc, and copper into gold.

They sought a universal solvent for metal transmutation and the elusive "Philosopher's Stone," believed to cure all diseases and grant eternal life. In their quest, alchemists gained valuable chemical knowledge. However, much of their work was done in secrecy, shrouded in mysticism, and few records survive from that period.

Though the alchemists lacked sound theoretical foundations and did not belong to the intellectual ranks of the Greek philosophers, they achieved something the philosophers had largely overlooked—they subjected materials to systematic treatments in what could be loosely described as early laboratory methods. These practices, carried out in alchemical workshops, revealed important natural facts and laid the groundwork for the experimental approach that defines modern science.

Alchemy began to wane in the 16th century when Paracelsus (1493–1541), a Swiss physician and revolutionary figure in chemistry, advocated for directing chemistry towards medical purposes, focusing on curing human diseases. He openly criticized the alchemists' mercenary pursuits of transforming base metals into gold, urging instead that the science of chemistry serve the needs of medicine.

7.6 The beginning of the Enlightenment era

Classical chemistry developed more slowly than astronomy and physics, only truly emerging in the 17th and 18th centuries. Pioneers like Joseph Priestley (1733–1804), who discovered oxygen in 1774, and Robert Boyle (1627–1691) helped usher in this new era by recording and publishing the results of their experiments and openly discussing their theories. Boyle, often considered the founder of modern chemistry, was one of the first to approach chemistry as a true science. He championed the experimental method and, in his influential book *The Sceptical Chymist*, clearly distinguished between "elements" and "compounds" or mixtures. Boyle is perhaps best known today for Boyle's Law, which describes the behavior of gases.

French chemist Antoine Lavoisier (1743–1794) further solidified chemistry as a science. He revolutionized the field by using a chemical balance to make precise, quantitative measurements of the substances involved in chemical reactions. This use of the balance in chemistry was as groundbreaking as the telescope was for astronomy. Thanks to Lavoisier, chemistry became a quantitative experimental science. He also made significant contributions to the organization of chemical data, established modern chemical nomenclature, and formulated the Law of Conservation of Mass, which states that matter is neither created nor destroyed in chemical reactions.

In the early 19th century, John Dalton (1766–1844), an English schoolteacher, advanced the atomic theory, which provided a rational basis for the atomistic concept of matter. Dalton's theory remains a foundational idea in modern science.

Since Dalton's time, the field of chemistry has made tremendous strides, especially in the late 19th and 20th centuries. Major achievements include breakthroughs in understanding atomic structure, the biochemical principles of life, advances in chemical technology, and the large-scale production of chemicals and related products.

Johann Becher (1635–c.1682) was a prominent and flamboyant German alchemist who accepted that air, water, and earth were elements, though he did not grant them equal importance. He viewed air as inert and uninvolved in transformative processes. Becher believed that the diverse dense substances found in the world originated from three distinct types of earth. **Terra fluida** was a fluid element responsible for the shininess and heaviness of metals. **Terra pinguis** referred to a "fatty earth," rich in organic matter from both animal and plant sources, which contributed to combustibility. Lastly, **Terra lapidea** was described as "vitreous earth," providing solidity to materials. Ultimately, these three earths can be understood as representations of mercury, sulphur, and salt, which played a foundational role in the development of modern chemistry.

In the seventeenth century, early chemists began to explore the nature of matter through practical experimentation. While alchemy always emphasized an experimental approach, it was during this time that a transitional group of natural philosophers emerged. Unlike traditional alchemists focused solely on the Great Work of transformation, these "chymists" aimed to study matter at a more practical level. One notable figure among them was Robert Boyle (1627–1691), the educated son of an Irish aristocrat.

Boyle became a prominent member of British science in the mid-seventeenth century, forming connections with influential figures like Isaac Newton, who secretly practiced alchemy. He was also involved in founding the Royal Society in 1661. Though Boyle shared a deep interest in alchemy, he was an independent thinker and a critical mind.

His seminal work, "The Sceptical Chymist" (1661), is often viewed as a critique of alchemy. However, its true aim was to differentiate learned alchemical adepts, like Boyle himself, from the "vulgar laborants" who blindly followed recipes in their pursuit of gold. Boyle's lasting contribution to chemistry lies in his challenge to existing theories about elements, arguing that they were incompatible with experimental evidence. He criticized the conventional four-element theory, which posited that all substances contained Aristotle's four elements. Boyle pointed out that some materials, like gold, could not be reduced to these classical components, stating, "Out of some bodies, four elements cannot be extracted." His insistence on experimentation as the path to understanding elements marked a significant shift in thought: "I must proceed to tell you that though the assertors of the four elements value reason so highly... no man had ever yet made any sensible trial to discover their number."

While Boyle's definition of an element was not particularly controversial for his time, he went further by questioning whether true elements existed at all. Although he did not propose a definitive replacement for the elemental frameworks he critiqued, he showed some interest in the Flemish scientist Johann Baptista van Helmont's idea that everything is made of water.

By the end of the seventeenth century, scientists had made little progress in identifying the elements compared to the Greek philosophers. However, a century later, British chemist John Dalton (1766–1844) outlined a modern atomic theory and presented a list of elements that, while still incomplete and occasionally incorrect, laid the groundwork for today's periodic table. This rapid advancement in understanding can be attributed to Boyle's emphasis on experimental analysis and the abandonment of outdated preconceptions about what elements should be.

For classical scholars, an element had to correspond to recognizable substances. In contrast, many elements today are ones most people will never encounter. Additionally, scientists began to recognize that substances could change their physical states—such as from solid to liquid to gas—without altering their elemental composition. For instance, ice is not water transformed into 'earth'; it is simply frozen water.

Ultimately, the nature of elements proved to be complex and elusive. Until the twentieth century, scientists struggled to understand why there were so many elements and why there weren't even more. Elements could not be discerned through casual observation; they required rigorous examination with advanced scientific tools. This complexity might explain why some prefer the simplicity of earth, air, fire, and water, which resonate more with human experience despite not being the elements recognized in chemistry.

Revolution: How Oxygen Changed the World

Antoine Laurent Lavoisier is often credited with revolutionizing chemistry in a manner akin to Isaac Newton's impact on physics and Charles Darwin's on biology. He transformed chemistry from a collection of disjointed facts into a cohesive science governed by unified principles.

Timing played a pivotal role in scientific evolution. Isaac Newton's work in the seventeenth century marked the dawn of the Enlightenment, fostering a belief in rationalism as a means to understand the universe and improve *the human condition*. As Charles Darwin's theories gained traction, the solid certainties of nineteenth-century science and culture began to dissolve in the face of modernism, altering the established norms in art, music, and literature simultaneously. However, Antoine Laurent Lavoisier's fate became emblematic of this brave new world; he was executed during Robespierre's Reign of Terror. The optimism of Enlightenment philosophers such as Voltaire, Montesquieu, and Condorcet crumbled under the passions and brutality of the French Revolution. Reason itself was upended, and chemistry evolved into a deeply Romantic science in the years that followed.

Lavoisier (1743–1794), much like Condorcet, found himself entangled in the political upheaval of the time. While science in England remained a pursuit of wealthy gentlemen with leisure, France's state-sanctioned Academy of Sciences had members who often occupied public office and played significant roles in political life.

Lavoisier, who had been a tax collector before his rise as a prominent scientist, became a target for the revolutionary fervor. His chemical expertise led him to the position of director of Louis XVI's Gunpowder Administration, and as treasurer of the Academy of Sciences, he staunchly opposed its dissolution by the anti-elitist Jacobin regime in 1793. This resistance painted a target on his back, and in 1794, he met the guillotine, which had already claimed his father-in-law.

Two centuries later, debate continues over whether Lavoisier was truly the discoverer of one of chemistry's most crucial elements: oxygen. Retro-Nobel prizes have been awarded for significant discoveries made before the prize was established in 1901, and the committee determined that the first chemistry prize should honor oxygen's discoverer, emphasizing its foundational role in the Chemical Revolution. While Lavoisier named the element, he was not the first to isolate it or to identify it as an essential substance. Yet, this is only part of the story. Oxygen serves as both a central organizing principle for modern chemistry and a link between the alchemical roots of Robert Boyle's "chymistry" and the marvels produced in today's chemical industries, marking a vital stage in the evolving understanding of elements.

As the "Isaac Newton of chemistry," Lavoisier, along with his wife and collaborator Marie Anne, delivered significant challenges to the Aristotelian concept of elements. Through his experiments on water, he concluded in 1783 that it was not a simple substance, as previously thought. Regarding air, he announced that it consists of two distinct fluids with different properties, which he termed "mephitic air" and "highly respirable air." Thus, neither water nor air qualified as true elements. He identified the components of water as hydrogen ("water-former") and oxygen, which combine in a two-to-one ratio, represented by the familiar chemical formula H_2O.

Air, as Lavoisier discovered, is a more complex substance. He recognized that the fraction termed "highly respirable air" is actually an element: oxygen. He derived the name from the Greek word for "acid-former," reflecting his belief that oxygen was present in all acids. For the other gas, which he called mephitic air, he proposed the name "azot" (from Greek), meaning life-inhibiting. Lavoisier found that isolating this component led to the death of animals forced to breathe it, leading him to label it noxious. In reality, it is not poisonous; rather, it is inert and cannot support life when separated from oxygen. Lavoisier noted that this gas contributes to the formation of nitric acid, giving rise to the name "nitrigen," although he favored "azot," which remains the French term for nitrogen today.

While Lavoisier did not seek to entirely dismantle tradition, he asserted, "We have not pretended to make any alteration upon such terms as are sanctified by ancient custom; and therefore… retain the word air to express that collection of elastic fluids which composes our atmosphere." His understanding of this "collection of fluids" was somewhat incomplete, albeit understandably so. Oxygen and nitrogen account for 99% of the atmosphere, but the remaining components create a diverse blend. Most notably, argon is an extremely unreactive element, and there are variable amounts of water vapor and carbon dioxide, among other trace gases like methane and ozone.

Until recent decades, many minor constituents of air remained undetected, but they play crucial roles in atmospheric and environmental chemistry. Some of these gases are greenhouse gases, while others are toxic pollutants, with origins ranging from natural processes to human activity.

In terms of elemental definitions, oxygen and nitrogen are classified as elements, while most of these other gases are compounds formed by the reaction of two or more different elements. For example, oxygen gas consists of diatomic molecules, where each atom is bonded to another oxygen atom, whereas carbon monoxide is formed when an oxygen atom bonds with a carbon atom. This leads to some confusion in chemistry: the term "element" may refer to a specific kind of atom (like oxygen in rust) or to a substance containing only one type of atom (like oxygen gas or a piece of copper metal). Some elements, particularly metals, are commonly found in compound forms, while others, like sulfur and gold, exist naturally in pure or "elemental" states.

Thus, just as one might view a cat both as an abstract concept with distinguishing features and as a tangible creature, air comprises primarily oxygen and nitrogen, while water consists of hydrogen and oxygen. However, the mixture of elements in air differs fundamentally from that in water. In water, chemical bonds link each oxygen atom to two hydrogen atoms, requiring a chemical reaction to separate them. In contrast, the elements in air are physically mixed, akin to sand and salt, allowing for separation without a chemical reaction. While Lavoisier utilized chemical reactions to separate these elements, modern techniques can achieve their physical separation.

Lavoisier's conclusions about air were not entirely novel; he was neither the first to synthesize water from its elements nor the first to deduce that air comprises different substances. What set Lavoisier apart was his interpretation of these observations. The latter half of the eighteenth century saw the rise of "pneumatic chemistry," which focused on the properties of gases, then commonly referred to as "airs." The invention of the pneumatic trough by English clergyman Stephen Hales facilitated this focus by enabling chemists to collect gases emitted from heated substances. In ancient times, "air" encompassed all gaseous substances, but Hales's apparatus allowed for the recognition that not all gaseous emanations were alike, indicating they could not be considered the same unadulterated element.

For instance, Scottish chemist Joseph Black (1728–1799) studied "fixed air," which he identified in the 1750s as a gas produced when carbonate salts were heated or treated with acid. He theorized that this gas was "fixed" in solids until liberated. Unlike common air, fixed air turned lime water (a solution of calcium hydroxide) cloudy due to the formation of insoluble calcium carbonate, essentially chalk. Black found that human breath, gases released during combustion, and those generated through fermentation all had the same effect on lime water. Fixed air is now recognized as carbon dioxide, which carbonates decompose into when heated. Daniel Rutherford (1749–1819), a student of Black, referred to this gas as "mephitic air," a name derived from a term for noxious emissions believed to cause pestilence. This label was fitting, as animals perished in the presence of this gas. Rutherford reported in 1772 that only about one-fifth of common air supports life, while the remaining gases suffocated animals and extinguished candles.

In the 1760s, two notable English pneumatic chemists, Henry Cavendish (1731–1810) and Joseph Priestley (1733–1804), made similar observations regarding gases, echoing findings from the time of Robert Boyle. However, it was Black who first (albeit marginally) proposed the idea that nitrogen, as it would later be named, was a distinct element.

Joseph Priestley's experiments with Hales's trough yielded remarkable results, isolating around twenty different gases, including hydrogen chloride, nitric oxide, and ammonia. Yet, neither he nor his contemporaries recognized these gases as unique compounds; instead, they viewed them as variations of 'common air' affected by different levels of purity. The lingering influence of Aristotle's elements shaped this perspective, as pneumatic chemists tended to interpret gases through the lens of impurities. Even Antoine Lavoisier struggled to break free from this mindset.

This viewpoint was not merely a nod to classical ideas; it was rooted in a chemical framework that explained gas reactions through the concept of phlogiston, a substance from alchemical theory. Alchemy evolved into modern chemistry through various stages, with phlogiston theory arguably representing the last phase. The notion of phlogiston can be traced back to Jabir ibn Hayyan's concept of sulfur, a supposed universal component of metals. Real sulfur, a yellow substance mined from the earth, was highly combustible and a key ingredient in gunpowder.

For some chemists, phlogiston was synonymous with fire, while others aligned with Georg Ernst Stahl's definition of 'terra pinguis': that metals contain an inflammable principle that escapes into the air when heated. It seemed logical that burning wood released some substance into the air, which was labeled phlogiston, the essence of flammability. To prove this, phlogiston theorists would burn a candle in a sealed container, claiming the flame extinguished because the air became saturated with phlogiston and could absorb no more.

Metals, unlike wood, do not burn brightly, but they can be transformed into dull substances when heated in air, a process known as calcination. It was believed that during calcination, metals released phlogiston, which charcoal could restore to recover the original metal. However, a significant problem arose: while burning wood loses mass, calcined metals actually gain weight. This contradiction led many chemists to evade the issue or suggest that phlogiston was weightless or had negative weight.

Stahl's phlogiston theory extended beyond combustion, attempting to account for various chemical and biological processes, including the behaviors of acids and alkalis, respiration, and the odors of plants. It provided a semblance of unity to the field of chemistry. By 1772, Lavoisier still adhered to phlogiston theory but began questioning its adequacy in explaining combustion. He suggested that metals absorbed 'fixed' air during calcination and released it when reduced back to their metallic forms with charcoal and heat. When he learned of Black's fixed air in 1773, he concluded that it was the substance metals combined with to form calxes, clarifying the weight gain.

Pierre Bayen, a French pharmacist, soon highlighted to Lavoisier that mercuric oxide, known as 'calx of mercury,' could revert to mercury simply through heating, without requiring charcoal rich in phlogiston. Moreover, the gas released in this process was distinct from Black's fixed air. The nature of this gas became clearer during a dinner with Joseph Priestley. Priestley, a nonconformist Presbyterian minister supported by the Earl of Shelburne, conducted an experiment in August 1774 similar to Bayen's. By heating mercuric oxide, he collected the gas produced, discovering that a candle flame burned even brighter in this gas than in common air, while smoldering charcoal became incandescent.

Priestley deduced that this gas lacked phlogiston, making it particularly eager to absorb it from burning substances. He dubbed the gas 'dephlogisticated air' and, in 1775, discovered it had remarkable properties: mice placed in a vessel filled with it survived far longer than those in normal air. He sensed a vital quality in the substance and even noted that inhaling it made him feel unusually light and invigorated. He envisioned its potential as a health-enhancing agent, though he joked that only he and his mice had the privilege of breathing it.

However, he was not entirely original; in 1674, John Mayow had claimed that a gas released by heating nitre (potassium nitrate) turned arterial blood red in the lungs. He suggested that metals gained weight during calcination due to the absorption of this 'nitro-aerial' gas, which was essentially oxygen. Around the same time, Swedish apothecary Carl Wilhelm Scheele isolated a gas that enhanced combustion, referring to it as 'fire air' and believing it combined with phlogiston during burning.

In October 1774, Priestley and Shelburne dined with Lavoisier in Paris, where Priestley shared his findings. Combined with Bayen's results, this convinced Lavoisier that metals did not combine with fixed air to form calxes. Bayen had only reported that the gas from mercuric oxide resembled common air. Lavoisier's account of this discovery was submitted for publication in 1775 but took two years to print. Scheele had also sent Lavoisier a letter in September 1774 detailing his findings, the outcome of which remains unknown; however, in Lavoisier's later work, oxygen became a central theme.

Lavoisier might have been dismissive regarding issues of priority in scientific discovery, but he advanced far beyond mere replication of results. Priestley would continue to view oxygen as merely a form of modified common air, influenced by phlogiston, while Scheele similarly interpreted it in these terms. Lavoisier, however, came to recognize that this 'pure air' was a substance in its own right, leading him to redefine air not as an element but as a mixture. His contributions transformed oxygen into an element, reflecting a shift in understanding.

Lavoisier's beliefs about elements remained somewhat traditional, likening them to colors or spices with intrinsic properties evident in mixtures. This notion was flawed; a single element could exhibit vastly different characteristics depending on its combination with other elements.

For example, chlorine is a toxic gas, yet when paired with sodium to form table salt, it becomes harmless. Similarly, while carbon, oxygen, and nitrogen are vital for life, carbon monoxide and cyanide are lethal.

Such ideas were challenging for chemists to accept. Lavoisier faced criticism for asserting that water consisted of oxygen and hydrogen, as water extinguishes flames (being the "most powerful antiphlogistic we possess," according to one detractor) while hydrogen is highly flammable. The process of burning ceases when oxygen is depleted, not when it becomes saturated with phlogiston, which added to the appeal of phlogiston theory.

In an essay from 1773, Lavoisier articulated three physical states of matter: solid, liquid, and gas, emphasizing the distinction between the physical and chemical natures of substances—a distinction that had confounded ancient thinkers and led to their minimal elemental schemes. He argued that "the same body can successively pass through each of these states" based on the quantity of fire matter it combines with.

Lavoisier revolutionized chemists' perceptions of elements. By the early eighteenth century, it was common to recognize just five elements. However, in 1789, he published his influential textbook, *Traité élémentaire de chimie* (Elementary Treatise on Chemistry), defining an element as any substance that could not be broken down into simpler components through chemical reactions, listing no fewer than thirty-three elements. While nineteenth-century physics would later reveal some of these to be fictitious (like light and caloric) and others compounds that had yet to be decomposed into their elemental forms, Lavoisier's message was clear: the elemental landscape was vast, and it was up to chemists to explore it.

Recently, scientists have glimpsed planets beyond our solar system, with the first extrasolar planet detected in 1996 through its gravitational effects on its star. By 1999, astronomers could even observe the light reflected from these planets, revealing a slight blue tint—though this does not imply Earth-like conditions, as the hue likely results from other atmospheric gases.

What if scientists were to find a planet with oxygen-rich reflected light, mirroring that of Earth? It would strongly suggest the presence of life. This raises the question: why does oxygen imply life? Until the 1960s, many believed Earth's oxygen-rich atmosphere—composed of roughly one-fifth oxygen and four-fifths nitrogen—was a consequence of geological processes, implying a planet could support life without necessarily hosting it.

In reality, the air's chemical composition is a product of biological activity. About two billion years ago, primitive organisms began transforming an atmosphere lacking in oxygen into one abundant with it. No known geological process can sustain elevated oxygen levels in our atmosphere indefinitely, as the gas will eventually react with rocks and become sequestered underground. Only biological processes can strip oxygen from its compounds and return it to the atmosphere. If all life were to cease on Earth, oxygen levels would eventually decline to negligible amounts. Therefore, an oxygen-rich atmosphere serves as a compelling indicator of life beneath its surface.

While all animals depend on oxygen, many bacteria are anaerobic, thriving in environments devoid of oxygen, such as mud in seabeds and marshlands, or deep within oilfields. These early bacteria likely emerged over 3.8 billion years ago, existing in an atmosphere that comprised nitrogen, carbon monoxide, water vapor, and perhaps methane. To fuel their biochemical processes, these primitive organisms required energy, which some researchers theorize they derived from the heat and chemical energy of hydrothermal vents. Over time, some bacteria evolved the ability to photosynthesize, utilizing sunlight to produce sugars while releasing oxygen as a byproduct. This marked a dramatic shift in the planet's atmosphere.

These photosynthetic organisms fed upon the carbon dioxide emitted by volcanic eruptions, gradually transforming Earth's atmosphere by saturating it with oxygen. At the same time, some of these bacteria developed cellular respiration methods that enabled them to harness oxygen to break down sugars and release energy, fueling the evolution of larger and more complex organisms.

After approximately two billion years, oxygen levels increased from negligible amounts to about 20% of the atmosphere. This biological revolution profoundly influenced the development of life on Earth, enabling the emergence of multicellular organisms, ultimately leading to the complexity and diversity observed today. Oxygen not only serves as a crucial fuel source for life but also shapes the very nature of our planet, paving the way for evolution to unfold.

The history of chemistry encompasses numerous pivotal discoveries, yet the advent of oxygen remains particularly significant. Its discovery heralded a profound transformation in our understanding of chemical processes and the relationship between life and the environment. Today, as scientists explore distant planets for signs of life, the presence of oxygen serves as a vital indicator, underscoring the profound interconnectedness between chemical principles and the biological processes that define life itself.

7.7 The effects of oxygen on the Earthly elements

In the New Testament, St. Matthew cautions that earthly treasures are fleeting, as "moth and rust doth corrupt." This warning holds particular significance when considering the relentless effects of oxygen, which until recently, provided no means to safeguard gleaming iron and steel from its corrosive nature. Oxygen tarnishes old paintings, transforming their varnish to a brown hue, and most metals develop an oxide layer within seconds of exposure to air. Humans and other animals inhale oxygen not out of necessity, but because we have evolved mechanisms to mitigate its harmful effects. If the air were toxic, life forms—including humans—would adapt to thrive in that toxicity. Enzymes work diligently to neutralize the harmful byproducts formed when oxygen is utilized to metabolize sugar in our cells' energy factories.

Among these byproducts are hydrogen peroxide, a common bleach, and the even more harmful superoxide free radical, both of which can damage vital biomolecules like DNA. While cells possess mechanisms to repair this damage, its gradual accumulation is a significant contributor to aging. Thus, living in an oxygen-rich environment is not ideal; it is merely the outcome of an evolutionary happenstance.

Oxygen is incredibly abundant, ranking as the third most plentiful element in the universe and constituting 47% of the Earth's crust. Remarkably, the biosphere has managed to stabilize atmospheric oxygen at levels optimal for aerobic organisms like us. If oxygen levels fell below 17%, we would suffocate, while concentrations exceeding 25% would make all organic matter highly flammable, leading to uncontrollable wildfires. A 35% oxygen concentration could have historically led to global fires that would obliterate most life on Earth. This is why NASA switched from using pure oxygen to normal air in their spacecraft following the tragic fire during the Apollo tests in 1967. The current level of 21% oxygen strikes a delicate balance.

The relative constancy of atmospheric oxygen supports the Gaia hypothesis, which posits that Earth's biological and geological systems collaboratively regulate the atmosphere and environment to sustain life. Although oxygen levels have fluctuated since the atmosphere became rich in this gas, the variations have been minimal. Today's atmospheric oxygen concentration is sufficient to facilitate the formation of the ozone layer in the stratosphere, which protects life from harmful ultraviolet radiation. Ozone, a UV-absorbing form of oxygen, consists of triplet oxygen atoms, unlike the diatomic molecules in regular oxygen gas.

How, then, is atmospheric oxygen maintained at such steady levels? It is produced during photosynthesis, a process by which organisms extract oxygen from water molecules. This includes all plants and many bacteria. Oxygen is consumed by animals and other aerobic life forms. While it may seem that a balance exists between these sources and sinks, the reality is more complex. The oceans serve as a buffer against significant fluctuations in atmospheric oxygen; the decomposition of marine organic matter slows if oxygen levels drop.

Oxygen is one of several essential elements that are continuously consumed and recycled through interactions involving the biosphere, Earth's rocks and volcanoes, and the oceans. These biogeochemical cycles are interconnected, meaning that changes in the cycle of oxygen, carbon, nitrogen, and phosphorus affect one another. The interlocking systems create a relatively stable environment on Earth. Alterations in the speed of one cycle—such as those resulting from industrial and agricultural practices that release carbon-rich gases into the atmosphere—can disrupt the entire system in unpredictable ways. This unpredictability contributes to the uncertainty surrounding the impact of human activities on global climate change.

The dynamic nature of Earth's chemistry is not one of equilibrium; rather, it is characterized by constant change. In a state of equilibrium, no change occurs, but the chemical stability of our planet is a result of ongoing processes. This distinction is akin to the difference between a person remaining stationary and someone staying in the same spot by walking on a treadmill. The imbalance in Earth's environment involves both inorganic processes in the sea and rocks, but it is ultimately sustained by the biosphere—by life itself. The movement of these interconnected systems is primarily powered by the energy of sunlight harnessed by photosynthetic organisms.

If life were to come to an end, the planet would gradually transition toward a static equilibrium vastly different from today's vibrant environment. This is evident when we examine the atmospheres of our neighboring planets. Venus and Mars, similar in size to Earth and formed from a comparable mix of elements, now exhibit dramatically different atmospheres. Both contain less than 1% oxygen and minimal nitrogen, with around 95% of their atmospheres composed of carbon dioxide. Venus's atmosphere is thick and dense, raising surface temperatures to about 750 °C, while Mars's tenuous atmosphere keeps temperatures at a frigid approximately −50 °C. In both cases, the lack of oxygen and the composition of their atmospheric gases signal the absence of life.

Ironically, gold is one of the most useless metals. Its very uselessness—its inert and non-reactive nature—is what makes it so precious. Gold does not react with atmospheric gases, meaning its surface remains untarnished, which is why it has been highly valued for fine jewelry. Chemists refer to gold as a "noble" metal, a term that reflects not only its lack of chemical reactivity but also its rich historical associations with excellence, magnificence, and nobility. In the late Middle Ages, the term "noble" referred to a gold coin in England.

The allure of gold goes deeper than mere aesthetics. Its resistance to aging ensures that it retains its beauty long after other metals lose their luster. This incorruptibility was thought by alchemists to symbolize spiritual purity, making the pursuit of gold a quest for more than just wealth; it was a religious aspiration. Chinese alchemists believed that because gold does not decay, it could prolong life, leading them to seek a vital, life-giving elixir as a means of securing the spirit of gold itself. Gold's yellow hue came to represent profound concepts, embodying human dignity and symbolizing the four cardinal directions. In ancient China, yellow was the color reserved for the emperor, just as purple was for the Roman elite. Among all metals, gold resonates with rich cultural associations and serves as an enduring symbol of eminence and purity.

Although platinum is rarer and often more expensive, efforts to elevate it above gold have failed due to the lack of legends and myths to support it. No other element possesses the same chemical characteristics that have embedded gold so deeply into our cultural fabric.

7.8 The branches of Chemistry

Chemistry can be broadly divided into two main branches: organic chemistry and inorganic chemistry. Organic chemistry focuses on compounds that contain carbon. The term "organic" originally referred to the chemistry of living organisms, including plants and animals. In contrast, inorganic chemistry encompasses all other elements and some carbon compounds. Substances categorized as inorganic primarily originate from mineral sources rather than from animal or plant sources.

Bernoulli

Other subdivisions of chemistry, such as analytical chemistry, physical chemistry, biochemistry, electrochemistry, geochemistry, and radiochemistry, are often viewed as specialized or auxiliary fields related to the two main branches. Chemistry and physics intersect, as both disciplines explore the properties and behavior of matter. Biological processes are inherently chemical; for example, the metabolism of food into energy for living organisms is a chemical process. Understanding the molecular structure of proteins, hormones, enzymes, and nucleic acids is aiding biologists in their studies of the composition, development, and reproduction of living cells.

7.9 The Scientific Method explained

Chemistry, as a science, deals with concepts and ideas related to the behavior of matter. While these concepts may seem abstract, their application has had a profound and tangible effect on human culture, particularly through modern technology, which began around 200 years ago and has advanced at an ever-increasing pace.

A key distinction between "science" and "technology" lies in their focus: science represents an abstract body of knowledge, while technology is the practical application of that knowledge in the world around us. Chemistry is an experimental science, and much of its progress has come from the scientific method applied through systematic research. Although significant discoveries sometimes occur by chance, most scientific achievements are the result of carefully planned experiments.

There is often confusion about the terms hypothesis, theory, and law. A hypothesis is a tentative explanation of certain facts that serves as a foundation for further experimentation. When a hypothesis is well-supported, it becomes a theory—a broader explanation of natural phenomena backed by substantial evidence. While hypotheses and theories offer explanations, scientific laws are simple statements of natural phenomena that hold true without exception under specific conditions.

7.10 Understanding Atoms

Atoms are complex, minuscule entities. For instance, a single grain of table salt contains about one quintillion sodium atoms (that's 1 followed by 18 zeros) and an equal number of chlorine atoms. As scientists sought to understand what these incredibly small particles are composed of, significant strides in the field of "atomic structure" began to emerge in the 17th century.

In 1738, Swiss mathematician Daniel Bernoulli introduced the concept that gases consist of small particles. He effectively explained how these gases exert pressure on their containers, attributing this pressure to the collisions of particles against the container walls. According to Bernoulli, the pressure of a gas in a container arises from the speed and frequency of molecular collisions with the walls. This speed is directly related to temperature; therefore, if volume, pressure, and temperature remain constant, the number of molecules also remains unchanged. Bernoulli's analysis of increased particle collisions in a reduced container volume (like a balloon) led him to derive the same mathematical principles outlined in Boyle's law. His work was groundbreaking, as it marked the first time the concept of atoms was used to calculate a physical property.

Eighty years prior to Bernoulli, English chemist Robert Boyle was already exploring atomic concepts in the 17th century, during the time of Newton. Boyle conducted experiments with gases, leading to the formulation of Boyle's law, which describes how gas pressure increases as its volume decreases, provided the temperature remains constant. For example, squeezing a balloon increases the pressure inside it; if you continue to squeeze, the pressure builds until the balloon eventually bursts.

The late 18th century saw atoms firmly integrated into scientific thought, particularly through the work of French chemist Antoine Lavoisier. He investigated the nature of combustion and identified various elements—pure chemical substances that cannot be broken down into simpler forms. Lavoisier recognized that burning substances results from the combination of oxygen from the air with other elements. Additionally, Isaac Newton contributed to the understanding of atomic theory by demonstrating that light is composed of particles, or atoms, further expanding our knowledge of the fundamental building blocks of matter.

Robert Boyle

7.11 Electrons Explained

The "cloud" of electrons forms the outer layer of the atom and dictates how it interacts with other atoms. The specific details of what lies deep within this electron cloud are largely irrelevant; what truly matters is the electrons themselves. The interactions between the electron clouds of multiple atoms are what drive chemical behavior.

Bohr's 20th-century model of the atom clarified the characteristics of the electron cloud and provided a scientific foundation for chemistry. Chemists had already observed that certain elements shared similar chemical properties despite having different atomic weights. When elements are organized in a table based on their atomic weights—and particularly when accounting for different isotopes—these similar elements appear at regular intervals.

Notably, elements with similar properties emerge every eight atomic numbers. This systematic arrangement of elements, where those with comparable properties are grouped together, is what gives the table its name: the "periodic" table.

7.12 The Elemental Periodic Table

In 1869, the Russian scientist Dmitri Ivanovich Mendeleyev created the "Periodic Table," which allowed him to make remarkable deductions about undiscovered elements, predicting not only their existence but also their properties, densities, and melting points. To appreciate the significance of the Periodic Table, it's essential to define what we mean by an element. Antoine Lavoisier provided a useful definition: a substance qualifies as an element if it cannot be broken down into simpler, more fundamental components. However, this definition is influenced by a chemist's expertise and the technology available at the time.

For instance, Lavoisier classified "lime" and "magnesia" as elements, but these are compounds—calcium oxide and magnesium oxide, respectively. The elements calcium and magnesium were first isolated by the English chemist Humphry Davy in 1808 using electrolysis, a method that separates compounds through electricity. Davy also discovered sodium and potassium this way in 1807.

The concept of atoms was long debated. Philosophers like Aristotle could remain skeptical, as discussions were largely theoretical. This changed in 1908 when French physicist Jean Perrin demonstrated that the random motion of tiny particles suspended in water aligned with Albert Einstein's theory that they were being struck by water molecules, which consist of hydrogen and oxygen atoms. This evidence convinced many, including chemist Wilhelm Ostwald, that atoms are indeed real.

John Dalton, a modest Quaker from Manchester, began illustrating atoms around 1800, drawing inspiration from Isaac Newton's earlier vision of solid, impenetrable particles. Dalton viewed atoms as eternal, unchangeable entities, aligning his ideas with the ancient Greek philosopher Democritus, from whom he borrowed the term "atomos." His drawings depicted circular particles, each distinguished by symbols, which combined in fixed ratios to form what we now refer to as molecules.

Dalton did not know what atoms were made of and considered the weights of atoms—their relative mass—far more important. He assumed that atoms of the same element had identical weights, while different elements had differing weights. For example, Dalton noted that hydrogen combined with eight times its weight of oxygen to form water, suggesting that the atomic weight of oxygen relative to hydrogen was 8. However, Dalton was mistaken; the water molecule actually contains two hydrogen atoms for every oxygen atom, meaning the true atomic weight of oxygen is 16. While his list of atomic weights was a mix of accurate and inaccurate values, subsequent chemists, including Jons Jacob Berzelius, refined the list, enhancing our understanding of atomic weights.

Dalton's atomic theory revolutionized chemistry, transitioning it into a precise science. The significance of quantitative measurements in chemical processes was recognized by contemporaries like Cavendish, Priestley, and Lavoisier, but without a theory explaining the elements, these figures were mere empirical observations. Dalton's assertion that atoms unite in simple ratios to form compound particles elucidated why chemical reactions consistently occur in specific proportions. This principle was later formalized in Louis Joseph Proust's Law of Definite

Proportions in 1788. However, disagreements remained, as early chemical analysis methods were often unreliable, leading to variability in observed compound ratios.

Dalton introduced his atomic theory in 1808 with the publication of *A New System of Chemical Philosophy*. His depictions of atoms and molecules bridged the microscopic and macroscopic worlds of chemistry, illustrating both observable phenomena—like the combination of hydrogen and oxygen to form water—and the unobservable union of atoms. According to historian William Brock, Dalton's symbols fostered belief in the reality of chemical atoms and helped chemists visualize complex chemical reactions. Together, Dalton and Lavoisier revolutionized the language of chemistry.

However, Dalton's symbolic system was cumbersome for typesetters, prompting Berzelius to propose an alphabetic notation for elements in the following years. Berzelius suggested using the first letter of each element's name, or two letters in cases of duplication. For instance, hydrogen became H, oxygen O, and carbon C, while cobalt was designated Co to differentiate it from carbon. He insisted that Latin names be used for certain elements, resulting in copper being represented as Cu (from cuprum), gold as Au (from aurum), and iron as Fe (from ferrum).

Berzelius also suggested using subscripts to denote the ratios of atoms in compounds. Thus, the two-to-one ratio of hydrogen and oxygen in water is written as H_2O. This notation provided a clearer representation of chemical elements and their combinations than Dalton's approach, though Dalton criticized Berzelius's symbols as perplexing and detrimental to the elegance of atomic theory.

While Dalton's atomic symbols were highly schematic, they visually indicated small, ball-like particles. In contrast, Berzelius's symbols lacked this mnemonic quality. Many 19th-century chemists viewed chemical formulas, such as C_6H_6 for benzene or C_2H_6O for dimethyl ether, merely as abbreviations for elemental analyses rather than representations of atomic models. They tended to disregard what these formulas implied about atomic arrangements, leading to limited inquiry into molecular shapes until the mid-19th century. By the end of the century, some chemists felt it was futile to ponder the nature or arrangement of atoms.

Dalton's theory provided a tangible reality for atoms and distinguished elements based on weight rather than color, fostering the belief that elements are fundamentally different and not interchangeable. However, some chemists, including the prominent Michael Faraday, remained skeptical about the idea of transmutation.

Others, like William Prout, suggested that under extreme conditions, elements might indeed transform into one another. Prout proposed in 1815 that all elements could be derived from hydrogen, invoking the ancient concept of *protyle*—the primal substance believed to compose all matter.

While Berzelius dubbed this idea "Prout's hypothesis," he did not fully endorse it. In the 1840s, French chemist Jean-Baptiste Dumas refined the concept by pointing out that some atomic weights did not neatly fit into integral multiples of hydrogen's weight. For instance, chlorine has an atomic weight of 35.5, challenging Prout's assumptions. Dumas speculated that the fundamental components of atoms might be smaller fractions of hydrogen, coining the term "protyle."

Despite these theories, no one had successfully subdivided a hydrogen atom or convincingly transformed one element into another. In the 1870s, astronomer Joseph Norman Lockyer posited that the right conditions for transmutation could be found in the intense environments of stars.

7.13 Explaining elements through science

The next crucial advancement in understanding the atom came with the realization that certain substances cannot be further reduced by chemical means. These fundamental substances, known as elements, combine in specific and predictable ways to form all the materials we encounter.

During the 19th century, while unrelated research was occurring, the self-taught English scientist Michael Faraday was investigating how electrical currents break apart water and other compounds. He proposed that electricity is not a fluid but consists of small particles that carry electric charge, a notion inspired by Ben Franklin. These particles later became known as "electrons."

As previously mentioned, the pioneering chemist John Dalton established a solid foundation for the role of atoms in chemistry in the early 19th century. He asserted that matter consists of atoms, which are indivisible; that all atoms of a given element are identical, while atoms of different elements vary in size and shape; that atoms cannot be created or destroyed but are rearranged during chemical reactions; and that a chemical compound, composed of two or more elements, consists of molecules, each containing a fixed number of atoms from the constituent elements. Thus, the modern atomic concept of the material world, as we learn it in textbooks today, emerged around two hundred years ago.

In 1808, Dalton demonstrated that the chemical rules governing element combination could be explained by three key assumptions:

1. Each element is made of specific atoms.
2. All atoms of the same element are identical and distinct from those of other elements.
3. These atoms combine in particular ways to create all the substances we observe in the universe.

In 1811, the Italian chemist Amedeo Avogadro formulated his famous hypothesis, stating that at a fixed temperature and pressure, equal volumes of gas contain the same number of molecules, regardless of the gas's chemical nature. This idea suggested that the chemical or physical properties of a gas do not affect the number of molecules in a given volume. Avogadro's theory inspired experiments that eventually led to the determination of what we now refer to as Avogadro's number, approximately 600 sextillion (6 followed by 23 zeros) molecules—an astonishingly large figure. Later experiments validated Avogadro's hypothesis, confirming that each liter of gas at one atmosphere of pressure and 0°C contains roughly 27,000 billion (2.1×10^{21}) molecules. However, it wasn't until the 1850s that Avogadro's compatriot, Stanislao Cannizzaro, refined the idea to a point where it gained wider acceptance among chemists.

Despite this progress, as late as the 1890s, many chemists still resisted the ideas of Dalton and Avogadro. By then, however, the developments in physics, particularly the detailed explanations of gas behavior using atomic concepts by Scottish physicist James Clerk Maxwell and Austrian physicist Ludwig Boltzmann, began to sway opinions.

Today, we can "indirectly" observe atoms. Modern scanning tunneling microscopes can map surfaces with extraordinary atomic resolution, and field ion microscopes from the 1950s allowed scientists to capture indirect images of atoms. In contrast, at the beginning of the 20th century, even though most physicists and chemists accepted the existence of atoms, some skepticism remained. To address this, Albert Einstein aimed to provide irrefutable proof of atomic existence in the early 20th century.

7.14 Measurements of Mass

The gram is a small unit of mass measurement, with a nickel weighing approximately 5 grams. In the SI system, the standard unit of mass is the kilogram, which is equal to 1,000 grams. The kilogram's mass is defined by international agreement to be exactly the same as the mass of a platinum-iridium weight known as the international prototype kilogram, stored in a vault in Sèvres, France.

When comparing mass units, 1 kilogram is equivalent to about 2.2 pounds, and 1 pound equals 454 grams (or 0.454 kilograms). Below is a summary of various measurements of gram weight:

- 1 g = 1,000 mg
- 1 kg = 1,000 g
- 1 kg = 2.2 lb.
- 1 lb. = 454 g

7.15 Heat and Temperature Scales

Heat is a form of energy associated with the motion of tiny particles in matter. It can refer to the total amount of energy within a system or to the energy added to or removed from that system. In this context, a "system" refers to the entity being heated or cooled. The amount of heat energy present determines whether the system feels hot or cold. Temperature measures the intensity of heat in a system, irrespective of its size. Heat always flows from areas of higher temperature to areas of lower temperature. In the SI system, the unit of temperature is the Kelvin.

It's important to note that the degree symbol is not used with Kelvin temperatures:

- Degrees Celsius = 0 °C
- Kelvin (absolute) = K
- Degrees Fahrenheit = 0 °F

Heat: Quantitative Measurement

The SI-derived unit for heat is the joule (pronounced "jool" and abbreviated as J). Another common unit for heat is the calorie (abbreviated as cal). The relationship between joules and calories is defined as follows: 4.184 J = 1 cal (exactly).

To illustrate the magnitude of these units, 4.184 J (or 1 cal) is the amount of heat energy needed to raise the temperature of 1 gram of water by 1 °C, typically measured from 14.5 °C to 15.5 °C.

Because joules and calories are relatively small units, kilojoules (kJ) and kilocalories (kcal) are often used to express heat energy in many chemical processes.

7.16 Matter Defined

The universe is made up of matter and energy, and each day, we interact with countless forms of matter. Air, food, water, rocks, soil, glass, and even this book are all examples of different types of matter. Broadly defined, matter is anything that has mass and occupies space.

Matter exists in three primary physical states: solid, liquid, and gas. Solids have a definite shape and volume, with particles that are rigidly bonded together. The shape of a solid is independent of its container; for example, a crystal of sulfur retains the same shape and volume whether it is placed in a beaker or laid on a glass plate.

Most common solids, such as salt, sugar, quartz, and metals, are crystalline, existing in regular, repeating three-dimensional geometric patterns. In contrast, solids like plastics, glass, and gels lack a regular internal structure, classifying them as amorphous solids—meaning they do not have a definite shape or form.

Liquids have a definite volume but take the shape of their container. The particles in a liquid are held together by strong attractive forces but can move freely, providing fluids with their characteristic flow. This particle mobility allows liquids to adapt to the shape of any container they occupy.

Gases, on the other hand, have no fixed shape or volume. The particles in a gas move independently and have enough energy to overcome the attractive forces that bind them in solids or liquids. As a result, a gas expands to fill its container, pressing continuously in all directions against the walls. The distance between gas particles is relatively large compared to those in solids and liquids, making gases easily compressible and expandable.

For instance, when a bottle of ammonia solution is opened in one corner of a laboratory, its familiar odor quickly spreads throughout the room. This phenomenon illustrates the rapid and free movement of gaseous particles, which tend to permeate the entire area around them.

Although matter appears continuous, it is made up of discrete particles held together by attractive forces. These forces are strongest in solids, giving them rigidity, while they are weaker in liquids, which still maintain a definite volume. In gases, attractive forces are so minimal that the particles act almost independently.

Table 7.1: Common Materials in the Solid, Liquid, and Gaseous States of Matter

Solids	**Liquids**	**Gases**
Aluminum	Alcohol	Acetylene
Copper	Blood	Air
Gold	Gasoline	Butane
Polyethylene	Honey	Carbon dioxide
Salt	Mercury	Chlorine
Sand	Oil	Helium
Steel	Vinegar	Methane
Sulfur	Water	Oxygen

A **substance** is a specific type of matter with a definite and fixed composition. Often referred to as a **pure substance**, it can be classified as either an element or a compound. Common examples of elements include copper, gold, and oxygen, while typical compounds include salt, sugar, and water.

Matter can be classified as **homogeneous** or **heterogeneous** based on its appearance and properties. Homogeneous matter appears uniform and exhibits the same properties throughout, while heterogeneous matter consists of two or more physically distinct phases.

A **phase** refers to a homogeneous part of a system separated from others by physical boundaries, and a **system** is simply the body of matter being examined. When visible boundaries exist between the components, the system is considered heterogeneous, regardless of whether the compounds are in solid, liquid, or gaseous states.

It's important to note that a pure substance, whether an element or compound, is always homogeneous in composition. However, it may exist in different phases within a heterogeneous system. For instance, ice floating in water represents a two-phase system composed of solid and liquid water. Each phase is homogeneous in composition, but the presence of two distinct phases makes the overall system heterogeneous.

A **mixture** consists of two or more substances and can be classified as either homogeneous or heterogeneous, with variable composition. For example, if we add a tablespoon of sugar to a glass of water, an immediate heterogeneous mixture forms, consisting of solid sugar and liquid water. Upon stirring, the sugar dissolves to create a homogeneous mixture or solution, where both substances are still present, and all parts of the solution are uniformly sweet and wet. The proportions of sugar and water can be adjusted by adding more sugar and stirring.

Some substances do not create homogeneous mixtures. For instance, mixing sugar with fine white sand results in a heterogeneous mixture. Careful observation may be required to determine that the mixture is heterogeneous, as both components are white solids. In fact, most ordinary matter exists as mixtures. If we examine materials like soil, granite, iron ore, or other naturally occurring minerals, we find they are typically heterogeneous mixtures. In contrast, seawater is a homogeneous mixture (solution) containing various substances, while air is a homogeneous mixture (solution) of several gases.

7.17 Properties of Substances

How do we identify substances? Each substance has a distinct set of properties that gives it a unique identity. These properties can be viewed as the "personality traits" of substances, classified into **physical** and **chemical** categories.

Physical Properties

Physical properties are inherent characteristics that can be observed without altering the substance's composition. They are linked to the substance's physical existence and include attributes such as color, taste, odor, and state of matter (solid, liquid, or gas). These properties describe a substance's ability to undergo reactions or decompose.

Physical Changes

Matter can undergo two types of changes: physical and chemical. **Physical changes** involve alterations in physical properties—such as size, shape, and density—or changes in state without affecting the composition of the substance. For example, the transformation of ice into water and then into steam illustrates physical changes across different states of matter, with no new substances formed.

When a clean platinum wire is heated in a flame, its appearance shifts from a silvery metallic luster to a glowing red. This change is classified as physical because the platinum can revert to its original metallic form upon cooling, and its composition remains unchanged throughout the heating and cooling processes.

Chemical Changes

In contrast, **chemical changes** result in the formation of new substances that possess different properties and compositions compared to the original material. These new substances need not resemble the initial material at all.

For instance, when a clean copper wire is heated, it changes from a metallic copper color to a glowing red. However, unlike platinum, the copper does not revert to its original form upon cooling; instead, it becomes a black material known as copper(II) oxide. This transformation occurs when copper reacts with oxygen in the air during the heating process. The unheated wire is nearly 100% copper, whereas copper(II) oxide consists of approximately 79.9% copper and

20.1% oxygen. One gram of copper will yield 1.252 grams of copper(II) oxide. Thus, while the platinum undergoes only a physical change, the copper experiences both physical and chemical changes when heated.

Conservation of Mass

The **Law of Conservation of Mass** states that there is no detectable change in the total mass of the substances involved in a chemical change.

7.18 Energy

Energy is the capacity of matter to perform work and exists in various forms, including mechanical, chemical, electrical, heat, nuclear, and radiant (light) energy. Matter can exhibit both potential and kinetic energy.

Potential Energy

Potential energy refers to stored energy, which an object possesses due to its position. For example, a ball located 20 feet above the ground has more potential energy than one at 10 feet and will bounce higher when allowed to fall. Water held behind a dam represents potential energy that can be converted into useful work, such as electrical or mechanical energy.

Gasoline serves as a source of chemical potential energy; when it burns (reacts with oxygen), the heat released indicates a decrease in potential energy, resulting in new substances that possess less chemical potential energy than the original gasoline and oxygen.

Kinetic Energy

Kinetic energy is the energy that matter possesses due to its motion. When the water behind a dam is released, its potential energy converts into kinetic energy, which can drive generators to produce electricity. All moving objects possess kinetic energy; for instance, the pressure exerted by a confined gas arises from the kinetic energy of rapidly moving gas particles. When two moving vehicles collide, their kinetic energy is converted into other forms of energy, such as heat, sound, and light (radiation) during the crash.

Energy can be transformed from one form to another, with some types converting more easily and efficiently than others. For example, mechanical energy can be converted into electrical energy with over 90% efficiency using an electric generator. In contrast, solar energy has thus far been directly converted into electrical energy at an efficiency of only about 15%.

7.19 Energy in Chemical Changes

In every chemical change, matter either absorbs or releases energy. These changes can produce various forms of energy. For instance, electrical energy is generated by chemical reactions in lead storage batteries, which start automobiles. Light energy is emitted as a flash during the chemical reaction in magnesium flashbulbs for photography. Similarly, heat and light energies are released during fuel combustion. All the energy necessary for our vital processes—such as breathing, muscle contraction, and blood circulation—is produced by chemical changes occurring within our cells.

Conversely, energy is also utilized to induce chemical changes. For example, electroplating metals involves a chemical reaction when electrical energy passes through a salt solution containing the submerged metal. Additionally, green plants utilize radiant energy from the sun in the process of photosynthesis, which is another example of a chemical change. As previously mentioned, heat can cause mercury (II) oxide to decompose into mercury and oxygen, representing yet another chemical change.

Often, chemical changes are primarily employed to generate energy rather than to produce new substances. In many cases, the heat or thrust generated by fuel combustion takes precedence over the formation of new materials.

7.20 Conservation of Energy

Energy transformations take place during every chemical change. If energy is absorbed during the reaction, the products will possess more chemical potential energy than the reactants. Conversely, if energy is released, the products will have less chemical potential energy compared to the reactants. For instance, in an electrolytic cell, water can be decomposed, absorbing electrical energy in the process. The resulting hydrogen and oxygen have higher levels of chemical potential energy than the original water. This energy is then released as heat and light when hydrogen and oxygen are burned to reform water. Thus, energy can be converted from one form to another or transferred between substances, ensuring that it is not lost.

Extensive research has been conducted on energy changes in various systems, revealing that no system can gain energy without it being sourced from another system. This principle is encapsulated in the Law of Conservation of Energy, which states that energy cannot be created or destroyed; it can only be transformed from one form to another.

7.21 Interchangeability of Matter and Energy

Matter and energy are intrinsically linked; addressing one invariably involves the other. This relationship puzzled scientists for centuries until the early 20th century. In 1905, Albert Einstein introduced one of the most groundbreaking concepts in science.

Einstein proposed that the energy (E) equivalent to a mass (m) can be calculated using the equation $E=mc^2$. In this equation, E is measured in ergs (where 1 erg = 1×10^{-7} J), m is in grams, and c represents the speed of light (approximately 3.0×10^{10} cm/s). According to this equation, whenever energy is absorbed by a substance, its mass increases, and when energy is released, its mass decreases. Although the energy changes observed in chemical reactions can be significant, the actual changes in mass are extremely small and often undetectable by even the most sensitive instruments.

For example, according to Einstein's equation, 9.2×10^7 J or (9.2×10^{14} ergs) of energy corresponds to a mass loss of just 0.0000010g (or 1.0 µg):

1 µg of mass = 9.0×10^7 J of energy (or 9.0×10^{14} ergs).

In practical terms, when *2.8×10^3 g* of carbon is burned to produce carbon dioxide, it releases *9.2×10^7 J* of energy. From this large quantity of carbon, only about *one-millionth* of a gram, specifically *3.6×10^{-8} g* of the original mass, is converted into energy. Consequently, in practical applications, we can consider the masses of reactants and products in chemical changes to remain constant. However, because mass and energy are interchangeable, we can combine the laws of conservation of matter into a single, more comprehensive statement: the total amount of mass and energy remains constant during a chemical change.

7.22 Elements and Compounds

All known substances on Earth—and likely throughout the universe—are composed of a "chemical alphabet" made up of 108 currently recognized elements. An element is a fundamental substance that cannot be broken down into simpler substances through chemical means. These elements serve as the building blocks for all materials.

The elements are arranged in order of increasing complexity, starting with hydrogen, which is element number 1. Out of the first 92 elements, 88 are naturally occurring. The remaining four—technetium (43), promethium (61), astatine (85), and francium (87)—either do not exist in nature or have only fleeting existences due to radioactive decay. With the exception of element 94, plutonium, the elements numbered above 92 are not known to occur naturally and are typically synthesized in laboratories, often in very small amounts. Recent reports have confirmed the discovery of trace amounts of plutonium in nature. Elements 107 and 109 were synthesized in 1981 and 1982, respectively, while there has been no reported synthesis of element 108. No elements beyond those found on Earth have been detected on other celestial bodies.

Most substances can be broken down into two or more simpler substances. For example, mercury (II) oxide can decompose into mercury and oxygen, water can be separated into hydrogen and oxygen, and sugar can break down into carbon, hydrogen, and oxygen. Table salt can also easily be decomposed into sodium and chlorine. In contrast, an element cannot be further decomposed into simpler substances through ordinary chemical changes.

If we take a small piece of an element, such as copper, and keep dividing it into smaller and smaller particles, we will eventually reach a point where we have a single unit of copper that cannot be subdivided further while still retaining its identity as copper. This smallest particle of an element that can exist independently is called an "atom."

An atom is also the smallest unit of an element that can participate in a chemical reaction. Atoms consist of even smaller components known as subatomic particles, but these subatomic particles do not exhibit the properties of the elements themselves.

7.23 Lockyer's discovery of Helium

In 1869, astronomer Joseph Norman Lockyer discovered a new element that had never been observed on Earth. He identified it by analyzing the light emitted from the sun, recognizing that atoms absorb light at specific wavelengths. This phenomenon results in the sunlight spectrum—when light is dispersed through a prism—showing narrow dark lines, akin to a barcode, where elements in the sun's atmosphere have absorbed certain wavelengths. Lockyer noticed an absorption line that didn't match any known elements, leading him to conclude it must be a new, unseen substance.

At the same time, French astronomer Pierre Janssen made the same observation from his observatory in Paris. This new element was named helium, derived from "Helios," the Greek word for the sun.

Helium is the lightest of the so-called noble gases, which are extremely unreactive, explaining why they had not been discovered earlier, despite their presence on Earth. Terrestrial helium was first identified twenty-seven years after Lockyer and Janssen's solar observations. Lockyer's studies of the solar spectrum led him to realize that the sun contains a vast array of chemical elements. He theorized in 1873, later elaborated in his book *Chemistry of the Sun* (1887), that in the hottest blue-white stars, stellar matter breaks down into atomic constituents—subatomic particles, referred to as "protyle" by Dumas. As stars cool, these particles combine to form regular elements, including some not yet known on Earth, like helium. Lockyer proposed that stars begin as loose aggregates of gas and dust, rich in various elements. Under gravity's influence, this material condenses, heating up until it becomes hot enough to dissociate atoms into protyle. As the star continues to contract and cool, protyle condenses into increasingly heavier elements, mirroring the process of biological evolution proposed by Darwin. Lockyer published this theory in the journal *Nature*, which he founded, in 1914. By that time, however, the questions of atomic composition and the possibility of atomic transmutation had become subjects of experimental inquiry on Earth. These experiments revealed that the proponents of protyle—Prout, Dumas, and Lockyer—were onto a profound truth about atomic structure.

Ernest Rutherford (1871–1937) aimed to "anatomize" the atom and chose gold for his experiments because, like medieval artists who used gold for decoration, it could be hammered into very thin sheets of nearly transparent material. This property allowed him to study a sample only a few atoms thick, essential for investigating atomic composition. Rutherford remarked, "I was brought up to look at the atom as a nice hard fellow, red or grey in color, according to taste." However, in 1907, he discovered that atoms were primarily "empty space."

While working at Manchester University with his students Hans Geiger and Ernest Marsden, they directed alpha particles from radioactive elements at thin gold foil and found that most particles passed through with little deflection. Geiger helped develop the instrument that detected the alpha particles, later turning it into the Geiger counter.

While it was expected for lighter alpha particles to pass through gold leaf, a small number of them bounced back. This unexpected result forced the researchers to reassess their understanding of atomic structure. Rutherford later recalled, "It was quite the most incredible event that has ever happened to me in my life...almost as incredible as if you fired a 15-inch shell at a piece of tissue paper and it came back and hit you." This experiment led to the discovery of the atomic nucleus. Rutherford concluded that while atoms are mostly empty space, they contain a dense central core where almost all their mass resides. This nucleus, about 10,000 times smaller than the overall atom, must be positively charged because it repels positively charged alpha particles. Surrounding it, he posited, is a cloud of negatively charged electrons, equal in charge to the nucleus.

Danish physicist Niels Bohr (1885–1962) refined Rutherford's vague description of the atom into a more precise and appealing model. By 1911, physicists were aware that atoms contained electrons, negatively charged particles discovered by Joseph John Thomson in 1897. Bohr, a young student, initially worked with Thomson at Cambridge but soon joined Rutherford's lab in Manchester. In 1912, he developed a model of the atom that he published the following year, earning him the Nobel Prize in 1922. Bohr's model depicted electrons orbiting a dense nucleus like planets around the sun, building on Rutherford's ideas. His crucial contribution was explaining how this arrangement could be stable; according to classical physics, orbiting electrons should emit light as they move and subsequently lose energy, spiraling into the nucleus. To resolve this dilemma, Bohr incorporated principles from quantum theory, emerging from the work of Einstein and Max Planck.

Bohr's description of the atom diverged significantly from Dalton's earlier model, which depicted it as an indivisible lump. Instead, atoms are composed of subatomic particles—the electrons and the nucleus—and are primarily empty space. The "size" of the atom is defined not by hard boundaries but by the extent of the electrons' orbits. Regarding the nucleus, Rutherford proposed that it consists of positively charged subatomic particles. He asserted that hydrogen, the lightest atom, contains a single positively charged particle, which he named the proton, marking the final realization of the prote hyle or protyle concept. Helium nuclei (alpha particles) possess twice the positive charge of hydrogen nuclei, indicating they contain two protons. This validated Prout's hypothesis: since their nuclei are made up of protons, all elements can be seen as derived from hydrogen.

However, Rutherford recognized there must be more to the story. Although helium nuclei have double the charge of hydrogen nuclei, they also have four times the mass. He thus proposed the existence of particles with the same mass as protons but lacking electrical charge. In 1932, Rutherford's student James Chadwick identified this neutral particle and named it the neutron.

In the commonly depicted Bohr model of the atom, electrons—much lighter than protons—orbit around a nucleus made up of protons and neutrons, forming ellipsoidal shapes similar to planetary orbits. While planets follow Newton's laws, electrons are governed by quantum mechanics, resulting in a less precise understanding of their positions.

It is impossible to pinpoint the location of an electron within an atom; we can only calculate the probability of finding it in a specific location at any given time. This fuzzy representation stems from the wavelike properties exhibited by very small objects, which behave as both particles and waves. As a result, it is more accurate to think of electrons forming a "cloud," akin to bees buzzing around a hive, moving too rapidly to observe any individual atom distinctly. Furthermore, these clouds do not adopt disc-like shapes, as a solar-system analogy might suggest; instead, they exhibit a variety of forms based on the energy levels of the electrons. Some clouds are spherical, while others resemble dumbbells or have multiple lobes centered around the nucleus, known as orbitals.

The development of the quantum atom enabled chemists to address a fundamental mystery: why elements exhibit their unique properties. For instance, why is helium so inert while sodium is highly reactive? Why do hydrogen atoms pair up in hydrogen gas, while carbon atoms bond with four others in diamond?

7.24 The atomic weight of an element

Before Rutherford proposed that atoms contain positively charged protons, the concept of atomic number did not exist, nor did the implications of such a number. Today, elements are arranged by atomic number, which increases sequentially by one for each subsequent element. This atomic number indicates the number of electrons in an atom, which equals the number of protons, ensuring that the atom remains electrically neutral.

Elements can bond in two primary ways: some atoms share electrons, effectively forming a kind of handshake, while others transfer electrons to become charged ions. For example, in methane, a carbon atom shares electrons with four hydrogen atoms, while in table salt, sodium atoms donate an electron to chlorine atoms, resulting in positively charged sodium ions and negatively charged chloride ions. The resulting electrostatic attraction binds the ions together. The bonding tendency, or valency, of each element depends on how many electrons its atoms can contribute. Generally, an atom cannot utilize all its electrons for bonding; typically, only the outermost electrons are involved.

Quantum theory reveals that electrons are arranged in shells around the nucleus. The first shell can hold two electrons, the second eight, and the third eighteen, reflecting the "magic numbers" of the Periodic Table. Beyond the first shell, electrons are grouped into sub-shells. For instance, the second shell has one sub-shell with two electrons and another with six; the third shell includes one sub-shell of two, one of six, and one of ten. The fourth shell contains sub-shells with 2, 8, 10, and 14 electrons.

These arrangements explain the block sizes in the Periodic Table, as they correspond to the filling of shells and sub-shells with electrons as atomic numbers rise. The process can be complex due to overlapping shells; for instance, the first sub-shell of the fourth shell is filled before the third sub-shell of the third shell. However, new blocks of elements emerge as one moves down the rows of the Periodic Table, with additional sub-shells becoming available for filling.

This periodicity arises because the filling pattern of each shell is similar to the one before it, causing a repetition in chemical properties. Each element aims to achieve a filled outer shell by either sharing electrons or gaining and losing them. For example, lithium, sodium, and potassium form ions by losing one electron, resulting in filled outer shells, while carbon and silicon achieve this through sharing. The noble gases, located at the end of each row, are inert because they already have filled outer shells and do not need to bond with other atoms.

Consequently, an element's position in the Periodic Table—its row and column—reveals much about its chemical behavior. Metals are generally found on the left, non-metals on the right, and the column number often predicts the element's valency. In general, chemical reactivity decreases as one moves down the rows. Thus, the Periodic Table serves as an invaluable reference for aspiring chemists preparing for exams.

7.25 Making new elements

So, how many elements are there? The answer remains uncertain. Scientists can identify the number of natural elements—those we expect to find throughout the universe—which ends at around uranium, the 92nd element. However, when it comes to the total number of possible elements, we have no definitive answer; it's merely a matter of speculation.

Since the mid-twentieth century, chemists and physicists have worked together to create new elements—substances previously unknown on Earth. They are gradually expanding the Periodic Table into uncharted territory, where predicting which elements may form and how they will behave becomes increasingly challenging. This area of study is known as nuclear chemistry. Unlike most chemists, who combine elements to create molecules and compounds, nuclear chemists manipulate subatomic particles, specifically protons and neutrons, to forge new combinations within atomic nuclei. This pursuit fulfills the ancient alchemists' dream of transmuting one element into another.

The alchemists were ultimately doomed to fail in their quest, as transmuting elements using "chemical energy"—the energy associated with making and breaking atomic bonds—is impossible. However, everything changed with the discovery of radioactivity at the end of the nineteenth century, marking the beginning of a remarkable and transformative era in chemistry. This groundbreaking work originated in a dilapidated wooden shed at the School of Chemistry and Physics in Paris, where Marie Curie and her husband, Pierre, conducted their experiments.

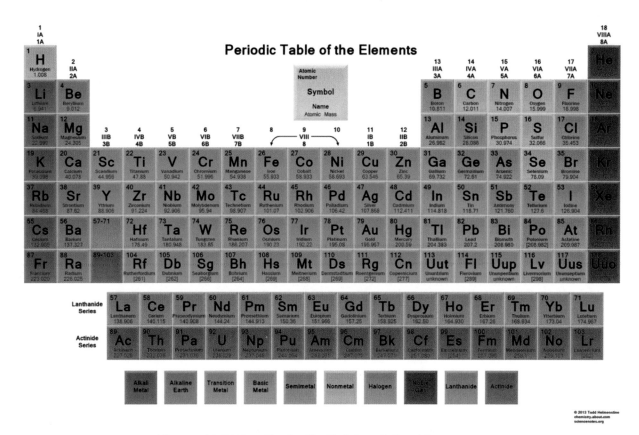

Figure 7-2 The Periodic Table of the Elements

7.26 Summary of the Chemistry of the Elements

The structure of matter has long fascinated thinkers throughout history. The foundations of modern atomic theory trace back to ancient Greek philosophers. Around 440 B.C., Empedocles proposed that all matter consisted of four fundamental "elements": earth, air, water, and fire. Approximately a century later, Democritus (c. 470-370 B.C.), an early atomistic philosopher, suggested that all matter was composed of indivisible particles he termed "atoms." He believed these atoms were in constant motion and could combine in various ways. However, his ideas were purely speculative and lacked scientific backing.

In contrast, Aristotle (384-322 B.C.) rejected Democritus's theory and instead championed the Empedoclean view. Aristotle's influence was so profound that his ideas dominated scientific and philosophical thought until the early 17th century. The term "atom" itself originates from the Greek word *atomos*, which means "indivisible."

Distribution of Elements

Elements are unevenly distributed throughout nature. At normal room temperature, two elements—bromine and mercury—are liquids, while eleven elements—hydrogen, nitrogen, oxygen, fluorine, chlorine, helium, neon, argon, krypton, xenon, and radon—are gases. All other elements are solids.

About 99% of the weight of the Earth's crust, seawater, and atmosphere is comprised of just ten elements, with oxygen being the most abundant at approximately 50% of this mass. This distribution covers the Earth's crust to a depth of about ten miles, as well as the oceans, freshwater, and the atmosphere. It does not account for the mantle and core of the Earth, which are thought to consist primarily of metallic iron and nickel. Additionally, since the atmosphere contains relatively little matter, its inclusion has minimal impact on the overall element distribution.

Names of the Elements

The names of the elements have various origins, many deriving from early Greek, Latin, or German terms that often describe specific properties of the elements. For instance, iodine comes from the Greek word *iodes*, meaning "violet-like," reflecting its violet color in vapor form.

Quantum Mechanics and Electron Behavior

In quantum mechanics, the concept of electrons in specific energy levels is maintained, but electrons are no longer viewed as orbiting the nucleus in fixed paths. Instead, they are found in orbitals—regions in space surrounding the nucleus where there is a high probability of locating a given electron.

Dalton's Atomic Theory

More than 2,000 years after Democritus, English schoolmaster John Dalton (1766-1844) revived the atomic concept and proposed a theory based on empirical evidence. This theory, articulated in a series of papers published between 1803 and 1810, asserted that each element is composed of a unique type of atom. The key points of Dalton's atomic theory include:

1. Elements consist of tiny, indivisible particles called atoms.
2. Atoms of the same element are identical in mass and size.
3. Atoms of different elements differ in mass and size.
4. Chemical compounds form when two or more atoms of different elements unite.
5. Atoms combine to create compounds in simple numerical ratios, such as one-to-one, two-to-one, or two-to-three.
6. Atoms from two elements may combine in varying ratios to produce more than one compound.

Dalton's atomic theory is a landmark in the evolution of chemistry. While its main principles remain valid today, some aspects require modification due to later discoveries, including that (1) atoms consist of subatomic particles; (2) not all atoms of a specific element have the same mass; and (3) under certain conditions, atoms can be decomposed.

Ch 8—Electricity and Magnetism

8.1 The Invisible forces of Nature

Newton's mechanics dominated the field of physics until the mid-19th century, with scientists creating sophisticated mathematical formulations to describe these principles. Meanwhile, other branches of physics, such as optics, electricity, magnetism, and the exploration of matter, evolved at a considerably slower rate. By the time Albert Einstein entered college in the early 20th century, these fields had largely matured, with electromagnetism being the most developed. This successful integration of electricity and magnetism had been established just a few decades prior, marking a significant advancement in the understanding of physical phenomena.

8.2 The forces of electricity and magnetism

Electricity and magnetism have been recognized since ancient times. The Greeks discovered that amber, a striking golden gem still used in jewelry today, could attract light objects like seeds or feathers when rubbed with cloth; they referred to amber as "elekron." They also noted that lodestone, or magnetite, attracted iron.

Knowledge of electricity and magnetism remained relatively stagnant until the late 16th century, when William Gilbert, the court physician to Queen Elizabeth I and a contemporary of Galileo and Johannes Kepler, began conducting meticulously designed experiments with magnets. He also examined the attractive properties of amber and introduced the term "electric" to describe anything that exhibited similar attraction. Gilbert published his findings in a comprehensive work titled *The Magnet*.

Despite Gilbert's contributions, electricity and magnetism continued to be seen as curiosities, primarily used for entertainment at social gatherings. Electric displays featuring sparks and magnet tricks were not yet considered worthy of serious scientific inquiry.

One of the main challenges in understanding electricity and magnetism lies in the fact that the sources of attraction and repulsion—both in magnets and electrified bodies—are invisible. In contrast to mechanics, where one can observe objects moving, accelerating, or colliding, and measure their mass and motion, the nature of these unseen forces remains elusive. What causes amber to attract pieces of straw? Why does a magnet attract iron regardless of which pole is used, yet either attract or repel other magnets depending on the side? Serious study of electricity and magnetism was not easy, but fortunately, a few determined individuals persisted in their investigations.

8.3 The opposing forces of nature

Benjamin Franklin, the American scientist and Renaissance man, delved into the nature of electricity. He was aware of experiments conducted in France by scientist Charles du Fay, who rubbed a glass rod with silk and used it to manipulate a gold leaf. Interestingly, the leaf was attracted to the glass rod before contact, but it repelled after they touched. Du Fay proposed that there were two types of electricity and theorized that like charges would repel each other, while opposite charges would attract.

Franklin

Franklin conducted similar experiments, including those with lightning, and observed that one type of electricity could neutralize the other. He proposed that there was only one kind of electricity and that objects typically contained a normal amount of it. When objects were brought together, some electricity would transfer from one to the other.

After this transfer, Franklin theorized that one object would have an excess of electricity, which he designated with a positive sign, while the other would have a deficiency, marked with a negative sign. For instance, when a glass rod is rubbed with silk and then pulled apart, the silk acquires negative electricity, leaving the glass positively charged. However, Franklin had no way of knowing which was which, so he assumed that rubbing the glass rod transferred electricity to it, making the rod positive and the silk negative.

In the 20th century, scientists identified that the carriers of electric charge, now known as electrons, are negatively charged. Rubbing glass with silk transfers these negative electrons from the rod to the silk, resulting in the glass losing electrons. Consequently, the silk ends up with an excess of electrons and becomes negatively charged, while the glass develops a deficiency of electrons, creating a charge imbalance in the oxygen and silicon atoms that make up the glass molecules. Typically, atoms are electrically neutral, with a core of positive charges and a surrounding cloud of negative electrons. If one electron is removed from an atom, it becomes positively charged. In the case of rubbing silk on a glass rod, the rod ultimately acquires a positive charge.

When rubbing a glass rod with silk multiple times, about a billion electrons can be transferred from the glass to the silk. While this number seems large, even vigorous rubbing typically removes electrons from only about one in a million atoms, assuming optimal conditions. Franklin believed that positive charges were transferred, whereas we now understand that it is the negatively charged electrons that move. Nonetheless, we still use Franklin's conventions of positive and negative signs, which allow us to state that like charges repel each other, while opposite charges attract.

8.4 Detecting Forces and Fields

Franklin taught us that electric charges attract or repel each other based on their signs. But what underlies this attraction or repulsion? What is the force that causes these charges to respond to one another?

There are two perspectives on this phenomenon, both predating Einstein. The first is the notion of a force acting directly between charges, prompting them to move toward or away from each other. The second, more nuanced and powerful idea, involves the concept of fields.

8.5 Studying Electric Force

Before Franklin and Du Fay, Newton had already established that any two objects in the universe attract each other with a force proportional to the product of their masses and inversely proportional to the square of the distance separating them. This principle is encapsulated in Newton's universal law of gravitation, which is classified as an inverse square law.

Curious about whether the force between electric charges also followed an inverse square law, Franklin speculated that there might be two types of forces: one attracting and the other repelling. He reached out to his friend Joseph Priestley in England to investigate further.

An inverse square law indicates that the strength of a force between two objects diminishes as the distance between them increases. Specifically, the force's strength decreases in proportion to the square of the distance. For example, if the distance is doubled, the force decreases to one-fourth of its original strength. Tripling the distance reduces the force to one-ninth, and quadrupling it reduces it to one-sixteenth.

Priestley reflected on the parallels between electric and gravitational forces, conducted thought experiments, and proposed that electric force was indeed an inverse square law. Two years later, French scientist Charles Augustin de Coulomb devised a clever method to measure this force, validating Franklin's and Priestley's insights.

Like gravitational force, the electric force between two charged objects depends on the inverse square of the distance between them and the product of their charges. Since there are both negative and positive charges, this product can yield positive or negative values. A positive force arises between like charges (either two positives or two negatives), resulting in a repulsive effect that pushes the charges apart. Conversely, a negative force occurs between unlike charges, leading to attraction that pulls the charges together, while simultaneously exerting a repulsive effect on like charges.

8.6 Defining Electric Fields

The second approach to understanding the interaction between electric charges involves the concept of a field. An electric field describes the influence of an electrically charged object on the space surrounding it. When a charged object is present, it alters or distorts the space around it, causing other charges within that field to experience a force of attraction or repulsion toward the original charge.

8.7 Coulomb's Balance Measuring Device

Calculating the force of gravity using Newton's universal law of gravitation is relatively straightforward, as both mass and distance can be measured with ease. However, measuring electrical charges presents challenges. Since electric charges are invisible, determining their separation and values (whether positive or negative) is not as simple.

To address this, Coulomb invented a device to measure the force required to twist a pair of small electrically charged spheres. These spheres were positioned at the ends of an insulating rod suspended by a thin wire. As the charged spheres either moved apart or came closer together, depending on their charge signs, the wire would twist. Coulomb calculated the repulsive force based on the angle of twisting, thereby confirming Priestley's insight: the electric force follows an inverse square law.

Coulomb also demonstrated that by placing uncharged spheres in contact with the charged ones, fractions of the original charge could be transferred. This experiment established that the electric force between two bodies is dependent on the product of their charges. Today, this electric force is known as the Coulomb force. The electric field around a charge can be visualized using a set of lines. Because like charges repel and unlike charges attract, the field lines include arrows indicating the direction a small positive test charge would follow if placed at various points within the field.

Augustin de Coulomb

The field lines of a small positively charged sphere radiate outward, as a test charge will always move away from a positive charge. Conversely, the field lines of a small negatively charged sphere point inward because a test charge is attracted to it from every position around it.

When you place two small spheres with equal and opposite charges close together, their field lines bend and intersect. These lines indicate the direction in which a positive charge will move when placed within the field, while a negative charge will move in the opposite direction.

There are numerous other fields in physics. For instance, the gravitational field surrounding the Earth is the property of the space around our planet, where any object experiences the Earth's gravitational attraction. Einstein utilized the concept of the gravitational field to illustrate that what he termed "space-time"—a four-dimensional combination of space and time—is curved, with gravity essentially being a manifestation of this curved space-time.

8.8 Magnetic Fields

One field that you can actually observe is the magnetic field. If you sprinkle iron filings on a piece of heavy paper and then place a small bar magnet underneath, you'll see the filings align along curved paths extending from one end of the magnet to the other. This alignment reveals the shape of the magnetic field surrounding the magnet. The presence of a magnet alters the space around it, and any other magnetized object within that region experiences a force due to this change. Like the gravitational force, the magnetic force follows an inverse square law.

As you may have noticed when handling small magnets, each magnet has two distinct sides, commonly referred to as the north and south magnetic poles. If you position two magnets with their north poles facing each other, they will repel one another. The same repulsion occurs when two south poles are brought close together. However, when you flip one magnet so that the north pole of one faces the south pole of the other, they attract.

8.9 The Attraction Between Electricity and Magnetism

Electricity and magnetism share many similarities. There are two types of electric charges: positive and negative, as well as two types of magnetic poles: north and south. Like charges repel one another, while unlike charges attract, just as like poles repel and unlike poles attract. Both magnetic and electric forces follow an inverse square law.

The electric field lines surrounding two equal and opposite charges mirror the shape of those created by the north and south magnetic poles in a magnet. However, there are notable differences: positive and negative charges can exist in isolation, whereas magnetic forces always exist in pairs.

8.10 Magnetism

In 1931, the esteemed English physicist Paul Dirac proposed the existence of single magnetic poles, or monopoles, to complete the symmetry between electricity and magnetism. If electric charges can exist separately, he questioned, why can't magnetic poles? Recent theories in particle physics and cosmology suggest that magnetic monopoles may have existed in the early Universe. If they exist today, they cannot be obtained by simply splitting a magnet in two, as this process results in two complete magnets, each retaining its own north and south poles.

8.11 Failing at a Demonstration and Changing Science

Encouraged by the parallels between electricity and magnetism, scientists sought to uncover the connection between the two. A promising starting point was the idea that electric currents could generate magnetic fields. An electric current is essentially the flow of electric charges, typically through metal wire. While electric currents can occur in a vacuum or in some non-metals, they are most commonly found in metals. Despite numerous attempts, scientists repeatedly failed to demonstrate this connection.

In 1819, a professor in Denmark named Hans Christian Ørsted prepared a demonstration for his students to illustrate that, despite their apparent similarities, electric and magnetic fields were not interconnected and that one could not be generated from the other. He had performed this demonstration numerous times before. Ørsted laid wires on the table in front of him and passed an electric current through them. Using a small magnetic compass, he demonstrated to his students that the compass needle consistently pointed north, regardless of how close he brought it to the wires.

Ørsted

After completing his demonstration, Ørsted picked up the compass and noticed that the needle twitched, pointing in a direction perpendicular to the wire. Intrigued, he continued experimenting. When he reversed the current, he observed that the compass needle also reversed direction while remaining perpendicular to the wire. In his attempt to prove that electric and magnetic fields were unrelated, Ørsted accidentally demonstrated their connection: electric fields could indeed generate magnetic fields. This link had gone unnoticed by others because they had positioned the compass directly next to the wire rather than above or below it.

8.12 Careless Experimenter

Ørsted's experiment marked the first documented instance of a force acting perpendicular to the motion of charged particles (in this case, the electrons in the electric current). He discovered that a magnetic compass would be deflected when placed directly above or below a current-carrying wire. When several compasses were positioned next to the wire lying on the table, they all pointed north. However, when one compass was lifted so that its magnetic needle was perpendicular to the wire, it became deflected.

Despite his significant discovery connecting electricity and magnetism, Ørsted was somewhat careless as an experimenter. One of his students noted that he was "a man of genius, but... he could not manipulate instruments."

If this description holds true, Ørsted's lack of skill as an experimenter may have actually contributed to his discovery. More skilled scientists likely turned off the current before dismantling their setups, thus missing the movement of the compass. Ørsted published his findings in 1820, and within months, scientists across Europe were attempting to replicate his results.

The young French physicist André-Marie Ampère successfully expanded upon Ørsted's discovery, formulating a mathematical description and demonstrating that all forms of magnetism arise from small electric currents. An electric current is simply the movement of one or more electric charges through space. Today, we refer to this breakthrough as Ampère's Law, which states that a moving electric charge generates a magnetic field.

Faraday

8.13 Discovering electrical current flow

Ørsted and Ampère had demonstrated that an electric current generates a magnetic field, prompting the crucial question: Could a magnetic field create an electric current? Many scientists sought to answer this when, in 1821, a self-taught English scientist named Michael Faraday entered his lab with several ideas in mind. Initially, none of Faraday›s concepts worked, just as no one else had. Yet Faraday remained persistent, spending ten years trying to show that a magnetic field could induce an electric current.

Finally, in 1832, he twisted an insulated copper wire around one side of an iron ring and connected the ends of the wire to a battery. He coiled a second insulated wire around the opposite side of the ring, monitoring any currents generated in this coil. Faraday understood that the electric current in the first wire would produce a magnetic field, as established by Ampère's law, and demonstrated in Ørsted's famous experiment. He expected the magnetic field created in the first coil to propagate through the iron ring to the second coil. While this propagation had been previously proven, Faraday was specifically looking for evidence that this magnetic field would induce a current in the second wire coil.

Despite numerous attempts, scientists had struggled to demonstrate this connection until Ørsted's experiment in 1819. He set up a demonstration to show his students that, despite their similarities, electric and magnetic fields were unrelated, claiming one could not produce the other.

After laying wires on the table and passing a current through them, Ørsted used a small magnetic compass to illustrate that the compass needle always pointed north, regardless of how close it was to the wires.

When he finished, he picked up the compass and noticed the needle twitching, pointing in a direction perpendicular to the wire. Continuing his experiments, Ørsted reversed the current and observed that the compass needle also reversed direction while remaining perpendicular to the wire. In his effort to prove the opposite, Ørsted inadvertently demonstrated that a connection existed after all: electric fields could generate magnetic fields. No one had previously recognized this link because they had been placing the compass directly next to the wire instead of above or below it. Ørsted's experiment marked the first documented instance of a force acting perpendicularly to the motion of a body—in this case, the electrons in the electric current.

Ørsted's discovery of the connection between electricity and magnetism was somewhat accidental; he was not known for meticulous experimentation. One of his students noted that Ørsted was "a man of genius but... he could not manipulate instruments."

Faraday's own experiment began with him connecting the first coil to a source of electricity while keeping it insulated from the iron ring. When he checked for current in the second coil, nothing happened. Repeating the experiment yielded the same result, but he noticed that each time he connected or disconnected the battery, the current meter on the second loop registered a small twitch. This observation puzzled him. Could the change in the magnetic field within the ring be causing the meter to respond?

Determined to investigate further, he tried varying the current in the first loop, and after numerous attempts, he finally succeeded. Faraday had made one of the century's major discoveries: he had produced electricity from magnetism.

8.14 Law of induction

After formulating his "law of induction," Michael Faraday invented the electric generator. Essentially, an electric generator consists of several loops of wire that rotate within a magnetic field. The movement of these wire loops can be powered by a paddlewheel turned by a waterfall or a motor. As the loops rotate, the magnetic field lines passing through them change, going from minimal when the loops align with the field lines to maximum when they are positioned at right angles. This changing magnetic field induces an electric current in the wire, demonstrating that it is indeed the variation in the magnetic field that generates the current.

For several months, Faraday experimented with other methods to produce current in isolated wires, discovering that iron was not necessary for induction. He demonstrated that a changing current in one coil creates a changing magnetic field, which in turn induces a current in a second nearby coil.

Thus, a moving magnet generates electric current, as the isolated coil reacts to the approaching or receding magnetic field. Today, we refer to this principle as Faraday's law of induction, which states that a changing magnetic field produces an electric current.

The discoveries of Ørsted, Ampère, and Faraday revealed the close relationship between electricity and magnetism, with Faraday providing a comprehensive understanding of their connection. However, due to his limited mathematical knowledge, he struggled to fully grasp Ampère's complex mathematical theories. Instead, he relied on Ørsted's experimental descriptions and Ampère's presentations to inform his work.

8.15 Integrating the laws of motion with the speed of light

All was proceeding smoothly for Galileo's concept of relativity until James Clerk Maxwell introduced his theory of electromagnetism in the 19th century. His four foundational equations revealed that light behaves as an electromagnetic wave. However, for a wave to propagate, it requires a medium through which to travel. Sound waves, for instance, move through air, water, or solids.

To account for the propagation of light, scientists proposed a mysterious substance known as the ether. Following Maxwell's developments, it became widely accepted that light traveled as a wave at approximately 300,000 kilometers (186,000 miles) per second through this ether. By 1882, measurements of the speed of light closely aligned with the modern value of 299,792.458 kilometers per second (186,282.397 miles per second).

Enters James Clerk Maxwell

Born in Edinburgh, Scotland, James Clerk Maxwell (1831-1879), the son of a Scottish landowner, was destined to become the foremost theoretical physicist of the 19th century. At just fifteen, he published his first paper on a geometrical method for tracing ovals. In 1855, he was awarded the Adams Prize by Cambridge University for demonstrating that Saturn's rings could not be solid but were instead composed of small, fragmented pieces of matter. By 1860, he played a pivotal role in advancing the kinetic theory of gases, which explained the properties of gases as a collection of particles in motion.

Maxwell's most significant contribution was to the theory of electromagnetism. A highly trained Scottish scientist, he possessed remarkable mathematical prowess. In his initial two papers on electromagnetism, he developed a mathematical model for Faraday's induction law. Through his model, he not only replicated Faraday's law—showing how a changing magnetic field produces a changing electric current—but also proposed that a changing electric field could similarly generate a magnetic field. This notion was in line with Ampere's findings, which established that a moving charge creates a magnetic field. Maxwell's formulation extended beyond Ampere's law by asserting that any changing electric field, regardless of whether charges were in motion, could produce a magnetic field. For instance, a changing electric field between two stationary metal plates would still generate a magnetic field, demonstrating the broader applicability of Maxwell's equations.

Maxwell

Maxwell chose to modify and extend Ampere's law to encompass this new possibility, leading to what we now refer to as the Ampere-Maxwell law. This extension states that a changing electric field produces a changing magnetic field. With this groundbreaking addition, Maxwell achieved one of the most significant scientific discoveries of all time, unifying electricity and magnetism into a single theory known as electromagnetism. The relationship between these two fields was now clear: one generates the other. Maxwell encapsulated this unity in mathematical form through his renowned equations, collectively known as Maxwell's field equations.

Building on Faraday's earlier work, Maxwell emphasized the concept of fields, contrasting with Newton's focus on the direct interaction of bodies across empty space (action at a distance). These force fields exert pressure on material objects, causing them to move. Both Faraday and Maxwell viewed the influence of an electrically charged body as producing stresses in its immediate environment, which in turn generated stresses that radiated outward in diminishing strength as the distance from the source increased. These stresses, thought to exist even in otherwise empty space, are referred to as "fields" or "force fields."

In essence, Maxwell synthesized the contributions of Coulomb, Ampere, Faraday, and others into a coherent and elegant theory, succinctly captured in just four equations. The first equation is a refined version of Coulomb's law, establishing the relationship between an electric charge and its electric field. The second describes magnetic field lines, highlighting the differences from electric field lines. The beauty of Maxwell's work lies in the last two equations: the third represents Faraday's law of induction, while the fourth is the Ampere-Maxwell law.

According to the third equation, a changing magnetic field induces a changing electric field, while the fourth asserts that this changing electric field produces a changing magnetic field. This reciprocal interaction means that when one field is initiated—regardless of whether it's the electric or magnetic field—the other is immediately generated, creating a dynamic interplay. The two interlinked fields merge into a single electromagnetic field that begins to expand through space. Maxwell unified his equations into a single expression, demonstrating that this electromagnetic field propagates through space as a wave at a speed of 288,000 kilometers per second (kps), a figure remarkably close to the measured speed of light, which was recorded at 311,000 kps at the time.

8.16 Confirming Maxwell's theory

James Clerk Maxwell passed away from cancer at the age of 48, never witnessing the validation of his groundbreaking discovery that light is an electromagnetic wave produced by the movement of charges. Had he lived just nine more years, he would have seen his theories confirmed.

Nine years after Maxwell's death, the German physicist Heinrich Hertz (1857-1894) set out to apply Maxwell's theory in his laboratory to generate and measure electromagnetic waves. Hertz, a brilliant physicist, earned his PhD at just 23 from the University of Berlin. By 1883, he immersed himself in Maxwell's writings to grasp the theory of electromagnetism, which was too new to be included in college curricula and poorly understood by many professors.

Hertz independently mastered electromagnetism within two years and aimed to generate the electromagnetic waves described by Maxwell. He built a version of the apparatus used by Michael Faraday in his famous induction experiment, which later became part of Maxwell's equations. However, Hertz made some modifications: instead of using two insulated wires wrapped around a metal ring, he opened one of the loops and placed two small metal balls at the ends of a wire, creating a small gap between them.

Drawing from Maxwell's theory, Hertz understood that connecting or disconnecting the battery in the first coil would produce a rapid change in current, resulting in a changing magnetic field that would generate a voltage in the second coil. In simple terms, the process could be summarized as: changing electric current → changing magnetic field → voltage in the second coil. When the voltage in the open coil became sufficiently high, a spark would leap across the two balls. According to Maxwell's theory, these sparks generated changing electric and magnetic fields that would propagate across the gap and travel through the surrounding space as electromagnetic waves.

8.17 The radio transmitter / receiver invention

Hertz realized that by adding a second open loop with metal balls at the ends, forming a small gap, he could capture the electromagnetic field generated by his first setup and produce a voltage in this new loop. In doing so, he effectively invented a radio transmitter (the first loop) and a radio receiver (the second loop). This innovation allowed Hertz to measure the speed of the electromagnetic waves produced by his spark apparatus, and he found it matched the speed of light, precisely as Maxwell had predicted.

Hertz's experiment demonstrated that light is an electromagnetic wave. Essentially, he created a form of light by running an electric current through a wire, producing what turned out to be an invisible radio wave. So, what connects radio waves and light? Both radio waves and light, along with other electromagnetic waves discovered later, are generated similarly—by accelerating electric charges. The primary distinction among them lies in the frequency of their oscillations. Once generated, all electromagnetic waves propagate through space in all directions at the speed of light, approximately 300,000 kilometers per second. You can visualize them as pulsating bubbles expanding outward.

As these electromagnetic waves emanate from their source, they resemble an expanding bubble. When observed with appropriate instruments, one can detect the electric and magnetic fields oscillating in unison. This oscillation travels through space without deforming; the wavelength remains constant for each specific electromagnetic wave. However, wavelengths vary across different types of waves. For example, radio waves, like those produced by Hertz, have wavelengths ranging from about 1 meter to several thousand kilometers. One hertz of electromagnetic radiation, named in honor of Heinrich Hertz, corresponds to a wavelength of about 1 meter.

Hertz

8.18 Applications of electromagnetism

The electromagnetic waves that heat your food in a microwave are measured in centimeters, while X-rays consist of much shorter wavelengths. In medical applications, X-rays can penetrate skin and muscle to produce images of a patient's skeletal structure for physicians to analyze.

Visible light has wavelengths longer than X-rays but shorter than those used in radios and TVs. Due to their small size, scientists use nanometers to denote these wavelengths—one nanometer (nm) is one-millionth of a millimeter. Visible light ranges from about 400 nm for red to 700 nm for violet.

The energy spectrum as defined by Newton, having various ranges of wavelengths discovered over time are listed. Each wavelength corresponds to a different energy level, with shorter wavelengths carrying more energy. Scientists have developed instruments to detect different ranges of electromagnetic wavelengths, including X-rays, gamma rays, and radio waves.

Following his groundbreaking discovery, Hertz published a paper to clarify the connection between light, radiant heat, and electromagnetic wave motion. He expressed confidence in the benefits that this identity would bring to the study of optics and electricity. Ironically, during these experiments, Hertz also uncovered the photoelectric effect, which later provided Albert Einstein with evidence that challenged the wave theory.

Einstein's concept of light quanta, or light as particles, contradicted the well-established wave theory that Hertz and others had supported. While the wave theory had proven remarkably successful, the idea of light quanta seemed absurd to many scientists. They reasoned that energy in a quantum of light was tied to its frequency, a property typically associated with waves, not with particle-like entities traveling through space.

Maxwell's field equations mathematically describe the relationships between electric charges, currents, and the resulting electric and magnetic fields. One of Maxwell's laws states that a changing magnetic field induces an electric field even in the absence of electric charges. This principle of electromagnetic induction is fundamental to electric generators, which generate electric voltages in their armatures by exposing them to varying magnetic fields.

From his research, Maxwell not only predicted that electromagnetic fields propagate at a finite speed but also calculated this speed as 186,000 miles per second, matching the speed of light. This led him to hypothesize that light is a form of electromagnetic phenomenon, a theory later confirmed. Hertz, inspired by Maxwell's predictions, conducted laboratory experiments that demonstrated the existence of electromagnetic waves, forming the basis for today's wireless communication technology.

Today, we are surrounded by electromagnetic waves produced by radio stations, remote controls, and cell phones. These waves all travel at approximately 300,000 kilometers per second, a value consistent with measurements taken when Maxwell was making his discoveries. We even send electromagnetic waves to our spacecraft on Mars, instructing rovers to climb intriguing hills visible due to the electromagnetic signals sent to us 20 minutes earlier.

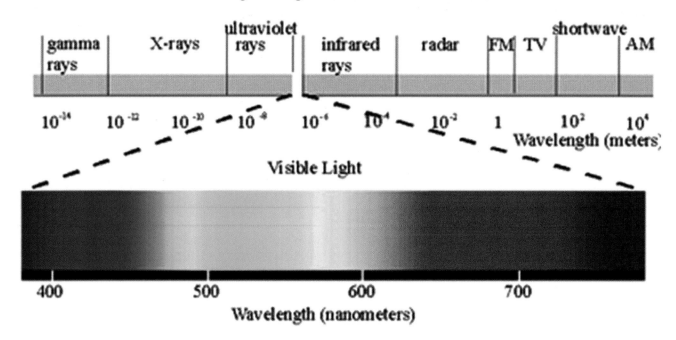

Figure 8-1 Electromagnetic Spectrum

8.19 Volume of Charge

When electric charge is distributed over a specific volume, each individual charge element contributes to the electric field at an external point. To find the total electric field, a summation or integration of these contributions is necessary. Although the smallest units of electric charge are electrons and protons, it is beneficial to treat charge distributions as continuous and define a charge density.

8.20 Summary of the discovery of Electromagnetism

During the 19th century, the field of electricity and magnetism was advanced by several prominent physicists, including the Danish scientist Hans Christian Ørsted (1777-1851), Englishman Michael Faraday (1791-1867), Scottish physicist James Clerk Maxwell (1831-1879), and German physicist Heinrich Hertz (1857-1894). Maxwell notably observed that electromagnetism could not be fully explained by Newtonian mechanics. While it was previously believed that only the distance between two objects determined the force they exerted on each other, it became evident that moving electric charges, such as those found in electric currents, produced effects that were absent when the charges were at rest.

In 1819, Ørsted discovered that an electric current flowing through a wire could deflect a compass needle. The following year, the French physicist François Arago found that a wire carrying an electric current acted like a magnet, attracting iron filings. Soon after, André-Marie Ampère demonstrated that two parallel wires would attract each other when carrying currents in the same direction but repel each other when the currents flowed in opposite directions. Intrigued by the ability of electricity to create magnetism, the British experimentalist Michael

Faraday sought to generate electricity using magnetism. He found that pushing a bar magnet in and out of a coil of wire generated an electric current, which ceased when the magnet remained still within the coil.

Celestial bodies only attract each other, whereas electric charges at rest can either attract or repel each other, exerting forces along the line connecting them. Ørsted discovered that an electric current, consisting of moving electric charges, exerted a force on a magnetic needle at right angles to this connecting line. Prior observations in astronomy suggested that the force between two bodies depended solely on their immediate configuration, but Hertz's experiments revealed that electromagnetic disturbances propagate as waves at a finite speed. Thus, the force experienced by one body could only be understood in relation to the historical state of the other.

Just as ice, water, and steam represent different states of H_2O, Maxwell showed in 1864 that electricity and magnetism are different manifestations of the same underlying phenomenon—electromagnetism. He encapsulated the diverse behaviors of electricity and magnetism into four elegant mathematical equations. Upon recognizing the significance of Maxwell's work, the Austrian physicist Ludwig Boltzmann could only quote Goethe in admiration: "Was it a God that wrote these signs?"

Using these equations, Maxwell made the groundbreaking prediction that electromagnetic waves travel at the speed of light through the ether, although he believed the ether itself was unnecessary. He demonstrated that light is a form of electromagnetic radiation. According to Faraday and Maxwell, material particles experience forces in the presence of fields, suggesting that these fields act as intermediaries between the particles, addressing Newton's concept of "action at a distance."

Maxwell did not live to witness the confirmation of his predictions through experimentation. He passed away from cancer at the age of 48 in November 1879, the same year that Albert Einstein was born. Less than a decade later, in 1887, Heinrich Hertz provided the experimental evidence that solidified Maxwell's unification of electricity, magnetism, and light as the pinnacle of 19th-century physics.

Ch 9— Radioactivity and the Atom

9.1 The discovery of the electron

In the late 19th century, scientists studying electricity were involved in a prolonged debate regarding the nature of the radiation emitted from a wire carrying electric current through a vacuum tube. The apparatus used in these experiments consisted of a sealed glass tube, evacuated of air, with two metal discs positioned at either end. These discs were connected to a battery, allowing researchers to observe the behavior of electricity in the vacuum environment of the tube.

Thomson

A "cathode ray tube" (CRT) is the predecessor of both the television tube and the computer CRT. The tube emitted a mysterious green glow, a phenomenon that was initially unexplained. Physicists referred to the metal disc producing the glow as the "cathode," while the opposite disc was named the "anode." The cathode was connected to the negative terminal of the battery, and the anode to the positive terminal. The rays emitted from the cathode were termed "cathode rays," and the entire apparatus became known as a cathode ray tube (CRT). At the time, the nature of cathode rays was debated—some scientists believed they were a form of radiation caused by vibrations in the ether, distinct from light and recently discovered radio waves, while others thought they were streams of tiny particles. German scientists mostly supported the ether wave theory, while British and French scientists favored the particle hypothesis. The discovery of X-rays by Wilhelm Röntgen in 1895 complicated the situation, although his work, for which he won the first Nobel Prize in Physics in 1901, proved unrelated to the cathode ray question, emerging before the atomic physics framework had fully developed.

In 1897, J.J. Thomson, director of the Cavendish Laboratory at Cambridge, England, conducted experiments to understand the mysterious green glow in CRTs. Working with a team of 20 researchers, Thomson succeeded in deflecting the beam of green light and discovered that it carried a negative charge. From these experiments, Thomson concluded that the green glow was produced by individual particles traveling through the vacuum in the tube. He was also able to measure the charge of these particles, which were later identified as electrons.

9.2 Envisioning the structure of the atom

J.J. Thomson conducted his groundbreaking work at the Cavendish Laboratory, a research center at Cambridge University founded in the 1870s by James Clerk Maxwell, the first Cavendish Professor of Physics. Thomson devised an experiment to investigate the nature of cathode rays, focusing on the electric and magnetic properties of charged particles. His apparatus was designed to balance the deflective effects of both electric and magnetic fields, allowing the cathode rays to travel in a straight path from a negatively charged metal plate (the cathode) to a detector screen. This setup only worked for electrically charged particles, which led Thomson to confirm that cathode rays are indeed negatively charged particles, now known as electrons. By analyzing the balance of electrical and magnetic forces, Thomson was able to calculate the ratio of an electron's charge to its mass (e/m). Regardless of the metal used to make the cathode, the result was always the same, leading Thomson to conclude that electrons are fundamental components of atoms and that all atoms contain identical electrons, even though different elements have different types of atoms.

Though the Cavendish Laboratory was originally established by Maxwell, it was under Thomson's leadership that it became a world-renowned center for experimental physics. Thomson's work, along with that of the researchers he mentored, revolutionized 20th-century physics. In addition to winning a Nobel Prize himself, seven of his collaborators at the Cavendish Laboratory were later awarded Nobel Prizes for their contributions to science. The Cavendish Laboratory remains a leading center for physics research to this day.

9.3 Ions

The "cathode rays" produced by the negatively charged plate in an evacuated tube were later discovered to be streams of negatively charged particles, now known as electrons. Since atoms are electrically neutral, it followed that there must be positively charged counterparts to electrons, which are atoms that have lost an electron. In 1898, Wilhelm Wien of the University of Würzburg conducted some of the earliest studies of these positive rays, finding that the particles were much heavier than electrons, as one would expect from atoms missing an electron. Following his work on cathode rays, J.J. Thomson took on the challenge of investigating these positive rays, or "canal rays," by conducting a series of complex experiments that extended into the 1920s. His work with modified cathode-ray tubes, which allowed a small amount of gas to remain, led to the study of ionized atoms or ions, created when electrons collided with gas atoms, knocking out additional electrons and leaving behind positively charged ions.

By 1913, Thomson's team was measuring the deflections of positive ions, including those of hydrogen, oxygen, and neon. His work with neon revealed a significant discovery: unlike electrons, which all have the same charge-to-mass ratio (e/m), neon ions had different masses despite having the same charge. This was the first indication that chemical elements can exist with atoms of different masses but identical chemical properties, a phenomenon now known as isotopes. Thomson's findings challenged earlier models of the atom, which viewed it as indivisible. He proposed a "plum pudding" model, where the atom was a mixture of positive and negative charges, with electrons embedded in a diffuse positive body. However, this model was soon rendered obsolete by discoveries in Germany and Paris, including Wilhelm Roentgen's identification of X-rays and Henri Becquerel's and Marie and Pierre Curie's research on radioactivity. These findings suggested that the atom's structure was far more complex than previously thought, and while Thomson's model was eventually disproven, it laid the groundwork for future atomic theories.

Thomson's model of the atom, with electrons embedded in a positive "pudding," provided scientists with a framework to test, but it couldn't explain the emission of certain rays from atoms. This limitation suggested that there were additional, undiscovered components within the atom. His work, though ultimately flawed, became a pivotal steppingstone that challenged scientists to further investigate atomic structure, leading to a more accurate understanding. To uncover the full answer to the atom's mysteries, we must look back into the history of scientific discovery and explore the experiments that followed.

Marie Curie

9.4 Radioactivity

The key to understanding atomic structure came with the accidental discovery of radioactivity in 1896. Much like the discovery of X-rays a few months earlier, this breakthrough was a fortunate accident. Physicists, including Wilhelm Röntgen, were experimenting with cathode rays when they noticed that electrons striking a material object produced a secondary, invisible radiation. Röntgen observed this radiation by its effect on a fluorescent screen and photographic plates. He named it "X" for its unknown nature. This secondary radiation, now known as X-rays, was soon understood to behave like waves, a form of electromagnetic radiation with much shorter wavelengths than visible light.

The discovery of X-rays in December 1895 created excitement within the scientific community. It reinforced the belief among many scientists, particularly in Germany, that cathode rays were also a form of wave-like radiation. Meanwhile, other researchers began to explore ways to produce similar forms of radiation, and Henri Becquerel in Paris was the first to succeed.

One of the most fascinating properties of X-rays was their ability to penetrate opaque substances, such as black paper, and leave an image on photographic plates, a revelation that opened the door to further discoveries in atomic and radiation physics.

In February 1896, Henri Becquerel conducted an experiment where he wrapped a photographic plate in a double layer of black paper, coated the paper with uranium and potassium bisulphate, and exposed it to sunlight for several hours. When he developed the plate, it revealed the outline of the chemical coating. Becquerel initially believed that sunlight had triggered the production of X-rays in the uranium salt, similar to how phosphorescence works. His interest lay in phosphorescence, which is the emission of light by a substance after it has absorbed light.

Ernest Rutherford

A fluorescent screen, as seen in the discovery of X-rays, emits light only when "excited" by incoming radiation. In contrast, a phosphorescent substance can absorb incoming radiation and slowly release it as light, even after being placed in the dark. Naturally, scientists sought a relationship between phosphorescence and X-radiation, but Henri Becquerel's discovery revealed something unexpected. Two days after preparing a photographic plate for a similar experiment, cloudy skies delayed his plans, and the plate remained in a cabinet. When Becquerel developed the plate on March 1, 1896, it once again showed the outline of the uranium salt. He realized that this new radiation had no connection to sunlight or phosphorescence. It was a previously unknown form of radiation, emitted spontaneously by the uranium itself—what we now call radioactivity.

Becquerel's discovery sparked further investigations into radioactivity. Marie and Pierre Curie, working at the Sorbonne, became pioneers in this emerging field. Their groundbreaking work led to the discovery of new radioactive elements, earning them the Nobel Prize in Physics in 1903. Marie Curie was later awarded a second Nobel Prize in Chemistry in 1911 for her continued research on radioactive materials. Their daughter, Irene, also contributed significantly to this field and received a Nobel Prize in Chemistry during the 1930s for her work on radioactivity, continuing the family's remarkable scientific legacy.

In the early 1900s, experimental discoveries in radioactivity advanced rapidly, far outpacing theoretical explanations. Among the prominent figures driving these investigations was Ernest Rutherford. A former student of J.J. Thomson, Rutherford, a New Zealander, made a key observation: the radioactivity discovered by French scientists consisted of two distinct types. By placing sheets of aluminum foil in front of a radiation detector, he found that one type of radiation, which he named alpha rays, could be easily stopped, while another, beta rays, required more layers of foil. Later, a third type, gamma rays, was also identified. Rutherford further demonstrated that alpha rays carried a positive charge, while researchers in Paris showed that beta rays were negatively charged.

Rutherford, who had worked under Thomson at the Cavendish Laboratory in the 1890s, became a Professor of Physics at McGill University in Montreal in 1898. There, alongside Frederick Soddy, he made a groundbreaking discovery in 1902, showing that radioactivity involved the transformation of one element into another. Rutherford named the two types of radiation produced by this radioactive "decay" alpha and beta radiation. When the third type was discovered, it was naturally called gamma radiation.

Beta radiation was eventually identified as fast-moving electrons, akin to cathode rays, while gamma rays were shown to be a form of electromagnetic radiation, similar to X-rays but with even shorter wavelengths. Alpha particles, however, were different— they had a mass approximately four times that of a hydrogen atom and carried a positive charge twice that of an electron, revealing a new understanding of atomic structure.

9.5 A new model of the atomic structure

Even before scientists fully understood what an alpha particle was or how it could be ejected at high speeds from an atom, transforming it into another element, researchers like Ernest Rutherford were already utilizing these particles. These "high-energy" particles, produced by atomic reactions, served as valuable probes for investigating atomic structure. In a fascinating twist, Rutherford would eventually discover the origins of the alpha particles he was using.

In 1907, Rutherford left Montreal to become a Professor of Physics at the University of Manchester, England. In 1908, he was awarded the Nobel Prize in Chemistry for his groundbreaking work on radioactivity, though he found this somewhat amusing. While the Nobel Committee considered the study of elements to be part of chemistry, Rutherford saw himself as a physicist and viewed chemistry as an inferior science.

Working with his assistant Hans Geiger, who would later invent the Geiger counter, Rutherford made a key discovery: when an alpha particle struck a screen coated with a specific substance, the coating emitted a flash of light. This provided Rutherford with a crucial tool. Although he couldn't directly observe the atom, he could detect the behavior of alpha particles as they interacted with atoms. By watching where the particles landed and how their trajectories changed, he could infer how atoms influenced them. This set the stage for Rutherford's future discoveries about atomic structure.

9.6 Creating a new model of the atom

With the advancements in our understanding of atoms and molecules through quantum physics, the old adage among physicists that chemistry is merely a branch of physics has become more than just a joke; it is now largely accurate. In 1909, Hans Geiger and Ernest Marsden, working in Rutherford's department at Manchester, conducted experiments where they directed a beam of alpha particles onto and through a thin sheet of metal foil.

The alpha particles originated from naturally radioactive atoms, as there were no artificial particle accelerators available at the time. The outcomes of the interactions between the alpha particles and the metal foil were measured using scintillation counters, which emitted sparks upon being struck by a particle. Some alpha particles passed straight through the foil, while others were deflected at angles to the original beam. Surprisingly, a few even bounced back from the foil, returning on the same side they had struck. Rutherford was astounded by these experimental findings.

Previously, J.J. Thomson had established that atoms contain negative electrons. However, Thomson and his contemporaries recognized that there must be a positive component within the atom to balance the negative charge of the electrons, as atoms are electrically neutral. Rutherford set out to identify this elusive positive element. He directed even more alpha particles at gold atoms to observe how they interacted with these unseen entities. With Geiger's assistance, he used the radioactive element radium as the source of alpha particles, aiming them at a thin sheet of gold foil. Behind the foil, he placed a screen coated with a special substance to detect the alpha particles (see Figure 9-2). Rutherford tasked Geiger with collecting data and adjusting the screen's position to better understand the deflection patterns of the alpha particles. By bombarding these invisible obstacles with projectiles, he aimed to make their presence known.

Figure 9-2 Particle detection apparatus (image No. 84)

The fast-moving alpha particles encountered a strong force within the nucleus that caused them to bounce back. After some calculations, Rutherford concluded that for this phenomenon to occur, the atom must contain a very small core of positive charge, which he termed the "positively charged nucleus." He deduced that this nucleus housed most of the atom's mass.

The flashes on Rutherford's screen provided crucial insights. As Geiger collected data in the lab, he observed most flashes when the screen was positioned directly behind the gold foil or within a few degrees. Typically, a fast alpha particle would pass through the electrons in a gold atom and continue on its path. Occasionally, however, one alpha particle would come close enough to an electron to experience a slight electrical repulsion.

This repulsion was minimal since the alpha particles were traveling rapidly. Curious about the results, Rutherford instructed a bright undergraduate named Ernest Marsden to move the screen farther out to around 45 degrees and take measurements. Marsden recorded a few flashes. Encouraged, he pushed the screen even farther and was able to detect flashes at angles of 75 and 80 degrees. To his surprise, he still observed some flashes even when the screen was placed perpendicular to the incoming alpha particles. He reported these findings to Rutherford, who was astonished by the results. "It was like firing a 15-inch shell at tissue paper and expecting to see the bullet bounce back at you," he later remarked.

When an alpha particle collides with an electron, it typically brushes it aside without any significant effect. The observed deflections were likely due to the positive charges within the atoms of the metal foil, which repel one another. If Thomson's "watermelon" model were accurate, no particles should bounce back; if the positive charge filled the atom, the alpha particles would pass straight through. If the "watermelon" allowed one particle to pass, it should allow them all.

However, if all the positive charge were concentrated in a tiny region much smaller than the overall atom, then an alpha particle striking this small concentration directly could be repelled, while most would pass through the empty space between the positively charged parts of the atoms. This arrangement could account for the varying responses of the alpha particles: some would be repelled, others slightly deflected, and some would remain almost undisturbed.

After extensive experimentation and analysis, Rutherford concluded that each alpha particle has a mass more than 7,000 times that of an electron (in fact, an alpha particle is identical to a helium atom minus two electrons) and can move at speeds close to that of light. These fast alpha particles encountered a significant force within the nucleus, leading them to bounce back. Rutherford's calculations suggested that for this to happen, the atom must possess a small central core of positive charge, which he referred to as the positively charged nucleus.

In 1911, Rutherford proposed a new atomic model that laid the groundwork for our modern understanding of atomic structure. He asserted that there must be a small central region, or nucleus, containing all the positive charge of the atom, equal in magnitude and opposite in charge to the electrons surrounding it. Together, the nucleus and the electrons form an electrically neutral atom. Subsequent experiments revealed that the nucleus is only about one hundred-thousandth the size of the entire atom.

9.7 The Predicted Collapse of the atom

Physicists soon determined that the nucleus of hydrogen consists of a single proton. Hydrogen's structure is straightforward: a proton at the center forms the nucleus, with one electron orbiting around it. Other atoms, however, are more complex and posed greater challenges.

This model revolutionized physics, prompting a key question: if unlike charges attract each other while like charges repel, why don't the negatively charged electrons spiral into the positively charged nucleus? The answer emerged from an analysis of atomic interactions with light, marking a significant advancement in the early development of quantum theory.

The calculations and measurements for these atoms were not aligning. The primary issue lay with the behavior of electrons orbiting the nucleus. According to James Clerk Maxwell's theory of electromagnetism, a moving electric charge in a curved path would radiate energy. This implied that electrons would lose energy while orbiting, spiraling inward and ultimately collapsing into the nucleus, a process physicists estimated would occur within about a microsecond.

The dilemma posed by Rutherford's atomic model hinged on the established fact that a moving electric charge emits energy in the form of electromagnetic radiation, such as light or radio waves. If an electron remained outside the nucleus, it should inevitably be drawn into the center due to the attractive force of the positively charged nucleus, rendering the atom unstable. As the atom collapsed, it would release a burst of energy.

One way to counteract this collapse was to envision electrons orbiting the nucleus like planets around the Sun. However, this orbital motion inherently involves continuous acceleration. Even if the electron's speed remains constant, its direction changes, which affects its velocity. This changing velocity should lead to energy loss, causing the electrons to spiral inward toward the nucleus. Thus, even the orbital model failed to prevent the predicted collapse of Rutherford's atom.

Yet, in reality, atoms are stable. The solution to this contradiction between theory and observation arose from two scientists, Max Planck and Albert Einstein, who were not directly investigating atomic physics at the time. Ultimately, a postdoctoral researcher named Niels Bohr would develop techniques that contributed to resolving this issue. Bohr started with the concept of electrons orbiting the nucleus and discovered a method to keep them in stable orbits without losing energy and spiraling inward. This approach fit well with the familiar analogy of the solar system, but it also opened the door to a different interpretation: envisioning the nucleus and electrons coexisting in space without the need for orbital motion.

This new perspective prompted theorists to reconsider why the attractive force between positive and negative charges does not cause the atom to collapse while radiating energy. In March 1911, this revised atomic model was shared with the scientific community, and by October 1912, Rutherford officially used the term "nucleus" for the first time.

Rutherford presented his new atomic model during a lecture at Cambridge in the fall of 1911, attended by J.J. Thomson. While Rutherford's alpha-scattering data supported a nuclear model, Thomson's did not. Niels Bohr, who had recently begun his postdoctoral research at Cambridge under Thomson's guidance, might have also been present.

Dissatisfied with the crowded atmosphere of the Cavendish Laboratory, Bohr decided to transfer to Manchester and work with Rutherford. With Rutherford's encouragement, he arranged to leave Cambridge and join the Manchester team. Bohr recognized that an electron in orbit around a nucleus possesses energy that can be expressed using Newtonian physics. However, he immediately imposed a quantum condition on this energy, asserting that only specific energy levels are permitted. He defined these allowed energies as states of the hydrogen atom.

Bohr's quantum condition restricted both the possible energies and the orbits of the electron. Larger energy states corresponded to larger orbits. This led to a conceptualization of Bohr's hydrogen atom as a series of discrete orbits surrounding a central nucleus: the smallest orbit having the lowest energy, with successively larger orbits representing higher energy levels.

By the time theorists began addressing these puzzles in the early 20th century, crucial discoveries had already been made regarding how matter interacts with radiation, forming the basis for a more refined model of the atom.

9.8 The interaction between material bodies and electromagnetic radiation

At the dawn of the 20th century, the prevailing scientific understanding of the natural world was rooted in a dualistic philosophy. While material objects were conceptualized as particles, or atoms, electromagnetic radiation—including light—was considered in terms of waves. This perspective led researchers to believe that studying the interaction between light and matter could provide a pathway to unify physics around 1900. However, it was precisely in attempting to explain how radiation interacts with matter that classical physics, which had excelled in many areas, encountered significant failures.

A straightforward way to observe the interaction between matter and radiation is by examining a hot object. A hot object emits electromagnetic energy, and the hotter it becomes, the more energy it radiates at shorter wavelengths (higher frequencies). For instance, a red-hot poker is cooler than a white-hot one, and even an object too cool to emit visible light can still feel warm, as it radiates lower-frequency infrared radiation.

By the late 19th century, it was increasingly clear that this electromagnetic radiation must be linked to the movement of tiny electric charges. Although the electron had just been discovered, it was evident that a charged component of an atom (now recognized as an electron) vibrating back and forth would produce a stream of electromagnetic waves, similar to how wiggling a finger in a pool creates ripples on the water's surface.

The challenge arose when combining the best classical theories—statistical mechanics and electromagnetism—which predicted a form of radiation that differed markedly from what was actually observed coming from hot objects.

9.9 The Blackbody absorption of radiation

To predict heat radiation, theorists often employed an idealized concept known as a "blackbody." This hypothetical object absorbs all incoming radiation, regardless of wavelength, and is also the most efficient emitter of electromagnetic radiation. Despite its name, a blackbody can appear red or white-hot; in fact, the Sun's surface behaves similarly to a *blackbody*.

Creating a makeshift blackbody is simple: one can use a hollow sphere or a closed tube with a small hole. Any radiation entering through the hole gets trapped inside, bouncing around the walls until it is absorbed, making the hole effectively act like a blackbody. This phenomenon gives rise to the German term "cavity radiation."

Our primary interest lies in the behavior of a blackbody when heated. Much like a poker, it first feels warm and then glows red or white-hot, depending on its temperature. Laboratory studies show that the emitted radiation spectrum—indicating how much energy is radiated at each wavelength—depends solely on the blackbody's temperature. There is minimal radiation at both very short and very long wavelengths, with most energy emitted in a middle range. As the temperature rises, the peak of the spectrum shifts toward shorter wavelengths (from infrared to red to blue to ultraviolet), but there is always a cutoff at very short wavelengths. This is where 19th-century measurements of blackbody radiation conflicted with theoretical predictions.

Strangely, classical theory predicted that a cavity filled with radiation should contain an infinite amount of energy at the shortest wavelengths. Instead of peaking and falling to zero at zero wavelength, the predictions indicated that measurements would escalate off the scale at the high-frequency end. This stemmed from the natural assumption that electromagnetic waves in a cavity could be treated similarly to waves on a violin string, allowing for waves of any size and wavelength. The vast array of wavelengths necessitated the application of statistical mechanics to predict the radiation's overall behavior, leading to the conclusion that energy radiated at any frequency is proportional to that frequency. Since frequency is the inverse of wavelength, this implied that blackbody radiation should produce enormous amounts of energy in the ultraviolet range and beyond. This discrepancy became known as the "ultraviolet catastrophe," highlighting flaws in the assumptions behind the predictions.

On the lower-frequency side of the blackbody curve, observations aligned well with classical theory, specifically the Rayleigh-Jeans Law, indicating that classical theory was partially correct. However, the puzzle remained as to why the energy at higher frequencies was not substantial but instead diminished to zero as the frequency increased.

This conundrum captured the attention of many physicists in the late 19th century, including Max Planck, a meticulous German scientist who leaned toward conservatism rather than revolutionary ideas. With a keen interest in thermodynamics, Planck aimed to resolve the ultraviolet catastrophe by applying thermodynamic principles.

By the late 1890s, two approximate equations had been established: one based on the Rayleigh-Jeans Law for long wavelengths, and another developed by Wilhelm Wien that approximated short wavelengths and predicted the wavelength at which the peak of the curve occurred at various temperatures.

In his quest to resolve the ultraviolet catastrophe, Planck worked on the problem from 1895 to 1900, publishing several key papers that linked thermodynamics and electrodynamics. Yet, he still struggled to solve the riddle of the blackbody spectrum. In 1900, he experienced a breakthrough—not through a calm, logical insight, but rather through a blend of luck and intuition combined with an understanding of the mathematical tools he was using. Recognizing that electric oscillators exist in every atom, Planck sought to connect thermodynamics with electrodynamics.

He proposed a revolutionary idea: modifying conventional electromagnetic laws so that electromagnetic wave energy always existed in discrete packets, or "quanta." He suggested that the energy in these packets, defined by the frequency of the wave, increased with higher frequencies (shorter wavelengths). More specifically, he introduced the concept that each quantum's energy equals the frequency multiplied by a constant now known as "Planck's constant," valued at approximately *6.6×10^{-23}Js*. This notion explained why, at lower temperatures, only low-frequency (long-wavelength) quanta were excited, while higher frequencies emerged only at elevated temperatures. Planck's theory not only aligned with observed patterns but also provided precise quantitative predictions of radiation at each wavelength for a given temperature.

Planck began by investigating how small electric oscillators should radiate and absorb electromagnetic waves, employing a distinct approach compared to Rayleigh and Jeans. His method ultimately produced the expected blackbody curve, including its ultraviolet catastrophe. While no one can definitively ascertain what Planck envisioned when he took the pivotal step that birthed quantum mechanics, his work has been meticulously analyzed by Martin Klein, a historian specializing in the era of *quantum theory's emergence*. Klein's reconstruction of the roles played by Planck and Einstein in this revolutionary period provides a compelling historical context.

In late summer 1900, Planck realized he could merge the two incomplete descriptions of the blackbody spectrum into a single mathematical formula that captured the entire curve. This innovative formula successfully reconciled Wien's Law and the Rayleigh-Jeans Law, marking a significant triumph. Planck's equation accurately matched experimental observations of cavity radiation, but unlike its predecessors, it lacked a physical foundation.

While Wien and Rayleigh—and even Planck himself over the previous four years—sought to construct a theory based on sensible physical assumptions, Planck had essentially conjured the correct curve without understanding the underlying physical principles that governed it. Ultimately, those principles turned out to be far from intuitive.

Ch 10—Light Quanta

10.1 Discovering the Quanta

Around the same time that J.J. Thomson was conducting his experiments with cathode ray tubes, Max Planck was in Germany tackling another significant and unresolved issue in physics: the way hot objects radiate energy. Planck's eventual solution would not only shed light on this phenomenon but also help scientists understand why electrons in atoms do not spiral into the nucleus. However, Planck's resolution required interpretation from Albert Einstein.

To grasp the problem that physicists were facing, consider a familiar example. When you turn on an electric stove, the heating element becomes warm before it visibly changes color. After a couple of minutes, it starts to glow red, and eventually, as it reaches a higher temperature, it emits an orange hue. This thermal radiation is a form of electromagnetic wave. While you may not see the radiation when it is in the infrared range, it can occasionally fall within the visible spectrum. Conversely, other times, the radiation may be in the shorter wavelengths of the ultraviolet region, where it remains invisible.

10.2 Max Planck

Max Planck was born in 1858 in Kiel, Germany, into an academic family. His father was a professor of law at the University of Kiel, and both his grandfather and great-grandfather had been professors as well. When Planck was nine, his father took a position at the University of Munich, where Planck attended high school. There, he developed a strong interest in math and physics, thanks to an excellent teacher, and quickly became a top student.

Planck enrolled at the University of Munich to study physics, but he found himself at odds with his professor, Philipp von Jolly, who famously discouraged him by saying that there was little left to discover in physics. Disillusioned, Planck transferred to the University of Berlin, where he was able to learn from renowned physicists Hermann von Helmholtz and Gustav Kirchhoff.

Like Albert Einstein would later do, Planck pursued topics outside the standard curriculum, studying thermodynamics through the original work of Rudolf Clausius. After earning his undergraduate degree, he wrote his doctoral thesis on the second law of thermodynamics, which earned him a PhD from the University of Munich at the young age of 21. As was common for physics PhDs at the time, Planck set his sights on an academic career.

Planck

In Germany at that time, aspiring professors typically began as instructors, or *Privatdozents*, an unpaid position with teaching responsibilities. *Privatdozents* earned small fees from students for administering exams, but this income was not enough to live on. From 1880 to 1885, Max Planck worked as a *Privatdozent* in Munich, relying on additional work to support himself.

In 1885, Planck was promoted to associate professor, a role that came with a regular salary. This financial stability allowed him to marry his childhood sweetheart, Marie Merck. In 1889, Planck moved to the University of Berlin, where he became a full professor, succeeding Gustav Kirchhoff, who was retiring.

Planck was a talented pianist and at one point considered pursuing a career in music before ultimately choosing physics. His decision led him to become one of the most influential scientists in history. In 1918, he was awarded the Nobel Prize in Physics for his groundbreaking discovery of the quantum of energy.

10.3 The source of thermal radiation

Physicists had long struggled to solve the issue of thermal radiation, particularly in the high-energy region where their equations were producing nonsensical results. The heat radiated by an object is generated by the energy changes of oscillating charged particles within the material. However, at short wavelengths, instead of increasing, the energy distributions actually decreased, approaching zero—especially in the ultraviolet part of the spectrum. This discrepancy was called the "ultraviolet catastrophe." The prevailing model incorrectly predicted that objects should radiate more energy at shorter wavelengths, but experimental observations showed the opposite—less energy was emitted.

This problem was especially concerning because the solutions physicists proposed were based on well-established theoretical foundations, yet the real-world data didn't match. The failure to reconcile theory with these new observations called into question the validity of thermodynamics, a key area of physics concerned with heat and energy transfer.

In 1900, while serving as a professor of physics at the University of Munich, Max Planck decided to address this radiation problem. He began by applying Maxwell's electromagnetism to create a theory that linked thermal energy and the charged oscillating "particles" of electromagnetic waves. Planck initially viewed electromagnetism as purely wave-like, but to develop his theory, he had to employ the statistical methods pioneered by Ludwig Boltzmann for understanding energy distribution in molecular collisions.

At the time, the dominant model assumed that thermal energy emitted by an object was due to continuous changes in the energy of oscillating charged particles. This model predicted that energy distributions at shorter wavelengths would increase indefinitely, leading to absurd results. However, this theory was in direct contradiction to Planck's and others' observations, which showed energy tapering off at short wavelengths instead of growing.

10.4 Energy comes in bundles

To derive his formula using Boltzmann's statistical methods, Planck first had to divide the total energy radiated by an object into discrete bundles or packets, all containing the same amount of energy. He then calculated the possible ways these energy packets could be distributed among the oscillating particles. Planck published his findings in a series of papers between 1897 and 1900. The resulting formula, now known as Planck's law, matched experimental observations perfectly. However, Planck himself was uneasy with the method he used, particularly the application of statistical methods in physics, which he wasn't fond of. Despite his reservations, the equations worked.

In James Clerk Maxwell's theory of electromagnetism, a charged electron orbiting the nucleus continuously emits energy. By analogy, Planck's model proposed that oscillating charges within hot objects radiate energy. However, if electrons were truly radiating energy as Maxwell's theory suggested, they should spiral into the nucleus within a microsecond, yet atoms are stable and don't collapse.

Planck had already addressed the ultraviolet catastrophe and explained the energy distribution of hot objects by 1900. In 1911, Rutherford proposed his model of the atom, but the riddle of the stable atom remained. During Einstein's "miracle year" in 1905, he expanded Planck's concept of energy quanta, applying it to light and radiation, laying the groundwork for a fuller understanding of atomic stability. All the necessary elements to solve the puzzle of the theoretically collapsing atom were now in place.

Enters Niels Bohr

In 1913, Danish physicist Niels Bohr introduced a new model of the atom that avoided the problem of electron collapse. His model was similar to Rutherford's planetary model but with a crucial difference: Bohr proposed that electrons orbit the nucleus in specific, stable orbits he called "stationary orbits." In these orbits, electrons do not radiate energy and remain stable. However, Bohr's theory allowed electrons to move between orbits. When an electron jumps to a lower orbit, it releases energy; once it reaches the new orbit, it becomes stable again in another stationary orbit. If the electron absorbs energy, such as from a photon, it jumps to a higher orbit. Importantly, electrons are only allowed to occupy these specific orbits—positions between them are forbidden. The energy released or absorbed during these jumps corresponds to Planck's quanta, meaning electrons can only exchange energy in discrete packets.

Niels Bohr

Bohr used Planck's theory to calculate the energy levels of these stationary orbits in the atom. His results matched experimental data perfectly, marking a significant step toward understanding atomic structure. However, when Bohr tried applying his model to more complex atoms, the calculations didn't align with experimental measurements. This inconsistency indicated that Bohr's model, though revolutionary, wasn't the final answer. It became clear that Newtonian mechanics, combined with Maxwell's electromagnetism and supplemented by the discoveries of Planck and Bohr, still fell short. A new framework was needed, and the 26-year-old Einstein would soon provide the key to this "new physics."

10.5 A Quantum Leap

Relativity wasn't Einstein's only groundbreaking theory. He also played a pivotal role in the development of quantum theory. In March 1905, during his "miracle year," Einstein published a paper titled "On a Heuristic Point of View Concerning the Production and Transformation of Light." The term "heuristic" refers to something that guides discovery or research, even if it isn't fully proven—a reflection of Einstein's cautious attitude toward quantum theory at the time.

Despite his early contributions, Einstein grew skeptical of the theory's implications. He famously remarked that "God does not play dice with the universe," expressing his discomfort with the randomness and uncertainty inherent in quantum mechanics. Throughout his life, Einstein believed that quantum theory was incomplete and that a more fundamental explanation of atomic behavior would eventually emerge. However, modern scientists widely accept quantum theory as a cornerstone of physics, and it remains central to our understanding of the atomic world.

10.6 Discovering the Quantum

In his March 1905 paper, Einstein began by explaining why the existing equations for thermal radiation fell short. While these equations worked well in the low-energy region, they failed in the high-energy range, predicting an infinite amount of energy. Einstein approached the problem from a fresh perspective, moving away from Planck's methods. He decided to start with basic physical principles, ultimately showing that the radiation from hot objects behaves as if it's made up of discrete packets, or quanta, of energy. He proposed that the energy of each quantum is tied to the wavelength of the emitted radiation—the shorter the wavelength, the higher the energy.

Up to this point, there was no radical departure from Planck's thinking, as the idea of energy quanta could be understood as a unique characteristic of radiation from heated objects. But Einstein took a daring leap that would eventually win him the Nobel Prize in 1922. He asserted that matter and radiation interact solely by exchanging these quanta of energy. Light, he said, wasn't just a wave as Thomas Young's experiments had demonstrated; light also behaves like a particle. These quanta, or photons, possess a fixed amount of energy and exert pressure on objects, interacting with matter as particles do. In other words, light is "lumpy."

Young's famous interference experiment had seemingly settled that light was a wave, and this wave theory of light is still demonstrated in classrooms today. Waves and particles were thought to be mutually exclusive—one could not exist if the other did. So, the question lingered: is light a wave or a particle? According to Einstein, light is both. Though this dual nature of light seemed nonsensical at the time, in the quantum world Einstein introduced, it made perfect sense. It would take another 20 years for scientists to fully reconcile this apparent contradiction.

10.7 Ascertaining quanta of various energies

Einstein's light-quantum idea offered a solution to the puzzling results physicists were encountering when studying radiation from hot objects. They had been measuring the energy emitted at different wavelengths and found that the theory worked well at long and intermediate wavelengths. However, at very short wavelengths, the experiments showed far less radiation than predicted by their equations, which suggested that energy should increase as the wavelength shortened. This discrepancy between theory and experiment left scientists baffled.

The key difference lay in the nature of photons. Low-energy photons have long wavelengths, while high-energy photons have short wavelengths. Experiments showed that very little short-wavelength energy was being radiated from hot objects, contrary to theoretical predictions. The light quantum idea clarified these results. Planck had previously suggested that the thermal energy emitted by an object results from charged particles oscillating within the object, with the energy of each photon tied to the wavelength of the oscillation. At short wavelengths, high-energy photons are involved, but few oscillators can produce such high-energy photons, which explains why little radiation is detected at these wavelengths.

Einstein later demonstrated that any classical approach to this problem would lead to what was known as the ultraviolet catastrophe—the prediction of infinite energy at short wavelengths. Planck avoided this problem by recognizing that energy must be emitted in discrete packets, or quanta, rather than being continuously distributed as classical physics assumed. He formulated a new equation, $E=h\nu$, where E represents the energy of a photon, the frequency, and h the constant now known as Planck's constant.

At longer wavelengths, photons have low energies, so many more of them can be emitted, though each contributes little to the total energy. At intermediate wavelengths, the emission of photons is higher, resulting in the greatest energy output, which matched experimental data. Planck's breakthrough was realizing that energy levels were tied to the wavelength of light emitted by oscillators, rather than being equally distributed, as classical theories suggested.

Einstein's major insight was recognizing that this quantization of energy wasn't just a curious feature of radiation but a fundamental property of nature. He demonstrated that light, and all electromagnetic radiation, is made up of discrete quanta. Photons at any given wavelength have the same energy and are identical to one another. Just as you can't split the steps of a staircase, you can't split photons.

With Einstein's new perspective on light, Planck's radiation law gained widespread acceptance as the explanation for thermal radiation. It showed that thermal energy comes from oscillating charged particles within an object, and these oscillations have specific, quantized energies. Einstein, once skeptical of Planck's law, now embraced it, seeing how his new quantum view made the law both sensible and revolutionary.

10.8 Solving the photoelectric effect

Another long-standing puzzle in physics, known as the photoelectric effect, had also defied explanation. Discovered by Heinrich Hertz in 1877, the phenomenon left physicists scratching their heads. When a beam of light is shone on certain materials, electrons are ejected from the surface, generating an electric current. This effect is the basis for modern solar cells that power everything from calculators to Martian rovers. Increasing the brightness of the light produces more electricity, but surprisingly, the speed of the emitted electrons remains unchanged. Additionally, if the wavelength of the light exceeds a certain threshold, no electrons are emitted, no matter how intense the light is.

The solution to this mystery came from Einstein's light-quantum theory. How did it explain the photoelectric effect? When light hits a material, photons—packets of energy—are sent at a specific energy level determined by their frequency. If the light is "monochromatic" (of a single color), all the photons have the same energy. When a photon strikes the material, it transfers all its energy to an electron. The electron then absorbs this energy, allowing it to escape from the material. Increasing the brightness doesn't increase the energy of each photon; it simply sends more photons of the same energy, which releases more electrons but doesn't affect their speed.

The solution to the second problem is more straightforward. For an electron to escape the material, it needs a minimum amount of energy. This energy is provided by the photons, which have a fixed energy determined by the wavelength of the light. If the wavelength is too long, the energy of each photon is too low to free the electron, no matter how bright the light is.

Increasing the brightness just adds more low-energy photons, which still aren't powerful enough to eject electrons. Thus, Einstein's theory explained both why increasing brightness doesn't change the electron speed and why long-wavelength light fails to emit electrons at all.

10.9 Waves of Matter

Physicists initially responded with skepticism to these two persistent and unresolved problems. In 1905, Einstein was relatively unknown, but the series of remarkable papers he published that year quickly brought him into the spotlight. Max Planck was one of the first to recognize Einstein's brilliance and became an early supporter of his special theory of relativity. However, the light-quantum theory posed a different challenge; even Planck hesitated to embrace it, despite its role in clarifying his own findings.

As late as 1913, when Einstein was acknowledged as one of the leading physicists in Europe—where most physics research was concentrated—there remained significant resistance to his quantum ideas. When he was nominated for membership in the Prussian Academy of Sciences that year, Planck and other esteemed physicists stated in their recommendations: "It can be said that there is hardly a major problem in modern physics to which Einstein has not made a remarkable contribution. However, he may have sometimes missed the mark in his speculations, as in his hypothesis of light-quanta, which should not be held against him too harshly."

For 15 years, Einstein remained isolated in his belief in the light-quantum concept. By 1918, he expressed his growing conviction, stating that he no longer doubted the reality of quanta, "even though I am still alone in this conviction."

Ch 11—Special Relativity (Space and Time)

About a century ago, an obscure civil servant in Switzerland recognized that the prevailing theories in physics were flawed and set out to rectify them. His contributions to science were so significant that Time magazine named him the Person of the 20th Century, surpassing kings, queens, presidents, artists, movie stars, and religious leaders. This remarkable individual was Albert Einstein. His equation, $E=mc^2$, has become the most famous formula in all of science and emerged from his groundbreaking theory of relativity.

Albert Einstein at age 5
In courtesy of the California Institute of Technology and The University of Jerusalem

The equation $E=mc^2$ states that mass and energy are fundamentally the same, with all objects possessing both. It illustrates that mass can be transformed into energy, and energy can be converted into mass. This is known as Einstein's mass-energy equation.

11.1 Electromagnetism - Einstein's Most Fascinating Subject

"Convert magnetism into electricity," wrote the self-taught English scientist Michael Faraday in his lab notebook in the early 1800s. His groundbreaking work, along with that of James Clerk Maxwell, established the foundation for electromagnetism. As a young secondary school student, Einstein immersed himself in the theories of mechanics and electromagnetism, which profoundly shaped his understanding of the universe. He found the theory of magnetism to be the "most fascinating subject" and often skipped college classes to read the original papers on the topic. By the time he graduated, he had become an expert in a field that was considered cutting-edge in physics. Later, Einstein identified an inconsistency between electromagnetism and Isaac Newton's concept of absolute time, which led him to develop his theory of relativity. Interestingly, Einstein conducted most of his scientific work at home or in the library of the patent office, rather than in a university or research lab. At this time, he was often referred to as an "amateur scientist." He earned his degree in physics in 1900.

11.2 Einstein's first scientific contribution

In his first two professional papers, published shortly after graduating from college, Einstein applied thermodynamic principles to explain various effects observed in liquids. His second paper focused on the atomic foundations of thermodynamics. In his groundbreaking 1905 paper on "Brownian motion," Einstein utilized Boltzmann's statistical techniques to elucidate the zigzag motion of smoke particles. The methods he employed in this work significantly advanced the field of statistical mechanics.

At the time, while this patent clerk was making his bold claims, scientists were just beginning to accept the reality of atoms. The phenomenon of radioactive decay, which involves the spontaneous emission of light and charged particles from atoms, had only been discovered five years earlier.

11.3 Proving that atoms exist

Einstein's Ph.D. thesis was titled "A New Determination of Molecular Dimensions," written during his remarkable year of creativity in 1905. This thesis marked his second attempt to earn a Ph.D. He had previously tried in 1902, a year and a half after graduating from the Polytechnic, submitting a thesis on molecular forces, which was ultimately rejected.

11.4 Imagining the atom

Living in the 21st century, we understand that matter is composed of atoms. However, a century ago, even physicists and chemists were not in agreement about the existence of atoms. The young Einstein began his scientific career with three pivotal papers that provided evidence for the existence of atoms. The foundation of our modern theories of matter can be traced back to John Dalton in the 19th century, though the concept of the atom was first introduced by Democritus 23 centuries earlier. In his Ph.D. thesis, Einstein demonstrated how to determine the actual dimensions of atoms. He did not develop his ideas in isolation; instead, he built upon the existing knowledge of atoms at the time as a springboard for his work.

11.5 Einstein's 2nd major scientific contribution

In the early 1900s, physics was facing a crisis. The two dominant branches at the time were electromagnetism, represented by James Clerk Maxwell's theory, finalized in 1873, which explained the nature of light among other phenomena, and Newtonian mechanics, Isaac Newton's well-established laws of motion that had been accepted for over 250 years by 1900. However, these two foundational theories were incompatible with one another.

One major flaw in the framework of physics was its inability to explain observations related to the heat emitted by objects. As objects heat up, they change color; for example, an electric stove burner glows red at first, then transitions to orange, and eventually bright yellow. Logically, one might expect that an object glowing in the ultraviolet spectrum—beyond visible light and detectable only with special instruments—would indicate an even higher temperature. Physicists' equations suggested this would be true, yet observations contradicted it: hot objects emitted less ultraviolet light and more light of other colors. This discrepancy became known as the "ultraviolet catastrophe."

Physicists had been conceptualizing the light emitted from hot objects as waves, in line with Maxwell's theory and supported by the famous experiments of English physicist Thomas Young in the early 19th century. In 1900, however, German physicist Max Planck proposed a groundbreaking idea: if the light emitted by hot objects could be thought of as being composed of discrete packets or "quanta" instead of continuous waves, he could derive a new equation that accurately described the observed phenomena. This approach explained why the radiated light peaked at certain colors and was nearly absent in the ultraviolet range.

Although Planck did not believe that these quanta represented the fundamental nature of light—consistent with Young and Maxwell's findings that light is continuous and wave-like—he noted that in certain situations, sound waves can exhibit lumps or beats when waves of slightly different frequencies overlap. He speculated that a similar effect might be occurring with the radiation of light quanta from hot bodies.

11.6 Absolute motion

A second significant problem within the flawed framework of physics was even more problematic, as it put mechanics and electromagnetism in direct contradiction: Newton and Maxwell disagreed on the existence of absolute motion.

According to Newton, the laws of physics should remain consistent regardless of whether you are at rest or moving at a constant velocity. In Newtonian mechanics, motion is always described relative to some object. He believed that no experiment could yield different results whether you were in motion or at rest.

Consider a familiar scenario: when you're on an airplane, it can be difficult to tell if you're moving (unless turbulence occurs). If you fall asleep before takeoff and wake up while the plane is cruising, you would need to look outside to confirm that you are not still parked at the gate.

This is because, in Newton's view, the laws of physics are the same in both situations: when the plane is at rest and when it is flying.

Maxwell's theory diverges from Newtonian mechanics. While Newton viewed light as a particle, Maxwell's electromagnetism conceptualized it as a wave. As a wave, light requires a medium to propagate, much like how water waves need a body of water and sound waves require air. On Earth, light travels through air, water, and glass, but what about when it travels through space? What medium does the light from the Sun or a star move through on its way to Earth and into our eyes? Nineteenth-century physicists referred to this hypothetical medium as "ether." They proposed that ether filled the entire universe, allowing planets, stars, and light to move through it. This would provide a means to determine whether you were moving at a constant velocity or at rest. You wouldn't need to look out the window on an airplane to know you were in motion; you would simply need to find a way to measure your movement through the omnipresent ether, which even permeated objects, including the airplane cabin.

11.7 Electric fields

When you move a magnet toward or away from a stationary wire, it generates a current in the wire by producing an electric field. This phenomenon is described by Faraday's law. Michael Faraday demonstrated that a changing magnetic field is necessary to create an electric field in the wire. As the magnet approaches the wire, the magnetic field strength increases; conversely, it decreases as the magnet moves away.

Now, consider what happens when the magnet remains stationary and the wire is moved. According to Maxwell, since the magnetic field around the stationary magnet isn't changing, there will be no electric field to induce a current in the wire. Although a current still appears in the wire, Maxwell explains that this occurs for an entirely different reason. Moving the wire toward the magnet does not generate an electric field.

In summary, when the magnet is in motion, an electric field is produced; when it is stationary, no electric field arises. Thus, according to Maxwell's electromagnetism, you can determine whether you're moving by bringing along a magnet and an electric field sensor.

11.8 Being at rest in the Universe

If you can perceive motion or determine when you are at rest, then motion cannot be considered relative. You could take your motion-detection device into space and gradually slow down until it indicates that you are at rest. Alternatively, you could use the device on the ground and move in the opposite direction of Earth's rotation and its revolution around the Sun, attempting to offset all the various motions of the Earth and the solar system until your indicator reads zero. At that point, you would be at rest in the universe. From this perspective, you would observe everything else moving around you in their true, "absolute motions," as scientists refer to them.

11.9 Trouble with Classical Physics

The kinematics formulated by Galileo and the mechanics established by Newton, which together lay the foundation of classical physics, achieved numerous successes. Among their significant accomplishments are the understanding of planetary motion and the application of kinetic theory to explain certain properties of gases. However, there were several experimental phenomena that these otherwise robust classical theories could not account for. While some scientists believed that physics was essentially complete with these two frameworks, by the time Einstein was a student, several problems were becoming evident. First, the two theories conflicted on certain points. Second, neither electromagnetism nor mechanics could adequately explain several new observations made by physicists. Einstein later remarked that physics at that time resembled a "botched-up" house, precariously close to collapse.

11.10 Problems with Electromagnetism

Einstein was uncomfortable with certain implications of electromagnetism as formulated by Maxwell. According to Maxwell's theory, electromagnetism applied only to objects that were at rest relative to the ether. This was why no electric field was produced in a moving wire; the wire was not stationary in the ether. However, Einstein rejected the concept of absolute motion. He agreed with Galileo that all motion is relative and that the laws of mechanics should hold true everywhere. Yet, he sought to extend this principle beyond just mechanics, asserting that the laws of physics should be consistent throughout the entire Universe.

11.11 Einstein's Paradox

Einstein proposed his special theory of relativity in 1905, motivated not by the need to explain the Michelson-Morley experiment concerning the ether, but rather by a thought experiment he devised. As a 16-year-old student, Einstein had studied Maxwell's theory of electromagnetism and pondered a paradox: if one were to move at the speed of light parallel to a light beam in a vacuum, they would observe "static" electric and magnetic field patterns. This was akin to observing a "static" disturbance on a string while moving alongside waves on that string. However, Einstein recognized that such static patterns in empty space contradicted Maxwell's theory. His intention was not to validate the Michelson-Morley experiment but to extend the principle of relativity beyond the realm of mechanics, aiming for a more elegant and simplified understanding of physics.

Galileo's theory of relativity posited that it is impossible to determine whether one is moving quickly, slowly, or not at all. This principle holds true whether one is on Earth or isolated in a smooth-moving train, blindfolded and ear-plugged, or in any other situation devoid of an external reference frame.

Unaware that Michelson and Morley had disproved the existence of the ether, Einstein used Galileo's principle of relativity as a foundational basis to explore the ether's existence. He recognized that the laws of mechanics involve only accelerations, while the laws of electromagnetism involve a universal velocity: the speed of electromagnetic wave propagation in a vacuum. This speed, as predicted by Maxwell's theory and consistently confirmed through experiments and observations, is constant across all wavelengths, including radio waves, light, and X-rays.

Faced with a dilemma, Einstein had two options to resolve the paradox: either Maxwell's theory was flawed, or the classical kinematics that allowed an observer to travel alongside a light beam was incorrect. Trusting his intuition, Einstein chose to uphold Maxwell's theory as fundamentally correct and sought a new kinematics to replace those of Galileo and Newton. This new framework, forming the basis of special relativity, asserts that no observer can catch up to a light beam. It also addresses the intricacies of time, length, and velocity.

In his seminal 1905 paper, Einstein universalized Galileo's principle of relativity. While Galileo and Newton had limited it to mechanics—the extent of their scientific knowledge—Einstein expanded it to encompass all areas of physics. He established this extended principle as a cornerstone of his new theory, articulating it as a postulate: the laws of physics are the same in all non-accelerated frames of reference.

11.12 Einstein's extended principle of relativity

The laws of physics are universal; they operate consistently regardless of the speed at which you are moving. It's impossible to differentiate between rest and motion, indicating that all reference frames are equivalent and that there is no concept of absolute motion. In the absence of a stationary reference point, all uniform motion is relative. With this postulate in hand, Einstein reformulated electromagnetism. In the second part of his 1905 paper, he demonstrated that all electric and magnetic effects remain unchanged across all reference frames in uniform motion. Einstein made a "slight" modification to electromagnetism, ensuring that it relied on relative motion.

What about the scenario involving a moving wire and a stationary magnet? According to Einstein, there is only relative motion between the magnet and the wire. If you remain stationary with the wire, the magnet appears to move toward you; conversely, if you stay with the magnet, the wire seems to approach you.

While Maxwell's electromagnetism suggested that a factor beyond the electric field generates a current when a wire is moved toward a resting magnet, Einstein argued that, in this specific case, Maxwell was mistaken; an electric field also produced the current in that situation. Modern physicists do not concern themselves with which object is moving; they simply recognize that the magnet and the wire are in relative motion. For today's physicists, this is clear-cut, and that clarity stems from Einstein's demonstration that absolute motion does not exist.

Einstein's correction brought electromagnetism and mechanics onto equal ground, unifying them into a single theory. The principle of special relativity applies to both. The motion-detection device that relies on the absence of an electric field in one scenario fails to work because there is no measurable difference in either case. Ultimately, you cannot distinguish between rest and motion. However, Einstein's contributions did more than resolve the inconsistencies between electromagnetism and mechanics; he expanded the principle of relativity to encompass all of physics.

11.13 Light Speed

The notion that the Universe lacks an absolute standard of motion aligns with the fundamental principle of relativity. No frame of reference holds an advantage regarding the speed of light or absolute motion; all perspectives in the Universe are equivalent. All forms of electromagnetic radiation, including light, travel at the same velocity. The theory of relativity further posits that exceeding the speed of light is impossible, and that all forces, effects, and material objects are constrained by this speed. There is no known method to transfer information through any medium faster than light.

How do we measure the speed of a light beam? Early experiments were conducted long ago by researchers using lanterns. By prior arrangement, one individual would uncover his lantern while the other would uncover his as soon as he saw the light from the first. The first person would then measure the time elapsed between uncovering his lantern and seeing the light from the second lantern. These experiments were conducted over large distances, such as between two hilltops. Unsurprisingly, the only significant delay in transit was due to the limitations in measuring the speed of light.

Later attempts to measure the speed of light proved to be more successful. Today, it is possible to accurately measure the speed of light using equipment commonly found in high school physics or electronics laboratories. Once scientists had the technology to demonstrate that light did not travel instantaneously, as was previously believed, they began to wonder if the speed of light would be affected by motion. Knowing that the speed of sound is influenced by motion, they theorized that light must also be similarly affected. They proposed that light must travel through some medium, much like sound does. This concept became known as the ether theory—the idea that a medium for light propagation existed everywhere, even in a vacuum.

The ether was thought to provide a standard for absolute motion in the Universe. If light traveled at a constant speed through this ether, then by measuring the speed of light in different directions, it should be possible to determine the speed of the Earth in relation to this ether. Such reasoning prevailed at the time.

11.14 Struggling with the speed of light

According to Maxwell's equations, light travels at a fixed speed relative to the ether. However, Einstein found no need for the ether. With his principle of relativity, he eliminated the concept of an absolute standard of rest. He proposed that light consists of independent electric and magnetic fields oscillating through empty space.

Maxwell also did not require the ether in his equations; it was physicists before him who invented this hypothetical substance to explain the propagation of electromagnetic waves through a vacuum. But without the ether, the question arises: with respect to what is light moving? Einstein's unexpected answer was that light travels at the same speed relative to everything. This simple insight contained the essence of relativity. Yet, Einstein did not arrive at this conclusion overnight; he grappled with it.

During the development of his theory, Einstein wrestled with the apparent contradiction between the constant speed of light and the varying speeds of ordinary objects. He came to understand that both time and space are relative. Unlike Newton, who believed they were fixed, Einstein recognized that time and space change when an object is in motion, adjusting themselves to ensure that the speed of light remains constant, no matter how fast one moves.

The postulate of the principle of light says:

1. The speed of light is a universal constant. All observers in uniform (nonaccelerated) motion measure the same value c for the speed of light.
2. Time no longer flows at the same rate for everyone.
3. Light moves at c with respect to anything.
4. Space and time is relative (varying) but the speed of light is absolute (constant)

11.15 Inertial frames of reference

As long as the laws of physics dealt only with acceleration, all inertial frames of reference were equally valid for describing nature. Within classical mechanics, no experiment could identify one specific frame as fundamental. However, since light travels at a universal speed of 186,000 miles per second (denoted by the symbol c), careful measurements of light's speed relative to laboratory instruments should reveal the velocity of those instruments relative to the rest of the Universe. There should be one frame of reference in which light always travels at c—a so-called "frame of absolute rest." Historically, this idea was linked to the "luminiferous ether," a medium thought to carry light waves similarly to how air carries sound.

To determine the state of absolute rest, scientists conducted precise experiments on electromagnetic phenomena. Among these experimenters were American physicists Albert Michelson and Edward Morley, who used Michelson's optical interferometer to detect variations in the speed of light. Their experiments sought to measure even tiny deviations, yet all attempts failed, yielding the surprising conclusion that no absolute motion of the Earth could be detected with respect to a privileged frame of reference.

The Dutch physicist Hendrik Lorentz and French mathematician Henri Poincaré suggested that the failure to identify an absolute frame of reference implied that no such frame existed. Newton's framework of multiple inertial frames, where no single one is special, seemed valid after all.

In 1905, Einstein combined Lorentz's and Poincaré's ideas into a new perspective on frames of reference. He offered a solution to the Michelson-Morley experiment's puzzling results without contradicting Maxwell's theory of electromagnetism. By altering the traditional concepts of space and time, Einstein could simultaneously preserve the equivalence of all inertial frames and the constancy of light's speed as described by Maxwell.

Einstein's special theory of relativity inadvertently resolved the Michelson-Morley problem, which had puzzled physicists for two decades. Hendrik Lorentz and George FitzGerald had proposed the concept of length contraction to explain the experiment's results, an idea met with reluctance by scientists. Lorentz also suggested that time could stretch or contract with motion, though he and others believed these effects were purely mathematical tools rather than physical realities.

Einstein's solution was even more radical. Like Lorentz's length contraction, time and space indeed change as an object moves, but for Einstein, these changes were not just mathematical tricks—they were real. His equations described the actual nature of the universe. The reason we don't notice these changes in everyday life is that they are incredibly small at normal speeds, becoming significant only when objects move at velocities close to the speed of light.

11.16 The beauty of Physics in describing nature

Einstein derived the equations needed to calculate speeds in his thought experiment, based on his theory of relativity. Remarkably, these equations turned out to be the same as the Lorentz transformations, which had been previously introduced to explain the results of the Michelson-Morley experiment. However, Einstein's interpretation of these equations differed significantly from Lorentz and FitzGerald's. While their transformations were seen as mathematical fixes, Einstein's equations represented a deeper, physical reality. Importantly, relativity applies only to uniform motion—the kind that cannot be distinguished from rest.

11.17 Accounting for the Ether

If light traveled through the ether, then Earth would also move through it. As a result, the speed of light should vary depending on Earth's motion. For example, when Earth moves in the same direction as the ether, it would "catch up" slightly with the light, leading to a lower measured speed. Conversely, when Earth moves in the opposite direction, you'd expect to measure a higher speed, as Earth would be "losing ground" to the light beam.

At the end of the 19th century, Albert Michelson and Edward Morley, working at what is now Case Western Reserve University, designed a precise experiment to detect changes in the speed of light as Earth moved through the ether. Surprisingly, their experiment found that the speed of light remained constant, regardless of Earth's movement.

This result baffled scientists. It was akin to measuring the speed of a boat in water moving at 20 kph and finding no difference in its speed whether it traveled upstream or downstream. The expected variation was missing, and the Michelson-Morley experiment introduced a new puzzle for physics.

11.18 The Michelson-Morley experiment

At the time, neither Weber nor Einstein were aware that Michelson and Morley had already conducted a groundbreaking experiment in 1886 with the same objective. Michelson and Morley had devised a clever way to detect Earth's motion through the ether by measuring the speed of light at different points during Earth's orbit around the Sun.

Michelson had developed the instrument for this experiment while working in Hermann von Helmholtz's laboratory in Germany. The idea itself was based on a suggestion made by James Clerk Maxwell in 1875. According to the ether theory, as Earth moved through the ether in its orbit, it would create an "ether wind" similar to the feeling of wind against your hand when you stick it out of a moving car window.

If you measured the speed of light in the direction of Earth's motion, you would be moving against the ether wind, and the speed of light should theoretically be the speed of light in the ether minus the ether wind's velocity. Conversely, if you measured light's speed in the opposite direction, it should be the speed of light in the ether plus the ether wind's speed. Michelson performed his first version of this experiment in 1881 in Germany, using a highly sensitive device called an interferometer.

He anticipated detecting a slight difference in the speed of light—300,000 kilometers per second (kps) minus the 30 kps speed of Earth's motion around the Sun. Instead, he measured the speed of light at 300,000 kps, showing no difference, as if there were no ether wind at all.

Later, Michelson accepted a position as a professor at the Case School of Applied Science in Cleveland, Ohio (now Case Western Reserve University). There, he met Edward Morley, a chemistry professor, and together they redesigned Michelson's original experiment with greater precision.

In their new setup, Michelson and Morley aimed to measure the speed of light in two directions simultaneously—one along the ether wind (opposite to Earth's motion in its orbit) and one across the ether wind. They used a two-way mirror to split a single light beam into two paths: one following the direction of Earth's movement through the ether, and the other perpendicular to it. The two beams were then recombined and superimposed on a screen. Because both beams originated from the same source, they were coherent, meaning their waves were in sync.

Einstein as a teenager
In courtesy of the California Institute of Technology and The University of Jerusalem

When the two beams met, they formed an interference pattern, which Michelson and Morley used to measure the speed of light with great precision. However, they were puzzled by the results. They expected the interference pattern to shift if the speed of light differed in the two directions, but no such shift occurred. The two scientists believed their experiment had failed because they saw no difference in the measurements.

Einstein was only seven years old when Michelson and Morley first conducted their experiment, and he learned about it only after finishing college. His special theory of relativity ultimately provided the solution to the puzzle, but he didn't develop it with their experiment in mind. His primary focus at the time was on electromagnetism and Galileo's principle of relativity.

The Michelson-Morley experiment, carefully executed, puzzled scientists when its results were published. It showed that the speed of light remained constant, regardless of the Earth's motion. This was unexpected. In response, two physicists, George FitzGerald of Trinity College in Dublin and Hendrik Antoon Lorentz of the University of Leyden in the Netherlands, independently proposed an explanation. They suggested that objects shrink in the direction of motion, with the amount of shrinkage precisely enough to maintain the speed of light as constant. This concept became known as the "Lorentz-FitzGerald contraction."

Lorentz further explained this phenomenon with his theory of matter, which was based on electrons. He proposed that as a body moves through the ether, its electrons compress, causing the object to shrink. He then modified Galileo's principle of relativity, creating an equation to compute this length contraction, now known as the "Lorentz-FitzGerald contraction."

In 1895, when Lorentz published his paper, a 16-year-old Einstein was already contemplating what it would be like to travel at the speed of light. Ten years later, Einstein provided the final solution to the mystery, using Lorentz's equation in his theory of relativity but interpreting it in a completely new way.

11.19 Jules Poincare and modern relativity

Jules Henri Poincaré in France took a different approach to the perplexing results of the Michelson-Morley experiment. While he agreed with Lorentz's ideas to some extent, he was uncomfortable with the concept of length contraction being introduced solely to explain the experiment's outcome. Poincaré sought a more fundamental explanation that didn't rely on ad-hoc adjustments like the contraction hypothesis.

11.20 Elastic time

Poincaré was disappointed that Lorentz had abandoned the principle of relativity. Since the laws of physics appeared to be the same in all frames of reference, Poincaré believed there had to be a more general principle of relativity at play. He urged Lorentz to expand his length contraction equation. Lorentz returned with a new set of transformation equations, which not only included length contraction but also introduced "time dilation." According to this theory, if you're moving relative to the Earth, not only do objects shrink in length, but time itself slows down—your clock ticks at a different rate. Time, Lorentz theorized, had an "elastic" property that stretched and contracted depending on your motion.

However, no one, not even Lorentz, thought these equations represented the real world. They were seen merely as mathematical tools to help with calculations. Poincaré suggested that clocks in different moving frames could be synchronized using light signals, but he emphasized that these clocks wouldn't show "true time." He argued that only the ether could serve as the standard for true time. Despite his belief in the ether, Poincaré began advocating for a new mechanics, one in which the speed of light couldn't be exceeded.

In his lectures, Poincaré spoke of this new mechanics as an "unrealized hope and conjecture." But just a year later, a 26-year-old Albert Einstein turned this unrealized hope into reality. Einstein's solution was even stranger than Lorentz and FitzGerald's. Like them, he agreed that time and space changed as you moved, but for Einstein, these changes were real, not just mathematical conveniences. His equations described the physical world. The reason we don't notice these changes is that they are incredibly small at ordinary speeds, becoming significant only at velocities approaching the speed of light.

Einstein's equations were, in fact, the same as Lorentz's transformations, but Einstein's interpretation was revolutionary. He argued that clocks throughout the universe tick at different rates depending on their motion. A clock in a spaceship might run slower or faster than one on Earth. Einstein extended Galileo's principle of relativity, which applied to mechanics, to all of physics. This eliminated the need for the ether and the concept of absolute motion. In Einstein's view, all motion is relative, and the speed of light is a universal constant.

This was a radical departure from the previous notion of absolute motion, where the ether was seen as a reference point for measuring the true motion of objects. In Einstein's universe, there was no such reference point. Motion could only be understood in relation to other objects, and the laws of physics remained the same for all observers in uniform motion.

In Einstein's famous equation, $E = mc^2$, the "c" represents the speed of light. But why c^2, not c^3 or another power of c? The answer goes back to the 17th century, when Christiaan Huygens and other scientists showed that the energy of an object is related to the square of its speed. Einstein's equation, as an energy equation, follows the same principle, but applies it to the speed of light.

11.21 Christian Huygens' famous energy equation

The Dutch scientist Christiaan Huygens was one of the most renowned figures in European science and a source of pride for his homeland. He came from an illustrious family; his father, Constantijn Huygens, was a major figure in Dutch literature. Influenced by his friend, the French philosopher and mathematician René Descartes, Huygens began developing ideas about space and mathematics early in life. By the time he was 30, he had already published several important mathematical papers, establishing his reputation.

In January 1669, Huygens presented a paper to the Royal Society of London that addressed a significant debate in physics at the time: the motion and collisions of objects. Specifically, scientists were puzzled by the behavior of two small metal balls that, after swinging, collided and then transferred their motion. A modern version of this experiment, often seen in novelty shops, features metal balls that collide and transfer their motion to one another. Physicists at the time couldn't understand why, after a collision, the first ball would stop completely while the second ball would swing back to the same height as the first ball had before the collision. Their existing knowledge of speed and energy couldn't explain why the first ball didn't simply bounce back with half the speed, while the second ball took off with the other half.

Huygens cleared up this confusion by introducing a new concept. He explained that if you multiply the mass of each ball by the square of its speed and then add these values together, the total remains the same before and after each collision.

This product of mass and the square of speed gives the ball's kinetic energy. Although the energy of each individual ball changes upon collision, the total energy in the system remains constant. Huygens' discovery laid the groundwork for the principle of the conservation of energy. This principle explains why the second ball always returns to the same height—the energy is transferred, not lost.

Huygens showed that the energy of motion is proportional to the object's mass and the square of its speed. This is the foundation of what we now call kinetic energy. Einstein's famous energy equation, $E = mc^2$, follows a similar structure: it multiplies an object's mass by the square of a speed, but in this case, it's the speed of light. In Einstein's equation, the speed of light represents the energy that exists in the form of particles, illustrating that light itself carries energy. Thus, Einstein's equation also reflects the conservation of energy, but on a cosmic scale.

11.22 Formulating $E = mc^2$

In a brief yet groundbreaking paper, Einstein introduced his most famous equation, $E = mc^2$, by demonstrating that when an object at rest, such as an atom undergoing radioactive decay, emits light, its energy decreases. This reduction in energy occurs because some of it is carried away by the emitted light. From the principle of energy conservation, Einstein knew that the light's energy had to come from the object itself, meaning the object's total energy diminished—a concept already understood.

However, Einstein took this idea further. He applied his special relativity equations to compare the energy of an atom emitting light while at rest in the laboratory versus while in motion relative to the laboratory. Through his calculations, he found that the atom's mass decreased after the emission of light.

What made Einstein unique was his ability to take a step beyond what was known. He expanded this observation, stating that any atom loses mass when it emits light. Then came his bold generalization: this principle applies to all matter. Einstein proposed that the mass of any object is a measure of its energy content and that this relationship holds universally.

In essence, Einstein showed that mass and energy are two forms of the same entity—interchangeable. These ideas are captured in the simple yet profound equation, $E = mc^2$.

11.23 Eliminating the ether

A central contradiction between mechanics and electromagnetism lay in the concept of absolute motion. Newton asserted that all motion is relative, meaning absolute motion cannot exist. In contrast, Maxwell's theory suggested it could. However, Einstein aligned himself with mechanics. In his fourth paper of 1905, commonly referred to as the relativity paper (despite the absence of the term in its title), he reformed electromagnetism to ensure it remained unchanged regardless of whether the observer was at rest or moving at a constant velocity. This meant he adapted electromagnetism so that its description depended solely on relative motion, eliminating the need for the ether. Light, according to Einstein, does not require a medium to propagate; it can travel through the empty space between stars.

With the publication of this paper, the concept of ether was eliminated from physics. Einstein asserted that absolute motion does not exist; when you're on an airplane, for example, you cannot determine if you're moving or stationary without looking outside. He posited that the laws of physics, including those governing mechanics and electromagnetism, are consistent throughout the Universe, irrespective of how one moves—as long as there is no acceleration. Furthermore, Einstein extended the idea of relative motion to light itself: anyone, anywhere in the Universe, regardless of their state of rest or constant motion, will always measure the same speed of light. This paper not only resolved issues in electromagnetism but also introduced a fundamentally new perspective on understanding the world.

11.24 Casting doubt on simultaneity

Einstein's special relativity paper (the fourth paper) raised questions about the concept of simultaneity. He argued that two events deemed simultaneous in one moving reference frame may not be simultaneous when viewed from another frame moving relative to the first. This insight into the relativity of simultaneity led to unexpected implications in our understanding of time and space.

11.25 Einstein conducts a thought experiment

Consider the following thought experiment involving two spaceships: You're aboard a spaceship on an interstellar journey when a crew member activates a light bulb in the center of the crew quarters. You observe that the light reaches both the front and back of the cabin simultaneously, as expected, since the speed of light remains constant regardless of motion, and the distance to each wall is the same. In this frame, the two events are simultaneous.

Meanwhile, an astronomer named Ellie is observing your transparent spaceship through a powerful telescope from her own ship. From her perspective, your ship is traveling toward the star Sirius at half the speed of light (0.5c). When the crew member turns on the light, Ellie notes that the light traveling to the front of the spaceship has a slightly longer distance to cover than the light going to the back. This is because the front wall of your ship has moved farther away from the light bulb since the light was emitted, while the back wall has moved closer, allowing the light to reach it a bit sooner. Ellie perceives the light beam moving diagonally as it bounces off the walls, suggesting that time flows more slowly for you, the space traveler, compared to her perspective.

To Ellie, the light does not reach the front and back of the ship simultaneously. Thus, these two events are not simultaneous when viewed from her frame. This thought experiment illustrates that events considered simultaneous in one moving frame may not be simultaneous in another. Only in an inertial reference frame are events simultaneous; in an accelerated frame, that is not the case.

11.26 Dilating Time

According to Einstein, you and Ellie have different perceptions of timing because you are moving relative to each other. Now, let's consider a second experiment: you send a laser pulse straight up to a mirror on the ceiling of the spaceship's cabin. From your perspective inside the spaceship, you observe the pulse travel straight up, bounce off the mirror, and return directly to the source.

However, Ellie, watching from her own spaceship, sees the pulse moving along a diagonal path. This is because, as the light travels upward, it also moves horizontally with the spaceship. When the pulse reflects off the mirror, Ellie observes it descending along another diagonal, again moving downward and sideways with the ship's motion.

Einstein states that both you and Ellie measure the light pulse traveling at the same speed. However, Ellie perceives the light pulse as covering a longer distance than you do. Consequently, the same event—the light pulse going up and down—takes more time to occur from her perspective than from yours.

11.27 Shortening Space

According to Einstein, time is relative; it varies based on the motion of the observer. However, time isn't the only aspect affected by movement—space is also altered. From the perspective of a stationary observer, space appears shortened or contracted, and time seems to progress more slowly in a moving frame. For example, if you are moving parallel to me, I would perceive your space as contracted, meaning that everything you carry, including yourself, appears shorter in the direction of motion.

From your viewpoint, everything seems normal: a yardstick still measures 36 inches, and you don't appear any thinner than usual. However, you see me moving at half the speed of light (c), and in your observation, my surroundings appear shortened by 13 percent in the direction of motion. This aligns with Einstein's Special Theory of Relativity.

This contraction of space is associated with the perceived lengthening of time for the moving observer, which manifests as time slowing down. Thus, both space and time are relative concepts.

Einstein arrived at this conclusion after determining that the speed of light is a "universal constant." This perspective fundamentally contrasts with Newton's view, where both space and time were fixed, and the speed of light could vary based on the observer's velocity.

11.28 Understanding length contraction

When you are moving relative to another observer, they will perceive your space as contracted, which means that anything you carry, including yourself, will appear shorter in the direction of motion. For instance, if your spaceship measures 300 meters and is traveling at half the speed of light (0.5c), the observer will measure its length as only 260 meters—a 13% contraction. Similarly, a yardstick lying flat in your moving ship will be seen as measuring 31 inches instead of 36. Although you look 13% thinner to the observer, your actual height remains unchanged.

According to Einstein, space is relative. This phenomenon pertains to uniform motion, not acceleration, meaning the frames of reference are inertial. In a moving frame, space is shortened, and these effects are genuine. Biological processes also slow down as you move faster through space-time. For example, your heartbeat and the rate of cell division in your body will decrease when you travel relative to Earth.

If you launch into space and reach a speed of 0.9c relative to Earth, your heart will beat normally, and your cells will continue to divide at their usual rate. However, from my perspective on Earth, your ship's clock will run more slowly due to your speed. For every hour that passes on my clock, only 26 minutes will elapse on yours. After one day for me, only ten and a half hours will have passed for you. A year later, you will be seven months younger than I am. In ten years, you will have aged only four years and four months.

This is a real difference in aging, not an illusion, as all your biological processes have genuinely slowed down. You will indeed age more slowly than those of us remaining on Earth. Relativity applies to uniform motion—motion that cannot be distinguished from rest.

However, when you take off in your spaceship, you must accelerate for some time to reach the constant speed you will travel at. Upon reaching your destination, you'll need to decelerate, stop, turn around, and then accelerate again to return home. When you arrive, you will slow down and finally stop to reunite with your younger friends. All these accelerations do not represent uniform motion, where special relativity is applicable. You will definitely feel the effects of acceleration; you won't need to look outside to know you are moving. If you encounter another spaceship while accelerating, you will have no doubt about your movement. This limitation troubled Einstein, as he felt his special theory did not encompass the full scope of relativity.

11.29 The mathematical origins of space-time

Einstein's mathematics professor at the Polytechnic, Hermann Minkowski, was not fond of him during his student years, famously calling him "a lazy dog." However, in 1907, as Minkowski was preparing for a seminar on the electrodynamics of moving bodies alongside the renowned mathematician David Hilbert in Göttingen, he came across Einstein's special relativity paper. Both men were highly impressed by the work, yet Minkowski found it hard to believe that Einstein had authored it. He remarked to a colleague that he never would have thought Einstein was capable of such a contribution.

Minkowski later went on to create an elegant mathematical framework for special relativity, merging the three dimensions of space with the one dimension of time into a single concept he termed "space-time." He stated, "From here on, space by itself and time by itself are doomed to fade away into mere shadow, and only a kind of union of the two will preserve an independent reality."

11.30 Mixing Space and Time

Einstein developed his theory of relativity from a single profound idea: the speed of light remains constant throughout the Universe, irrespective of your location or motion. However, the consequences of this notion are far from straightforward. When you and I are not moving relative to each other—essentially moving parallel—our clocks remain synchronized, and we experience time at the same rate. We both recognize that a day has passed when we wake up the next morning.

As soon as you start moving relative to me, part of your time-based motion transforms into spatial motion. If I remain stationary, I retain all my motion through time. The faster you move, the more of your motion through time shifts into motion through space.

If you could entirely convert your motion through time into spatial motion, you would reach the speed of light, at which point time would effectively stand still for you. This illustrates why Einstein claimed that nothing other than light can travel at light speed; the flow of time cannot be entirely halted, though it can be nearly frozen at speeds approaching light.

According to relativity, the total of your motion through time and your motion through space must equal the speed of light. This relationship between the three spatial dimensions and the single dimension of time is what later came to be known as "space-time," a four-dimensional construct that reveals how space and time are interwoven, contrasting with Newton's view of them as separate entities.

No one had recognized this connection before Einstein, as these effects are only observable at speeds close to that of light. At slower, everyday speeds, only a minuscule amount of motion through time is converted into motion through space, making it difficult to detect.

In September 1905, just three months after he published his groundbreaking paper on special relativity, Einstein completed another significant work during his miraculous year. This concise, three-page paper, titled "Does the Inertia of a Body Depend on Its Energy Content?" was published in November of that year. In it, Einstein concluded that mass and energy are equivalent, distilling this profound insight into the simple yet powerful equation $E = mc^2$, which means energy (E) equals mass (m) multiplied by the speed of light (c) squared (^{2}).

11.31 Mass is the measure of inertia

Mass measures the resistance encountered when attempting to change an object's motion, a property known as "inertia." Essentially, mass quantifies inertia. However, it's important to note that there are two distinct ways to measure mass: through inertia and through weight, which refers to the gravitational attraction an object experiences towards Earth. Physicists differentiate these two measurements by referring to the first as inertial mass and the second as gravitational mass. Therefore, the terms mass and weight should not be used interchangeably.

The term inertial mass is quite descriptive; an object with a large inertial mass exhibits significant resistance to changes in its motion. Inertial mass does not depend on drag or friction, as it applies to objects moving freely through space. Additionally, gravity does not influence an object's inertial mass. When we stand still, we are moving together through time. If you begin to move relative to me, I would observe that you are converting part of your motion through time into motion through space. As your speed increases, more of your time-based motion shifts into spatial motion.

If you could fully convert your motion through time into motion through space, you would reach the speed of light, effectively halting time. However, completely stopping time is impossible, implying that you cannot achieve light speed. For light itself, there is no motion through time; it travels solely through space at the speed of light.

As an object's speed increases, its inertial mass—the resistance to acceleration—also rises, making it increasingly difficult to accelerate an object closer to the speed of light. If an object were somehow to reach light speed, its inertial mass would become infinite. Since infinite mass would require infinite force for acceleration, this necessitates an infinite amount of energy, leading us to conclude that no object can travel at light speed. The equation $E=mc^2$ encapsulates this concept, where E represents energy, m denotes mass, and c^2 serves as the conversion factor between mass and energy.

While writing about special relativity, Einstein began to consider its limitations. His principle of relativity states that the laws of physics remain consistent for anyone in uniform motion, meaning they behave the same whether at rest or in uniform motion, making it impossible to detect uniform motion.

This framework is limited because it does not account for "accelerated motion," which is easily detectable. For instance, during an airplane's takeoff or landing, one can clearly sense acceleration. In contrast, if you fall asleep while listening to music and awaken later, you might not realize that the plane has left the gate until you look out the window.

Only uniform motion is relative; accelerated motion is not. Recognizing this, Einstein expanded his special theory of relativity to encompass all forms of motion, not just uniform motion. In doing so, he developed a theory of gravity that replaced Isaac Newton's universal law of gravitation, which he called the general theory of relativity.

11.32 What is a dimension

We often use the term "dimensions" without providing a clear definition. While we can agree that our Universe is three-dimensional and that space-time can be viewed as four-dimensional, we should clarify what it means to have a certain number of dimensions rather than attempting to define the term "dimension" itself.

In our Universe, we can draw three lines through any given point such that they all intersect at that point and are mutually perpendicular. A helpful way to visualize this is to imagine a corner of a room where two walls meet the ceiling. One line is formed by the intersection of the two walls and runs vertically, while the other two lines, formed by the intersections of the walls with the ceiling, run horizontally. At the point where they intersect, these lines are mutually orthogonal, each forming a right angle with the other two. This property holds true for every point in our Universe; we can always find exactly three mutually perpendicular lines.

If it were possible to find four mutually perpendicular lines at a point, we would be dealing with a four-dimensional space, or hyperspace. Conversely, if we could find only two such lines, we would have a two-dimensional space; for example, the surface of the Earth represents a two-space.

On the surface of a sphere, only two lines can be found that are perpendicular at any one point. Thus, we can define the concept of dimensionality based on the number of mutually perpendicular lines that can be drawn in a given space. A space is considered n-dimensional if, through any point in that space, we can find exactly n lines that are mutually perpendicular.

Two observations arise from this definition. First, n will always be a whole number; we will never have a space with fractional dimensions. Second, we will assume that a space does not vary in dimensionality from one location to another; if one point accommodates n mutually orthogonal lines, then we assume this property holds for all points in the Universe.

Another way to define the dimensionality of a Universe is by considering how many coordinates are necessary to pinpoint a given point. For instance, to identify a point on the surface of the Earth, we need two coordinates: latitude and longitude.

When considering all of space surrounding the Earth, we must introduce a third coordinate to uniquely determine every point in the Universe. A two-dimensional space requires two coordinates for precise location, while a three-dimensional space requires three. Generally, an n-dimensional space will necessitate exactly n different numbers to establish the unique position of a point. Consequently, different points will correspond to different combinations of coordinate values, ensuring that each point has a unique set of numbers identifying it. Both the orthogonal-line approach and the coordinate scheme can be used to define dimensionality.

11.33 The Cartesian System

There are various types of coordinate systems, but the simplest and most commonly used is the **Cartesian coordinate system**. Named after the French mathematician René Descartes, this system is created by constructing sets of mutually perpendicular number lines.

In a two-dimensional Cartesian plane, any point is defined by an ordered pair of numbers, such as (3, -2). In a three-dimensional Cartesian coordinate system, three number lines—referred to as the x, y, and z axes—intersect perpendicularly at the origin point, defined by the ordered triple (0, 0, 0). The angle between the x and y axes is exactly 90 degrees, which also holds true for the angles between the y and z axes and the x and z axes. Consequently, any point in three-dimensional space can be represented by a unique ordered triple.

The Cartesian system can be extended to any number of dimensions. In n-dimensional hyperspace, the system requires exactly n mutually orthogonal lines, known as the $x_1, x_2, ..., x_n$ axes. Any point in this n-dimensional space is represented by an ordered n-tuple $(x_1, x_2, ..., x_n)$.

11.34 Polar Coordinates

There are alternative methods for establishing the uniqueness of a point beyond the Cartesian system. For example, on the spherical surface of the Earth, using a Cartesian coordinate system is impractical due to "geometric distortion." While it works well for small areas, it becomes complicated when applied across the entire surface of a sphere.

To pinpoint geographic locations on the Earth's surface, we utilize a system based on angles instead of number lines, specifically latitude and longitude coordinates. To define points in three-dimensional space, we introduce a radius coordinate. This allows any point in space to be uniquely represented by two angles and a distance vector. Another type of three-dimensional coordinate system, known as the polar coordinate system, employs one angle and two distance values. To locate a given point *P*, we identify the angle *Theta* or Θ with respect to a reference line and a radius *r*, which defines a point *P'* on the "image plane" directly above or below *P*. We then measure the distance *d* from the "image plane" to *P*.

From either the polar or Cartesian three-dimensional systems, we can create a four-dimensional coordinate system by adding an additional distance vector. This process can continue indefinitely, allowing for the construction of five or more dimensions through the addition of more distance vectors.

While visualizing a four-dimensional coordinate system is essentially impossible for us as three-dimensional beings, we can grasp some properties of hyperspace by considering time as the fourth dimension. Any two-dimensional space that deviates from perfect "flatness" is termed non-Euclidean, as the principles of plane geometry no longer hold. Two-dimensional creatures confined to their own space, unable to conceive of a three-dimensional existence, would struggle to determine whether their universe is Euclidean. In contrast, we have discovered that our three-dimensional universe is non-Euclidean. Generally, this deviation from the "flatness" of Euclidean geometry is minor, but under certain conditions—such as intense gravity or vast distances—our universe can be significantly non-Euclidean. It is gravity that causes this curvature of space.

11.35 Time as the Fourth Dimension

We briefly explored some models of coordinate systems that use time as the fourth dimension. Theoretically, time fits this role perfectly. We can envision time as continuously "flowing" or "moving" along a line, to which we can assign coordinates. This approach is often employed in history textbooks to help us compare significant events, such as the Civil War and the decline of the Roman Empire or the evolution of the solar system.

We have the flexibility to choose intervals of any length for the coordinates on a timeline. In historical examples, we might use years or centuries, while for other contexts, we may prefer days, hours, or even minutes. For more precise applications, we might opt for microseconds or nanoseconds (units of 10^{-6} or 10^{-9} seconds) as our base units. Time is indeed a continuum, allowing us to specify units in whatever magnitude suits our needs.

The constancy of the speed of light provides theoretical justification for correlating time with spatial distance. For instance, an interval of one second corresponds to a spatial distance of 300,000 kilometers. Similarly, an interval of one year equates to a distance of one light year, and this relationship extends to other time intervals as well.

11.36 Centrifugal forces

We are all familiar with the so-called "force that does not exist." This refers to the apparent outward force experienced by a rotating or revolving mass, often termed centrifugal force. While this "force" is not considered theoretically real, it does produce observable effects. A passenger in a centrifuge experiencing 10 g's might challenge the claim that centrifugal force is nonexistent.

Rapidly rotating planets, like Jupiter and Saturn, exhibit noticeable oblateness due to this effect. For instance, Jupiter, which has a rotational period of less than 10 hours and a predominantly gaseous outer layer, is distinctly flattened—a feature that is evident even in photographs of the planet.

11.37 Summary of Einstein's architecture of Space and Time

Sound travels through the atmosphere by conduction at a fixed speed. When a sound source is stationary relative to the surrounding air, sound waves propagate equally in all directions. However, if the source moves through the air, sound waves travel away from the source more quickly in the direction opposite its motion and more slowly in the direction of its motion. The speed of sound remains constant with respect to the air, regardless of the motion of the source. This raises the question: does the same apply to light? Experiments were conducted to investigate this.

One experiment aimed to measure the potential effect of Earth's motion relative to distant stars. Light from a star was observed at two different times: first, when Earth was moving toward the star in its orbit around the Sun, and six months later, when it was moving away. The hope was that this and similar experiments would reveal Earth's absolute motion within the Universe, assuming the ether was at rest and provided an absolute standard for motion.

However, every attempt to detect Earth's motion relative to the hypothesized ether yielded negative results. In response, scientists created complex explanations to maintain the idea of a conductive medium in the Universe, but these were inconsistent with observed facts. The ether theory's primary aim was to establish an absolute standard of motion; as it consistently failed experimental validation, it was ultimately abandoned. Consequently, researchers needed to find a new explanation for the constancy of the speed of light.

11.38 Theoretically faster than Light

One significant consequence of the fundamental axiom of relativity, as we will explore in later chapters, is that no material object can achieve the speed of light. Regardless of the amount of energy expended to accelerate an object, its speed will always fall short of light speed. Even with the most advanced spacecraft engines, we will never reach this ultimate velocity (speed); the speed of light remains the fastest possible speed in our Universe.

Scientific theories typically start with general statements known as "postulates," which form the foundation of the theory. From these postulates, we derive mathematical laws in the form of equations that relate various physical variables. These predictions are then tested in laboratory settings. A theory remains valid until experimental evidence contradicts it, prompting modifications or replacements of the initial postulates, thereby restarting the cycle.

For nearly two centuries, the mechanics established by Galileo and Newton withstood rigorous experimental scrutiny. These mechanics were based on the absolute nature of space and time. However, through a thought experiment involving the pursuit of a light beam, Einstein recognized the necessity of revising the Galilean laws of relative motion. In his 1905 paper, "On the Electrodynamics of Moving Bodies," Einstein introduced two postulates that serve as the foundation for his special theory of relativity:

1. **The principle of relativity:** The laws of physics are the same in all inertial reference frames.
2. **The principle of the constancy of the speed of light:** The speed of light in free space is constant at c in all inertial reference frames.

The first postulate asserts that the laws of physics are absolute and universal, applying equally to all inertial observers. Laws that hold true for one inertial observer cannot be violated by any other inertial observer. The second postulate is more challenging to accept, as it contradicts our intuitive understanding, which is grounded in Galilean kinematics learned from everyday experiences.

To illustrate, consider three observers—A, B, and C—each at rest in different inertial reference frames. When observer A emits a flash of light, they observe it traveling at speed c. Observer B, moving away from A at a speed of c/4, would, according to Galilean kinematics, measure the light's speed as c−c/4=3c/4. Meanwhile, observer C, moving toward A at c/4, would calculate the speed of light as c+c/4=5c/6. However, Einstein's second postulate asserts that all three observers measure the same speed ccc for the emitted light pulse.

11.39 The Special Theory of Relativity

Since Newton's era, it has been widely accepted that all frames of reference rely on individual frameworks of rods, lines, or other systems of markers to identify locations in space. However, it was assumed that all observers naturally utilize the same universal time. This assumption is so intuitive that it is often taken for granted and not explicitly stated.

Einstein challenged this notion by highlighting that comparing clocks, regardless of their motion, is straightforward only when all velocities involved are significantly smaller than the speed of light—approximately 186,000 miles per second (or 300,000 kilometers per second). Because no method of signaling is known to transmit information faster than light, new complications arise when comparing clocks moving at, for instance, 100,000 miles per second. To compare the rates of two clocks, at least two readings from each clock are necessary; without this, one cannot be certain that their rates are equal.

One of the key insights from Einstein's analysis was the concept of the relativity of simultaneity: two events occurring at widely separated locations may seem simultaneous to one observer but may not be perceived as such by an observer in a different state of motion.

Einstein at age 25
In courtesy of the California Institute of Technology and The University of Jerusalem

Next, Einstein established a new framework for relating time and distance measurements between two distinct inertial frames of reference. This framework is known as the Lorentz transformation. He further demonstrated that if these newly proposed relationships are valid, Maxwell's laws of the electromagnetic field can be fully applicable in both inertial frames. In doing so, he resolved the paradox between the system of equivalent inertial frames in Newtonian mechanics and the concept of an absolute rest frame that seemed necessary in Maxwell's theory.

The special theory of relativity once again treats all inertial frames of reference as equivalent, thereby reinstating the principle of relativity established by Newton in the 17th century. However, in this framework, the concept of universal time, along with the idea of a universal rate of clocks and a universal length for yardsticks, had to be discarded. Initially, the term "principle of relativity" referred to the relative equivalence of inertial frames, but it soon evolved to encompass discussions about the "relativity of time" and the "relativity of length."

When two complete frames of reference are in relative motion, each observer perceives the clocks and yardsticks of the other as contracted in the direction of motion, while remaining unaffected in the perpendicular direction.

In essence, the special theory of relativity replaces the notions of "absolute length" and "absolute time," which were abandoned, with a new concept often referred to as the invariant or proper "space-time interval." For two events occurring at distant locations, their spatial separation is not absolute, even within the Newtonian framework; only the time lapse between them can be considered absolute. If the two events are not simultaneous, an observer moving with their entire frame of reference might cover a distance between the occurrences of the two events, leading them to perceive them as happening at the same location, while they may appear far apart to an observer traveling in a different direction.

11.40 Minkowski's Four-Dimensional World

Before the advent of relativity, space and time were considered entirely separate domains—space consisting of three dimensions and time just one. This distinction was justified because time was thought to be independent of any spatial relationships. Most observational data available to scientists before the late 19th century involved motions that were so slow compared to the speed of light that the latter could be regarded as almost infinite. The main exception to this was the study of the eclipses of Jupiter's satellites, which led the Danish astronomer Olaus Roemer (1644-1710) to discover that light has a finite speed. By observing that the periodicity of these eclipses was delayed or advanced depending on whether Earth and Jupiter were on opposite sides of the Sun or on the same side, Roemer deduced that light takes nearly 20 minutes to travel a distance equal to the diameter of Earth's orbit around the Sun. This travel time, when combined with his best estimate of the speed of light, was less than 20 percent off the value accepted today, based on more precise experiments.

Roemer's discovery dates back nearly as far as Newton's formulation of a comprehensive theory of mechanics. For the following two centuries, astronomers corrected their observations to account for the finite time it takes for light to travel from the event to the observer on Earth. During this period, there was no concern about whether the speed of light should be measured relative to its source, the Earth, or some other reference point like the Sun. This was because the speed of light is so much greater than any velocities encountered within the solar system—approximatelimage)y 10,000 times faster than Earth's orbit around the Sun—that any minor uncertaintie9s or errors in accounting for light's transit time were insignificant compared to other sources of observational error. Roemer's findings certainly did not challenge the notion that each event in the universe could be assigned a specific time; time was regarded as universal, independent of the observer's motion and any specific frame of reference.

However, the theory of Special Relativity undermined the universality of time and space measurements, replacing them with the concept of an invariant interval. This invariant interval provides a consistent numerical relationship for any two events separated by both space and time, yielding the same value across all frames of reference. In 1908, Hermann Minkowski (1864-1909), a Russian-born mathematician, proposed that space and time should no longer be treated as separate continua but should instead be merged into a single four-dimensional framework known as the "space-time continuum." In this unified continuum, the invariant interval serves a role analogous to ordinary distance in three-dimensional space.

Minkowski

In making this proposal, Minkowski did not aim to alter the special theory of relativity, which was only three years old at the time. Instead, he offered a fresh interpretation of the theory that has proven invaluable in the years since. Formally, it is always possible to combine two continua (referred to as manifolds by mathematicians; this term suggests the connectedness of the points within the continuum) into a single entity. This is achieved by considering a point from the first manifold alongside a point from the second manifold as a point in the new ("product") manifold.

To view space and time not as separate continua but as dimensions of a unified continuum or manifold—often referred to as space-time, the world, or the Minkowski Universe—provides a coherent framework, particularly since none of these axes remain constant during transitions between inertial frames of reference. Minkowski's approach proves useful because it encourages the search for concepts that are "natural" within four-dimensional space-time, independent of any separation into distinct space and time. In any space-time diagram, the history of a particle and its movements over time is depicted by a curve. From Minkowski's perspective, this curve represents the only natural depiction of that history, with essential characteristics that remain unchanged across different inertial frames. This is known as the world curve of the particle, which must be time-like throughout; thus, in any inertial frame, the particle cannot attain or exceed the speed of light.

Minkowski's geometry fundamentally differs from conventional Euclidean geometry in that the primary relationship between any two points can be time-like, space-like, or light-like. Consequently, the square of the invariant interval between these points can be positive, negative, or zero.

Similar to Euclidean distance, the values of Minkowski's invariant interval are independent of the choice of coordinate system (or frame of reference) and represent intrinsic geometric relationships. The world points in the Minkowski Universe are described using a coordinate system with four axes: one representing the conventional time scale and the other three corresponding to ordinary spatial coordinates. In conventional space, the preferred coordinate systems are known as Cartesian coordinates, named after the French mathematician René Descartes (1596-1650), where all three axes are straight and mutually perpendicular.

In Minkowski geometry, there are also preferred four-dimensional coordinate systems for observers in uniform motion, using Cartesian coordinates for spatial reference and standard clocks for timing events. These are known as Lorentz coordinate systems or "Lorentz frames of reference." The space-time interval formulas discussed in the preceding chapter are based on Lorentz frames. In ordinary geometry, transitioning from one Cartesian coordinate system (x and y) to another (x-prime and y-prime) involves various types of transformations. One type, known as a translation, shifts the origin of the coordinate system (where all values are zero, marked as 0 or 0-prime) without altering the directions of the axes. Another involves rotating the coordinate axes around the unchanged origin, with the angle of rotation denoted as theta. Combining a translation with a rotation results in a general type of coordinate transformation that leads from one Cartesian system to another, referred to as an orthogonal transformation.

In the Minkowski Universe, transitioning from one inertial frame to another—i.e., moving from one Lorentz frame to another—is called a Poincaré transformation. The most general Poincaré transformations include translations of the time axis origin, the space coordinate origins, and changes in the observer's (unaccelerated) state of motion.

If only the latter occurs, with the time and space origins unchanged, it is termed a Lorentz transformation. In this context, Lorentz transformations are a special subset of Poincaré transformations. A quantity or relationship that remains unchanged under a Poincaré transformation is referred to as Poincaré invariant, while those primarily concerned with changes in the observer's velocity are called Lorentz invariant. Mathematically, physicists seek Poincaré-invariant (or Lorentz-invariant) relationships.

In the subsequent development of relativity theory and its application across various areas of physics, this mathematical framework has demonstrated considerable heuristic value. Modern physics would be unthinkable without Minkowski's four-dimensional perspective. Nonetheless, discussions about relativity can also be framed in a way that separates the three dimensions of space from the time axis. Einstein's early papers were written in such a manner; these works (1905-1908) laid the groundwork for the relativistic theory of electricity, magnetism, and much of relativistic mechanics. Given that spatial and temporal measurements require different instruments, separating space and time is crucial for designing any experimental or observational methods. The three-plus-one and four-dimensional representations in relativity complement one another.

At every world point in the Minkowski Universe, the full range of directions in space and time forms a "pencil" of arrows radiating from that point. Some of these directions are light-like, while others are space-like. The light-like directions form two cones, known as light cones, associated with that world point. Each cone has its interior, composed entirely of time-like directions.

The directions within one light cone point toward the future, forming what is called the "future light cone," while the other is referred to as the "past light cone."

A Poincaré transformation alters a space-time direction into another direction without changing its fundamental nature. It consistently transforms a time-like direction into another time-like direction, while preserving the characteristics of light-like and space-like directions. Since all light-like directions remain light-like, the integrity of both light cones is maintained; each light-like direction is merely shifted within its own light cone. These cones of directions, which are independent of the chosen frame of reference, have no equivalents in ordinary Euclidean space. They exist solely within "Minkowski geometry" because the invariant interval, which otherwise resembles the distance between points in Euclidean spaces, is zero between certain pairs of points, making the direction from one point to another "special" and light-like.

Light cones encompass all the directions along which light pulses can travel. Specifically, the past light cone includes all directions in space-time from which light-borne information can reach an observer located at the world point where the cone is centered. Consequently, each direction on the past light cone can be associated with a point on the celestial sphere, representing our perspective of the sky and stars, much like a globe depicts our view of Earth.

11.41 The Twin Paradox

The previous examples of relativistic time distortion illustrate applications of the special theory of relativity. For some, the concept that time can be "stretched" is difficult to grasp. They may dismiss it as an illusion or merely a clever construct of a mathematician's imagination. However, this phenomenon has been experimentally observed, and we cannot ignore the results of these experiments. The idea of time distortion also leads to an intriguing paradox known as the "Twin Paradox" or "Clock Paradox."

11.42 Mass, Energy and Momentum

Relativity emerged from the foundations of electromagnetic theory, and since the electromagnetic field interacts with electrically charged particles, relativity had to address particle behavior while integrating concepts from Newtonian mechanics.

According to Newton's laws, two interacting bodies exert equal and opposite forces on each other, resulting in changes to their velocities that are inversely proportional to their respective masses. If one of the interacting particles is moving at a velocity close to the speed of light, relativity states that nothing can exceed this limit, implying that some inherent mechanism must prevent any speed from surpassing light. This necessity for a speed limit requires modifications to Newtonian mechanics for compatibility with relativistic principles.

Furthermore, Newtonian mechanics posits that when two bodies interact, their velocities adjust in such a way that their common center of mass—often referred to as the center of gravity—maintains a constant velocity. The law of inertia, which states that an object will not change its velocity unless acted upon by external forces, applies not only to elementary particles but also to larger bodies made up of numerous interacting particles. For this law to hold in relativistic physics, the motion of an unaccelerated center of mass in one Lorentz frame must have a constant velocity across all Lorentz frames. However, because velocities transform differently under Poincaré transformations compared to Newtonian physics, the common center of mass will not automatically satisfy the requirement of unaccelerated motion across all Lorentz frames. Einstein discovered that to reconcile this issue, he had to assume that a particle's mass is dependent on its velocity and thus varies across different Lorentz frames; the faster a particle moves, the greater its mass becomes. The mass of a moving particle is found to exceed its rest mass by an amount equal to its kinetic energy divided by the square of c.

This new velocity-dependent mass ensured that no particle could be accelerated to a speed greater than that of light, hinting at the profound equivalence of mass and energy captured in the equation $E=mc^2$, a formula that has since been confirmed far beyond its original theoretical prediction. Elementary particles, the fundamental building blocks of matter, can be transformed into radiant energy directly or through intermediary processes. The small discrepancies between the masses of atomic nuclei and their constituent particles represent energy—the so-called binding energy of atomic nuclei—which can be released during nuclear reactions, either gradually in nuclear reactors or explosively in nuclear weapons.

The mass of an object measured in the Lorentz frame where it is at rest is known as its rest mass. This intrinsic property of a physical body differs from its relativistic mass, which depends on the relative motion between the object and the observer. Although the sum of the relativistic masses of several interacting bodies remains constant throughout the interaction, the rest masses can change. For instance, when an atomic nucleus undergoes radioactive decay into smaller structures while emitting a gamma ray (a high-energy electromagnetic pulse with a wavelength shorter than visible light or even X-rays), the rest masses of the resulting products will be less than that of the original nucleus. However, the total of the relativistic masses, including the relativistic mass of the emitted gamma ray, will equal that of the nucleus before decay.

A quantity that remains unchanged over the course of a physical system's history—when isolated from external influences—is said to be conserved. In Newtonian physics, both mass and energy are conserved quantities, and any principle asserting that a quantity, such as mass or energy, remains constant over time is termed a "conservation law."

In relativistic physics, energy and relativistic mass are equivalent, though measured in different units. Specifically, the mass unit is larger by a factor of c^2 than the energy unit. Consequently, a relatively small amount of mass in a nuclear bomb (less than one percent of the active material in a thermonuclear device) corresponds to an enormous release of explosive energy when it is unleashed.

Under certain conditions, a body can perform work by converting part or all of its rest mass into available energy, independent of any potential or kinetic energy it may possess. Therefore, it is reasonable to view an object's energy as comprising potential, kinetic, and rest energy, the latter being c^2 times its rest mass. The total energy of a physical object is referred to as its total "relativistic" energy, which equals c^2 times the object's relativistic mass.

Is there a law relating the energy measured in one frame of reference to that in another? In both Newtonian and relativistic mechanics, the answer is yes. A mathematical relationship exists between the energy values in the two frames, involving both the relative velocity between the frames and another quantity: linear momentum. Linear momentum is defined as the product of an object's mass and velocity, serving as a measure of the impact that object would have when colliding with a stationary body. Linear momentum is a vector quantity, possessing both direction (the direction of the particle's velocity) and magnitude, and it can be conveniently expressed in its three components aligned with the three coordinate axes in a Cartesian coordinate system.

Given the conservation of mass, Newton's law of inertia can also be articulated as the linear momentum of an isolated system remaining constant over time. This constancy persists despite internal interactions, as the changes in linear momentum induced by any two constituents cancel each other out due to the principle that the forces causing these changes are equal and opposite.

In standard three-dimensional space, the three components of a vector, denoted as a, b, and c, combine to determine its magnitude V using the equation:

$$V^2 = a^2 + b^2 + c^2,$$

as described by Pythagoras' theorem. When comparing two different Cartesian coordinate systems, the vector's components will vary between them; however, the sum of their squares will remain constant. In mathematical terms, this means that the magnitude V is *invariant* under orthogonal coordinate transformations—those transformations that convert one Cartesian coordinate system into another.

11.43 Time is Distance

The constant ccc allows us to define time in relation to spatial separation, and vice versa, through the simple equation

$$d = t \cdot c,$$

where d denotes distance and t denotes time. For instance, we can interpret a distance of 300,000 kilometers as corresponding to a time interval of one second, and conversely, one second can be viewed as a spatial separation of 300,000 kilometers. This concept helps us visualize time as a fourth dimension, through which our Universe moves at velocity c into the future.

This model effectively illustrates certain space-time phenomena: any two events occurring 300,000 kilometers apart—regardless of the orientation of the connecting line—are separated by a time interval; similarly, any two events occurring one second apart in time are also exactly 300,000 kilometers apart in distance. Thus, the speeds of time and light can be considered equivalent.

11.44 Addition of Velocities

The special theory of relativity provides a framework for mathematically evaluating the addition of large velocities. When dealing with velocities that are much smaller than the speed of light, we can simply add them together.

11.45 Time Dimensions

The inability to accelerate a material object to or beyond the speed of light arises from the fact that its mass increases infinitely as it approaches that speed, requiring an infinite amount of energy to reach it. While constructing a spacecraft capable of exceeding light speed is unlikely in our three-dimensional universe, it is intriguing to consider other universes where speeds greater than light are possible. Although this remains a mathematical concept rooted in intuition, we cannot dismiss the possibility of additional "time dimensions." Our universe comprises three spatial dimensions, and there is compelling cosmological evidence suggesting that space may have four or more dimensions. So, why couldn't there be multiple, or even infinite, dimensions of time? Even if we can't observe or detect these other time dimensions, envisioning them is fascinating, particularly as they offer an escape from the absolute limitation imposed by the speed of light.

Renowned British physicist Stephen Hawking explored whether time had a beginning or if it extends infinitely backward. He pondered whether the cosmic clock had a first tick or had been ticking forever. In collaboration with James Hartle from the University of California at Santa Barbara, Hawking proposed a model suggesting that time could be activated quantum-mechanically at the moment of the Big Bang.

Their approach combined Einstein's concepts of time and space with quantum physics in a compelling way. Hawking was upfront about the speculative nature of the Hartle-Hawking theory, acknowledging its somewhat tenuous foundations, yet it represents a sincere attempt to address what may be the ultimate scientific challenge.

A central aspect of their theory is what Hawking refers to as "imaginary time." Unfortunately, many misunderstand this term to imply something mystical, akin to "the time of the imagination." Others interpret it as a type of time that exists only in thought, distinct from the "real" time of experience. In reality, "imaginary" is used in a technical mathematical sense that is entirely unrelated to imagination.

In school, we learn to square numbers: for instance, the square of 2 is *2×2=4* and the square of 3 is *3×3=9*. Taking the square root reverses this process; the square root of 4 is 2, and the square root of 9 is 3. More advanced students encounter squaring negative numbers, where multiplying two negative numbers yields a positive result (e.g., *(−3)×(−3)=9)*. Consequently, there are two numbers whose squares yield *9: 3* and -3. Conversely, the square root of 9 must be either 3 or -3.

A challenge arises when attempting to take the square root of a negative number, such as -9. Since both negative and positive numbers squared produce a positive number, no standard number can be squared to yield a negative number. To address this, mathematicians created new numbers in the 16th century, termed "imaginary." This label is not due to their lack of reality compared to "ordinary" numbers but rather because they do not appear in everyday arithmetic, such as counting sheep or money. The term "imaginary" is merely mathematical jargon. There are also irrational, transcendental, real, complex, rational, transfinite numbers, and vulgar fractions. These names are historically significant.

So, what does this have to do with time? The connection traces back to Hermann Minkowski's work, which established that a single "space-time" continuum emerges from Einstein's special theory of relativity. Minkowski treated time as a fourth dimension, similar to space but with distinctions in how time and space are integrated into the space-time description.

To understand this difference, we need to examine the concept of distance in space-time. The distance between two points in space is well-defined as the length of a ruler connecting them in a straight line. The interval in time between two events is equally straightforward, represented by the time difference recorded on a clock at rest within the relevant reference frame. However, when space and time are unified in a single space-time framework, the question arises: how do we reconcile the two?

11.46 The Light Barrier

Some electrically charged particles can travel very close to the speed of light. According to the special theory of relativity, a subatomic particle cannot exceed the speed of light in a vacuum. However, light travels more slowly in air, allowing a nuclear particle to potentially exceed light speed within the atmosphere, albeit slightly slowed by the medium. When such a charged particle moves faster than the speed of light in air, it generates an electromagnetic shock wave similar to a sonic boom, but with light instead of sound. This phenomenon is known as "Cherenkov radiation," named after its Russian discoverer. The distinctive angle of its beam makes Cherenkov radiation easy to identify.

In essence, the concept of traveling faster than light can imply "backwards" in time, introducing various puzzles and paradoxes. It's important to note that the theory of relativity does not state that "nothing can go faster than light," as is often claimed. While it prohibits objects from traveling at superluminal speeds in a vacuum, it does not prevent them from existing or operating at speeds that would otherwise be considered faster than light. According to Einstein's theory, nothing can breach the "light barrier" by accelerating past or decelerating below the speed of light.

Physicists have coined the term "tachyons" for hypothetical superluminal particles, derived from the Latin word for "speed." Roger Clay and his colleagues believed they had discovered tachyons. Although the theory of relativity does not categorically rule out the existence of tachyons, they present a challenge for physicists because they could potentially be used to send signals into the past. While Ms. Right cannot physically travel into the past without violating relativity, she might be able to manipulate tachyons to send a message backward in time.

Physicists often describe light-speed particles in terms of their energy rather than their velocity. This approach arises from the fact that, due to the light-speed barrier, all very fast particles move at nearly the same speed—just a fraction below the speed of light in a vacuum. Therefore, one particle may possess ten times the kinetic energy of another similar particle while moving only slightly faster.

11.47 Expanding Space

All stars in the universe emit a unique spectrum. When the light from any star passes through a prism, it spreads into a rainbow, revealing specific spots or frequencies where no light appears. These gaps are known as absorption lines, created by the atoms of certain elements that prevent the emission of discrete frequencies. The pattern of these absorption lines is consistent across all stars and is easily recognizable.

Sometimes, this pattern can shift either upward or downward in frequency. An upward shift, or blue shift, indicates that the star is moving toward us, while a downward shift, or red shift, signifies that it is moving away. Most stars and galaxies are in motion relative to us, resulting in some degree of spectral shift based on their movement along our line of sight.

In very distant galaxies, the spectra predominantly show a red shift, suggesting they are receding from us. Generally, the farther away a galaxy is, the more pronounced the red shift.

Now, imagine if we could travel in four dimensions, beyond our three-dimensional continuum. In this hypothetical "four-space" or "hyperspace," which is difficult to visualize, we could arrange four straight rods so that each one is perpendicular to the other three.

11.48 Time Distortion

We have observed that the apparent rate of time can be altered by Doppler effects when an object is moving toward or away from us. However, at sufficiently high speeds—specifically, at a significant fraction of the speed of light—a different kind of time distortion occurs, known as "relativistic time distortion" or "time dilation." Although this effect is not noticeable until extreme speeds are reached, it theoretically exists at any speed, even when simply walking down the street. This phenomenon fundamentally challenges the concept of "simultaneity" on a cosmic scale, particularly among objects that are not perfectly stationary relative to one another.

Relativistic time distortion increases as the relative speed between two objects rises. At high velocities, this distortion can be significant, with factors of 2, 3, or even 100 possible. In theory, there is no upper limit to how great this factor can become; a thousand years could be compressed into a mere second. The time-distortion factor approaches infinity as the relative speed nears the speed of light. At light speed, according to the equations of relativity, time effectively comes to a standstill, meaning the speed of light is equivalent to the speed of time itself.

The special theory of relativity, introduced in 1905, marks the conclusion of the classical period and the dawn of a new era. It builds upon established classical concepts of matter spread across space and time, as well as deterministic laws of nature. Yet, it also introduces revolutionary ideas about space and time, fundamentally critiquing traditional notions as defined by observable phenomena. This critique represents Einstein's most significant achievement, setting his work apart from that of his predecessors and differentiating modern science from classical science.

Even prior to Einstein's contributions, investigations into the physical world had begun to push the boundaries of human perception. Scientists were aware of invisible forms of light, such as ultraviolet and infrared, and inaudible sounds. They worked with electromagnetic fields in empty space, which were imperceptible to the senses and could only be indirectly observed through their effects on matter.

These generalizations became necessary as the limitations of direct sensory impressions were recognized. For example, the sensations of hot and cold were not precise enough for a comprehensive theory of heat, leading to the use of thermometers to measure thermal differences as changes in a mercury column.

There are countless instances where one sense has been replaced or complemented by another, highlighting a web of interconnections within science. This is why geometric structures perceived through vision or touch are often favored; they are deemed the most reliable observations, as independent of the individual observer as possible. For example, electromagnetic fields, which cannot be directly perceived, have been understood by translating them into measurable mechanical quantities in terms of space and time.

The overarching question of whether a direction or point in space, or a moment in time, is absolute was addressed by Newton's famous axioms. His statements unequivocally affirm this idea. However, his equations of motion present a contradiction: they reveal that there exist certain equivalent reference frames in relative motion, each of which can justifiably be considered at absolute rest. Therefore, Newton's concept of space is absolute only in a limited sense. Further research, particularly in the fields of electromagnetism and optics, has exposed additional and more profound challenges to the Newtonian perspective.

Einstein shattered existing paradigms through a critical evaluation of contemporary ideas about space and time. Finding them inadequate, he introduced more satisfactory concepts, adhering to the foundational principles of scientific research: objectivization and relativization. He also incorporated another principle that had been recognized prior but was primarily used for logical criticism rather than for scientific construction. This principle, championed by physicist and philosopher Ernst Mach—who greatly influenced Einstein—argued that concepts and statements lacking empirical verification should have no place in a physical theory.

Einstein analyzed the simultaneity of events occurring at different locations in space and identified it as a non-verifiable notion. This realization led him, in 1905, to reformulate the fundamental properties of space and time. About a decade later, he applied the same principle to motion under gravitational forces, which guided him in developing his theory of general relativity.

The principle advocating for the elimination of the unobservable has sparked considerable philosophical debate. Often referred to as "positivistic," it aligns with the philosophy championed by Mach. Positivism maintains that only immediate sensory impressions are real, viewing everything else as a mental construct, which fosters a skeptical attitude toward the existence of an external world. This stance was far removed from Einstein's beliefs; in later years, he explicitly opposed positivism. Instead, his method should be seen as a heuristic principle highlighting the weaknesses of traditional theories that proved empirically unsatisfactory. This approach has emerged as a crucial method in fundamental research in modern physics, especially in the development of quantum theory.

Consequently, Einstein's way of thinking not only marked the peak of the classical period but also ushered in a new era in physics.

Now, consider two bodies, S_1 and S_2, made of the same deformable (fluid) material and of equal size, positioned in astronomical space at a distance where their ordinary gravitational effects on one another are negligible. Each body exists in equilibrium due to the gravitational interactions among its components and other physical forces, resulting in no relative motion between its parts. However, the two bodies are in a state of relative rotational motion with constant velocity around the line connecting their centers. From the perspective of an observer on body S_1, body S_2 appears to rotate uniformly relative to his reference frame, and vice versa. Now, suppose each observer assesses the shape of the body they occupy; it turns out that S_1 is a sphere, while S_2 takes the form of a flattened ellipsoid of rotation.

Newtonian mechanics would suggest that the differing shapes of the two bodies imply that S_1 is at rest in absolute space while S_2 is undergoing an absolute rotation. The flattening of S_2 can be attributed to "centrifugal forces." This example illustrates how absolute space is introduced as a (fictitious) cause. S_1 cannot be responsible for the flattening of S_2 since both bodies are in the same condition relative to each other and thus cannot deform differently. Relying on space as a cause does not satisfy logical requirements regarding causality. Since we only have centrifugal forces to indicate its existence, we support the hypothesis of absolute space solely based on the very phenomena it was meant to explain. Sound epistemological criticism rejects such artificially constructed hypotheses, as they are overly simplistic and contradict the goal of scientific inquiry, which aims to establish criteria for distinguishing scientific results from mere fanciful notions. This critical selection of "admissible causes" sets apart a reasoned approach to understanding the world, as pursued by physical research, from mysticism, spiritualism, and other manifestations of unchecked imagination. In fact, the concept of absolute space can be seen as almost "spiritualistic." When we ask, "What causes centrifugal forces?" the answer is simply, "Absolute space." However, if we question what absolute space is or how it manifests, we find that no one can provide an answer beyond stating that absolute space is the cause of centrifugal forces without any further properties. This consideration indicates that space, as a cause of physical occurrences, must be excluded from our understanding of the world.

The notion that the totality of distant masses must account for centrifugal forces was first articulated by philosopher-physicist Ernst Mach, whose ideas significantly influenced Einstein. There is no experience to contradict this view, as the reference system used in astronomy for determining the rotations of celestial bodies is deliberately chosen to be at rest relative to the entire stellar system. More accurately, this reference frame is configured so that the apparent motions of fixed stars relative to it appear irregular and lack a preferred direction. The flattening of a planet increases with its rotational velocity concerning this reference frame aligned with distant masses.

Consequently, we must require that the laws of mechanics—and physics in general—depend solely on the relative positions and motions of bodies. No reference system should be favored a priori, as seen with the inertial systems of Newtonian mechanics and Einstein's special theory of relativity. Otherwise, absolute accelerations relative to these preferred reference systems, rather than just the relative motions of bodies, would become part of physical laws. Thus, we arrive at the postulate that the laws of physics must be applicable in precisely the same way across reference systems that are in arbitrary motion, marking a significant extension of the principle of relativity.

11.49 The Rotational Acceleration Field

We are all familiar with the inverse-square gravitational field that surrounds all material objects. The force exerted on a given mass diminishes as the distance from the source increases. This relationship is expressed by the following equation:

$$f = ma = mk/r^2,$$

where m represents the mass of the object being affected, r is the distance from the gravitational source, and k is a constant that varies based on the mass of the gravitational source. This relationship is referred to as the "inverse square."

11.50 Acceleration Distorts Space

When two different coordinate systems are stationary relative to each other, or when their relative velocity is constant, the transformation between them is quite straightforward. In the thought experiment, as long as Jim and Joe maintain a constant relative velocity, straight lines in Jim's universe will also appear straight in Joe's universe, and vice versa. While objects may seem spatially distorted, a straight rod in Jim's ship will still look straight to Joe, regardless of its orientation, though its length may not appear the same. When all straight lines in one coordinate system remain straight in another, we say the two systems are connected by a linear transformation; otherwise, the transformation is nonlinear.

Now, imagine that the velocity between the vessels is not constant. In this case, straight lines in one universe may not always appear straight in the other reference frame. This means Jim's straight rod might appear bent to Joe, and Joe's straight rod might appear bent to Jim. In this scenario, the transformation becomes nonlinear, but there is still a homeomorphism between the two spaces.

11.51 The Principle of Equivalence

The fulfillment of this postulate necessitates a completely new formulation of the law of inertia, as this is what gives inertial systems their privileged status. Inertia should no longer be viewed as an effect of absolute space but rather as a consequence of interactions with other bodies.

Currently, we recognize only one type of interaction between material bodies: gravitation. Moreover, experiments have revealed a remarkable relationship between gravitation and inertia, encapsulated in the law stating that gravitational mass is equal to inertial mass. Therefore, these two phenomena—attraction and inertia—that appear distinct in Newtonian physics must share a common origin. This insight represents Einstein's significant discovery, transforming the general principle of relativity from an "epistemological" postulate into a precise scientific law.

We can characterize the subject of the forthcoming investigation as follows: In classical mechanics, the motion of a massive body (not subjected to electromagnetic or other forces) is governed by two factors: (1) its inertia, which resists acceleration relative to absolute space, and (2) its gravitational attraction to other masses. Our goal is to find a formulation of the law of motion in which inertia and gravitation merge into a higher-order concept, such that motion is dictated solely by the distribution of masses in the universe. However, before we establish this new law, we must navigate several conceptual challenges. Despite the differing masses of two freely falling bodies, they fall or accelerate at the same rate.

Classical mechanics differentiates between the motion of a body left to itself, experiencing no forces (inertial motion), and the motion of a body influenced by gravity. The former is characterized by uniform, rectilinear motion in an inertial system, while the latter follows curvilinear paths and is non-uniform. According to the principle of equivalence, this distinction must be eliminated. By transitioning to an accelerated reference frame, we can convert the uniform, rectilinear motion of inertia into a curved, accelerated motion indistinguishable from that caused by gravity. The reverse is also true, at least for limited segments of the motion, as will be elaborated in the next chapter. Since then, we have referred to any motion of a body that is unaffected by electrical, magnetic, or other forces, but solely influenced by gravitational masses, as "inertial motion."

11.52 A thought experiment using the Principle of Equivalence

Imagine Jim and Joe, along with the little man from the earlier example, experiencing an acceleration of approximately 9.9 m/s². They would recognize this acceleration as "one gravity," the same strength as that felt on the surface of the Earth. Although this level of acceleration results in a very slight curvature of the "geodetic line," it could potentially be detected with sufficiently sensitive instruments. If Jim's ship had opaque walls, preventing him from observing his surroundings, he might feel completely at home, indistinguishable from standing on his home planet.

Similarly, the little man inside the hollow ball would feel at ease, assuming his planet has the same gravitational field strength as Earth. If the ball were also opaque, he might conclude that it is resting on the surface of his planet or on Earth itself.

Are there any experiments Jim or the little man could conduct to determine whether they are accelerating through space or at rest on a planet? Unfortunately, the answer is no. We currently know of no way to distinguish between a gravitational field created by a nearby planet or star and the acceleration resulting from a change in velocity. By the laws of physics, when confined to a small space, there is no method to differentiate between these two situations. Therefore, the two forces are considered equivalent.

This is known as the "principle of equivalence," which has profound implications. All phenomena occurring in a "true" gravitational field are equivalent to those in an accelerating reference frame. This equivalence applies not only to massive objects but also to light. For instance, rays of light should travel along curved paths in a gravitational field near a planet or star. At Earth's surface, a beam of light emitted parallel to the ground should not follow a perfectly straight line through space but rather a path that curves slightly downward. Although this effect is minuscule in Earth's gravitational field, it does occur; a ray of light passing close to Earth will bend slightly toward the planet.

As gravitational fields intensify, these effects become more pronounced. For example, a photon's path would bend significantly near the Sun. If a gravitational source is massive and dense enough to create a strong gravitational acceleration, it might even cause a photon to return to its source. In theory, if gravity becomes powerful enough, a photon could be captured and never escape, leading to the fascinating phenomenon of a "black hole," where objects can enter but cannot exit.

In Newtonian gravitational physics, the deflection of a light ray passing near a massive object is attributed to the idea that photons are projectiles with a certain mass and velocity. Thus, the gravitational pull of a nearby celestial body should alter the photon's path, similar to how a spacecraft would be affected. While it's possible to calculate the degree of light deflection in a gravitational field using this Newtonian model, these calculations have proven inaccurate, yielding results that are approximately half of what actual observations indicate.

In contrast, the relativistic model predicts the curvature of light passing near the Sun with high precision, as we will soon explore. There are other observed phenomena predicted by Einstein's general theory of relativity that Newton's theory does not account for. According to Einstein, gravitation is simply the curvature of space, manifested by the bending of geodesic lines. A light beam traveling between two points is "bent" by a nearby mass because the space itself is curved. Consequently, the straightest path is actually curved; in the presence of gravity, there are no true straight lines.

11.53 Red and Blue Shift

Motion toward or away from a light source does not affect the apparent speed of the light reaching us. Specifically, if we were to measure the speed of a beam of light arriving at our spaceship, it would consistently appear to come in at the same speed, ccc, regardless of whether we were moving toward or away from the source, and independent of our speed relative to it. This phenomenon is known as the "red shift" and "blue shift," which refer to decreases and increases in frequency, respectively, caused by moving away from or toward a source of electromagnetic radiation.

In the vastness of interstellar and intergalactic space, such relative motion is quite common. The red and blue shifts are examples of "Doppler effects," the same phenomenon that leads to discrepancies in the number of ticks counted by observers on Earth and those on a spacecraft during a lunar trip. These shifts in light and other electromagnetic waves arise from continuously changing distances between the source and observer, leading to fluctuations in frequency. Each cycle of the electromagnetic field can be thought of as one "tick" of a clock.

Ch 12—Early Quantum Physics

12.1 The discovery of the quantum

Quantum mechanics stands as one of the greatest intellectual achievements of the 20th century, and its development is a fascinating story. The history can be divided into two key periods: the "Old Quantum Theory" from 1900 to 1925, and the period from 1925 to around 1935, during which the foundations of modern quantum mechanics and electrodynamics were laid out.

The journey began in 1900 when Max Planck discovered that he could only explain the spectrum of thermal radiation—which clashed violently with classical electrodynamics—by proposing that the energy of material sources was discrete, or "quantized." Planck recognized that this introduced a profound contradiction with classical physics, and it deeply troubled him.

In 1905, Albert Einstein pushed the idea of quantization further by applying it to the electromagnetic field's thermodynamics. He introduced what would later be known as the photon and accurately predicted the key feature of the photoelectric effect. Although Einstein's work had a significant impact, his suggestion that light had "corpuscular," or particle-like, aspects was only widely accepted after the discovery of the Compton effect in 1923.

The next major breakthrough came in 1913 with Niels Bohr's quantum theory of the hydrogen atom, based on Ernest Rutherford's atomic model. This marked a major advancement in understanding atomic structure and spectra. The first phase of quantum theory culminated in Wolfgang Pauli's exclusion principle and the discovery of electron spin by Samuel Goudsmit and George Uhlenbeck in 1925. The uncertainty relations, central to quantum mechanics, do not determine whether one can measure one of the pair (position or momentum) with absolute precision while giving up knowledge of the other.

12.2 The formal framework

Any physical theory relies on fundamental concepts that are more basic than the ones it aims to clarify. In classical mechanics, these primitive ideas include time and points in three-dimensional Euclidean space. Using these, Newton's equations provide clear definitions of momentum and force—concepts that had previously been vague and qualitative. Similarly, Maxwell's equations and the Lorentz force law, alongside Newton's equations, define electric and magnetic fields in terms of the motion of test charges.

However, the transition from classical to quantum mechanics is not as straightforward. While classical physics often helps in formulating the Schrödinger equation to describe specific phenomena, the statistical interpretation of quantum mechanics is not directly derived from the

Schrödinger equation itself. This has made the interpretation of quantum mechanics a subject of ongoing debate, despite its consistent empirical success.

In quantum mechanics, all systems with the same degrees of freedom share a common "Hilbert space." While the theory typically provides statistical predictions, this does not mean that quantum uncertainty is as overwhelming as it is sometimes portrayed. In fact, in certain ways, quantum mechanics offers a deeper level of insight than classical mechanics.

12.3 The Quantum Revolution

Max Planck introduced his "quantum" formula at a meeting of the Berlin Physical Society in October 1900. Over the next two months, he worked tirelessly to uncover the physical basis for this formula, experimenting with different assumptions to match the mathematical equations. Planck later described this period as the most intense work of his life. Many of his attempts failed, and eventually, he was left with one reluctant option.

Planck was a physicist rooted in classical ideas. He had been hesitant to embrace the molecular hypothesis and was especially uncomfortable with the statistical interpretation of entropy, introduced by Ludwig Boltzmann into thermodynamics. Entropy, a fundamental concept in physics, is closely linked to the passage of time. While Newton's laws of mechanics are time-reversible, the real world does not behave that way.

Consider a stone dropped to the ground. When it hits, its kinetic energy is converted into heat, increasing the disorder of the system. But if we were to heat the stone, it would not rise back up into the air. This demonstrates the second law of thermodynamics, which states that natural processes always tend toward increasing disorder, or entropy. This law suggests that once energy becomes disordered, it cannot be easily reorganized into orderly motion.

Boltzmann offered a variation on this idea, proposing that while highly unlikely, a reversal of entropy could happen. He argued that just as air molecules randomly move, there is a slim chance they could all gather in a corner of a room. However, this possibility is so remote that it can be practically ignored.

Planck had long argued against Boltzmann's statistical interpretation of the second law, believing that entropy always increased with absolute certainty. Yet, by the end of 1900, after exhausting all other possibilities, he reluctantly applied Boltzmann's statistical methods to his blackbody radiation problem. To his surprise, the calculations worked. Ironically, due to his unfamiliarity with Boltzmann's equations, Planck applied them inconsistently, arriving at the correct solution but for the wrong reasons. It wasn't until Albert Einstein revisited Planck's work that the true significance of his findings became clear.

Planck's reluctant acceptance of Boltzmann's statistical interpretation was a major step forward in science. His work demonstrated that while entropy tends to increase, it is not an absolute certainty. This has profound implications for cosmology, where vast stretches of time and space mean that improbable events might still occur. In theory, the entire Universe could represent a massive statistical fluctuation—a rare instance of low entropy that is now gradually running down.

Planck's "mistake," however, revealed something even more profound about the universe. Boltzmann's statistical approach involved dividing energy into discrete chunks and handling them mathematically before integrating them back into a continuous form. But before Planck reached this final step, he realized he already had the mathematical formula he needed. He chose to stop there, a bold and completely unjustified move in the context of classical physics, yet one that fundamentally changed our understanding of the Universe.

12.4 What is the meaning of *h*

The resolution to the ultraviolet catastrophe becomes clearer through Planck's approach. At high frequencies, the energy required to emit a single quantum of radiation is substantial, meaning that only a few oscillators have enough energy to emit these high-energy quanta. In contrast, at lower frequencies (with longer wavelengths), many low-energy quanta are emitted, but their collective energy remains minimal. The middle range of frequencies presents an ideal balance, where enough oscillators possess the energy to emit moderate-size quanta, which together create the peak in the blackbody radiation curve.

However, Planck's discovery, announced in December 1900, raised as many questions as it answered. His early papers on quantum theory were not particularly clear, possibly due to his struggle to integrate the quantum concept into thermodynamics. For some time, many physicists considered his work a mere "mathematical trick" to bypass the ultraviolet catastrophe, without deeper physical meaning. Planck himself felt uncertain. In a letter to Robert William Wood in 1931, he described his 1900 work as "an act of despair" and admitted that a theoretical interpretation had to be found "at any cost." Despite this, Planck realized he had stumbled onto something significant. According to Heisenberg, Planck's son recalled his father comparing his discovery to Newton's during a walk in the Grunewald forest outside Berlin.

In the early 20th century, physicists were more focused on breakthroughs in atomic radiation, and Planck's quantum explanation of the blackbody curve did not seem immediately transformative. It wasn't until 1918 that Planck received the Nobel Prize for his work—an unusually long delay compared to the swift recognition received by figures like the Curies or Rutherford. This slow acknowledgment likely stemmed from the fact that theoretical breakthroughs often take longer to be fully accepted, requiring confirmation through experiments to stand the test of time.

Planck's new constant, h, introduced something unusual. Though small—6.6×10^{-34} joule-seconds—the size itself wasn't the strange part, as a larger value would have revealed itself earlier in physics. What was peculiar about h were the units: energy (measured in ergs) multiplied by time (measured in seconds), which are known as units of "action." In classical mechanics, action wasn't a commonly discussed quantity—there is no "law of conservation of action" comparable to the laws of conservation of mass or energy. However, action has an intriguing property: like entropy, it is a constant across all observers in space-time, meaning it remains the same for all regardless of their frame of reference. This four-dimensional constancy only became significant when Einstein introduced the theory of relativity.

This leads us to Einstein, whose next major contribution would build on Planck's work. His special theory of relativity treats the three dimensions of space and the one dimension of time as a single four-dimensional space-time continuum. Observers moving at different speeds through space perceive things differently—they may disagree on the length of an object or the duration of an event. But when considering the object as existing within four-dimensional space-time, these disagreements about individual components (length and time) balance out. For instance, the "area" of the object in space-time—measured as length multiplied by time—remains constant for all observers, even if they disagree about each individual measurement. Similarly, action (energy x time) is also a four-dimensional quantity, and its value is conserved in special relativity, just like energy.

Planck's constant, *h*, seemed odd only because it was discovered before relativity, which altered the Newtonian view of space and time. Of Einstein's three great contributions in 1905—special relativity, Brownian motion, and the photoelectric effect—the special theory of relativity may seem the most different. However, all these works share a common foundation in theoretical physics. Despite the fame of his theory of relativity, Einstein's most significant contribution might have been his advancement of quantum theory, which grew from Planck's work through his study of the photoelectric effect.

Planck's work in 1900 revolutionized physics by revealing a fundamental limitation of classical physics—specifically, that some phenomena could not be explained by classical mechanics alone. This discovery marked the beginning of a new era in physics, even though Planck's initial formulation was more limited than modern interpretations suggest.

Despite inventing the concept of the quantum and moving physics beyond classical constraints, Planck initially proposed that only the *oscillators* within atoms were quantized, meaning they could emit energy only in specific "packets" or quanta. He didn't initially suggest that radiation itself was quantized. Throughout his career, he worked to reconcile these new ideas with classical physics, seeking a middle ground rather than fully embracing the quantum revolution he had set in motion. Planck didn't doubt his discovery of the quantum, but he struggled to see just how far removed his blackbody equation was from classical physics. His derivation relied on thermodynamics and electrodynamics—both classical theories.

Rather than abandoning his ideas, Planck's efforts to find a bridge between classical and quantum physics reflected a shift in his thinking. However, his deep grounding in classical physics made it difficult for him to fully embrace the new quantum ideas. Ultimately, it took a younger generation of physicists—less tied to classical concepts and more enthusiastic about the discoveries in atomic physics—to carry the quantum revolution forward.

12.5 The meaning of Planck's constant in terms of space and time

It is possible to measure a "smallest distance." Physicists have theorized the existence of such a limit, called the Planck length, named after Max Planck. This length, about 1.6×10^{-35} meters—approximately 10^{-20} times the size of a proton—is considered the smallest meaningful division of space. Below this scale, the concept of distance becomes meaningless according to our current understanding of the universe. But what about time?

There may also be a "smallest time" interval, known as Planck time. Similar to how the Planck length represents the smallest measurable distance, Planck time is the smallest measurable unit of time. It is defined as the time it takes for light to travel one Planck length, which is about 1.6×10^{-35} meters. Given the speed of light as the fastest possible speed, the shortest measurable time is the Planck length divided by the speed of light. Because the Planck length is extremely small and the speed of light is extremely fast, Planck time is incredibly brief—approximately 5.3×10^{-44} seconds. Beyond this, the very concept of time begins to break down. Some theories even suggest that time itself may be divided into discrete units called "chronons," with each chronon lasting exactly one Planck time.

Planck Time =

$1.6 \times 10\text{-}35$ m$/ 3.0 \times 108$ m/s = $5.3 \times 10\text{–}44$ second.

12.6 Einstein, Light and Quanta

In March 1900, Einstein was just twenty-one years old. He began his well-known position at the Swiss patent office in the summer of 1902 and spent the early years of the 20th century focused primarily on thermodynamics and statistical mechanics. His initial scientific publications were traditional in style and approached the same problems as the previous generation, including those tackled by Max Planck. However, in his first paper referencing Planck's ideas on the blackbody spectrum, published in 1904, Einstein started to forge a new path, developing a distinctive approach to solving physical problems.

Martin Klein notes that Einstein was the first to take the physical implications of Planck's work seriously, recognizing them as more than mere mathematical tricks. Within a year, this acknowledgment led to a significant breakthrough—the revival of the corpuscular theory of light. Alongside Planck's work, Einstein's 1904 paper also drew on the investigations into the photoelectric effect conducted independently by Phillip Lenard and J.J. Thomson at the end of the 19th century.

Lenard, born in 1862 in a region of Hungary that is now part of Czechoslovakia, received the Nobel Prize in Physics in 1905 for his research on cathode rays (electrons). In one of his notable experiments from 1899, he demonstrated that light shining onto a metal surface in a vacuum could produce cathode rays. The energy from the light caused electrons to be emitted from the metal.

Lenard's experiments utilized monochromatic light, meaning all the light waves had the same frequency (color). He investigated how the intensity of the light affected the ejection of electrons from the metal and discovered an unexpected result. When he increased the brightness of the light (by moving the light closer to the metal, which had a similar effect), he expected that the greater energy hitting each square centimeter of the metal surface would result in electrons being knocked out more quickly and with higher velocity. However, Lenard found that as long as the wavelength of the light remained constant, all the emitted electrons were ejected at the same velocity.

Moving the light closer to the metal increased the number of ejected electrons, but each electron still emerged with the same velocity as those produced by a weaker beam of light of the same color. In contrast, when Lenard used light with a higher frequency—like ultraviolet instead of blue or red—he observed that the electrons moved faster.

This phenomenon can be simply explained if one is willing to set aside classical physics and accept Planck's equations as physically meaningful. The significance of this acceptance is highlighted by the fact that, in the five years following Lenard's initial work on the photoelectric effect and Planck's introduction of the concept of quanta, no one took that seemingly straightforward step. Essentially, Einstein applied the equation $E=h\nu$ to electromagnetic radiation rather than to the oscillators within atoms. He proposed that light is not a continuous wave, as scientists had believed for a century, but rather comes in discrete packets known as "quanta." Each packet of light at a specific frequency ν (or color) carries the same energy E. When a light quantum, or photon, strikes an electron, it imparts the same amount of energy, resulting in the same velocity for the ejected electrons. Increased light intensity simply means more photons are striking the surface but changing the light's color alters its frequency and thus the energy carried by each photon. This groundbreaking work eventually earned Einstein the Nobel Prize in 1921.

Once again, a theoretical breakthrough required time for full recognition. The concept of photons did not gain immediate acceptance. While Lenard's experiments broadly aligned with Einstein's theory, it took over a decade for the precise relationship between the electrons' velocity and the light's wavelength to be rigorously tested and validated.

This work was accomplished by American scientist Robert Millikan, who also achieved a highly accurate measurement of Planck's constant hhh. Millikan received the Nobel Prize in Physics in 1923 for this research and for his precise measurements of the electron's charge.

One paper led to a Nobel Prize for Einstein, another confirmed the existence of atoms, and a third established the theory for which he is most famous: relativity. Almost incidentally, during this time in 1905, Einstein was completing another significant piece of research concerning molecular sizes, which he submitted as his doctoral thesis to the University of Zurich. He was awarded his PhD in January 1906. Although a PhD did not guarantee a path to active research as it does today, it is remarkable that the three pivotal papers of 1905 were published by someone who could still only sign himself "Mr. Albert Einstein."

In the subsequent years, Einstein continued to work on integrating Planck's quantum ideas into other areas of physics. He discovered that these concepts offered explanations for long-standing puzzles related to the theory of specific heat—the amount of heat required to raise the temperature of a given material by a certain degree, which depends on how atoms vibrate within the substance, a process that is also quantized. While this aspect of science is often overlooked in discussions of Einstein's contributions, the "quantum theory of matter" gained acceptance more rapidly than his "quantum theory of radiation," persuading many traditional physicists that quantum ideas deserved serious consideration.

Einstein refined his theories on quantum radiation until 1911, establishing that the quantum structure of light was an inevitable consequence of Planck's equation. He pointed out to a skeptical scientific committee that understanding light better required merging wave and particle theories, which had been at odds since the 17th century. By 1911, however, he had shifted his focus to gravity and, over the next five years, developed his General Theory of Relativity, his most significant work. It wasn't until 1923 that the quantum nature of light was conclusively established, sparking renewed debate over particles and waves that helped transform quantum theory into modern quantum mechanics.

The first major advancements in quantum theory occurred during the decade in which Einstein turned his attention elsewhere. This progress stemmed from a synthesis of his ideas with Rutherford's atomic model and was largely driven by Danish scientist Niels Bohr, who had worked with Rutherford in Manchester. After Bohr introduced his atomic model, the value of quantum theory as a description of the physical world at the smallest scales became undeniable.

12.7 Bohr's Atom

By 1912, the fundamental components of quantum theory were beginning to align. Einstein had affirmed the widespread validity of the concept of quanta and had introduced the notion of photons, despite its limited acceptance at the time. He asserted that energy is truly quantized, existing only in discrete packets of specific sizes. Meanwhile, Rutherford had proposed a novel model of the atom, characterized by a small central nucleus enveloped by a cloud of electrons, although this innovative idea also had not yet gained widespread endorsement.

Rutherford's model of the atom, however, could not maintain stability under the classical laws of electrodynamics. The solution lay in applying quantum principles to explain the behavior of electrons within atoms. Once again, a young researcher with a novel perspective emerged to tackle the problem, highlighting a recurring theme in the development of quantum theory.

Niels Bohr, a Danish physicist, earned his doctorate in the summer of 1911 and subsequently went to Cambridge that September to work with J.J. Thomson at the Cavendish Laboratory. As a junior researcher who was shy and spoke English imperfectly, Bohr struggled to find his place at Cambridge. However, during a visit to Manchester, he met Rutherford, who was both approachable and genuinely interested in Bohr's work. In March 1912, Bohr moved to Manchester to join Rutherford's team, focusing on the complexities of atomic structure. After six months, he returned to Copenhagen briefly but continued his collaboration with Rutherford's group in Manchester until 1916.

12.8 Energetic Electrons

Bohr's genius was precisely what atomic physics needed over the next decade or so. Rather than striving for a complete theoretical framework, he was open to piecing together various ideas to create a functional "model" that aligned roughly with the observations of real atoms. Once he established a basic understanding, he could refine it, making adjustments to improve its coherence and accuracy. He envisioned the atom as a miniature "solar system," where electrons orbited the nucleus in accordance with classical mechanics and electromagnetism. He proposed that electrons could not spiral inward while emitting radiation, as classical theory suggested, because they were limited to emitting whole energy packets—quanta—rather than continuous radiation. These "stable" orbits corresponded to specific energy levels, each a multiple of a fundamental quantum, and there were no intermediate orbits, which would require fractional energy values. To extend the solar system analogy, it was akin to stating that while the orbits of Earth and Mars are stable, no stable orbit exists between them.

The concept of orbits stems from classical physics, while the idea of electron states tied to fixed energy levels originates from quantum theory. By combining elements of classical and quantum theories, Bohr created a model of the atom that, while lacking true insight into atomic behavior, provided a practical framework for further progress. Although his model was ultimately incorrect in many aspects, it served as a vital steppingstone toward a genuine quantum theory of the atom. Unfortunately, due to its straightforward blend of classical and quantum concepts and the appealing imagery of the atom as a miniature solar system, this model has lingered in both popular literature and educational texts. If you learned about atoms in school, you likely encountered Bohr's model, regardless of whether it was explicitly named. However, at this stage in our exploration, it's important to recognize that Bohr's model does not capture the entire truth. It's essential to move beyond the notion of electrons as tiny "planets" orbiting the nucleus. Instead, think of an electron as a particle that resides outside the nucleus, possessing a specific amount of energy and other properties. Its movement, as we will explore, is governed by more complex principles.

Bohr's significant breakthrough came in 1913 when he successfully explained the spectrum of light emitted by hydrogen, the simplest atom. Spectroscopy, a field dating back to the early 19th century when William Wollaston discovered dark lines in the solar spectrum, gained prominence with Bohr's insights. Like Bohr's innovative approach, we too must step back from Einstein's ideas about light quanta to understand spectroscopy effectively. In this context, light is best viewed as an electromagnetic wave. As Newton established, white light comprises all colors of the rainbow, which we refer to as the spectrum. Each color corresponds to a specific wavelength, and by using a glass prism to separate white light into its colored components, we effectively disperse the spectrum, laying the different frequencies side by side on a screen or photographic plate. Short-wavelength blue and violet light appears at one end of the spectrum, while long-wavelength red is at the opposite end, extending beyond the visible spectrum on both sides.

When sunlight is dispersed in this manner, the resulting spectrum reveals sharp, dark lines at precise points, corresponding to specific frequencies. Researchers in the 19th century, including

Joseph Fraunhofer, Robert Bunsen (whose name is memorialized in the Bunsen burner), and Gustav Kirchhoff, established through experimentation that each element produces its unique set of spectral lines. For example, when sodium is heated in a Bunsen burner flame, it emits light of a characteristic color (yellow) visible as bright emission lines in the spectrum. Conversely, when white light passes through a gas or liquid containing sodium, it produces dark absorption lines at the same frequencies, even if the sodium is part of a compound. This phenomenon helped explain the dark lines seen in the solar spectrum.

With spectroscopy, astronomers can analyze distant stars and galaxies to determine their composition, while atomic physicists can investigate the inner structure of atoms using the same technique. The hydrogen spectrum is particularly straightforward because hydrogen is the simplest element, consisting of just one positively charged proton in its nucleus and one negatively charged electron.

The spectral lines that uniquely identify hydrogen are known as the Balmer lines, named after Johann Balmer, a Swiss schoolteacher who derived a formula describing their pattern in 1885, the same year Bohr was born.

12.9 The Balmer lines

Balmer's formula establishes a relationship between the frequencies of the spectral lines of hydrogen. Beginning with the frequency of the first hydrogen line, found in the red portion of the spectrum, the formula allows for the calculation of the next hydrogen line, located in the green. Using the frequency of the green line, the same formula can be applied to find the frequency of the subsequent line in the violet, and so forth. Although Balmer originally identified only four visible hydrogen lines when he derived his formula, other lines had already been discovered that conformed precisely to it. When additional hydrogen lines were later detected in the ultraviolet and infrared regions, they too adhered to this straightforward numerical relationship. Clearly, Balmer's formula indicated something significant about the structure of the hydrogen atom.

By the time Bohr entered the field, Balmer's formula was widely recognized among physicists and featured in every undergraduate physics course. However, it existed amid a sea of complex spectral data, and Bohr was not an expert in spectroscopy. Initially, he did not see the Balmer series as an obvious key to the mystery of the hydrogen atom's structure. But when a colleague specializing in spectroscopy highlighted the simplicity of the Balmer formula, despite the complexities present in the spectra of other elements, Bohr quickly recognized its significance. A simplified version of the formula indicates that the wavelengths of the first four hydrogen lines can be calculated by multiplying a constant (36.456×10^{-5} by 9/5, 16/12, 25/21, and 36/32). In this formulation, the numerators correspond to a sequence of squares (3^2, 4^2, 5^2, 6^2), while the denominators are differences of squares ($3^2 - 2^2$, $4^2 - 2^2$, and so on).

In early 1913, Bohr was already convinced that part of the answer to the atomic structure puzzle lay in incorporating Planck's constant, h, into the equations governing the atom. Rutherford's model of the atom included only two fundamental numbers: the electron's charge, e, and the masses of the particles involved. However, no combination of mass and charge could yield a number with the dimensions of "length," meaning that Rutherford's model lacked a "natural" unit of size. With the addition of an "action" constant like h, it became possible to construct a value with dimensions of length, offering a rough estimate of atomic size. The expression h^2/me^2 corresponds numerically to a length of about *20 x 10⁻⁸* cm, which aligns well with the properties of atoms inferred from scattering experiments and other studies.

For Bohr, it was evident that h was integral to atomic theory, and the Balmer series revealed exactly how it fit in. An atom produces a sharp spectral line by emitting or absorbing energy at a precise frequency. Energy is related to frequency through Planck's constant ($E = hv$), meaning that if an electron in an atom emits a quantum of energy hv, the electron's energy must change by exactly that amount E. Bohr proposed that electrons "in orbit" around the nucleus remain in place because they cannot radiate energy continuously. Instead, they can radiate or absorb a complete quantum of energy—one photon—allowing them to jump from one energy level (or orbit, in the classical view) to another. This seemingly simple concept marks a profound departure from classical ideas.

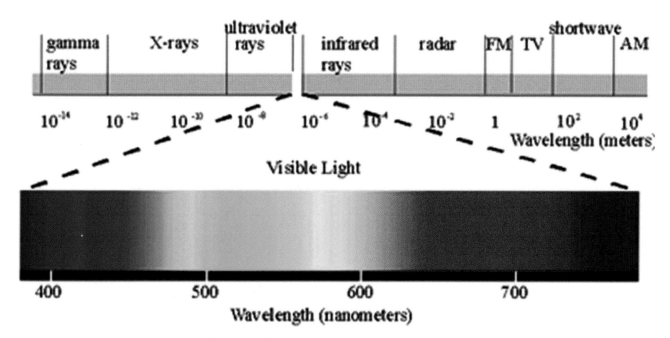

Figure 12-1 Balmer series – spectral lines of radiation

Imagine if Mars suddenly vanished from its orbit and reappeared in Earth's orbit, simultaneously radiating a pulse of energy into space, which in this context would be gravitational radiation. This thought experiment illustrates the inadequacy of the solar system atom model in explaining atomic behavior. It's far more effective to think of electrons existing in different states that correspond to various energy levels within the atom.

Transitions between these states can occur in either direction along the energy ladder. When an atom absorbs light, the energy quantum hv is used to elevate the electron to a higher energy level. If the electron subsequently returns to its original state, it will emit precisely the same energy hv as radiation. The constant 36.456×10^{-5} found in Balmer's formula can naturally be expressed in terms of Planck's constant, allowing Bohr to calculate the permissible energy levels for the single electron in a hydrogen atom. The observed frequencies of spectral lines then represent the energy differences between these levels.

In everyday applications, traditional energy units are often too large, so the electron volt (eV) is a more convenient measure. An electron volt is the energy an electron gains when moving across an electric potential difference of one volt, equivalent to 1.602×10^{-19} joules. To put this in perspective, a 100-watt light bulb consumes energy at a rate of 100 joules per second, which can be expressed as 6.24×10^{20} eV per second. While it sounds impressive to say that a light bulb emits six and a quarter hundred million trillion electron volts each second, it conveys the same energy as the 100-watt lamp. The energy involved in electron transitions that create spectral lines typically falls within a few electron volts—specifically, it requires 13.6 eV to completely remove an electron from a hydrogen atom. In contrast, energies associated with particles produced by radioactive decay are often in the millions of electron volts (MeV).

Thirteen years after Planck's groundbreaking incorporation of quantum theory into the realm of light, Bohr introduced quantum concepts into atomic theory. However, it would take another thirteen years before a cohesive quantum theory emerged. Progress during this period was slow, characterized by frequent setbacks.

Bohr's model of the atom was a "hodgepodge," merging quantum ideas with classical physics, employing whatever elements seemed necessary to patch up the model. This approach allowed for the prediction of many more spectral lines than could be observed, leading to arbitrary rules that deemed some transitions between energy states as "forbidden." New quantum numbers were assigned ad-hoc to align with observations, lacking a solid theoretical basis to justify their necessity or the reasons for the forbidden transitions.

In the midst of these scientific developments, the outbreak of World War I disrupted the European scientific community, just a year after Bohr proposed his initial atomic model. While Holland and Denmark remained scientific havens during this time, Bohr returned to Denmark in 1916 to become Professor of Theoretical Physics in Copenhagen and founded the research institute that now bears his name in 1920. Despite the turmoil, news from German researchers like Arnold Sommerfeld—who refined Bohr's model to the extent that it was sometimes referred to as the "Bohr-Sommerfeld atom"—could flow through neutral Denmark to reach Rutherford in England. Although progress continued, the dynamic of scientific advancement had changed.

After the war, German and Austrian scientists were excluded from international conferences for many years, and Russia was engulfed in revolutionary turmoil, leading to a loss of some of science's international character and a generation of young men. It was left to a new generation of physicists to elevate quantum theory from Bohr's halfway model—the "hodgepodge" atom—to a fully developed quantum mechanics. This generation includes renowned figures such as Werner Heisenberg, Paul Dirac, Wolfgang Pauli, and Pascual Jordan. Born during the years following Planck's significant contribution, these scientists began their research careers in the 1920s. Unlike Bohr, they did not carry the burden of classical physics, allowing them to create theories of the atom that did not require classical concepts.

Interestingly, the span from Planck's discovery of the blackbody radiation equation to the maturation of quantum mechanics was just twenty-six years, coinciding with the emergence of a new generation of physicists into the research field. This generation inherited two vital legacies from their still-active predecessors: first, Bohr's model of the atom underscoring the necessity of incorporating quantum ideas into any comprehensive atomic theory; second, the groundbreaking contributions of Einstein, who, in 1916, introduced the concept of probability into atomic theory. Although Einstein's ideas were initially seen as a temporary fix to align Bohr's model with experimental observations, they ultimately became foundational for true quantum theory, despite Einstein later famously asserting, "God does not play dice."

As previously mentioned in the early 1900s, Rutherford and his colleague Frederick Soddy were exploring the nature of radioactivity when they uncovered a fundamental property of the atom—specifically, its nucleus. Radioactive decay involves a fundamental transformation of an individual atom (now understood as the disintegration of the nucleus and the emission of nuclear particles), yet it appears unaffected by external factors. Whether atoms are heated or cooled, placed in a vacuum, or submerged in water, the process of radioactive decay continues undisturbed.

Although it is impossible to predict when a specific atom will undergo decay, experiments demonstrate that a consistent proportion of atoms from a large sample of a radioactive element will decay over a certain timeframe. Each radioactive element has a unique characteristic known as its "half-life," the period during which half of the atoms in a sample decay. For example, radium has a half-life of 1,600 years, carbon-14 has a half-life of just under 6,000 years—making it useful for archaeological dating—and radioactive potassium has a half-life of 1,300 million years.

Unaware of the specific mechanisms driving the decay of individual atoms, Rutherford and Soddy established a statistical theory of radioactive decay, akin to the actuarial techniques used by insurance companies. Just as insurers understand that while some policyholders may die young, leading to significant payouts, others will live longer, balancing the books through statistical tables, physicists can similarly manage the decay of radioactive substances by examining large collections of atoms.

One peculiar characteristic of radioactive decay is that it never completely ceases within a sample of radioactive material. Out of millions of atoms present, half will decay over a specific timeframe. In the subsequent half-life, half of the remaining atoms will decay, and this process continues indefinitely. Consequently, while the quantity of radioactive atoms diminishes progressively toward zero, each step only takes the total halfway there. In those early years, physicists like Rutherford and Soddy anticipated that a breakthrough discovery would eventually elucidate the reasons behind individual atom decay, thus clarifying the statistical methods applied to the Bohr model of atomic spectra. Bohr also expected that future discoveries would eliminate the need for "actuarial tables," but they were mistaken. The electron can occupy multiple energy levels and transition between them, resulting in numerous lines in each element's spectrum. Each spectral line corresponds to a transition between energy levels with distinct quantum numbers. For example, all transitions culminating in the ground state generate a series of spectral lines similar to the Balmer series, while transitions leading to the second energy level yield another distinct set of lines.

In a hot gas, atoms frequently collide, exciting electrons to higher energy levels, which then fall back, emitting bright spectral lines in the process. Conversely, when light passes through a cold gas, ground-state electrons are elevated to higher energies, absorbing light and creating dark lines in the spectrum. If the Bohr model of the atom holds any validity, this mechanism of hot atoms radiating energy should align with Planck's law, implying that the blackbody spectrum of cavity radiation arises from a multitude of atoms radiating energy as their electrons shift between energy levels.

By 1916, after completing his General Theory of Relativity, Einstein redirected his focus back to quantum theory, likely inspired by Bohr's success and the growing acceptance of his own corpuscular theory of light. Robert Andrews Millikan, an American physicist at the California Institute of Technology (Caltech), had previously been one of Einstein's strongest critics regarding the photoelectric effect interpretation presented in 1905.

After a decade of meticulously testing the data, he sought to disprove Einstein's explanation but ultimately confirmed its accuracy in 1914, validating Einstein's view of light quanta, or photons. In the process, Millikan derived a highly precise experimental value for Planck's constant h and, in 1923, received the Nobel Prize for his groundbreaking work and his measurement of the electron's charge.

Einstein recognized a similarity between the decay of an atom transitioning from an "excited" energy state—where an electron is at a higher energy level—to a state of lower energy, where the electron occupies a lower energy level, and the process of radioactive decay. He employed the statistical methods developed by Boltzmann to analyze individual energy states, calculating the probability that a specific atom would be in an energy state corresponding to a particular quantum number n. By applying the probabilistic "actuarial tables" of radioactivity, initially utilized by Rutherford, he determined the likelihood of an atom in state n decaying to a lower energy state (i.e., one with a lower quantum number).

This line of reasoning naturally led to Planck's formula for blackbody radiation, which was derived entirely from quantum principles. Utilizing Einstein's statistical insights, Bohr was able to enhance his atomic model by explaining why certain lines in the spectrum are more pronounced than others; some energy state transitions are simply more probable than others. While Bohr could not provide a definitive explanation for this phenomenon, it was not a significant concern at the time. Like others studying radioactivity, Einstein believed that the actuarial tables were not the ultimate answer, and that future research would clarify why specific transitions occur at particular moments rather than at others.

At this juncture, quantum theory began to diverge from classical concepts, and no fundamental explanation for why radioactive decay or atomic energy transitions happen at specific times has been discovered. It appears that these processes occur purely by chance, based on statistical probabilities, which raises profound philosophical questions.

In the classical framework, every event has a cause, allowing for a linear tracing of causality back in time to find the root cause—extending potentially to the Big Bang in cosmological contexts or to the moment of Creation in religious views. However, in the realm of quantum mechanics, this direct causality starts to fade, particularly when examining radioactive decay and atomic transitions.

An electron doesn't transition from one energy level to another at a specific time for any particular reason. Instead, the lower energy level is statistically more favorable for the atom, making it likely (and the likelihood can even be quantified) that the electron will transition sooner rather than later. There is no external force pushing the electron, nor is there any internal mechanism that times the jump. It simply occurs without any specific reason, at one moment instead of another. While this idea may not represent a significant departure from strict causality, many 19th-century scientists might have found it unsettling. However, it's unlikely that readers of this book are particularly troubled by this concept.

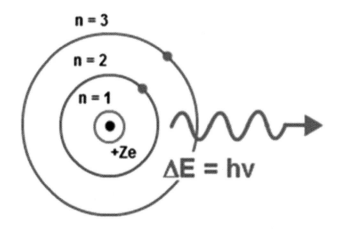

Figure 12-2 An atom's electron energy states

However, this was just the tip of the iceberg, the first indication of the true "strangeness" of the quantum world, and its significance was not fully appreciated at the time. This insight came in 1916 from Einstein.

In June 1922, Bohr visited the University of Göttingen in Germany to deliver a series of lectures on quantum theory and atomic structure. Göttingen was poised to become one of the three key centers for the development of a comprehensive version of quantum mechanics, led by Max Born, who had become Professor of Theoretical Physics there in 1921. Born, born in 1882 to the Professor of Anatomy at the University of Breslau, was a student during the early 1900s when Planck's ideas first emerged. Initially studying mathematics, he shifted to physics after completing his doctorate in 1906, even spending some time at the Cavendish Laboratory. This background would prove invaluable in the years to come. As an expert on relativity, Born's work was always marked by rigorous mathematical precision, standing in stark contrast to Bohr's more improvisational theoretical frameworks, which, while insightful and intuitive, often left others to grapple with the mathematical complexities. Both types of genius were crucial for advancing our understanding of atoms.

Bohr's lectures in June 1922 were a significant milestone in the revitalization of German physics post-war, as well as in the history of quantum theory. Scientists from across Germany attended, and the event became known as the "Bohr Festival."

In those lectures, after meticulously laying the groundwork, Bohr introduced the first successful theory of the periodic table, a framework that remains largely intact today. His theory proposed that electrons are added to an atom's nucleus in a specific manner. Regardless of the atomic number, the first electron occupies an energy state equivalent to the ground state of hydrogen. The second electron follows suit, creating an atom resembling helium, which contains two electrons.

However, Bohr noted that no additional electrons could occupy that energy level, necessitating that the next electron transition to a different energy level. Thus, for an atom with three protons and three electrons, two electrons would be closely bound to the nucleus, leaving one electron more loosely attached, behaving similarly to a one-electron atom, like hydrogen.

This three-proton element is lithium, which indeed shares some chemical similarities with hydrogen. The subsequent element in the periodic table with properties akin to lithium is sodium, with an atomic number of 11, eight places beyond lithium. Bohr argued that there must be eight available positions in the energy levels outside the inner two electrons. Once these were filled, the next electron—the eleventh—would need to occupy another energy state, even less tightly bound to the nucleus, again mimicking a one-electron atom. These energy states, termed "shells," form the basis of Bohr's explanation of the periodic table, which involved sequentially filling these shells as atomic number increased. One could visualize the shells as concentric layers, similar to onion skins; for chemistry, the key factor is the number of electrons in the outermost shell, as the dynamics of the inner layers play a secondary role in determining atomic interactions.

Through an exploration of electron shells and drawing from spectral evidence, Bohr elucidated the relationships among elements in the periodic table based on atomic structure. While he couldn't explain why a shell containing eight electrons is considered full or "closed," he left his audience convinced that he had uncovered an essential truth. As Heisenberg later remarked, Bohr "had not proven anything mathematically... he just knew that this was more or less the connection."

Chemistry focuses on how atoms interact and combine to form molecules. For instance, why does carbon bond with hydrogen to create methane, consisting of four hydrogen atoms attached to a single carbon atom? Why do hydrogen atoms exist as diatomic molecules, while helium does not form molecules at all? The shell model provides elegant answers: each hydrogen atom possesses one electron, whereas helium has two. The innermost shell is complete with two electrons, and (for reasons yet unknown) filled shells exhibit greater stability—atoms prefer to have full shells. When two hydrogen atoms form a molecule, they share their electrons, enabling each to achieve a closed shell. Helium, already possessing a full shell, shows no interest in forming bonds and remains chemically inert.

Carbon, which has six protons and six electrons, contains two electrons in its closed inner shell and four in the next, which is only half full. Four hydrogen atoms can each share an outer carbon electron while contributing their own, resulting in each hydrogen atom acquiring a semi-closed "pseudo-closed" shell of two electrons, while the carbon atom achieves a semi-closed second shell of eight electrons.

Bohr posited that atoms combine to get as close as possible to achieving a closed outer shell. In some instances, such as the hydrogen molecule, it's useful to visualize two nuclei sharing a pair of electrons; in other scenarios, like with sodium, an atom with an unpaired electron in its outer shell might donate that electron to an atom with seven electrons and one vacancy, such as chlorine.

Each atom finds satisfaction in this arrangement—the sodium atom becomes a positively charged ion by losing an electron, while the chlorine atom, by gaining one, fills its outermost shell. The resulting attraction between these oppositely charged ions leads to the formation of an electrically neutral molecule of sodium chloride, or table salt.

All chemical reactions can be understood as the sharing or transferring of electrons between atoms in pursuit of the stability that comes with filled electron shells. Energy transitions involving outer electrons give rise to the distinct spectral fingerprints of elements, while transitions in deeper shells, which require significantly more energy (found in the X-ray region of the spectrum), are consistent across all elements. Like all great theories, Bohr's model was validated by a successful prediction. Even in 1922, the periodic table contained several gaps corresponding to undiscovered elements with atomic numbers 43, 61, 72, 75, 85, and 87. Bohr's model predicted the properties of these "missing" elements, notably suggesting that element 72 should resemble zirconium. This prediction was confirmed within a year with the discovery of hafnium, element 72, which exhibited spectral properties that aligned perfectly with Bohr's predictions.

This marked the pinnacle of the old quantum theory. Within three years, however, this framework was largely replaced, although for most practical chemistry, the ideas of electrons as small particles orbiting atomic nuclei in shells that strive to be full (or empty, but not partially filled) suffice. Nineteenth-century physics remains adequate for everyday applications, and the physics from 1923 serves well for most chemistry. The physics of the 1930s delves further into the search for "ultimate truths." There hasn't been a major breakthrough since; the rest of science has been catching up with the insights of a select few geniuses. The successful Aspect experiment in Paris in the early 1980s concluded this period of catching up, providing the first direct experimental evidence that even the most peculiar aspects of quantum mechanics accurately describe reality.

Now is the time to explore just how strange the quantum realm truly is. The additional insights required to explain more complex molecules emerged in the late 1920s and early 1930s, drawing on the complete development of quantum mechanics. The majority of this work was done by Linus Pauling, who is better known today as a peace activist and vitamin C advocate. He received the first of his two Nobel Prizes for his research into the nature of the chemical bond and its application in elucidating the structures of complex substances in 1954. Those complex substances, illuminated through quantum theory by Pauling, a physical chemist, paved the way for the exploration of life's molecules. The significance of quantum chemistry in molecular biology is acknowledged by Horace Judson in his monumental work, *The Eighth Day of Creation*; however, detailing that story lies beyond the scope of this book.

12.10 Photons and electrons

Despite the achievements of Planck and Bohr in guiding the development of a physics for the microscopic realm distinct from classical mechanics, the modern understanding of quantum theory truly began with the acceptance of Einstein's concept of the light quantum. This breakthrough led to the recognition that light must be described as both particles and waves. Although Einstein first introduced the idea of the light quantum in his 1905 paper on the photoelectric effect, it wasn't until 1923 that it gained widespread acceptance and respectability.

Einstein himself approached this idea with caution, fully aware of its revolutionary implications. At the first Solvay Congress in 1911, he stated, "I insist on the provisional character of this concept, which does not seem reconcilable with the experimentally verified consequences of the wave theory." Although Millikan demonstrated by 1915 that Einstein's equations regarding the photoelectric effect were correct, many found it difficult to accept the reality of light as particles. Reflecting on his work in the 1940s, Millikan noted, "I was compelled in 1915 to assert its unambiguous verification in spite of its unreasonableness... it seemed to violate everything we knew about the interference of light."

At the time, he expressed his skepticism more emphatically. When reporting the experimental confirmation of Einstein's equation for the photoelectric effect, he stated, "The semi-corpuscular theory by which Einstein arrived at his equation seems at present wholly untenable." This was written in 1915. In 1918, Rutherford remarked that there appeared to be "no physical explanation" for the connection between energy and frequency that Einstein had elucidated thirteen years earlier with his light quantum hypothesis. It wasn't that Rutherford was unaware of Einstein's suggestion; he simply lacked conviction in it. Given that all experiments at the time were designed to support the wave theory of light, how could light also be composed of particles?

In 1909, around the time Einstein transitioned from being a patent clerk to his first academic role as an associate professor in Zurich, he made a notable advancement by referring to point-like quanta with energy $h\nu$ for the first time. In classical mechanics, particles such as electrons are depicted as "point-like" entities, which significantly contrasts with wave descriptions. However, the frequency of the radiation ν still indicates the energy of the particle.

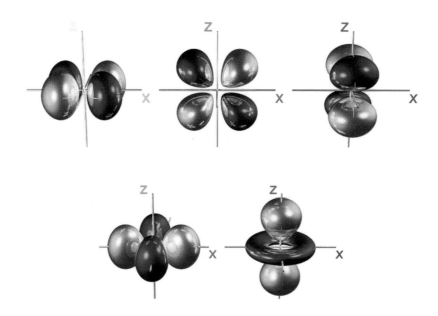

Figure 12-3 Atoms with various geometrical configurations

In 1909, Einstein remarked, "It is my opinion that the next phase in the development of theoretical physics will bring us a theory of light that can be interpreted as a kind of fusion of the wave and the emission theory." This statement captures the essence of modern quantum theory. By the 1920s, Bohr articulated this new foundation of physics through the "principle of complementarity," which posits that the wave and particle theories of light are not mutually exclusive but rather complementary. A comprehensive description necessitates both concepts, particularly evident in the need to measure the energy of a light "particle" in relation to its frequency or wavelength.

However, shortly after making these observations, Einstein shifted his focus away from quantum theory to develop his General Theory of Relativity. When he returned to quantum discussions in 1916, he brought with him a further logical advancement regarding light quanta. His statistical insights refined the understanding of the Bohr atom and enhanced Planck's model of blackbody radiation. These calculations regarding how matter absorbs or emits radiation also illustrated how momentum is transferred from radiation to matter, given that each quantum of radiation $h\nu$ carries a momentum of $h\nu/c$.

This work echoes his influential 1905 paper on Brownian motion. Just as pollen grains are jostled by gas or liquid atoms, proving the reality of atoms, the atoms themselves are similarly affected by the "particles" of blackbody radiation. Although the Brownian motion of atoms and molecules cannot be observed directly, the resulting buffeting produces statistical effects measurable in terms of properties like gas pressure. Einstein explained these statistical effects through the lens of blackbody radiation particles, which carry momentum.

However, the expression for the momentum of a light "particle" derives straightforwardly from special relativity. In relativity theory, the energy (E), momentum (p), and rest mass (m) of a particle are related by the equation:

$$E^2 = m^2c^4 + p^2c^2$$

Since a light particle has no rest mass, this equation simplifies quickly to:

$$E^2 = p^2c^2$$

which can be expressed as:

$$p = E/c$$

12.11 Particles of Light

In 1909, around the time Einstein left his position as a patent clerk and began his first academic role as an associate professor in Zurich, he made a small yet pivotal advancement. For the first time, he referred to "point-like" quanta with energy hv. In classical mechanics, particles like electrons are described as "point-like" objects, a concept that significantly differs from wave-based descriptions. However, the frequency of the radiation, v, provides a crucial link, as it indicates the energy of the particle.

It may seem surprising that Einstein took so long to make this connection, but he was preoccupied with other projects, such as general relativity. Once he did link the concepts, the alignment between statistical arguments and relativity theory significantly strengthened the case. From another perspective, since statistics indicate that *p=E/c*, one could argue that the relativistic equations imply the particle of light has zero rest mass.

This work ultimately convinced Einstein of the reality of light quanta. The term "photon," which refers to the particle of light, was not introduced until 1926 by Gilbert Lewis in Berkeley, California, and it became more widely accepted following the 1927 Solvay Congress titled "Electrons and Photons." Although Einstein stood alone in his belief in the existence of what we now call photons in 1917, this is an appropriate moment to introduce the term. It wasn't until six years later that incontrovertible, direct experimental proof of the existence of photons was obtained by American physicist Arthur Compton.

Compton had been studying X-rays since 1913, working at various American universities and the Cavendish Laboratory in England. Through a series of experiments in the early 1920s, he reached the conclusion that the interaction between X-rays and electrons could only be explained by treating X-rays as particles—photons. The key experiments focused on how X-ray radiation scatters when interacting with an electron. When an X-ray photon collides with an electron, the electron gains energy and momentum, moving off at an angle. Simultaneously, the photon loses energy and momentum, also changing direction, which can be calculated using the laws of particle physics.

The collision resembles a moving billiard ball striking a stationary one, with momentum transfer occurring in a similar manner. However, in the case of the photon, the loss of energy results in a change in the frequency of the radiation, corresponding to the amount $h\nu$ transferred to the electron. A complete explanation of the experiment requires both particle and wave descriptions. When Compton conducted his experiments, he found that the interactions behaved exactly as predicted—scattering angles, wavelength changes, and electron recoil all aligned perfectly with the notion that X-ray radiation consists of particles with energy $h\nu$.

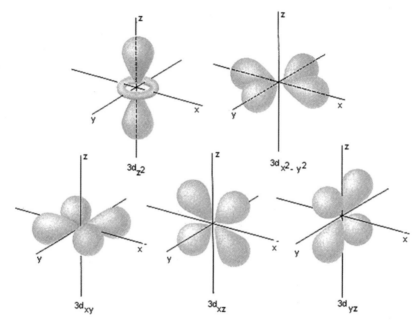

Figure 12-4 Atoms with various dimensional configurations

This phenomenon is now known as the "Compton effect," and in 1927, Compton was awarded the Nobel Prize for his discovery. By 1923, the reality of photons as particles carrying both energy and momentum had been firmly established. However, Niels Bohr initially resisted this idea, striving to find an alternative explanation for the Compton effect. Bohr hesitated to accept the necessity of incorporating both particle and wave descriptions into a complete theory of light, viewing the particle theory as a rival to the wave theory central to his atomic model. Nevertheless, evidence for the wave nature of light remained undeniable. As Einstein remarked in 1924, "there are therefore now two theories of light, both indispensable... without any logical connection."

The attempt to reconcile these two theories laid the foundation for the development of quantum mechanics in the years that followed. Progress was made rapidly across multiple areas, but the discoveries and ideas did not emerge in the precise order needed to build a cohesive new physics. To present a clear narrative, it's necessary to impose a bit more structure on the account than science itself had at the time. One approach is to explain the key concepts before diving into quantum mechanics, even though the theory started taking shape before all those concepts were fully understood. The implications of particle-wave duality, for example, were not completely grasped when quantum mechanics began to develop. However, in any logical account of quantum theory, the next step after recognizing the dual nature of light is the discovery of the dual nature of matter.

12.12 Particles/Wave duality

The discovery originated from a simple yet profound insight by a French nobleman, Louis de Broglie. His idea was straightforward: "If light waves can behave like particles, why shouldn't electrons also behave like waves?" This seemingly speculative thought could have easily been dismissed. But instead, de Broglie followed through with mathematical rigor, revolutionizing the understanding of particle-wave duality. His work led him to receive the Nobel Prize in 1929 and solidified his role as a pioneer of quantum theory.

Similar speculations had been made before, notably concerning X-rays as early as 1912, when W.H. Bragg, another Nobel Laureate, remarked, "The problem is not to decide between two theories of X-rays, but to find one that can encompass both." De Broglie's true contribution was to advance this particle-wave duality into the realm of matter and offer a clear mathematical framework to describe it. His older brother, Maurice, a respected experimental physicist, played a pivotal role by encouraging Louis to explore the dual aspects of particles and waves.

De Broglie's

Born in 1892, de Broglie entered the University of Paris in 1910, where he became fascinated by quantum mechanics, partially due to Maurice, who shared insights from the first Solvay Congress. However, his physics studies were interrupted by compulsory military service from

1913 until 1919 due to World War I. After the war, he returned to his studies and began to develop his theory of particle-wave duality, culminating in a breakthrough in 1923 with the publication of three papers on light quanta.

These early papers, though not immediately impactful, paved the way for his doctoral thesis at the Sorbonne in 1924, where he laid out the full theory of matter waves. His thesis presented the idea that electrons, thought of at the time as mere particles, also exhibited wave-like behavior, particularly evident in their existence at distinct energy levels in atoms. He noted that electrons, like waves, seemed to "fit" into these energy levels, much like how certain wavelengths fit a vibrating string.

De Broglie further posited that particles like photons are guided by associated waves, a concept he supported with a thorough mathematical description. Despite the initial skepticism of his examiners, de Broglie confidently suggested that the matter waves could be observed experimentally by diffracting an electron beam through a crystal. He had derived a simple relation for this—momentum multiplied by wavelength equals Planck's constant ($p\lambda = h$)—and predicted that electrons, due to their small mass and momentum, would exhibit wave-like behavior when diffracted through small gaps between atoms in a crystal lattice.

Unbeknownst to de Broglie, electron diffraction effects had already been observed as early as 1914 by American physicists Clinton Davisson and Charles Kunsman. Yet de Broglie continued to promote his hypothesis, and eventually, his work reached Einstein, who saw its significance and relayed it to Max Born in Göttingen. By 1925, despite existing evidence, the concept of matter waves was still considered speculative until Erwin Schrödinger incorporated de Broglie's ideas into his atomic theory, spurring experimental verification.

In 1927, experiments by two groups—Davisson and Lester Germer in the U.S. and George Thomson (son of J.J. Thomson) and Alexander Reid in England—proved de Broglie right. Electrons, when diffracted by crystal lattices, behaved as waves. Remarkably, J.J. Thomson had received the Nobel Prize in 1906 for proving that electrons are particles, and his son, George, received the same award in 1937 for demonstrating their wave-like nature. Both were correct: electrons are both particles and waves.

This breakthrough led to the discovery that other particles, such as protons and neutrons, also exhibit wave properties. In the late 1970s and 1980s, Tony Klein and colleagues at the University of Melbourne revisited classic wave-theory experiments, this time using neutron beams instead of light, further confirming de Broglie's revolutionary idea.

Ch 13—Quantum Physics

13.1 Revolution of the new physics

One of the challenges of atomic spectroscopy that the simple Bohr model of the atom could not explain was the splitting of spectral lines, which should be closely spaced "multiplets." Each spectral line corresponds to a transition between energy states, revealing how many energy levels exist within the atom and how deep each level is on the quantum staircase. Physicists studying spectra in the early 1920s proposed several possible explanations for this multiplet structure.

The most effective explanation came from Wolfgang Pauli, who introduced four distinct quantum numbers for electrons in 1924. At that time, physicists still viewed electrons primarily as particles and sought to explain their quantum properties using familiar concepts. Three of these quantum numbers were already included in the Bohr model, describing an electron's angular momentum (the speed of its orbit), the shape of its orbit, and its orientation. The fourth quantum number was needed to account for another property of the electron, which could manifest in two forms, thus explaining the observed splitting of spectral lines.

It quickly became apparent that Pauli's fourth quantum number represented the electron's "spin," which could be visualized as pointing either up or down, resulting in a convenient double-valued quantum number. This concept was first proposed by young physicist Ralph Kronig, who suggested that the electron possessed an intrinsic spin of one-half in natural units ($h/2\pi$) and that this spin could align either parallel or antiparallel to the atom's magnetic field.

To his surprise, Pauli himself opposed this idea, primarily because it conflicted with the notion of the electron as a particle within relativistic theory. Just as a classical electron orbiting the nucleus was deemed unstable according to classical electromagnetic theory, a spinning electron also seemed unstable according to relativity. Pauli's reluctance to embrace the idea ultimately led Kronig to abandon it, and he never published his findings.

Less than a year later, however, George Uhlenbeck and Samuel Goudsmit from the Institute for Theoretical Physics in Leyden independently proposed the same concept. They published their suggestion in the German journal *Die Naturwissenschaften* in late 1925 and in *Nature* in early 1926.

The theory of the "spinning electron" was soon refined to provide a comprehensive explanation for the previously unaccounted-for splitting of spectral lines, and by March 1926, even Pauli was convinced of its validity. But what exactly is this phenomenon known as spin? Attempting to explain it in everyday language reveals the challenge of grasping such a quantum concept; understanding often eludes us. For instance, one explanation might state that electron spin differs from the spin of a child's top, as the electron must complete two full rotations to return to its original position. However, this raises the question: how can an electron wave "spin" at all?

Pauli

Nobody was happier than Pauli when Bohr demonstrated in 1932 that electron spin cannot be measured through any classical experiment, such as deflecting electron beams with magnetic fields. This property only manifests in quantum interactions, like those responsible for the splitting of spectral lines, and holds no "classical" significance whatsoever. How much simpler it might have been for Pauli and his colleagues grappling with atomic theory in the 1920s if they had referred to the electron's "gyre" instead of its "spin" from the outset! Today, however, we are left with the term "spin," and any efforts to eliminate classical terminology from quantum physics are unlikely to succeed. While no one truly comprehends the intricate workings of atoms, Pauli's four quantum numbers do illuminate several essential features and properties of the atom.

13.2 Pauli and the Exclusion principle

Wolfgang Pauli was among the most distinguished scientists who contributed to the foundation of quantum theory. Born in Vienna in 1900, he enrolled at the University of Munich in 1918, already known as a precocious mathematician with a completed paper on general relativity that caught Einstein's attention, leading to its publication in January 1919.

Pauli absorbed knowledge from university classes and the Institute of Theoretical Physics, coupled with his extensive reading. His mastery of relativity was so profound that, in 1920, he was tasked with writing a comprehensive review article on the topic for a prominent mathematics encyclopedia. This remarkable article, penned by the twenty-year-old student, garnered widespread acclaim in the scientific community, including praise from Max Born, who welcomed Pauli as his assistant in Göttingen in 1921.

From Göttingen, Pauli moved on to Hamburg and then to Bohr's Institute in Denmark. Born did not lament Pauli's departure; his new assistant, Werner Heisenberg, was equally talented and played a vital role in developing quantum theory. Before the term "spin" was coined for Pauli's fourth quantum number, he had already leveraged these four numbers in 1925 to resolve a significant puzzle in the Bohr model of the atom.

In hydrogen, the single electron occupies the lowest energy state at the bottom of the quantum staircase. When excited, it can jump to a higher energy level and then return to the ground state, emitting a quantum of radiation in the process. However, in more massive atoms with additional electrons, these electrons do not all settle into the ground state; instead, they distribute themselves across higher energy levels. Bohr described electrons as occupying "shells" around the nucleus, with new electrons filling the lowest energy shell first until it was full, before moving to the next higher shell, thus constructing the periodic table of elements.

However, he did not clarify how or why a shell fills or why the first shell can accommodate only two electrons.

Each of Bohr's shells corresponds to a set of quantum numbers, and in 1925, Pauli recognized that by introducing his fourth quantum number for the electron, the number of electrons in each full shell perfectly matched the number of unique sets of quantum numbers associated with that shell. He formulated what is now known as the "Pauli Exclusion Principle," which states that no two electrons can occupy the same set of quantum numbers. This principle provided a rationale for how the shells fill in more massive atoms.

The exclusion principle and the discovery of electron spin were ahead of their time, only fully integrated into the framework of new physics in the late 1920s, after modern physics had been established. Due to the rapid advancements in physics during 1925 and 1926, the significance of the exclusion principle was sometimes overlooked. However, it is as fundamental and far-reaching as the core concepts of reality and has extensive applications across physics. The Pauli Exclusion Principle applies to all particles with a "half-integer" spin, such as $(1/2)\hbar$, $(3/2)\hbar$, $(5/2)\hbar$, and so forth. In contrast, particles with no spin (like photons) or integer spins such as $\hbar$, $2\hbar$, $3\hbar$, etc., follow a different set of rules

The statistical rules governing "half-spin" particles are known as Fermi-Dirac statistics, named after Enrico Fermi and Paul Dirac, who developed them in 1925 and 1926. These particles are referred to as "fermions." In contrast, the rules for "full-spin" particles are called Bose-Einstein statistics, named after the two physicists who formulated them, with these particles being termed "bosons." In 1924, Einstein became intrigued by the work of Indian physicist Satyendra Nath Bose.

Einstein was so impressed by Satyendra Bose's work that he translated it into German himself and forwarded it with a strong recommendation, facilitating its publication in August 1924. In his research, Bose developed a straightforward derivation of the law for massless particles that adhere to a specific type of statistics. He sent a copy of his findings in English to Einstein, asking him to help get it published in the *Zeitschrift für Physik*.

Using an innovative method for counting particles, which he invented, Bose was able to derive Max Planck's formula for the radiation of bodies. Einstein built upon Bose's work, applying it to atoms and molecules. This approach became integral to the modern development of "quantum statistical mechanics." Together, Einstein and Bose predicted the existence of a new state of matter known as the Bose-Einstein condensate, which was later confirmed through experimentation.

Satyendra Bose was born in Calcutta in 1894, and by 1924, he was a Reader in Physics at the newly established Dacca University. Following the works of Planck, Einstein, Bohr, and Sommerfeld from a distance, and recognizing the limitations of Planck's law, he aimed to derive the blackbody law anew. He began with the assumption that light consists of photons, as they are now called. By discarding classical theory elements and deriving Planck's law from a combination of light quanta—considered relativistic particles with zero mass—and statistical methods, Bose successfully disentangled quantum theory from its classical roots. This approach allowed radiation to be treated as a quantum gas, focusing on counting particles rather than wave frequencies.

Bose-Einstein statistics were being developed concurrently with the excitement surrounding de Broglie waves, the Compton effect, and electron spin during 1924-1925. These statistics marked Einstein's final significant contribution to quantum theory and represented a complete departure from classical concepts.

As mentioned earlier, Einstein expanded on these statistics, applying them to what was then a theoretical scenario involving a collection of atoms—whether in gas or liquid form—that followed the same rules. While these statistics were not suitable for real gases at room temperature, they perfectly accounted for the unusual properties of superfluid helium, a liquid cooled to near absolute zero (-273 °C). With the introduction of Fermi-Dirac statistics by 1926, it took some time for physicists to discern which rules applied in different contexts and to appreciate the implications of half-integer spin.

All familiar matter, including electrons, protons, and neutrons, are classified as fermions. Without the exclusion principle, the diversity of chemical elements and the characteristics of our physical world would not exist. In contrast, bosons are more elusive particles, such as photons, and the blackbody law arises from the competition among photons to occupy the same energy state. Under the right conditions, helium atoms can exhibit bosonic properties, becoming superfluid due to the arrangement of their two protons and two neutrons, which results in a net half-integer spin of zero. Fermions are conserved in particle interactions, meaning that the total number of electrons in the universe cannot increase, while bosons can be generated in large quantities, as anyone who has turned on a light can attest.

Although this framework appears coherent from a 2011 perspective, quantum theory was in a state of "disarray" by 1925. Progress was not linear; rather, many researchers were navigating their own paths through the complexities of quantum theory. Leading scientists were acutely aware of this chaos and expressed their concerns publicly. However, a significant breakthrough was on the horizon, primarily from a new generation of researchers who emerged after World War I and were perhaps more open to innovative ideas.

In 1924, Heisenberg, following an unsuccessful attempt to calculate the structure of the helium atom, remarked to Pauli in early 1923, "What a misery!" Pauli echoed this sentiment in a letter to Sommerfeld in July of that year, lamenting the difficulties with theories concerning atoms that contain more than one electron. By May 1925, Pauli wrote to Kronig that "physics at the moment is again very muddled," and even Bohr expressed his concerns about the numerous challenges facing his atomic model.

As late as June 1926, Wilhelm Wien, whose blackbody law had served as a crucial foundation for Planck's groundbreaking work, expressed his frustrations to Schrödinger regarding the "morass of integral and half-integral quantum 'discontinuities' and the arbitrary application of classical theory." By 1925, all the prominent figures in quantum theory were acutely aware of these issues. Established scientists like Henri Poincaré, Lorentz, Planck, J.J. Thomson, Bohr, Einstein, and Born remained influential, while a new generation, including Pauli, Heisenberg, Dirac, and others, was beginning to make significant contributions.

Heisenberg

The two leading figures in quantum theory, Einstein and Bohr, began to diverge significantly in their scientific perspectives by 1925. Initially, Bohr was a staunch opponent of the concept of light quanta, while Einstein grew increasingly concerned about the role of "probability" in quantum theory, leading Bohr to become its most prominent advocate. Ironically, the statistical methods that Einstein had introduced became foundational to quantum theory. However, as early as 1920, Einstein confided to Born, "that business about causality causes me a lot of trouble, too… I must admit that… I lack the courage of my convictions." This initiated a dialogue between Einstein and Bohr on this topic that would continue for thirty-five years, lasting until Einstein's death.

13.3 Matrices and Waves

Werner Heisenberg was born on December 5, 1901, in Würzburg. In 1920, he began studying physics at the University of Munich under the guidance of Arnold Sommerfeld, a prominent physicist who had contributed to the development of the Bohr model of the atom. Heisenberg immediately dove into quantum theory research, tasked with finding quantum numbers that could explain the splitting of spectral lines into doublets. Within a few weeks, he proposed a solution using half-integer quantum numbers. Despite the simplicity and accuracy of his solution, it was met with resistance. His supervisor, Sommerfeld, along with his colleagues, was deeply committed to the Bohr model, where integral quantum numbers were accepted as fundamental. As a result, Heisenberg's idea was quickly dismissed.

The main concern was that introducing half-integers might lead to a cascade of fractional values, threatening the foundational principles of quantum theory. However, a few months later, the more senior physicist Alfred Lande arrived at the same conclusion and published the idea. It was later discovered that half-integer quantum numbers are essential in quantum theory, particularly in describing electron spin. Particles with integer or zero spin, like photons, follow Bose-Einstein statistics, while particles with half-integer spin (e.g., 1/2, 3/2) obey Fermi-Dirac statistics. This discovery linked electron spin to the structure of atoms and the periodic table of elements. Although quantum numbers only change in whole increments, a shift from 1/2 to 3/2 or 5/2 to 9/2 is just as valid as changes between whole integers.

While Heisenberg missed the opportunity to receive credit for this early insight, he soon compensated with groundbreaking contributions. In 1924, Heisenberg succeeded Wolfgang Pauli as Born's assistant in Göttingen, allowing him to work with Bohr in Copenhagen. By 1925, Heisenberg was well-positioned to make significant strides in quantum theory. His breakthrough stemmed from a profound idea circulating within the Göttingen group: that a physical theory should only concern itself with phenomena that can be observed through experiments. This seemingly simple concept had deep implications. For instance, experiments that "observe" electrons in atoms don't show tiny particles orbiting a nucleus. Instead, observations of spectral lines reveal how electrons move between energy states. Heisenberg discarded everyday analogies like orbits and focused on the mathematics of associations between pairs of energy states, leading to his revolutionary formulation of quantum mechanics.

13.4 Breakthrough in Quantum Theory

The often-repeated story of Werner Heisenberg's breakthrough begins with a severe case of hay fever in May 1925, prompting him to retreat to the remote island of Helgoland. There, with no distractions and his hay fever alleviated, Heisenberg intensely focused on the task of interpreting quantum behavior. After his time on the island, Heisenberg returned to Göttingen, where he spent three weeks refining his ideas into a publishable form. Unsure of his work's readiness, he shared the paper with his friend Wolfgang Pauli, who was enthusiastic. However, Heisenberg, exhausted by his efforts, left the paper with Max Born, uncertain about publication, before departing for a lecture tour in Leyden and Cambridge. Ironically, he did not discuss his new work with his audiences, leaving them to learn of it through other channels.

Born, recognizing the significance of Heisenberg's paper, submitted it to *Zeitschrift für Physik*. He quickly realized that Heisenberg's mathematics—describing the interactions between two quantum states—couldn't be expressed using ordinary numbers. Instead, it required arrays of numbers, or "tables," as Heisenberg called them, analogous to a chessboard with rows and columns. Each element in these arrays described a pair of states. However, a curious result puzzled Heisenberg: when multiplying these arrays, the outcome depended on the order of multiplication, unlike ordinary arithmetic where, for example, 2 × 3 equals 3 × 2.

This non-commutative behavior baffled Heisenberg, but Born, after much contemplation, had an epiphany. The arrays Heisenberg had constructed were actually known in mathematics as "matrices," a branch of math Born had studied decades earlier in Breslau. Matrices are unique because their multiplication is order-dependent, or "non-commutative," a property that struck Born as fundamentally important in quantum mechanics.

Born's recognition of matrices brought clarity to Heisenberg's work. Just as each square on a chessboard can be identified by a pair of coordinates (e.g., b4 or f7), quantum mechanical states are defined by pairs of numbers. The state of a chessboard square is determined by the piece occupying it, and changes in the board's state (such as a pawn moving from e2 to e4) can be described algebraically. Quantum transitions, similarly, are described by linking initial and final states without detailing how the transition occurs. Heisenberg, Paul Dirac, and Erwin Schrödinger later developed different mathematical notations—akin to different chess notations—to describe the same quantum events, highlighting the deep connection between mathematics and quantum theory.

13.5 Quantum Mathematics

In the summer of 1925, Max Born, collaborating with Pascual Jordan, laid the foundation for what would later be known as "matrix mechanics." When Werner Heisenberg returned to Copenhagen in September, he joined their efforts via correspondence, culminating in a significant scientific paper on "Quantum Mechanics." This paper, more detailed and explicit than Heisenberg's initial publication, emphasized the central importance of the non-commutativity of quantum variables. In an earlier paper with Jordan, Born had already uncovered the relation $pq - qp = \hbar/i$, where p and q are matrices representing quantum variables, such as momentum and position, or where p and q are matrices representing quantum variables, the equivalent in the quantum world of momentum and position. Planck's constant $\hbar$ and the imaginary unit i (the square root of -1) appeared in this equation, which the Göttingen team highlighted as the "fundamental quantum-mechanical relation."

But what did this equation mean in physical terms? By 1925, physicists were familiar with Planck's constant, and equations involving iii, which hinted at oscillations or waves. However, matrices were unfamiliar to most mathematicians and physicists at the time, and the concept of non-commutativity seemed as puzzling as Planck's introduction of $\hbar$ had been back in 1900.

Heisenberg remarked on how matrix mechanics replaced Newtonian equations with similar ones involving matrices, yet surprisingly, many of the familiar results of Newtonian mechanics, such as the conservation of energy, could still be derived. In essence, matrix mechanics absorbed Newtonian mechanics within itself, much like Einstein's theory of relativity had previously done. It offered a deeper, more comprehensive framework, with classical mechanics emerging as a special case within the broader quantum theory.

Enters Paul Dirac

Paul Dirac, born on 8 August 1902, was only a few months younger than Werner Heisenberg, and is often regarded as one of the greatest English theorists since Isaac Newton. Dirac would go on to develop the most complete form of what we now call quantum mechanics. Yet, his path to theoretical physics was unconventional. After graduating with an engineering degree from Bristol University in 1921, Dirac couldn't find a job in his field. Instead, he was offered a studentship to study mathematics at Cambridge. However, financial constraints delayed his plans. Living with his parents in Bristol, Dirac compressed a three-year mathematics course into two years, earning a BA in applied mathematics by 1923. With this degree, he finally made his way to Cambridge, supported by a grant from the Department of Scientific and Industrial Research. It was only upon arriving at Cambridge that Dirac first learned about quantum theory.

As a new and unknown research student, Dirac attended a talk by Heisenberg in July 1925. Although Heisenberg didn't publicly discuss his latest work on quantum mechanics, he mentioned it to Ralph Fowler, Dirac's supervisor. Fowler, in turn, received a proof of Heisenberg's groundbreaking paper in mid-August, before it was published in *Zeitschrift für Physik*. He passed it to Dirac, who became one of the first people outside Göttingen, aside from Pauli, to study the new theory.

In his paper, Heisenberg had identified the non-commutativity of quantum variables but had not fully developed the idea. When Dirac began working through the equations, he quickly recognized the importance of the fact that:

(a * b) is not-equal-to (b * a) or (a∗b)≠(b∗a).

Unlike Heisenberg, Dirac was already familiar with mathematical objects that behaved this way, thanks to his knowledge of the work of William Hamilton, who had developed such mathematics a century earlier. Ironically, Hamilton's equations, originally created to solve problems involving planetary orbits, now provided the framework for quantum theory, which dispensed entirely with the notion of electron orbits.

Dirac independently discovered that the equations of quantum mechanics shared the same structure as those of classical mechanics. Moreover, classical mechanics was revealed to be a special case of quantum mechanics, applicable when quantum numbers are large, or Planck's constant is set to zero. Following his own path, Dirac further advanced quantum mechanics by introducing "quantum algebra," a system involving the addition and multiplication of quantum variables, or "*q* numbers." In this mathematical framework, concepts like comparing the size of two numbers—whether *a* is larger than *b*—had no meaning, reflecting the strange and non-intuitive nature of quantum systems. Yet, these rules perfectly described the observed behavior of atomic processes. Quantum algebra, in fact, encompassed matrix mechanics and extended beyond it.

In early 1926, Dirac published a series of four papers, culminating in his doctoral thesis. Meanwhile, Wolfgang Pauli used matrix mechanics to correctly predict the Balmer series for the hydrogen atom, and by the end of 1925, the phenomenon of spectral line splitting into doublets was attributed to the new property of electron "spin." The pieces of quantum mechanics were fitting together, and the different mathematical approaches—whether Dirac's quantum algebra or Heisenberg's matrix mechanics—were revealed to be different expressions of the same underlying reality.

Dirac

The game of chess offers a useful analogy for understanding different ways to describe quantum mechanics. There are various ways to represent a chess game. One could use a printed chessboard with all the pieces' positions marked, but this would be cumbersome for recording an entire match. Another method is to describe the moves: "King's pawn to King's pawn four." A more concise approach is algebraic notation, where the same move is written as simply "d2-d4." All three formats convey the same information about a real event—the transition of a pawn from one state to another. Similarly, in quantum mechanics, we have different formulations that describe the same underlying phenomena.

Dirac's "quantum algebra" is mathematically elegant and efficient, while the matrix methods developed by Born and others after Heisenberg's work are more cumbersome but equally effective. Some of Dirac's early breakthroughs came when he incorporated special relativity into his quantum mechanics. He was thrilled by the idea of light behaving as a particle, or photon, and delighted in discovering that by treating time as a quantum variable in his equations, he could predict that atoms would recoil when emitting light—just as they would if light carried its own momentum. He then applied this framework to develop a quantum-mechanical interpretation of the Compton effect.

Dirac's approach involved two steps: first, the mathematical manipulation of quantum variables, and second, the interpretation of the resulting equations in terms of observable physical events. This process mirrors how nature seems to work, presenting us with an observed event—like an electron transition—after some internal "calculation." However, instead of fully embracing quantum algebra, physicists in the years after 1926 were drawn away by another mathematical technique that offered solutions to the challenges posed by quantum mechanics, despite initially picturing the electron as a particle transitioning between quantum states.

What about de Broglie's idea that electrons and other particles should also be thought of as waves? In Dirac's formulation of quantum mechanics, a crucial term in the Hamiltonian equations is replaced by the quantum mechanical expression:

$$(ab-ba)/i\hbar.$$

This is simply another form of what Born, Heisenberg, and Jordan referred to as the "fundamental quantum-mechanical relation" in their collaborative "three-man paper." Although this paper was written before Dirac's first quantum mechanics paper, it was published after Dirac's work appeared.

Schrodinger

With his characteristic modesty, Dirac recounted how straightforward it became to make progress once it was understood that the correct quantum equations were simply classical equations reformulated in Hamiltonian form. For many of the challenges that arose in quantum theory, the solution was clear: set up the equivalent classical equations, convert them into their Hamiltonian form, and solve the problem. "It was like a game," Dirac said, "and a very interesting one to play."

13.6 Schrodinger's Theory

While matrix mechanics and quantum algebra were making their relatively quiet debuts in the scientific world, significant activity was unfolding in the broader realm of quantum theory. By late 1925, de Broglie's theory of electron waves had emerged, though the key experiments proving the wave nature of the electron had yet to be conducted. Separate from the work of Heisenberg and his colleagues, this wave idea led to the development of another quantum mathematics.

The concept originated from de Broglie and was brought to wider attention by Einstein. De Broglie's work might have remained a theoretical curiosity, dismissed as an interesting but untested mathematical hypothesis, had Einstein not recognized its potential. Einstein alerted Max Born to de Broglie's idea, which ignited a series of experiments that ultimately confirmed the reality of electron waves. It was through one of Einstein's papers, published in February 1925, that Erwin Schrödinger first encountered a crucial comment: "I believe that it involves more than merely an analogy."

In those days, Einstein's words carried great weight, and this subtle endorsement inspired Schrödinger to explore the implications of de Broglie's theory. Schrödinger was somewhat of an outsider among the younger physicists leading the quantum revolution. Born in 1887, he was 39 years old when he made his most significant scientific contribution—an advanced age for groundbreaking work. With a doctorate obtained in 1910 and a professorship at Zurich since 1921, Schrödinger was seen as a respected, if traditional, figure in physics.

Yet his approach to quantum theory differed fundamentally from that of the Göttingen group and Dirac. Where they had made quantum theory more abstract, detaching it from classical physical concepts, Schrödinger sought to reintroduce tangible, familiar ideas. He envisioned quantum physics in terms of waves—observable features of the physical world—and resisted the emerging notions of indeterminacy and electrons jumping between states instantaneously. While Schrödinger provided an immensely valuable tool with wave mechanics, conceptually, it represented a step back to 19th-century ideas.

De Broglie had suggested that electron waves in orbit around an atomic nucleus must fit a whole number of wavelengths into each orbit, making certain orbits "forbidden." Schrödinger applied wave mathematics to calculate the energy levels allowed under this model, but his initial results didn't match known atomic spectra. The issue was not with his method, but rather that he hadn't accounted for electron spin—an idea that hadn't yet been introduced in 1925.

Schrödinger temporarily shelved the work, missing the chance to be the first to publish a comprehensive mathematical treatment of quantum mechanics. However, when he was later invited to present a colloquium on de Broglie's work, he revisited his calculations. By ignoring relativistic effects, he found that his results aligned well with atomic observations where those effects weren't significant. As Dirac would later demonstrate, electron spin is fundamentally a relativistic phenomenon, unrelated to everyday notions of spinning objects. Schrödinger's wave mechanics, published in a series of 1926 papers, followed closely after contributions from Heisenberg, Born, Jordan, and Dirac. Despite its 19th-century roots, Schrödinger's work became a cornerstone of quantum theory.

Schrödinger's variation of quantum theory introduced equations that were strikingly familiar to those used to describe real-world waves—like ocean waves or sound waves traveling through the atmosphere. These wave equations were met with enthusiasm by the physics community because they felt intuitive and accessible. This was in sharp contrast to Heisenberg's approach, which abandoned any visual representation of the atom and focused solely on quantities measurable through experiments. Heisenberg's theory revolved around electrons as particles, while Schrödinger's centered on the concept of electrons as waves. Both approaches, despite their differences, produced equations that accurately described measurable quantum phenomena.

What was initially surprising soon became a point of clarity when Schrödinger, along with Carl Eckart and later Dirac, mathematically proved that the various sets of equations—whether based on waves or matrices—were actually equivalent. These were simply different ways of viewing the same mathematical reality. Schrödinger's equations incorporated the same non-commutative relations and the crucial factor $\hbar/i$ (Planck's constant divided by i, the square root of minus one), just as they appeared in Heisenberg's matrix mechanics and Dirac's quantum algebra. The realization that all these approaches led to the same fundamental results bolstered physicists' confidence in quantum theory as a whole.

Of the various methods, Dirac's quantum algebra was the most comprehensive, as it encompassed both matrix mechanics and wave mechanics as special cases. However, physicists in the 1920s gravitated towards Schrödinger's wave equations because they aligned with familiar concepts from classical physics—optics, fluid dynamics, and other well-understood fields.

Ironically, the success of Schrödinger's wave mechanics, with its easy-to-grasp physical interpretations, may have delayed a deeper understanding of the quantum world for many years. While it provided a powerful tool for solving quantum problems, the wave-based view, rooted in classical physics, somewhat masked the more abstract and counterintuitive nature of quantum mechanics.

13.7 The original founder of Quantum Mechanics

Historically, Dirac did not invent or discover wave mechanics. The quantum equations that became crucial in quantum mechanics actually trace their origins to a 19th-century effort to unify wave and particle theories of light. This work was pioneered by Sir William Hamilton, a Dublin-born mathematician considered by many to be one of the greatest of his time. Hamilton's most significant achievement, though not fully appreciated in his day, was uniting the laws of optics and dynamics into a single mathematical framework. This set of equations could describe both wave and particle motion, making it a groundbreaking development.

Hamilton's work, published in the late 1820s and early 1830s, influenced researchers in both mechanics and optics throughout the latter half of the 19th century. However, few recognized the full potential of Hamilton's system, which linked mechanics and optics at a deeper level. His revolutionary implication—that just as light rays had to be understood as waves in optics, particle motion might also be better described by wave motions in mechanics—was too radical for 19th-century physics. Even Hamilton himself did not explicitly propose it. The idea was so foreign that it never occurred to anyone at the time; it wasn't rejected as absurd but simply unimaginable given the framework of classical physics. It was only after classical mechanics failed to explain atomic processes that such an idea could emerge.

In this sense, Sir William Hamilton can be seen as a forgotten founder of quantum mechanics. He also developed what is now known as matrix algebra, another vital tool for the quantum revolution. Had Hamilton lived in the quantum era, he might have quickly grasped the connection between matrix mechanics and wave mechanics. Although Dirac later built upon these ideas, it's not surprising that he initially missed the link. As a young researcher focused on abstract concepts and following Heisenberg's lead in distancing quantum theory from the traditional view of electrons orbiting atomic nuclei, Dirac was not looking for a simple, intuitive physical model of the atom. He was deeply immersed in his first major research project, and even for a mind as brilliant as his, there are limits to how much one can accomplish in a short period.

What people did not initially recognize was that wave mechanics, despite Schrodinger's expectations, did not offer a comforting picture of the quantum world. Schrodinger believed he had removed the concept of quantum jumps between states by incorporating waves into quantum theory. He imagined the "transitions" of an electron from one energy state to another as akin to the change in vibration of a violin string moving from one note to another (or one harmonic to another). In this framework, he viewed the wave described by his wave equation as the matter wave proposed by de Broglie.

Bohr, Heisenberg and Pauli – Courtesy of the University of Copenhagen

As researchers delved deeper into the equations, the hopes of restoring classical physics to a central role began to fade. Bohr, for instance, found the wave concept perplexing. How could a wave, or a set of interacting waves, cause a Geiger counter to click as if it were detecting a single particle? What was actually "waving" within the atom? Most importantly, how could blackbody radiation be explained using Schrodinger's waves? In 1926, Bohr invited Schrodinger to Copenhagen to address these questions together, but their solutions were not particularly palatable for Schrodinger.

Upon closer examination, it became evident that the waves were as abstract as Dirac's q numbers. The mathematics indicated that they could not be real waves in space, like ripples on a pond; instead, they represented a complex form of vibrations within an imaginary mathematical realm known as "configuration space." Moreover, each particle, such as an electron, required its own three-dimensional space. Thus, a single electron could be described by a wave equation in three-dimensional configuration space, while two electrons needed a six-dimensional space, and three electrons would require nine dimensions, and so on.

Regarding blackbody radiation, even when reformulated in wave-mechanical terms, the necessity for discrete quanta and quantum jumps persisted. This revelation left Schrodinger disillusioned, leading him to famously remark, with various translations, "Had I known that we were not going to get rid of this damned quantum jumping, I never would have involved myself in this business."

As Heisenberg noted in his book *Physics and Philosophy*, the paradoxes arising from the duality of wave and particle representations were not resolved; instead, they were obscured within the mathematical framework. The appealing image of "physically" real waves orbiting atomic nuclei, which inspired Schrodinger to formulate the wave equation that bears his name, is fundamentally flawed. Wave mechanics does not offer a clearer insight into the reality of the atomic world than matrix mechanics; however, it provides an "illusion" of familiarity and comfort that matrix mechanics lacks.

This illusion has persisted into the present, obscuring the reality that the atomic world is vastly different from our macroscopic experience. Generations of students, many of whom have since become professors, might have developed a deeper understanding of quantum theory had they confronted the abstract nature of Dirac's approach rather than being led to believe that their knowledge of everyday wave behavior accurately represented atomic phenomena. Consequently, despite significant advancements in applying quantum mechanics to various intriguing problems, we find ourselves, over eighty years later, not much better off than the physicists of the late 1920s regarding our fundamental understanding of quantum physics. The very success of the Schrodinger equation as a practical tool has hindered deeper contemplation about how and why it functions as it does.

13.8 Developed Quantum physics

The foundations of quantum theory and practical quantum physics since the 1920s are rooted in concepts developed by Niels Bohr and Max Born during that era. Bohr provided a philosophical framework that reconciles the dual particle-wave nature of the quantum realm, while Born established essential principles for constructing our quantum models. Bohr posited that both particle physics and wave physics are equally valid and complementary descriptions of the same reality.

Neither perspective is complete on its own; specific situations may warrant the use of the particle model, while others are better suited to the wave model. An electron, for instance, is neither strictly a particle nor a wave; it exhibits wave-like behavior in some contexts and particle-like behavior in others. However, no experiment can simultaneously demonstrate both characteristics. This notion that wave and particle representations are two complementary aspects of an electron's complex nature is known as complementarity.

Born developed a new interpretation of Schrödinger's waves, focusing on a key element of Schrödinger's equation known as the "wave function," typically represented by the Greek letter Ψ. While working in Göttingen with experimental physicists who were regularly confirming the particle nature of electrons, Born found it difficult to accept that this wave function represented a "real" electron wave. Despite this skepticism, like many physicists of his time, he recognized the convenience of wave equations for solving various problems.

In seeking to associate a wave function with particle existence, Born refined an idea previously discussed in the context of light. He argued that while particles are real, they are, in a sense, influenced by the wave. The strength of the wave, specifically the value of $Ψ^2$ at any given point in space, serves as a measure of the probability of locating the particle at that point. While we can never determine an electron's exact position, the wave function allows us to calculate the likelihood of finding the electron in a particular location during an experiment.

The most intriguing aspect of this idea is that any electron could theoretically be anywhere; however, it is much more likely to be found in certain locations than others. This is reminiscent of statistical principles, which suggest that while all the air in a room could, in theory, gather in the corners, such an occurrence is extremely improbable. Born's interpretation of Ψ thus added an element of uncertainty to the already ambiguous quantum world.

Born's ideas aligned well with Heisenberg's discovery in late 1926 that uncertainty is inherent in quantum mechanics. The mathematical principle that states $pq \neq qp$ also implies that we cannot be certain about the values of p and q. If we define p as the momentum of an electron and q as its position, we can measure either p or q with high precision. The measurement uncertainties can be denoted as Delta qΔq, with the Greek letter Δ or (Delta) representing small variations in quantities. The amount of "error" in our measurement might be called Δ-p or Δ-q, since mathematicians use the Greek letter Δ, to symbolize small pieces of variable quantities.

Heisenberg demonstrated that if we attempt to measure both the position and momentum of an electron simultaneously, we will never achieve complete accuracy, as Δ-p - Δ-q must always exceed $\hbar$ (where $\hbar$ is Planck's constant divided by 2π or 2pi).

Thus, increased precision in determining an object's position results in decreased certainty about its momentum, and vice versa. The uncertainty principle has profound implications, but it's crucial to understand that it does not reflect a limitation in the experimental methods used to measure the properties of an electron.

It is a *cardinal rule of quantum mechanics* that in principle it is impossible to measure precisely certain pairs of properties, including position/momentum, simultaneously. **There is no absolute truth at the quantum level**.

13.9 Heisenberg's uncertainty relation

The Heisenberg uncertainty principle quantifies the extent to which the complementary descriptions of fundamental entities, like electrons, overlap. Position is a characteristic of particles, which can be precisely located, while waves lack a fixed position but possess momentum. The more accurately you measure one aspect, such as a particle's position, the less you know about the other, like its wave-related momentum, and vice versa.

In experiments, the nature of the setup determines the outcome: experiments designed to detect particles always reveal particles, while those set to detect waves always reveal waves. However, no experiment can show an electron behaving as both a particle and a wave simultaneously. Niels Bohr emphasized that our understanding of the quantum world relies entirely on experiments, each of which essentially poses a question to quantum reality. These questions are framed in terms of familiar concepts, such as "momentum" or "wavelength," which shape the results we observe. Even though classical physics doesn't accurately describe atomic processes, our experiments are still based on its principles. Moreover, observing atomic processes requires interference, leading Bohr to conclude that asking what atoms are when unobserved is meaningless.

As Max Born noted, all we can do is calculate the probability of a particular outcome for a given experiment. This collection of ideas—uncertainty, complementarity, probability, and the disturbance caused by observation—is known as the "Copenhagen interpretation" of quantum mechanics. While no formal statement was ever written down as the definitive "Copenhagen interpretation," and key elements like the statistical interpretation of the wave function came from Born in Göttingen, this framework became a central way of thinking about quantum mechanics.

The Copenhagen interpretation remains a flexible concept. Bohr first presented it publicly at a conference in Como, Italy, in September 1927.

This marked the completion of a coherent quantum mechanics framework, enabling any skilled physicist to solve problems involving atoms and molecules without needing to deeply question the theory's foundations. Instead, they could simply follow the established methods and derive accurate results.

In the following decades, figures like Paul Dirac and Wolfgang Pauli made significant contributions to the field, and many of the pioneers of quantum theory were honored with

Nobel Prizes, though the awards didn't always follow a clear logic. Heisenberg won the prize in 1932, but he felt that his colleagues Max Born and Pascual Jordan were unfairly overlooked.

Born, in particular, remained bitter for years, often pointing out that Heisenberg didn't even know what a matrix was until Born explained it to him. In a 1953 letter to Einstein, Born lamented that Heisenberg "reaped all the rewards of our work together, such as the Nobel Prize." Schrödinger and Dirac shared the physics prize in 1933, but Pauli didn't receive his, for the exclusion principle, until 1945. Born was finally recognized in 1954 with a Nobel Prize for his work on the probabilistic interpretation of quantum mechanics.

In truth, there is no "real" model of what atoms and elementary particles are like, or any clear understanding of what happens when we aren't observing them. However, the equations of wave mechanics—especially popular and widely used—allow us to make statistical predictions. If we measure a quantum system and obtain result A, the equations can predict the likelihood of getting result B (or C, or D, etc.) in a later measurement.

Quantum theory doesn't describe the nature of atoms or their actions when unobserved. Unfortunately, many who use wave equations today don't fully appreciate this, merely paying lip service to the probabilistic nature of the theory. As Ted Bastin noted, students often learn a "crystallized form of the ideas" from the late 1920s, without ever questioning the foundational aspects. They tend to think of quantum waves as real, and most finish their studies with a mental image of the atom, despite the theory's abstract nature.

People often use the probabilistic interpretation without fully grasping it, yet the effectiveness of Schrödinger's and Dirac's equations, combined with Born's interpretation, allows physicists to successfully work with quantum systems even without fully understanding why the theory works so well.

The first true "quantum designer" was Paul Dirac. Just as he had been the first person outside of Göttingen to fully grasp the new matrix mechanics and expand upon it, he similarly advanced Schrödinger's wave mechanics by placing it on firmer ground and further refining it. In 1928, while adapting the equations to align with the principles of relativity—introducing time as the fourth dimension—Dirac discovered that he needed to incorporate a term now understood to represent electron spin. This addition unexpectedly solved the puzzle of the "doublet splitting" of spectral lines, a problem that had confounded theorists for years.

Furthermore, Dirac's improvement of the equations led to another unforeseen discovery, one that would pave the way for the development of modern particle physics.

13.10 Antimatter

According to Einstein's equations, the energy of a particle with mass mmm and momentum ppp is given by:

$$E^2 = m^2 * c^4 + p^{-2} * c^2$$

This simplifies the famous equation $E=mc^2$ when the momentum is zero. However, this doesn't tell the full story. Since this equation involves taking the square root, mathematically, E can be either positive or negative. Just as *2×2=4 and −2×−2=4*, the energy can be expressed as *E=+/− mc²*.

In many cases, when negative roots appear in equations, they can be dismissed as irrelevant, with the "obvious" solution being the positive value. But Dirac, always thinking deeply, didn't take this shortcut. He contemplated the implications of both the positive and negative roots. In the relativistic version of quantum mechanics, energy levels can be calculated in two sets: one positive, corresponding to mc^2, and one negative, corresponding to $-mc^2$.

According to the theory, electrons should fall into the lowest available energy state. Since the negative energy states are lower than the positive ones, why didn't all electrons collapse into these negative states and vanish?

Dirac's brilliant insight lay in the fact that electrons are fermions, which means that only one electron can occupy each possible state (or two per energy level, one with each spin). He reasoned that the negative energy states must already be fully occupied, preventing electrons from falling into them.

What we think of as "empty space" is actually a sea of negative-energy electrons, according to Dirac's theory. But he didn't stop there. If an electron in this negative sea gains enough energy, it can jump up into the positive energy states and become visible as a normal electron.

To transition from the state $-mc^2$ to the state $+mc^2$, an electron needs an energy input of $2mc^2$, which, for an electron's mass, is about 1 MeV. This energy can easily be supplied during atomic processes or particle collisions. When a negative-energy electron is promoted into the real world, it would behave like an ordinary electron. However, it would leave behind a "hole" in the negative energy sea—essentially the absence of a negatively charged electron. Dirac theorized that this hole should act like a positively charged particle. Much like a double negative forming a positive, the absence of a negative charge in the sea would manifest as a positive charge.

Initially, Dirac speculated that this positively charged particle could be the proton, the only other known particle at the time. But in hindsight, he realized this was incorrect. The positive particle should have the same mass as the electron, not the proton. As he later reflected in his book "Directions in Physics," he regretted not boldly predicting the discovery of a new particle: the positron.

At first, Dirac's work didn't receive widespread acceptance. The idea of a positive counterpart to the electron being the proton was dismissed, but no one knew what to make of his theory. That changed in 1932 when Carl Anderson, an American physicist, observed a positively charged particle while studying cosmic rays. Cosmic rays, energetic particles from space, had been discovered by Victor Hess before World War I, and Hess later shared the 1936 Nobel Prize with Anderson for their contributions.

Anderson's experiments, using a cloud chamber to track charged particles, revealed particles whose tracks bent in a magnetic field just like electrons, but in the opposite direction. These particles had the same mass as electrons but carried a positive charge. Anderson called them "positrons." For this discovery, Anderson won the Nobel Prize in 1936, just three years after Dirac received his for quantum theory.

The discovery of the positron reshaped the way physicists understood the particle world. In 1932, James Chadwick discovered the neutron (earning him the 1935 Nobel Prize), completing the picture of an atomic nucleus composed of positive protons and neutral neutrons, surrounded by negative electrons. Dirac's positron provided the missing piece, introducing the idea of antiparticles and transforming the field of particle physics.

Positrons initially had no place in the established framework of particle physics, and the idea that particles could be created from energy transformed the concept of what a fundamental particle is. According to Dirac's theory, any particle can, in principle, be generated from energy through his process, provided that it is accompanied by its antiparticle, the "hole" in the negative energy sea. Although today's particle-creation theories are more sophisticated, the basic rules remain the same. One key rule is that whenever a particle encounters its antiparticle, they annihilate each other, releasing energy ($2mc^2$) in the form of gamma rays.

Before 1932, many physicists observed particle tracks in cloud chambers, which in hindsight must have been due to positrons. However, it was always assumed that these tracks came from electrons moving toward atomic nuclei rather than positrons moving outward. The idea of new particles was not readily accepted by the scientific community at the time. Today, the situation has reversed. As Dirac remarked in *Directions in Physics*, "People are only too willing to postulate a new particle on the slightest evidence, either theoretical or experimental" (p. 18). The result is a "particle zoo" with over 200 particles, compared to the two fundamental particles known in the 1920s. These particles can be created in particle accelerators given enough energy, although most are highly unstable and quickly decay into other particles and radiation. Among this plethora of particles, the discovery of the antiproton and antineutron in the mid-1950s stands as important confirmation of Dirac's original theory, though they are often overlooked in the crowded field of particle physics.

Entire books have been written about the particle zoo, and many physicists have focused their careers on classifying these particles. However, it seems unlikely that such a vast number of particles could all be truly fundamental. The situation mirrors that of spectroscopy before quantum theory, when scientists could catalog relationships between spectral lines but lacked an understanding of the underlying causes. Something more fundamental likely governs the creation of these particles, as Einstein suggested to his biographer, Abraham Pais, in the 1950s. Einstein believed that the answers would eventually emerge through a unified field theory. Sixty years later, it appears he may have been right. The explosion of particle physics since the 1940s can be traced back to Dirac's pioneering work on quantum theory, laying the foundation for the continued search for deeper, unified principles.

13.11 Inside the nucleus

The nucleus is about 100,000 times smaller than the atom in terms of radius, but because volume is proportional to the cube of the radius, it's more accurate to say that the atom is a *million billion* (10^{15}) times larger than the nucleus. Basic properties like the mass and charge of the nucleus can be measured, leading to the concept of isotopes—nuclei with the same number of protons, and thus the same number of electrons (and similar chemical properties), but differing numbers of neutrons, which result in different masses.

Since protons are all positively charged and naturally repel each other, a stronger force, called the strong nuclear force, must be acting to hold them together. This force operates only over the short distances found within the nucleus. Neutrons also play a role in the stability of the nucleus, and physicists have found a pattern similar to the electron shells surrounding the nucleus when counting the protons and neutrons in stable nuclei. Uranium, with 92 protons, is the largest naturally occurring element, though physicists have artificially created elements with up to 106 protons. However, these heavier nuclei are unstable (except for certain isotopes of plutonium, with atomic number 94) and break down into smaller nuclei. Altogether, there are around 260 known stable nuclei, although our understanding of nuclear structure is still less complete than the Bohr model of the atom.

Certain numbers of nucleons (neutrons or protons)—2, 8, 20, 28, 50, 82, and 126—result in particularly stable nuclei. These are referred to as "magic numbers," and elements with these numbers of nucleons are more abundant in nature than those with slightly different numbers. While the structure of the nucleus is dominated by protons, the number of neutrons in stable isotopes tends to be slightly larger than the number of protons, particularly in heavier elements. Nuclei with "magic numbers" of both protons and neutrons are especially stable.

Theorists predict that superheavy elements with around 114 protons and 184 neutrons should be stable, but these massive nuclei have not yet been observed in nature or successfully created in particle accelerators.

The most stable nucleus is iron-56. Lighter nuclei tend to gain nucleons in an effort to become iron, while heavier nuclei aim to lose nucleons to move toward this most stable form. Inside stars, light nuclei like hydrogen and helium undergo nuclear fusion, merging to form heavier elements such as carbon and oxygen, eventually leading to iron. These fusion reactions release energy.

In the explosive environment of a supernova, immense gravitational energy drives fusion beyond iron, producing heavier elements like uranium and plutonium. As heavy elements decay, they release energy by shedding nucleons, such as alpha particles, electrons, positrons, or neutrons—this energy is a remnant of the supernova that created them. An alpha particle, which consists of two protons and two neutrons, reduces the mass of a nucleus by four units and the atomic number by two when ejected. This process follows the principles of quantum mechanics, particularly Heisenberg's uncertainty relations.

Within the nucleus, nucleons are bound by the strong nuclear force, while an alpha particle just outside the nucleus would be repelled by the electric force. The interplay of these forces creates a "potential well," acting as a barrier. However, due to quantum mechanics, there's a small probability that an alpha particle can escape from the nucleus.

One of Heisenberg's uncertainty relations involves energy and time, stating that a particle's energy can fluctuate within a range ΔE over a short time period Δt, such that $\Delta E \times \Delta t$ is greater than Planck's constant. This allows a particle to "borrow" energy momentarily to overcome the potential barrier before returning to its normal state, but outside the barrier, allowing it to escape. Alternatively, in terms of position, the uncertainty in the particle's location might cause it to appear just outside the barrier, even though it "belongs" inside. This process, known as quantum tunneling, explains how particles escape the nucleus in radioactive decay. However, to understand nuclear fission, a different model of the nucleus is required.

Let's set aside the idea of individual nucleons in their shells and instead think of the nucleus as a liquid droplet. Just as a water droplet wobbles and changes shape, some of the nucleus's collective properties can be understood by its shifting shape. A large nucleus can "wobble" in and out, transitioning from a spherical form to a dumbbell-like shape and back again.

When energy is added to such a nucleus, the oscillations can become so extreme that the nucleus splits in two, producing smaller nuclei along with a spray of particles like alpha particles, beta particles, and neutrons. For certain nuclei, this splitting—known as nuclear fission—can be triggered by a fast-moving neutron colliding with the nucleus. If each fission event produces enough neutrons to cause more nearby nuclei to undergo fission, a chain reaction can occur.

In uranium-235, for example, which has 92 protons and 143 neutrons, fission typically produces two smaller nuclei with atomic numbers between 34 and 58, along with a few free neutrons. Each fission event releases around 200 MeV of energy, and if the uranium sample is large enough to prevent the escape of neutrons, a runaway chain reaction occurs—this is the principle behind an atomic bomb. In contrast, when the reaction is moderated by materials that absorb neutrons, the process can be controlled, resulting in a nuclear reactor that generates heat to produce steam and electricity.

Fusion, on the other hand, replicates the energy production seen in stars like the Sun. On Earth, we've only managed to replicate the first step in the fusion process—fusing hydrogen into helium—and so far, this reaction has only been uncontrolled, as seen in the hydrogen bomb. Unlike fission, where large nuclei are encouraged to break apart, fusion requires forcing small nuclei together. This means overcoming the electrostatic repulsion of their positive charges and getting them close enough for the strong nuclear force to take over and pull them together. Once fusion begins, the heat generated can cause an outward rush of energy that blows apart other nuclei before they have the chance to fuse, halting the process.

The promise of unlimited energy from nuclear fusion depends on finding a way to hold enough nuclei together for long enough to extract a useful amount of energy. It's also crucial that the fusion process produces more energy than is consumed in forcing the nuclei together.

This is relatively easy in a bomb, where the fusion fuel is surrounded by uranium, and a fission explosion triggers the fusion process. But for controlled fusion power, that challenge remains unsolved.

The inward pressure from a surrounding explosion in a fusion reaction brings hydrogen nuclei close enough together to trigger a second burst of energy, usually from laser beams that physically compress the nuclei. Lasers, of course, are based on principles derived from quantum theory.

At the core of an atomic nucleus lies a "potential well." A particle at point A remains trapped inside the well unless it gains enough energy to "jump" over the top to point B, at which point it escapes, rushing "downhill." However, quantum uncertainty allows particles to sometimes "tunnel" through the barrier from A to B (or vice versa) without having the energy required to scale the barrier. This tunneling effect is a key feature of quantum mechanics.

A similar process occurs when nuclei fuse together. When two light nuclei are pressed together, such as in the immense pressure inside a star, they can only fuse if they overcome the potential barrier that keeps them apart. The energy that each nucleus has depends on the temperature inside the star. In the 1920s, astrophysicists were surprised to find that the calculated core temperature of the Sun was slightly too low for its nuclei to overcome the potential barrier and fuse, at least according to classical mechanics. The explanation lies in quantum mechanics—some of the nuclei are able to tunnel through the barrier even with less energy than classical theory would predict.

Quantum mechanics, in fact, explains why the Sun shines when classical mechanics says it shouldn't be able to. On Earth, one method of generating energy from fusion is by combining two hydrogen isotopes—deuterium (one proton and one neutron) and tritium (one proton and two neutrons). This fusion creates a helium nucleus (two protons and two neutrons), releases a free neutron, and generates 17.6 MeV of energy. Stars, however, follow more complex processes, involving nuclear reactions between hydrogen and trace elements like carbon. The net effect in stars is to fuse four protons into a helium nucleus, releasing two electrons and 26.7 MeV of energy, with carbon cycling back to catalyze further reactions.

On Earth, fusion experiments primarily focus on reactions involving deuterium and tritium, as scientists aim to harness the immense energy potential of nuclear fusion.

13.12 Lasers and masers

When an atom absorbs a quantum of energy, an electron jumps to a higher orbit. If left undisturbed, the excited atom will eventually return to its ground state, releasing a precisely defined quantum of radiation with a specific wavelength. This process is known as "spontaneous emission," which contrasts with "absorption."

In 1916, while exploring these phenomena and laying the statistical foundations for quantum theory—which he later found troubling—Einstein recognized another possibility. An excited atom can be prompted to release its excess energy and return to its ground state if nudged by a passing photon. This process, called stimulated emission, occurs only if the incoming photon has the exact wavelength corresponding to the energy level of the atom.

Similar to the chain reaction in nuclear fission, imagine a group of excited atoms. When a photon of the right wavelength stimulates one atom to emit radiation, that atom produces a new photon. These two photons can then stimulate two more atoms, leading to four photons, which in turn stimulate four more, and so forth. This cascading effect results in a burst of radiation, all at the same frequency. Moreover, because of the synchronization in the emission process, the waves move in perfect harmony—each crest and trough aligns—producing a highly coherent beam of radiation. Since none of the peaks and troughs cancel each other out, all the energy released by the atoms is concentrated in the beam, allowing it to be directed onto a small area of material.

When a collection of atoms or molecules is heated, they fill a range of energy levels and subsequently emit radiation of various wavelengths in a disorganized manner, resulting in less effective energy release. However, techniques exist to preferentially fill a narrow band of energy levels and then trigger the excited atoms to return to their ground state. This trigger involves a weak input of radiation at the correct frequency, leading to a stronger, amplified output beam with the same frequency.

These techniques were first developed in the late 1940s by teams in the U.S. and the U.S.S.R., focusing on radiation in the microwave band (1 cm to 30 cm). The pioneers of this work were awarded the Nobel Prize in 1954. This radiation, known as microwave radiation, and the amplification process based on Einstein's 1917 ideas led to the term "microwave amplification by stimulated emission of radiation," abbreviated as MASER.

It took another decade before researchers successfully adapted this principle to optical frequencies. Today, several types of lasers exist, with the simplest being optically pumped solid lasers. In this design, a rod of material, such as ruby, is polished with flat ends and surrounded by a bright light source, often a gas discharge tube that rapidly pulses to produce light strong enough to excite the atoms in the rod.

The entire apparatus is kept cool to minimize interference from thermal excitation of the atoms in the rod, while bright flashes from the lamp stimulate (or pump) the atoms into an excited state. When the laser is activated, a pulse of pure ruby light, carrying thousands of watts of energy, emerges from the flat end of the rod.

There are various types of lasers, including liquid lasers, fluorescent-dye lasers, and gas lasers, all sharing the same fundamental characteristics— incoherent energy is inputted, and coherent light is emitted in a concentrated pulse that carries substantial energy. Some lasers, like gas lasers, produce a continuous, pure beam of light that serves as the ultimate "straight edge" for surveying, finding widespread application in rock concerts and advertising. Others generate brief but powerful energy pulses capable of drilling through hard materials and potentially having military applications.

Laser cutting tools are utilized in a range of industries, from clothing manufacturing to microsurgery. Additionally, laser beams transmit information more efficiently than radio waves, as the data capacity increases with the radiation frequency. For example, barcodes on many supermarket products (including the cover of this book) are scanned by laser scanners, and the video and compact audio discs introduced in the early 1980s also use laser technology. Lasers can even create genuine three-dimensional photographs, known as holograms.

The list of applications is virtually endless, especially when considering the use of masers for amplifying weak signals from communications satellites and radar. All of these advancements owe their foundation to Albert Einstein and Niels Bohr, who established the principles of stimulated emission over eighty-five years ago.

13.13 Chance and Uncertainty

Heisenberg's uncertainty principle is now regarded as a fundamental aspect—arguably the central aspect—of quantum theory. However, it didn't gain immediate recognition among his peers and took nearly a decade to attain this prestigious status. Since the 1930s, its prominence may have been overstated.

The principle emerged from Schrödinger's visit to Copenhagen in September 1926, during which he famously remarked to Bohr about "damned quantum jumping." Heisenberg realized that one of the primary reasons for the disagreements between Bohr and Schrödinger was a clash of concepts. Terms like "position" and "velocity" (or later "spin") do not hold the same meanings in microphysics as they do in everyday life.

So, what meanings do they possess, and how can the two realms be connected? Heisenberg revisited the fundamental equation of quantum mechanics:

$$pq - qp = \hbar/i$$

From this, he demonstrated that the product of the uncertainties in position $(\Delta\text{-}q)$ and momentum $(\Delta\text{-}q)$ must always exceed $\hbar$ (not h). This uncertainty principle also applies to any pair of what are known as conjugate variables—variables that, when multiplied, yield units of action, such as $\hbar$. The units of action are energy multiplied by time, and another crucial pair of these variables is energy (E) and time (t).

Heisenberg asserted that classical concepts from the everyday world still exist in the micro-world but can only be applied in the limited context revealed by the uncertainty relations. The more precisely we determine a particle's position, the less accurately we can know its momentum (which includes both speed and direction), and vice versa.

13.14 The meaning of uncertainty

These groundbreaking conclusions were published in *Zeitschrift für Physik* in 1927. While theorists like Dirac and Bohr, well-versed in the new equations of quantum mechanics, quickly recognized their significance, many experimentalists perceived Heisenberg's assertion as a challenge to their abilities. They believed he was implying that their experiments were inadequate for simultaneously measuring both position and momentum, leading them to devise experiments aimed at disproving him. However, this was a misguided effort, as that was not Heisenberg's intention at all.

This misunderstanding persists today, partly due to the way the concept of uncertainty is often taught. Heisenberg himself illustrated his point using the example of observing an electron. We can only perceive objects by looking at them, which involves bouncing photons of light off these objects and into our eyes. A photon has minimal impact on a large object, like a house, so we don't expect the act of looking to significantly affect it. In contrast, observing an electron presents a different challenge. Because an electron is so tiny, we must use electromagnetic energy with a short wavelength to see or interact with it (with the aid of experimental equipment) at all.

Gamma radiation is highly energetic, and any photon of gamma radiation that scatters off an electron can significantly alter the electron's position and momentum. If the electron is part of an atom, observing it with a gamma-ray microscope might even eject it from the atom entirely.

While this illustrates the challenge of precisely measuring both the position and momentum of an electron, the uncertainty principle goes further: it asserts that, according to the fundamental equations of quantum mechanics, no electron can have both a well-defined momentum and a well-defined position simultaneously.

This principle has profound implications. As Heisenberg noted at the conclusion of his paper in *Zeitschrift*, "We cannot know as a matter of principle the present in all its details." This marks a departure of quantum theory from the determinism inherent in classical physics. For Newton, if one knew the position and momentum of every particle in the universe, one could predict the entire future. In contrast, modern physicists contend that such perfect prediction is unattainable since we cannot precisely know the position and momentum of even a single particle.

The same conclusion emerges from various formulations of quantum mechanics, including wave mechanics, the Heisenberg-Born-Jordan matrices, and Dirac's q numbers, with Dirac's approach appearing particularly fitting as it avoids direct comparisons with everyday experiences. Notably, Dirac nearly derived the uncertainty relation before Heisenberg. In a December 1926 paper for the *Proceedings of the Royal Society*, he indicated that in quantum theory, it is impossible to answer questions concerning numerical values of both position and momentum, though one might expect to answer questions based on either variable alone.

It wasn't until the 1930s that philosophers began exploring the implications of these ideas for the concept of causality—the notion that every event is caused by a specific prior event—and the challenge of predicting the future. Meanwhile, although the uncertainty relations were derived from the fundamental equations of quantum mechanics, some influential figures began teaching quantum theory by starting with the uncertainty relations themselves. Wolfgang Pauli was a key figure in this trend; he wrote a significant encyclopedia article on quantum theory that opened with the uncertainty relations and encouraged his colleague, Herman Weyl, to begin his textbook, *Theory of Groups and Quantum Mechanics*, in a similar manner.

This situation is a peculiar quirk of history. The fundamental equations of quantum theory naturally lead to the uncertainty relations; however, if one begins with the concept of uncertainty, it becomes impossible to derive those basic quantum equations. Even more problematic, the only way to introduce uncertainty without referring to the equations is through examples like the gamma-ray microscope used for observing electrons. This approach often leads people to perceive uncertainty as merely an issue of experimental limitations rather than a "fundamental truth" about the nature of the universe.

As a result, traditional methods of learning quantum theory often require students to grasp one concept, backtrack to understand something else, and then revisit the initial topic with a clearer perspective. Science is not always "logical," and neither are scientists or their professors. This has led to generations of confused students and misconceptions about the uncertainty principle—misconceptions that you, as the reader, do not have, as you have learned these concepts in the correct order.

13.15 The Copenhagen interpretation

An important aspect of the uncertainty principle that often goes unnoticed is its asymmetrical relationship with time. Unlike many concepts in physics, which remain indifferent to the direction of time, the universe exhibits a clear "arrow of time," marking a distinction between the past and the future. The uncertainty relations indicate that we cannot simultaneously know an object's position and momentum, making the future inherently unpredictable and uncertain.

However, quantum mechanics allows us to conduct experiments that enable us to calculate backward, determining the exact position and momentum of an electron, for instance, at a specific point in the past. While the future remains uncertain—leaving us unsure of where we are heading—the past is well-defined; we have a clear understanding of where we originated.

To paraphrase Heisenberg, "We can know, as a matter of principle, the past in all its details." This aligns with our everyday experience of time, as it moves from a known quantum world at its most fundamental level. This concept may be connected to the arrow of time that we perceive in the broader universe, and its more bizarre implications will be explored in the next chapter.

As philosophers began to ponder the intriguing implications of the uncertainty principle, Bohr found these ideas illuminating, shedding light on concepts he had been struggling to articulate. The notion of "complementarity," which asserts that both wave and particle descriptions are essential for understanding the quantum realm (even though, in reality, an electron is neither purely a wave nor a particle), gained mathematical support through the uncertainty principle, which states that position and momentum cannot be precisely known. Instead, they represent complementary and, in a sense, mutually exclusive aspects of reality.

Between July 1925 and September 1927, Bohr published little on quantum theory. However, during a lecture in Como, Italy, he introduced the concept of complementarity and the "Copenhagen interpretation" to a broader audience. He emphasized that in classical physics, systems of interacting particles function like clockwork, operating independently of observation.

In contrast, in quantum physics, the observer interacts with the system to such an extent that the system cannot be viewed as having an independent existence. By choosing to measure position accurately, we increase the uncertainty in momentum and vice versa; if we focus on a particle's wave properties, we negate its particle characteristics. No experiment can reveal both aspects simultaneously. In classical physics, we can precisely describe particle positions in space-time and predict their behavior accurately. In quantum physics, this level of precision is unattainable, making even relativity a "classical" theory.

It took considerable time for these ideas to develop and for their significance to be recognized. Today, the core features of the Copenhagen interpretation can be more readily explained in terms of the experimental observation process. First, we must acknowledge that the act of observation alters the observed object, meaning we, the observers, are intrinsically part of the experiment; there is no clockwork that operates independently of observation. Second, our knowledge is limited to the outcomes of experiments. For instance, we might observe an electron in energy state A, then later in energy state B, leading us to infer that the electron transitioned from A to B—possibly as a result of our observation.

What we learn from experiments, and from quantum theory, is the probability that if we observe a system once and obtain result A, we might get result B upon a subsequent observation. We cannot comment on what occurs when we are not observing, nor can we explain how the system transitions from A to B, if indeed it does. The "damned quantum jumping" that troubled Schrodinger reflects our interpretation of receiving two different results from the same experiment, and this interpretation is fundamentally flawed.

Particles can sometimes be found in state A and at other times in state B, rendering the question of what exists in between or how they transition from one state to another entirely meaningless. This represents a fundamental aspect of the quantum world. While it's intriguing to consider the limits of our understanding regarding an electron's behavior when we observe it, it's even more astonishing to realize that we have no clue what it does when we are not observing it. Since then, neutrinos have indeed been discovered in three different varieties (along with their three corresponding anti-varieties), and other types have been postulated. Can we genuinely take Eddington's doubts at face value? Is it conceivable that the nucleus, the positron, and the neutrino did not exist until scientists uncovered the appropriate tools to reveal their forms? Such speculations challenge the very foundations of our sanity and our conception of reality, yet they are entirely reasonable inquiries within the quantum realm.

If we follow the quantum equations correctly, we can conduct an experiment that yields a set of readings indicating the presence of a specific type of particle. Nearly every time we apply the same equations, we obtain consistent results. However, the interpretation of these results in terms of particles resides in our minds and may be nothing more than a coherent delusion. The equations provide no insight into what particles do when we are not observing them. Before Rutherford's time, no one had ever observed a nucleus, and prior to Dirac, the existence of the positron was unimaginable. If we cannot articulate a particle's behavior when it is not being observed, we cannot assert its existence in those moments either. It is reasonable to claim that nuclei and positrons did not exist before the 20th century because no one had observed one before 1900.

In the quantum realm, what you see is what you get; nothing is truly real. The best we can hope for is a collection of delusions that align with one another. Unfortunately, even this hope is shattered by some of the simplest experiments. Consider the double-slit experiments that "proved" the wave nature of light—how can these be reconciled with the concept of photons?

One of the most renowned educators in quantum mechanics over the past fifty years has been Richard Feynman from the California Institute of Technology. His three-volume work, "Feynman Lectures on Physics," published in the early 1960s, has become a benchmark for other undergraduate texts. He has also participated in popular lectures on the subject, including his 1965 BBC series titled "The Character of Physical Law."

Born in 1918, Richard Feynman reached the height of his theoretical physics career during the 1940s, a period in which he helped establish the equations of quantum electrodynamics (QED), the quantum version of electromagnetism. His contributions to this field earned him the Nobel Prize in 1965. Feynman holds a unique place in the history of quantum theory as part of the first generation of physicists who grew up with a solid foundation in quantum mechanics, complete with all the essential principles and established ground rules. In contrast to earlier figures like Heisenberg and Dirac, who navigated a rapidly evolving landscape where new ideas often emerged out of order and the logical connections between concepts (such as spin) were not always clear, Feynman's generation was fortunate to have all the pieces of the puzzle available. This allowed them to understand the logical framework of quantum mechanics, even if it required some thoughtful consideration and intellectual effort to fully grasp.

13.16 Matter Waves

The conventional view of light as a wave, along with its particle-like properties, led the French physicist Louis de Broglie to propose that objects typically considered particles might also exhibit wave characteristics. For instance, a beam of electrons, usually envisioned as a stream of tiny bullet-like particles, can behave as a wave under certain conditions.

This groundbreaking idea was first confirmed in the 1920s by Davidson and Germer, who directed an electron beam through a crystal of graphite. They observed an interference pattern similar to that produced when light passes through a set of slits. This experiment provided direct evidence that the wave model of light could also apply to electrons.

Subsequent research revealed similar wave properties in heavier particles, such as neutrons, leading to the conclusion that wave-particle duality is a universal characteristic of all particles. Even everyday objects like grains of sand, gloves, or cars possess wave properties, although these waves are practically unobservable. This is partly due to their extremely small wavelengths, but also because classical objects are made up of atoms, each associated with its own wave, resulting in constantly fluctuating wave patterns.

As noted earlier, in the case of light, the vibration frequency of the wave is directly proportional to the energy of the quantum. In contrast, defining and measuring the frequency of matter waves is much more complex. Instead, there is a relationship between the wavelength of the wave and the momentum of the object: the greater the momentum of the particle, the shorter the wavelength of its associated matter wave.

In classical waves, there is always a medium that exhibits some form of "waving." For instance, in water waves, the surface of the water moves up and down; in sound waves, air pressure oscillates; and in electromagnetic waves, electric and magnetic fields fluctuate. But what corresponds to this in the case of matter waves? The conventional answer is that there is no physical quantity that directly represents this. We can use quantum physics principles and equations to calculate wave functions and predict measurable quantities, but we cannot directly observe the wave itself. Therefore, it is unnecessary to define it physically, and we should refrain from doing so.

To highlight this distinction, we use the term "wave function" instead of just "wave," underscoring that it is a mathematical construct rather than a physical entity. Another significant technical difference between wave functions and classical waves is that while classical waves oscillate at a specific frequency, the wave function for matter waves remains constant over time. Despite not being a physical entity, the wave function is crucial for applying quantum physics to real-world situations. For instance, when an electron is confined within a specific region, the wave function generates standing waves, similar to those previously discussed, resulting in the wavelength—and thus the particle's momentum—adopting one of a set of discrete quantized values.

13.17 The experiment with two holes

When we conduct experiments to detect the presence of an electron at a specific location, we are more likely to find it in areas where the wave function is large compared to those where it is small. This concept was quantitatively formulated by Max Born, whose rule states that the probability of locating the particle near a particular point is proportional to the square of the wave function's magnitude at that point.

Electrons within atoms are confined to a small region of space due to the electric force that pulls them toward the nucleus. As previously discussed, we can anticipate that the corresponding wave functions will create a standing-wave pattern, which will help us understand important properties of atoms.

13.18 Varying potential Energy

Particles move in regions where the potential energy remains constant. Since total energy is conserved, this means the kinetic energy, and consequently the particle's momentum and speed, must also remain constant throughout its motion. In contrast, when a ball rolls up a hill, it gains potential energy, loses kinetic energy, and slows down as it ascends.

The de Broglie relation links a particle's speed to its wavelength, so if the speed remains constant, the wavelength will also be consistent across different locations, which we have implicitly assumed.

In situations where potential energy varies, a more complex analysis is necessary, involving the mathematical equation that describes the wave's behavior in general. This equation is known as the Schrödinger equation. In the examples we discussed earlier, where the potential is uniform, the solutions to the Schrödinger equation take the form of traveling or standing waves, validating our simpler approach. However, fully understanding more complex scenarios is mathematically challenging and beyond the scope of this book.

13.19 Quantum Tunneling

Based on our earlier discussion of matter waves, we expect particles approaching the step to be represented by traveling waves moving from left to right. After bouncing back, the wave will travel from right to left. Since we generally cannot determine the particle's specific position at any given moment, the wave function to the left of the step will be a combination of these traveling waves. This expectation is confirmed when the Schrödinger equation is solved mathematically.

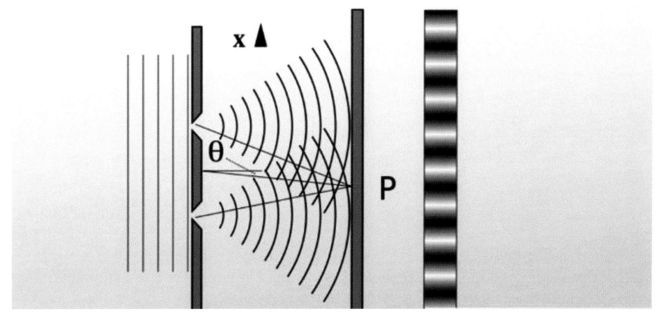

Figure 13-1 Traveling probability waves

However, a traveling wave with a relatively small but finite amplitude appears to the right of the barrier. This can be interpreted physically as indicating a small probability that a particle approaching the barrier from the left will not bounce back but will instead emerge on the other side. This phenomenon is known as "quantum mechanical tunneling," as the particle seems to tunnel through a barrier that is classically considered impenetrable.

Quantum tunneling is observed in various physical phenomena. For instance, in many radioactive decays, "alpha particles" are emitted from the nuclei of certain atoms. The probability of this occurring for a particular atom can be very low—so low that a specific nucleus may take millions of years, on average, before decaying. This process is understood in terms of the alpha particle being trapped inside the nucleus by a potential barrier, similar to the one discussed earlier. A very low amplitude wave exists outside the barrier, indicating a tiny (but non-zero) probability of the particle tunneling out.

13.20 A Quantum Oscillator

In this scenario, we examine a particle moving within a parabolic potential. The size, or "amplitude," of its oscillation is determined by the particle's energy: at the bottom of the well, all of this energy is kinetic, while the particle comes to rest at the boundaries of its motion, where all the energy is potential. The wave functions are derived by solving the Schrödinger equation. The wave function for the lowest energy state is characterized by a single peak that reaches its maximum at the center; the next highest state features two peaks, one positive and one negative, with the wave function crossing the axis, and this pattern continues for higher energy states.

13.21 The hydrogen atom

The simplest atom is that of hydrogen, consisting of a single negatively charged electron bound to a positively charged nucleus by the electrostatic (or "Coulomb") force. This force is strong when the electron is near the nucleus and gradually weakens as the electron moves further away. Consequently, the potential energy is significantly negative near the nucleus, approaching zero as the distance increases. A key simplifying aspect of the hydrogen atom is that the Coulomb potential is "spherically symmetric," meaning it depends solely on the distance between the electron and the nucleus, regardless of the direction of separation.

As a result, many of the wave functions corresponding to the allowed energy levels exhibit the same symmetry. The Coulomb potential confines the electron near the nucleus, while the oscillator potential restricts the particle's motion. When an atom transitions between energy levels, it absorbs or emits energy in the form of a photon, with its frequency related to the energy change by the Planck relation. The frequencies calculated from this energy level pattern align perfectly with those observed experimentally when electrical discharges pass through hydrogen gas. A comprehensive solution to the Schrödinger equation predicts a value for the constant RRR based on the electron's charge, mass, and Planck's constant, and this value matches exactly with that derived from experimental measurements. Thus, we achieve complete quantitative agreement between the predictions of quantum physics and the experimental data regarding the energy levels of the hydrogen atom.

While we have employed the principle of wave-particle duality to derive these quantized energy levels, how should we interpret the wave function associated with each level? The answer lies in the Born rule, which states that the square of the wave function at any point indicates the probability of finding the electron near that point. In this context, a model of the atom suggests that the electron should not be viewed as a point particle but rather as a continuous distribution spread throughout the volume of the atom. We can picture the atom as a positively charged nucleus surrounded by a cloud of negative charge, where the concentration at any point corresponds to the square of the wave function at that location.

This model is effective in many scenarios but should not be taken too literally. If we actually search for the electron within the atom, we will always locate it as a point particle. Conversely, it is equally inaccurate to think of the electron as a point particle when we are not observing its position. In quantum physics, we utilize models but avoid interpreting them too rigidly.

13.22 Other Atoms

Atoms other than hydrogen have more than one electron, leading to additional complexities. To address these complexities, we must first consider another key quantum principle known as the **Pauli exclusion principle**, named after its creator, Wolfgang Pauli. This principle states that no more than one particle of a given type, such as an electron, can occupy a particular quantum state simultaneously. Although this principle is straightforward to articulate, proving it requires advanced quantum analysis, which we will not attempt here.

Before we can effectively apply the exclusion principle, we need to understand another key property of quantum particles known as *spin*.

While we know that the Earth spins on its axis as it orbits the Sun, if we were to think of the atom as a classical object, we might expect the electron to spin in a similar fashion. This analogy has some merit, but there are significant differences between classical and quantum scenarios. Two fundamental rules govern the properties of spin: first, for any given type of particle (such as an electron, proton, or neutron), the rate of spin is always constant; second, the spin direction can only be either clockwise or counterclockwise around a defined axis.

This means that an electron within an atom can exist in one of two possible spin states. Consequently, a quantum state described by a standing wave can accommodate two electrons, provided their spins are opposite.

To illustrate the application of the **Pauli exclusion principle**, consider placing multiple electrons in a container. To achieve the state with the lowest total energy, all electrons must fill the lowest available energy levels. For instance, if we add them one by one, the first electron occupies the ground state, followed by the second electron, which has an opposite spin. Once this level is filled, the third electron must occupy the next highest energy level, along with a fourth electron, which must also have an opposite spin. This process continues, allowing up to two electrons in each energy state until all are accommodated.

Let's apply this method to atoms, starting with helium, which has two electrons. If we initially overlook the repulsive electrostatic force between the electrons, we can calculate the quantum states in a manner similar to that used for hydrogen, while also considering the doubled nuclear charge. This doubling results in all energy levels being significantly reduced (i.e., made more negative). Despite this change, the set of standing waves remains quite similar to those found in hydrogen, and the pattern is only slightly modified when interactions between the electrons are taken into account. Therefore, in helium, the lowest energy state consists of both electrons spinning in opposite directions within the lowest energy level.

In the case of lithium, which has three electrons, two of these will occupy the lowest energy state, while the third must occupy the next higher energy level. This latter state can hold up to six electrons: two will be in a spherically symmetric state, while the others fill three separate non-spherical states.

A collection of states sharing the same principal quantum number *n* is referred to as a **shell**. When all the states within a shell are filled with electrons, it is termed a **closed shell**. For instance, lithium has one electron outside its closed shell, while sodium, which has eleven electrons, has two electrons in the *n=1* closed shell, eight in the *n=2* closed shell, and one electron in the *n=3* shell. It is observed that many properties of sodium resemble those of lithium, and similar relationships among the properties of various elements are the foundation of the **periodic table**.

13.23 Summary of Quantum properties

Classical waves, such as water waves, sound waves, and light waves, are characterized by two key properties: frequency, which defines how many times per second a point on the wave oscillates, and wavelength, which measures the distance between repeating points along the wave. Waves can take the form of either traveling or standing waves. Traveling waves move at a speed determined by their frequency and wavelength, while standing waves are confined to a specific region in space, restricting their wavelengths and frequencies to a set of allowed values, much like the notes produced by musical instruments.

Although light behaves as a wave, it can also act like a stream of particles, known as photons, depending on the situation. Similarly, particles such as electrons exhibit wave-like behavior in certain contexts. When an electron is confined by a potential, such as in a "box," its matter waves form standing waves with specific wavelengths, leading to quantized energy levels. Transitions between these energy levels involve absorbing or emitting photons. The wave-like nature of quantum particles also allows them to tunnel through barriers that would be insurmountable in classical physics.

The precise agreement between calculated and measured energy levels in the hydrogen atom provides strong evidence supporting quantum physics. The Pauli exclusion principle states that no two electrons can occupy the same quantum state. Since an electron can exist in one of two spin states, each standing wave can hold up to two electrons, provided they have opposite spins. A change in an electron's speed results in a change in its momentum and wavelength, which is only allowed if it adheres to the exclusion principle, ensuring that both electrons can occupy the ground state if their spins are opposite.

First, the energy increases due to the electrostatic repulsion between the two positively charged protons. Second, it decreases because each electron is now attracted by both protons. Third, it increases again due to the repulsion between the two negatively charged electrons. Additionally, the kinetic energy of the electrons decreases as they move between the two nuclei, effectively enlarging the "box" that confines them. The overall effect of these changes depends on the distance between the atoms: when far apart, the total energy remains mostly unchanged, while when they are very close, the electrostatic repulsion between the nuclei dominates. However, at intermediate distances, the total energy decreases, reaching its lowest point when the protons are about 7.4×10^{-10} meters apart. At this distance, the difference between the energy of the molecule and that of two widely separated hydrogen atoms is about one-third of the hydrogen atom's ground-state energy.

So, where does this surplus energy go? Part of it increases the kinetic energy of the moving molecule, while the rest is released as photons. Both forms of energy manifest as heat, which leads to a rise in temperature—just as we would expect from a fuel.

This example demonstrates how energy can be released by bringing atoms together to form molecules. However, in the case of hydrogen, it is not a practical energy source on Earth because the hydrogen gas we have is already in molecular form.

13.24 The Higgs particle - God particle

Quantum electrodynamics (QED) is the theory that explains the interaction between matter and light, as described by Richard Feynman, a brilliant 20th-century American physicist often regarded as a true genius in the scientific community. According to QED, when two particles become entangled, their connection remains instant and powerful, no matter the distance between them, enabling seemingly impossible feats. "Quanta" are the fundamental building blocks of reality, representing tiny packets of energy or matter. A quantum is typically a minuscule, uniform unit, whether it's a photon of light, an atom of matter, or a subatomic particle like an electron.

Working with quanta means dealing with fixed, measurable quantities, unlike continuous quantities. This is analogous to the difference between digital information, which is based on discrete units (0s and 1s), and analog information, which can vary continuously across a range.

In the physical world, a quantum refers to a very small unit, much like how a "quantum leap" is actually a tiny change, contrary to how it's often used in everyday language. The central phenomenon of this discussion is the connection between the unimaginably small particles that make up our universe. At the quantum level, particles such as photons, electrons, and atoms can be linked so intimately that they effectively become part of the same entity, even when separated by vast distances. This is the concept of quantum entanglement. If two entangled particles are placed on opposite sides of the universe, they remain interconnected, and any change made to one instantly affects the other, which seems as bizarre to physicists as it does to the rest of us.

Albert Einstein, despite being instrumental in the early development of quantum theory, was uncomfortable with the idea of entanglement. He famously described it as "spooky action at a distance," reflecting his unease with the notion that entangled particles could influence one another without any physical link. This defiance of locality—where things influence each other only when close—was a major point of contention for Einstein, challenging his understanding of reality.

The term "entanglement" was introduced by physicist Erwin Schrödinger. In English, the word often implies chaos or disorder, as in a knotted string. However, the original German term has a more neutral meaning, suggesting an orderly crossing or connection. For Einstein, entanglement symbolized the nonsensical nature of quantum mechanics, as it appeared to break the fundamental rule of locality.

Locality is a principle so ingrained that we rarely think about it. It dictates that to act on something distant, we need some form of connection or intermediary. For example, if you want to knock a can off a fence, you can't simply will it to move—you must throw a stone to push it.

The stone travels across the gap, makes contact with the can, and, assuming your aim is true, the can falls. But quantum entanglement seems to bypass this need for an intermediary, creating a mysterious connection across vast distances without a physical link.

Similarly, if you want to speak to someone across the room, your vocal cords vibrate, pushing against nearby air molecules. These molecules, in turn, create sound waves that travel through the air, causing molecules to ripple across the space between you and the listener. Eventually, these vibrations reach the other person's ear, causing their eardrum to vibrate and allowing your voice to be heard. In this case, the sound wave is the intermediary, just as the thrown ball was in the earlier example. In both cases, something physically moves from point A to point B, illustrating the principle of locality: that action on a distant object requires some form of physical connection or transfer.

Scientific studies suggest that we are hardwired from birth to find the concept of influencing distant objects without contact unnatural. Research with infants shows they expect objects to interact only when in direct contact. Our brains instinctively reject the notion of "action at a distance." However, just because it seems unnatural doesn't mean it's impossible. Many things that defy our basic intuitions become clearer with greater understanding.

For instance, gravity operates across vast distances with no apparent connection between the objects it attracts, challenging the principle of locality. This notion of gravitational attraction was introduced with Newton's view of the universe, yet even ancient Greeks, long before the idea of gravity, were aware of other examples of action at a distance. Amber, when rubbed with a cloth, would attract lightweight objects like paper. Natural magnets, or lodestones, would pull metals toward them and, when set to float on water, would align themselves in a specific direction. In all of these cases, the interaction seemed to occur without any visible link between the objects.

The ancient Greeks had various explanations for these phenomena. One school of thought, the atomists, believed that everything was made up of different types of atoms and that a void existed between them. Since nothing could act across a void, they argued there must be a continuous chain of atoms connecting cause and effect. Other Greek philosophers believed in a more metaphysical explanation, suggesting that objects influence each other much like people attract one another. This idea bordered on the supernatural, with some Greeks attributing these mysterious forces to occult or divine intervention. This notion endured through antiquity, forming the basis of astrology—a belief that the planets exerted mystical influence over human lives.

Nearly two thousand years after the Greeks, Newton demonstrated remarkable genius in describing the effects of one form of apparent action at a distance—gravity. However, like the Greeks, he couldn't explain how one mass could influence another without anything physically connecting them. In his groundbreaking work published in 1688, Newton essentially stated that gravity exists but refrained from speculating on the underlying mechanism, avoiding any "non-empirical" guesses.

Some continued to believe that gravity had an "occult" mechanism, similar to the forces believed in astrology, but for the most part, the mystery of how gravity worked remained unaddressed until Einstein's revolutionary insights.

One key outcome of Einstein's work was the understanding that nothing could travel faster than the speed of light. It had been known since 1676, when Danish astronomer Ole Roemer made the first effective calculation of light's speed, now measured at approximately 186,000 miles per second. Einstein showed that no action, not even gravity, could surpass this universal speed limit. This became the fundamental constraint of the cosmos.

13.25 Light Quanta's entanglement property

Planck was not particularly enthusiastic about the idea of light quanta (light particles) and even openly criticized Einstein for his promotion of the concept. When Planck recommended Einstein for membership in the Prussian Academy of Sciences in 1913, he condescendingly noted that the Academy should not hold it against Einstein if he occasionally "missed the mark in his speculations, such as in the theory of light quanta." Ironically, that same year, Einstein's concept would gain traction and be further developed by Niels Bohr, who would later become Einstein's main intellectual rival over quantum theory, especially regarding "quantum entanglement."

Meanwhile, New Zealand-born physicist Ernest Rutherford had made a name for himself by discovering the atomic nucleus, further advancing atomic theory. However, Einstein struggled with the randomness at the core of quantum theory. He believed that there must be a strict, underlying causal process governing the behavior of particles, arguing that if all the relevant information were known, the electron's path and timing could be predicted. In contrast, quantum theory posited that such precise predictions were impossible—an electron's behavior could only be described probabilistically, and it wasn't until a measurement was made that properties like position would shift from a probability to an actual value. The same applied to all other properties of quantum particles.

The allure of instant communication over any distance is a thought that crosses the minds of many when they learn about quantum entanglement. After all, that's precisely what entangled particles achieve—changes in one particle are instantly mirrored in another, regardless of the distance separating them. However, even on the relatively small scale of Earth, the speed of light—the fastest means of transmitting information at approximately 300,000 kilometers (or 186,000 miles) per second in a vacuum—introduces delays. This speed creates frustrating pauses during satellite phone calls and presents significant challenges for engineers designing large-scale communication networks. The farther information must travel, the more pronounced the lag becomes.

Sound, while useful in navigating obstacles, dissipates quickly and is heavily influenced by wind direction, making it unreliable over distances greater than about half a mile. This limitation necessitates a vast number of stations to relay sound messages across a country.

Einstein raises a critical question: how can we truly ascertain that two lightning strikes are simultaneous? What does "simultaneous" mean for two events that occur in different locations?

We cannot occupy both spaces simultaneously, nor can we guarantee that clocks at the two sites are perfectly synchronized. According to Einstein, the only way to verify simultaneity is to place an observer midway between the two strikes, equipped with mirrors that allow them to view both events at once.

When we refer to light or an electron as a wave or a particle, we're employing models that help us conceptualize their behavior—comparable to the actual ripples we see on the surface of water for waves, or a stream of tiny dust particles for particles. However, it's crucial to understand that this doesn't mean light or electrons are strictly waves or particles. Although these models are familiar and helpful, we must remember they are just that—models, not the essence of the phenomena themselves.

So, what are light and electrons? Light is light, and an electron is an electron. They exhibit behaviors akin to waves and particles at times, which aids us in predicting their actions, but that doesn't define their true nature. Acknowledging this perspective can help contextualize the oddities of quantum theory. The peculiarities of phenomena like a photon seemingly passing through two slits simultaneously or the instantaneous nature of entanglement seem strange only because we have allowed our models to overshadow the realities they aim to describe.

This realization doesn't imply that anything is possible. Our models are incredibly useful, enabling us to refine our predictions about the world's behavior and characteristics (as opposed to its intrinsic nature). However, we should not expect absolute understanding to emerge solely from working with these models.

This creates a trap for would-be speculators. Because we're working with models of reality, it's easy to propose ideas that lack direct experimental support. For example, some have suggested that entanglement could account for various phenomena, including consciousness as a quantum phenomenon involving entanglement. However, the biological evidence regarding brain mechanisms indicates that we don't need to invoke such intricate and delicate phenomena to explain what seems to be a robust and consistent capability.

Max Planck's initial efforts a century ago to quantify the energy radiated by hot bodies initiated a process of capturing the mysterious and elusive world of quantum mechanics in mathematical form. Despite its 'spooky' interactions at a distance, this field has led to significant inventions like the laser and transistor. Today, quantum theory is being successfully integrated with theories of information and computation. These advancements might even provide insights into processes that remain elusive within conventional science, such as "telepathy," an area where Britain is leading research.

A true understanding of quantum mechanics remains out of reach; we tend to adopt too localized a perspective on nature. Quantum theory, as noted, is our most precise description of atomic behavior and forms the foundation of all that we know about chemistry. In turn, chemistry provides the mechanisms that underpin biology, including those that drive our metabolism and reproduction. Vedral poses the intriguing question of whether quantum effects not only govern the behavior of inanimate matter but also play a vital role in the existence of life itself, concluding that "our very existence could be made possible by entanglement."

According to Penrose, the reason entanglement doesn't spread uncontrollably is that observation destroys it. Other physicists, like Professor Tony Sudbery of York University, suggest that it's not the act of observation that destroys entanglement; rather, our involvement in the entangled world limits our ability to perceive it fully. Bosons, like photons, are one of the two types of quantum particles (the other being fermions, such as electrons). Bosons can share quantum states, while fermions cannot. The Higgs boson, a hypothetical particle, remains unobserved. Despite numerous attempts to detect it and occasional declarations that the concept is outdated, conclusive evidence remains elusive.

The idea of the Higgs boson was conceived by Peter Higgs of Edinburgh University in the early 1960s to explain the origin of mass and the variability of mass among different particles. Each fundamental force has a corresponding field—a sort of environmental texture in which it operates—communicated through bosons.

The photon is the carrier of the electromagnetic field. Higgs proposed the existence of another field—the "Higgs field"—responsible for mass. According to this theory, a particle's mass arises from its interactions with the Higgs boson, the Higgs equivalent of a photon. However, Higgs bosons remain a topic of controversy. While they effectively explain current theories and observations, no one has yet detected these so-called "God particles," named for their fundamental role in imparting mass to other particles. In 2001, speculation arose about their non-existence due to experiments failing to reveal them, yet many scientists continue to believe in their reality, arguing that we simply lack sufficiently powerful particle colliders to capture the elusive Higgs. Unfortunately, governments often cancel collider projects because of their immense scale.

Despite more than seventy years of skepticism, entanglement remains a persistent phenomenon. Each experiment brings us closer to understanding the bizarre nature of the quantum world. Even a groundbreaking thinker like Albert Einstein initially rejected the validity of quantum theory, but this approach has proven untenable. Quantum theory functions and reveals a complex web of secrets. Entanglement establishes a mysterious connection—an unfathomable bond—between two particles. It is perhaps unsurprising that, once practical applications of entanglement emerged, it caught the interest of those involved in secure communication. The eerie connection between entangled particles suggested that if entanglement could transmit a message, it would be completely secure against interception.

13.26 Force Carriers: Particles that make things happen

Here's a concise way to understand the essence of physics: it revolves around what exists (objects) and what occurs (events). The particles we can observe—such as leptons, baryons, and mesons—along with those we study but cannot directly see, like quarks, make up the fundamental components of existence. Additionally, there is a category of particles known as force carriers that dictate how these components interact and influence events. It's also worth noting that the absence of certain processes is just as intriguing as the presence of others. Many phenomena do not happen (and we believe cannot happen), such as the spontaneous creation of electric charge from nothing, the emergence or disappearance of energy, or potentially even the radioactive decay of a proton.

13.27 Gravitational Interaction

There are four types of force carriers, each corresponding to different interactions, ordered by increasing strength. Gravity is the weakest of these forces. Even in the tiniest gravitational interactions we can measure, countless gravitons are involved, meaning we only observe their collective effect in large quantities, never as individual particles. Consequently, we currently have no hope of directly detecting the graviton.

Interestingly, gravity, despite being the weakest force in nature, holds us firmly on Earth and keeps our planet in orbit around the Sun for two main reasons. First, gravity is always attractive, while the much stronger electric forces can be both attractive and repulsive, leading to a sort of equilibrium. For instance, Earth is electrically balanced with both positive and negative charges, so even if you build up a charge by shuffling across a rug on a dry day, you won't feel any significant electric force pulling you down. However, if all the negative charge were to vanish, leaving only the positive charge on Earth, and you had a modest negative charge from shuffling, you would be crushed instantly by the immense electric pull. Conversely, if the Earth's positive charge were stripped away, leaving only its negative charge, you would be launched into space at speeds far exceeding any rocket. This fine balance of attractive and repulsive forces exists throughout the universe, making gravity the dominant force.

The second reason gravity feels so pronounced, despite its weakness, is the immense mass that exerts it. We are anchored to the ground by the gravitational pull of six thousand billion billion tons of matter. Every piece of matter attracts every other piece through gravity. However, when we consider ordinary-sized objects, gravity's weakness becomes more apparent. In the subatomic realm, gravity plays no significant role; for example, between the proton and the electron in a hydrogen atom, the electric force overpowers gravity by an astonishing factor of over 10^{39} (to put that into perspective, if you stacked that many atoms end to end, they would stretch to the edge of the universe and back a thousand times). Yet, even though gravity is weaker than the size of a proton, it still intertwines with quantum theory in ways we can only begin to understand.

13.28 Electromagnetic Interaction

The photon has an intriguing history, beginning with its "invention" by Albert Einstein in 1905. During the 1920s, it existed in a somewhat ambiguous state as a "corpuscle," not fully recognized as a real particle. However, in the 1930s and 1940s, its importance became clear as physicists associated it with electrons and positrons, contributing to the development of the robust theory known as quantum electrodynamics. Today, we understand the photon as a fundamental particle with truly zero mass and size, acting as the force carrier for electromagnetism.

In fact, we encounter photons nearly every moment of our waking lives. While we can see many of them, such as those from visible light, there are countless others that go unnoticed, including those that transmit radio and television signals, radiate heat from warm walls, and deliver X-rays through our bodies. The universe is also permeated with low-energy photons, known as cosmic background radiation, which is a remnant from the Big Bang. Remarkably, there are about a billion photons for every material particle in the universe.

13.29 Feynman Diagrams

American physicist Richard Feynman, known for both his brilliant contributions to physics and his engaging writing and wit, developed a method for diagramming events in the subatomic realm. These "Feynman diagrams" serve as a valuable tool for visualizing the interactions taking place in this world.

Specifically, Feynman diagrams illustrate what we believe to be occurring when a force carrier is exchanged between two particles, helping to explain their interaction. For theoretical physicists, these diagrams offer more than just a visual representation; they enable the cataloging of potential reactions among particles and allow for calculations of the probabilities of various interactions. In this context, however, I will focus on using them solely as a visual aid to clarify how the forces generated by the exchange of force carriers operate.

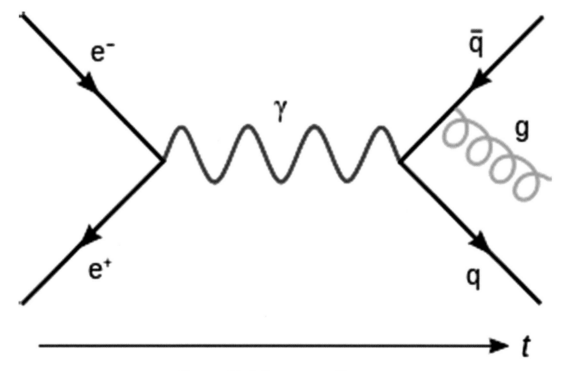

Figure 13-2 Feynman diagram

A Feynman diagram acts as a compact "space-time map." To grasp this concept, let's begin with a standard two-dimensional map, like one you'd find in a road atlas. Typically, such a map has north at the top and east to the right, with lines indicating paths through space—or, as a mathematician might describe it, a projection of a route through the physical world. To incorporate time alongside space, we would need to expand into four dimensions, entering the complex realm of space-time, which exceeds our ability to visualize.

Feynman

Physicists who are well-versed in relativity are no better than the average person at visualizing four dimensions. A space-time map provides information about both the "where" and "when" of an event. Each segment of a world line encompasses all that is known. However, when one segment transitions into another at a point of interaction, the situation changes. In relativity theory, an "event" is defined as something that occurs at a specific space-time point—indicating a precise location in space at an exact moment in time. Therefore, in the particle world, events happen at distinct space-time points rather than being spread out over space or time.

Experiments suggest that all occurrences in the subatomic realm stem from tiny explosive events at specific space-time points—events in which nothing persists. What enters a point differs from what exits.

Two key features appear in all Feynman diagrams: one is obvious, while the other is less apparent. The obvious feature is that at the interaction points A and B, three particle lines converge. These points, known as vertices, specifically refer to three-prong vertices, marking the locations of interactions. If you examine the other diagrams, you'll notice that they all contain these three-prong vertices.

Moreover, these vertices are of a specific type, where two fermion lines and one boson line intersect. The fermions depicted in these diagrams can be either leptons or quarks, while the bosons act as force carriers, such as photons, W bosons, or gluons. This leads to the remarkable generalization that physicists now accept as true.

Every interaction in the universe ultimately arises from the emission and absorption of bosons (the force carriers) by leptons and quarks at specific space-time points. Central to every interaction are three-prong vertices, marking truly catastrophic events where particles are either annihilated or created. At point A, an incoming electron is destroyed, a photon is emitted, and a new electron is generated. The electron that flies upward to the left in the diagram is not the same as the one that entered from the lower left; while they are both identical electrons, claiming that the outgoing electron is the same as the incoming one is meaningless.

Figure 15.5 illustrates how an electron and its antiparticle, a positron, can collide and annihilate, resulting in the creation of two photons, represented by the equation:

$$e- + e- \rightarrow 2\ gamma.$$

Here, there are again two interaction vertices, A and B, where two fermion lines and one boson line converge. At vertex A, an incoming electron emits a photon and creates a new electron that travels to vertex B, where it encounters an incoming positron and emits another photon.

To visualize this process progressing through time, you could imagine a horizontal ruler slowly moving upward, ignoring the arrows for a moment. You might wonder why the arrows are present if you're meant to disregard them—they serve as labels, indicating whether a line represents a particle or an antiparticle.

The line on the right, with a downward-pointing arrow, signifies a positron moving forward in time—upward in the diagram. This is where the Wheeler-Feynman perspective comes into play: the forward-moving positron is equivalent to a backward-moving electron. Thus, it's also possible to interpret the diagram as it appears visually. An electron approaches from the left, moves forward in time, emits photons at A and B, and then seemingly reverses its path through time.

Wheeler and Feynman demonstrated that both descriptions—one involving a forward-in-time positron and the other a backward-in-time electron—are "correct" because they are mathematically equivalent and indistinguishable. However, you might object that we lack the option to move backward in time; we progress inexorably forward, like the horizontal ruler sliding upward across the diagram.

What we observe in Figure 9.1 is a positron moving left toward a collision with an electron, yet, trained in the principles of the quantum world, we are open to believing that this can also be described as an electron moving right while backtracking through time as it unwinds.

13.30 Quantum Lumps

Max Planck did not set out to become a revolutionary figure. When he presented his theory of radiation to the Prussian Academy in Berlin in December 1900, introducing his now-famous constant h, he believed he was simply refining classical theory by correcting a minor flaw in a robust framework. However, as the quantum revolution he sparked gained momentum in the subsequent years, Planck distanced himself from it, unable to fully embrace the implications of his own work.

Planck's contributions addressed a problem that arose from the merging of electromagnetism and thermodynamics. Electromagnetism encompasses electricity, magnetism, and light, while thermodynamics involves "radiation," temperature, and the flow and distribution of energy within complex systems. These foundational theories of 19th-century physics struggled to explain "cavity radiation," which refers to the radiation emitted from a closed container at a constant temperature.

When Planck introduced the concept of quantization in 1900, he was already forty-two years old, well past the age at which many theoretical physicists make their most significant contributions—often considered to be around twenty-six. Some of the physicists who would later develop the complete theory of quantum mechanics between 1924 and 1928 had not yet been born when Planck initiated this revolutionary shift.

As Niels Bohr reportedly remarked, "If your head doesn't swim when you think about the quantum, you haven't understood it." Similarly, Richard Feynman, a brilliant American physicist renowned for his deep understanding of quantum mechanics, stated, "My physics students don't understand it either. That is because I don't understand it."

One enduring legacy from that December day in Berlin is Planck's constant, which remains a fundamental constant in quantum theory, with implications far beyond its original purpose of relating radiated energy to frequency. It serves as a crucial factor that defines the scale of the subatomic world and distinguishes it from the "classical" realm of everyday experience. This chapter explores the concept of quantum "lumps," of which I have only discussed one so far: the quantum of radiated energy that manifests as the particle of radiant energy known as the photon.

There are two types of granularities in nature: the granularity of matter itself and the discrete properties of those materials. To begin with the granularity of matter, it's widely understood that matter cannot be subdivided indefinitely. If you continue to divide it, you eventually reach atoms (a term originally chosen to mean "indivisible"). Further separation of atoms reveals electrons and atomic nuclei, leading down to the fundamental particles known as quarks and gluons.

As far as we currently understand, this is the limit of our knowledge. We have yet to find any size or structure for electrons and quarks. You might wonder, "Isn't that simply because we haven't developed the means to probe deeper? Why couldn't there be layers upon layers of worlds within worlds?" Scientists have a couple of compelling reasons to believe that the structure of reality has a finite core, and that we are either at or very near that core.

One reason for this belief is that only a few quantities are needed to completely characterize a fundamental particle. For example, an electron is defined by its mass, charge, flavor, and spin, as well as its strength of interaction with the force-carrying bosons of the weak interaction. That's essentially it. Physicists are confident that if there are additional properties of the electron waiting to be discovered, they are few in number. This allows for a concise list that fully describes everything about an electron.

Another reason supporting the notion that we are close to an essential core of matter, closely tied to the simplicity of description, is the identity of particles. Unlike even the most precisely manufactured ball bearings, which can never be truly identical, all red up-quarks, for instance, are genuinely indistinguishable from one another. The Pauli exclusion principle, which states that no two electrons can occupy the same state of motion simultaneously, makes sense if we assume that electrons are identical. However, it would not hold true if electrons varied in any way. If there were an infinite number of layers of matter to uncover, we would expect electrons to be as complex as ball bearings, with each one requiring a vast amount of information for precise description. This is not the case. The simplicity and indistinguishability of fundamental particles strongly suggest that we may be nearing the ultimate "reality" of matter.

13.31 Charge and Spin

One of the "quantized" properties of matter that has been discussed is electric charge. Some observed particles are neutral, while all others possess a charge that is an integral multiple (either positive or negative) of the proton's charge, denoted as e. Another quantized property is spin, which can either be zero or an integral multiple of the electron's spin, represented as $\hbar$ in angular momentum units. For particles, including composite ones, the multipliers (the multipliers e and $1/2\hbar$ are typically 0, 1, or 2) for charge and spin are typically 0, 1, or 2, although everyday objects can exhibit charges and angular momenta significantly greater than e and $1/2\hbar$.

The concept of lumpiness, or quantization, indicates that there is a finite difference between allowed values of a property. However, it does not imply that there are only a finite number of possible values. Similar to how even numbers like 2, 4, and 6 have a finite separation yet extend infinitely, charges and spins also have finite separations but an infinite range of possible values. We currently do not understand the specific magnitude of the charge quantum e, which is relatively small by particle physics standards. This value quantifies the strength of the interaction between charged particles and photons, known as the electromagnetic interaction, which is about one hundred times weaker than the quark-gluon interaction, aptly termed the strong interaction. Thus, while we label the electron's charge as small, both of these interactions are enormously strong compared to the weak interaction. It's important to note that the size of the charge lump is simply a measured value, and we do not know why it has its specific value.

Similarly, the magnitude of $\hbar$ (which determines the size of the spin lumps) is also a measured quantity lacking theoretical explanation. The theory of quantum mechanics, developed in the 1920s, explains the existence of spin and angular momentum quantization, but it does not address the magnitude of the quantum unit.

1. Fermions, such as leptons and quarks, possess half-odd-integer spins in units of $\hbar$ (e.g., 1/2, 2/3, 2/5), while bosons, such as photons and gluons, have integer spins (0, 1, 2, etc.).
2. Orbital angular momentum is always an integral multiple of $\hbar$ (0, 1, 2, etc.).
3. Angular momentum, whether spin or orbital, can only point in specific directions, and the projections of angular momentum along any chosen axis differ by exactly $\hbar$ (one unit).

13.32 Mass

In many ways, mass is the most evidently quantized property, as each particle has its own specific mass. Every composite entity, such as an atomic nucleus or a protein molecule, also has a definite (and thus quantized) mass. However, due to the energy contribution to mass, the mass of a composite particle does not equal the sum of the masses of its constituent particles.

For instance, a single neutron has a mass that is significantly greater than the combined masses of the three quarks (and any massless gluons) that compose it. Here, energy plays a role in contributing to the mass.

Consider the case of a deuteron (the nucleus of heavy hydrogen), which consists of a proton and a neutron. Its mass is slightly less than the combined masses of the proton and neutron due to binding energy, which acts as a negative contributor to mass. To separate a deuteron into its constituent particles, energy must be added—specifically, enough to counteract the binding energy.

Charge and angular momentum are more straightforward. The neutron's charge (zero) is the total of the charges of the quarks it contains. The deuteron's spin is the vector sum of the spins of its proton and neutron, accounting for both direction and magnitude. This principle applies to all combinations of particles. The fact that the component masses do not simply add together serves as a reminder that, at the deepest level, a composite entity is not merely a straightforward combination of parts; it forms a new entity entirely.

Why is the mass of the muon more than two hundred times that of the electron? Why is the mass of the top quark roughly fifty thousand times that of the up quark? These questions remain unanswered. Quantized mass exists, but we have yet to explain it.

Albert Einstein, widely regarded as the greatest physicist of the 20th century—and ironically, one of the architects of quantum theory—was not fond of quantum probability. He often expressed skepticism, famously stating that he did not believe God played dice. In 1953, he remarked, "In my opinion, it is deeply unsatisfying to base physics on such a theoretical outlook, since relinquishing the possibility of an objective description cannot help but cause one's picture of the physical world to dissolve into fog." Another famous quote of his, inscribed in stone at Princeton University, reads, "God is subtle but not malicious." Einstein accepted the idea that the laws of nature are not immediately apparent and require significant effort to understand. However, he believed that no grand designer of the universe would intentionally incorporate unpredictability into the fundamental laws. If a reason for quantum unpredictability is ever discovered, it may clarify this fog of uncertainty.

Here's a quiz: Is a sodium-23 atom, with a nucleus containing 11 protons and 12 neutrons, a boson or a fermion? You might think it's a fermion because its nucleus has 23 fermions—an odd number. Or, if you count quarks—three for each proton and neutron—the nucleus contains 69 fermions, still an odd number. But wait. The 11 electrons orbiting the nucleus add another layer. Counting quarks plus electrons results in 80, an even number (or 34 when counting protons, neutrons, and electrons). Therefore, a sodium-23 atom is a boson.

Bosons and fermions share many properties that do not differentiate them. For example, they can be either fundamental or composite, charged (positive or negative) or neutral. They can interact strongly or weakly and can possess a wide range of masses, including zero mass for particles like photons.

However, bosons and fermions differ significantly when it comes to spin. Bosons possess integral spin values (0, 1, 2, etc.), while fermions have half-odd-integral spin values (1/2, 3/2, 5/2, etc.).

The most notable distinction between these two classes of particles is their behavior in groups. Fermions are "antisocial," adhering to the exclusion principle, which dictates that no two identical fermions (such as electrons) can occupy the same state of motion simultaneously. In contrast, bosons are "social." They can not only coexist in the same state of motion but actually prefer to do so.

Why do these two types of particles exhibit such differing behaviors? The explanation lies in a subtle yet relatively straightforward aspect of quantum mechanics that I will elaborate on at the end of the chapter. This feature is unique to quantum mechanics and has no equivalent in classical physics, yet it has profound implications for the large-scale world we live in.

13.33 Fermions

Wolfgang Pauli, at just twenty-five years old, introduced the exclusion principle in 1925. Born in Austria and educated in Germany, he later settled in Switzerland and spent considerable time in the United States. After earning his Ph.D. at Munich in 1921, he published a comprehensive survey of relativity theory that impressed Albert Einstein with his deep understanding of the subject. In 1926, shortly after presenting the exclusion principle, Pauli was instrumental in applying Werner Heisenberg's newly developed quantum theory to atomic structures. By 1930, at the age of thirty, he proposed the existence of the neutrino and had become a professor at Zurich.

In his later years, Pauli gained a reputation for intimidating physicists presenting their research. He would sit in the front row, shaking his head and scowling at their work. His initial response to their correspondence on particle theory was always the same: "Alles Quatsch" (total nonsense). However, after further discussions, he would often acknowledge that their ideas had some merit, eventually congratulating them on their insights.

Pauli developed the exclusion principle during a tumultuous twelve-year period for physicists. Following Niels Bohr's formulation of a quantum theory for the hydrogen atom in 1913, scientists recognized that Planck's constant (h) was crucial to atomic behavior. They assumed that Bohr's concepts—where electrons occupy stationary states and transition between them by emitting and absorbing photons—applied to all atoms.

However, a comprehensive quantum theory was still lacking until Werner Heisenberg (at thirty-three) and Erwin Schrödinger (at thirty-eight) connected the dots and established the quantum framework we use today. Pauli's exclusion principle played a vital role in this scientific revolution.

Before delving into the implications of the exclusion principle, it is essential to clarify the concepts of "state of motion" and "quantum number." When we say an electron is in a specific state of motion or possesses certain quantum numbers, we refer to a description of its movement rather than its position. For instance, an automobile traveling westward at a constant speed on a straight road represents a particular state of motion, defined by its speed and direction rather than its location. If another car travels alongside it at the same speed and direction, it shares the same state of motion, regardless of distance apart. Similarly, an Indianapolis race car that maintains a consistent speed and acceleration during a race is also in a defined state of motion.

Two vehicles are considered to be in the same state of motion if they exhibit identical patterns of speed and acceleration, regardless of their distance from one another. A satellite orbiting the Earth illustrates this concept; its state of motion is determined by its energy and angular momentum, not its precise location at a given moment. A second satellite, although it may possess the same energy, could have less angular momentum, placing it in a different state of motion. Therefore, the state of motion is a comprehensive property that reflects the totality of an object's movement, rather than any specific detail.

For electrons within an atom, physicists cannot trace the precise details of their motion due to nature's constraints. Instead, they can only gather global information, akin to observing a race car moving too quickly to discern its specific position. While it's evident that the car is confined to the track and moving at an average speed, its exact location remains unknown. An electron's state within an atom similarly represents a blur of probabilities, indicating where it might be located at any given time.

Nonetheless, certain attributes of the electron can be precisely defined, including its energy, angular momentum, and orientation of its orbital motion. This precision allows for the assignment of numbers to characterize the electron's state—one number for energy, another for angular momentum, and a third for the orientation of angular momentum. Since these physical quantities are quantized—meaning they can only take specific discrete values—the corresponding numbers are also quantized, known as quantum numbers. For example, the principal quantum number (n) is designated as 1 for the lowest-energy state, 2 for the next higher state, and so forth, indicating the position of the state within the hierarchy of allowed energies.

The angular momentum quantum number, denoted as l, quantifies angular momentum in units of $\hbar$ and can be zero or any positive integer. Meanwhile, the orientation quantum number, mmm, can assume both negative and positive values, spanning from $-l$ to $+l$.

13.34 Bosons

In 1924, Satyendra Nath Bose, a thirty-year-old physics professor at the University of Dacca, sent a letter to Albert Einstein in Berlin. Enclosed was his paper titled "Planck's Law and the Hypothesis of Light Quanta," which had been rejected by the prominent British journal *Philosophical Magazine*. Undeterred by the rejection, Bose decided to reach out to the world's most celebrated physicist, possibly encouraged by the fact that he had translated Einstein's relativity text from German to English for the Indian audience or perhaps due to his confidence that his paper held significant insights.

The "Planck's Law" mentioned in Bose's paper refers to the mathematical formula introduced by Max Planck in 1900, which describes the distribution of energy among different frequencies of radiation within an enclosure at constant temperature, known as blackbody or cavity radiation. It's important to note that, in Planck's formulation, the equation $E=hf$ represented the minimum energy that a material object could absorb from or emit to radiation, rather than the energy of a photon of frequency f. For nearly a quarter of a century, Planck and most physicists believed that it was the energy transfer to and from radiation that was quantized, not the radiation itself—despite Einstein's 1905 proposal of the photon (which was named later) and Arthur Compton's evidence of photon-electron scattering in 1923. When Bose wrote his paper in 1924, he still regarded "light quanta" (photons) as hypothetical entities. His work would be pivotal in shifting the photon from a mere hypothesis to an accepted reality.

The term "photon" was coined by Gilbert Lewis in 1926. At the time, Bose was unaware of the exclusion principle, which was introduced the following year and applied by Pauli to electrons rather than photons. In the paper he submitted to Einstein, Bose demonstrated that he could derive Planck's law by positing that radiation consists of a "gas" of "light quanta" that do not interact with one another and can occupy any energy state regardless of whether another "light quantum" is already in that state. Einstein quickly recognized that Bose's derivation represented a significant advancement over Planck's original approach and provided strong, albeit indirect, evidence for the reality of light quanta.

Einstein took a personal interest in Bose's work, translating his paper from English into German and submitting it to the prominent German journal *Zeitschrift für Physik*, recommending its publication. The paper was published without delay.

Intrigued by Bose's findings, Einstein shifted his focus from his ongoing quest to unify electromagnetism and gravitation to consider how atoms might behave if they adhered to the same principles as photons. His paper was published shortly after Bose's, and this collaboration led to what we now refer to as Bose-Einstein statistics. A few years later, Paul Dirac proposed that particles following these statistical rules be called bosons.

Einstein explored what would happen to a gas of atoms at extremely low temperatures, assuming that these atoms conformed to Bose-Einstein statistics—a condition that applies to about half of all atoms. Using the apartment-house analogy, he suggested that there is no limit to the number of bosons that can occupy the same energy state. While one might expect all bosons to congregate in the lowest-energy state, this tendency is fully realized only at extremely low temperatures.

Einstein noted that the bosons would not only occupy the same energy state but would also be in identical states of motion. As a result, each boson would completely overlap and interpenetrate with the others, effectively occupying the entire energy state. Every atom would spread out according to a probability distribution that is the same as that of every other atom, now referred to as a "Bose-Einstein condensate." It took seventy years for experimentalists to produce a Bose-Einstein condensate in the lab, primarily due to the challenges in achieving the extraordinarily low temperatures required. Unfortunately, neither Bose nor Einstein lived to witness this remarkable confirmation of boson behavior.

Dirac, known for his modesty, also introduced the term "fermions" to describe particles whose properties he and Enrico Fermi discovered. One key distinction between fermions and bosons is their numbers: evidence suggests that the total number of fermions in the universe remains constant (with anti-fermions assigned a negative particle number), while the number of bosons can fluctuate. The fermion rule is evident in every individual particle reaction. For example, in the decay of a negative muon into an electron, a neutrino, and an antineutrino, there is one fermion before the decay and one afterward (counting the antineutrino as negative). Similarly, in the decay of a neutron, one fermion exists before the decay and one net fermion afterward. In the annihilation of an electron and a positron, the fermion count is zero both before and after the event.

Similarly, when a proton collides with another proton in an accelerator, various bosons can be produced. For instance, three bosons may emerge from the collision where there were none previously, while the number of fermions remains constant at two.

Another example can be seen in the decay of a negative pion into a muon and an antineutrino. In this case, the boson count decreases from one to zero, while the fermion count remains at zero (with the antineutrino assigned a negative particle number).

However, fundamental questions remain unanswered: Why do fermions maintain their numbers, while bosons can appear and disappear in arbitrary amounts?

13.35 Why Fermions and Bosons?

How does Nature determine whether each particle is social or antisocial—either inclined to cluster in the same state of motion or resistant to doing so? Classical theory fails to provide an answer, even an approximate one, which makes this question particularly intriguing. To find an answer, we must explore a mathematical feature of quantum theory that is, hopefully, comprehensible. This inquiry also hinges on the existence of identical particles in nature.

The exclusion principle does not state that no two fermions can occupy the same state of motion. Instead, it specifies that no two identical fermions (such as two electrons, two protons, or two red up quarks) can share the same state. Similarly, it is only identical bosons (like two photons, two positive pions, or two negative kaons) that prefer to occupy the same state of motion. If every particle in the Universe were slightly different from every other, it wouldn't matter whether a particle was a fermion or a boson, as there would be no barriers preventing bosons from clustering together. Particles would then behave like baseballs—each slightly distinct, with no incentives or disincentives to aggregate. Thus, the existence of truly identical entities in the subatomic realm has profound cosmic implications. For instance, if electrons were not precisely identical, they would not fill sequential shells in atoms, resulting in no periodic table—and ultimately, no you or me.

One distinguishing feature of quantum theory is its focus on unobservable quantities. One such unobservable is the wave function, or wave amplitude. The probability of a particle being in a specific location or moving in a particular manner is proportional to the square of the wave function. Therefore, while the wave function multiplied by itself relates to observable outcomes, the wave function alone does not. This means that whether the wave function is positive or negative has no observable consequences, as the squares of both positive and negative numbers are positive.

The story is somewhat more complex. The unobservable wave function can be a complex number, which combines real and imaginary numbers, moving it even further from direct observability. The "absolute square" of a complex number is always a positive quantity, and this is what we can observe.

13.36 Quantum Mechanics and Gravity

I previously mentioned *quantum foam*, which describes the turbulent nature of space-time at scales of 10^{-35} meters and times of 10^{-45} seconds. In the realm of scientific inquiry conducted so far, even at dimensions much smaller than an atomic nucleus, gravity and quantum theory appear unrelated. However, as Wheeler noted, if we delve even deeper into the fabric of space-time, we encounter a point where the fluctuations and uncertainties inherent to particles begin to influence space-time itself.

The smooth uniformity of the space we perceive and the orderly progression of time dissolve into bizarre convolutions that are challenging to visualize. This is the realm of Planck-scale physics, a domain that would have likely astonished Max Planck himself, and it is now being extensively explored by theoretical physicists.

13.37 Revisiting Heisenberg's uncertainty principle

Heisenberg began with the quantum state of the system in question, such as a single electron, an atom, or a molecule, and proposed that the most logical way to formulate mechanics for the system was by modeling the act of observation. Here, "observation" refers to any interaction experienced by the system, like the scattering of light off an electron. In the absence of interaction, the system remains entirely isolated from the outside world, rendering it irrelevant. Only through some form of interaction or observation does the system exist in a "definite state."

Heisenberg represented observations of a system as mathematical operations on its quantum state. This approach enabled him to formulate equations governing the behavior of a quantum system, yielding results that aligned with the wave mechanics developed by Schrödinger, particularly in predicting the energy levels of the hydrogen atom. The equivalence of the two methods becomes clear when one recognizes that Heisenberg's expressions for representing observations are differential operators acting on the quantum state, which is described by the wavefunction of the system.

Consequently, this approach leads to a differential equation for the wavefunction, identical to the wave equation that Schrödinger derived by analogy with the wave equation for light. Heisenberg's uncertainty principle emerges from the understanding that any observation of the quantum system will "disturb" it, thus precluding perfect knowledge of the system for the observer. This concept is best illustrated by considering what happens when we attempt to observe an electron's position in an atomic orbit by scattering a photon of light off it. The relationship between a photon's wavelength and momentum is given by the equation $\lambda = h/p$.

To determine the electron's position as accurately as possible, we should use a photon with the highest possible momentum (shortest wavelength), as we cannot resolve distances shorter than the wavelength of the light used. However, employing a high-momentum photon will provide a good estimate of the electron's position at the moment of measurement, but the electron will be significantly disturbed by the photon's high momentum (energy), leading to a high uncertainty in its momentum. This illustrates the essence of Heisenberg's uncertainty principle: knowing one parameter introduces uncertainty in its conjugate parameter.

A long-wavelength (low-momentum) photon can only provide a rough estimate of the electron's position without causing significant disturbance to the atom. Conversely, a short-wavelength (high-momentum) photon accurately localizes the electron but causes considerable disruption. Note that a particle's wavefunction reflects its localization. The relationship is expressed as:

delta-omega or (Δ) (λ) = h/(delta-p).

This spread in wavelengths (or frequencies) results in a localized wave packet in the wavefunction, reflecting the electron's rough localization. When the electron is very specifically localized, for instance, during a high-energy collision with another particle like a photon, the uncertainty in its momentum increases (along with the spread in the wavelength components of the wavefunction), resulting in a very localized wave packet. In such cases, it is reasonable to treat the electron as a particle. This wave description complicates the simplistic Bohr model of orbiting electrons. The dimensions of the electronic wavefunction are comparable to those of the atom itself. Until a measurement localizes the electron more precisely, assigning a more detailed position to the electron is meaningless. However, this explanation leaves the electron with a poorly defined role within the atom.

In 1926, German physicist Max Born proposed that the square of the wavefunction's amplitude at any point correlates with the "probability" of finding a particle at that location. The wavefunction itself does not have a direct physical interpretation beyond that of a "probability wave." When squared, it provides the likelihood of locating the particle at a specific point during measurement. Therefore, the probability density for finding the particle at position xxx at *time t* is given by:

Hence, the probability density for finding the particle at the position x at time t is probability density = $|omega(x, t)|^2$.

This means that the position of the electron within the atom is not entirely indeterminate. The solution to Schrödinger's equation for an electron in the electric field of a proton yields an amplitude for the wavefunction as a function of the distance from the proton, along with the energy levels. Squaring this amplitude reveals the probability of finding the electron at any specific location. Consequently, we can only assign probabilities to its position within an orbit and the likelihood of locating it in the space between orbits. Notably, there is even a small probability of the so-called orbital electron existing inside the nucleus.

Schrödinger's wavefunction associates each point in space (and time) with two real numbers: the amplitude (or size) of the wavefunction and its phase. Generally, the phase of a wave indicates its position within a cycle relative to an arbitrary reference point, measuring how far one is from a wave crest or trough. This phase is usually represented as an angle. Unlike the amplitude of the wavefunction, which relates to probability, the phase is inherently unobservable; only differences in phase can be detected, such as through interference patterns in optics.

13.37 Electron spin

Having just developed a sophisticated understanding of the electronic wavefunction, we will now revert to the familiar concept of Bohr's orbital atom to illustrate the next significant advancement in quantum theory. By 1925, physicists studying atomic spectra realized that not everything aligned with Bohr's model. Specifically, where Bohr's theory predicted a single spectral line, two lines were sometimes observed very close together.

To address this and other similar anomalies, Dutch physicists Sam Goudsmit and George Uhlenbeck proposed that electrons spin on their axes as they orbit the nucleus, much like the Earth spins around its north-south axis while orbiting the Sun. This electron spin introduces magnetic effects within the atom that can explain the spectral line splitting.

The electron's orbit around the nucleus creates a small loop of electric current, generating a magnetic field; thus, the orbiting electron acts like a tiny magnet. Additionally, the electron's spin generates its own magnetic moment, known as the "magnetic moment of the electron." This magnetic moment interacts with the orbital magnetic moment, resulting in variations in energy depending on the electron's spin direction. Consequently, this leads to slight differences in energy levels for the various spins of the electron, ultimately causing the observed splitting of the spectral lines associated with the Bohr orbit.

13.38 The Pauli exclusion principle

A cursory examination of the Bohr model of the atom reveals a fundamental principle that seems to be missing. There appears to be nothing preventing all the electrons in an atom from occupying the same orbit. However, we know that a typical atom has its electrons distributed across several different orbits. If this were not the case, transitions between these orbits would be rare, contradicting observations of atomic spectra. Thus, there must be a rule that keeps electrons spread out among the atom's orbits.

In 1925, Austrian physicist Wolfgang Pauli formulated the principle stating that no two electrons can occupy the same quantum state simultaneously. This means that electrons cannot have identical values of momentum, charge, and spin in the same region of space. Pauli arrived at this conclusion by carefully analyzing the atomic spectra of helium, where he found that certain transitions were consistently absent. For example, the lowest orbit (or ground state) of helium, where both electrons have the same spin, is not observed. In contrast, the state with opposing spins is seen.

The significance of this principle in atomic physics cannot be overstated. Since no two electrons can exist in the same state, adding extra electrons results in the gradual filling of outer electron orbits, preventing overcrowding in the lowest orbit. Only two electrons are allowed in the ground state, as they can differ solely by their two available spin values.

Higher orbits can accommodate more electrons because their quantum states can vary across a broader range of orbital angular momenta around the nucleus, which is also quantized. The Pauli exclusion principle is crucial for defining the chemical identities of all atoms of the same element, as it determines the allowed configurations of atomic electrons.

While our discussion has centered on atoms, the exclusion principle applies to any quantum system, defined primarily by the wavefunctions of its component particles.

In the case of isolated electrons with definite momentum whose wavefunctions extend throughout all space, the exclusion principle dictates that only two electrons with opposite spins can share the same momentum. For electrons confined within a crystal, where their wavefunctions span the dimensions of the crystal, this rule applies to all electrons present in that crystal.

Pauli's exclusion principle can also be expressed in terms of the behavior of a quantum system's wavefunction. While we've previously discussed the wavefunctions of individual particles, these can be combined to represent the entire quantum system. For instance, the total wavefunction of a helium atom describes the behavior of both electrons simultaneously. Just as a single electron's wavefunction reflects its localization as a wavepacket, a two-electron wavefunction will have two wavefunction peaks, each representing the localization of one of the electrons.

The exclusion principle arises from the requirement that a multi-electron wavefunction changes sign when any two electrons are interchanged. This means that wherever the wavefunction is positive, it must become negative, and vice versa. Thus, the wavefunction is anti-symmetric with respect to the interchange of two electrons. Considering a two-electron helium atom, if we denote the wavefunction for one electron at position $1x_1$ and the other at position $2x_2$, the probability of both electrons occupying the same position becomes zero, leading to the exclusion principle.

Since any two electrons are indistinguishable, swapping them is merely relabeling, which should not affect physical results such as energy levels and probability densities. The anti-symmetry of the wavefunction accommodates this. Since all physical quantities are proportional to the square of the wavefunction, changing its sign does not alter the outcomes.

Particles such as electrons and protons, which have a spin of ½ h (as well as more exotic particles with half-integral spins like *3/2 h 5/2 h*, etc.), obey the exclusion principle. Their wavefunctions are anti-symmetric when two identical particles are interchanged, classifying them as fermions. The dynamics of these fermions are governed by statistics formulated by Italian physicist Enrico Fermi and Englishman Paul Dirac, known as Fermi-Dirac statistics. These statistics describe how momentum is distributed among particles in an ensemble. Due to the exclusion principle, there is a limit to the number of particles that can occupy a specific momentum value, resulting in a diverse range of momenta among the particles.

In contrast, particles such as the photon with spin h (and other particles we shall meet with integral spins such as 0, h, 2h, 3h, . . . do not obey the exclusion principle and are called bosons.

The wavefunction of bosons remains unchanged when two particles are interchanged. An ensemble of bosons follows dynamical statistics initially developed by Indian physicist Satiendranath Bose and Albert Einstein. In Bose-Einstein statistics, there is no upper limit to the number of particles that can share the same momentum value. This property enables bosonic assemblies to behave coherently, as seen in laser light.

Ch 14— Many Worlds Theory & Philosophies and Paradoxes in Physics

14.1 A complete break from classical physics

A complete departure from classical physics occurred with the understanding that all particles and waves are not distinct entities, but rather a blend of both, known as particle-wave duality. In our everyday world, the particle aspect of this duality dominates interactions with material objects, making the wave component almost negligible. However, the wave aspect is still present, as described by the relation $p \cdot \lambda = h$, although it holds little significance at the macroscopic level. In the realm of the very small, where particles and waves are equally important, the behavior defies anything we can grasp from our daily experiences. The Bohr model, with its electron "orbits," was an incorrect depiction—and, in fact, no classical analogy can accurately describe atomic behavior. Atoms simply behave like atoms, without a direct comparison to familiar concepts.

The concept of a "field" was first introduced by Faraday but gained widespread acceptance only after Maxwell demonstrated that all electric and magnetic phenomena could be described by just four equations involving electric and magnetic fields. These equations, known as Maxwell's equations, form the foundation of electromagnetism. They are as fundamental to electromagnetism as Newton's laws are to mechanics—perhaps even more so, as Maxwell's equations are compatible with the theory of relativity, whereas Newton's laws are not. This set of equations encapsulates all of electromagnetism and stands as one of the great achievements of human intellect.

Electromagnetic waves, like radio and TV signals, are transmitted and received via antennas, and all EM waves, regardless of frequency, travel at the speed of light. Light itself is an electromagnetic wave. While we won't delve into the mathematical form of Maxwell's equations, which involve calculus, we will summarize their key principles in non-mathematical terms.

The four key principles, known as Maxwell's equations, are as follows: (1) A generalized form of Coulomb's law, known as Gauss's law, relates electric fields to their sources—electric charge. (2) A similar law applies to magnetic fields, except that since no magnetic monopoles (single magnetic charges) have been observed, magnetic field lines are continuous and do not begin or end as electric field lines do. (3) A changing magnetic field produces an electric field—this is known as Faraday's law. (4) A magnetic field is generated by an electric current or by a changing electric field.

The first part of (4), which states that an electric current produces a magnetic field, was discovered by Oersted and is mathematically expressed by Ampere's law. However, the second part of (4), which predicts that a changing electric field generates a magnetic field, was a new concept introduced by Maxwell. He proposed this based on the principle of symmetry in nature, hypothesizing that if a changing magnetic field can create an electric field (as Faraday's law indicates), the reverse must also be true. Although the effect is usually too small to detect easily, Maxwell's insight laid the groundwork for significant advances in physics.

These scientific discoveries ultimately contributed to the development of relativity and quantum theory, leading to new philosophical interpretations and paradoxes within modern science.

14.2 Revisiting the Twin Paradox

Soon after Einstein introduced the special theory of relativity, a seeming paradox emerged, known as the twin paradox. It goes like this: Imagine two 20-year-old twins. One twin embarks on a journey to a distant star and back in a spaceship traveling at a very high speed, while the other twin stays on Earth. According to the Earth-bound twin, time passes more slowly for the traveling twin due to the high speed. So, while 20 years might elapse for the Earth twin, only about a year could pass for the traveling twin, depending on the spacecraft's speed. Upon the traveler's return, the Earth twin would be 40 years old, while the spacefaring twin would only be 21.

But what about the perspective of the traveling twin? Since all inertial frames of reference are considered equally valid in special relativity, couldn't the traveling twin claim the reverse—that the Earth is moving away at high speed and therefore, time should pass more slowly for the twin on Earth? If both twins make these opposing claims, they cannot both be correct, because when the spacecraft returns, their ages can be directly compared. On the surface, this contradiction seems to present a paradox.

However, there is actually no paradox. The implications of special relativity—specifically, time dilation—apply only to observers in inertial (non-accelerated) reference frames. The Earth represents such a frame. During the periods of acceleration, the twin on the spacecraft is not in a steady inertial reference frame, rendering the traveling twin's predictions based on special relativity invalid. In contrast, the twin on Earth remains in an inertial frame and can make accurate predictions. Therefore, the notion of a paradox is unfounded. The traveling twin's perspective is incorrect, and the Earth twin's prediction—that the traveling twin will return having aged less—is the valid outcome. This conclusion is further supported by Einstein's general theory of relativity, which addresses situations involving accelerating reference frames.

14.3 Probability versus Determinism

The classical Newtonian perspective of the universe is fundamentally deterministic. A key principle of this view is that if the position and velocity of an object are known at a specific moment, its future position can be accurately predicted, provided the forces acting on it are also known. For instance, if a stone is thrown multiple times with the same initial velocity and angle, and the forces—like gravity and air resistance—remain constant, its trajectory will consistently follow the same path. This "mechanistic" outlook suggests that the unfolding of the universe, composed of discrete particles, is entirely predetermined.

However, this classical deterministic view has been profoundly transformed by quantum mechanics. As demonstrated in the double-slit experiment, electrons prepared in identical conditions do not all converge at the same location. Quantum mechanics introduces the concept of probabilities, indicating that an electron could arrive at various points, which starkly contrasts with the classical notion where a particle's path is entirely predictable based on its initial position, velocity, and the forces acting upon it.

Furthermore, quantum mechanics asserts that the position and velocity of an object cannot be simultaneously measured with complete accuracy, a concept encapsulated in the uncertainty principle. This limitation arises because fundamental entities, such as electrons, are not merely particles; they also exhibit wave-like properties. In quantum mechanics, we can only calculate the likelihood of finding an electron (when regarded as a particle) at different locations. Ultimately, quantum mechanics reveals an inherent unpredictability in nature that challenges the deterministic view of classical physics.

Since matter is composed of atoms, even everyday objects are believed to be influenced by probability rather than strict determinism. For instance, there exists a finite (albeit very small) chance that when you throw a stone, its trajectory might unexpectedly curve upward instead of following the expected downward parabola of projectile motion. Quantum mechanics predicts with high probability that ordinary objects will behave in accordance with classical physical laws, but these predictions are framed as probabilities rather than certainties.

The reason macroscopic objects tend to conform to classical laws is that when large numbers of particles are involved, any deviations from the average behavior are negligible. It is the collective behavior of vast numbers of molecules that adheres to the established laws of classical physics with high certainty, creating an illusion of "determinism." However, when examining small quantities of molecules, deviations from classical predictions become observable. Thus, while quantum mechanics does not offer precise "deterministic" laws, it does provide "statistical" laws grounded in probability.

It is essential to differentiate between the probabilities inherent in quantum mechanics and those utilized in the 19th century for understanding thermodynamics and the behavior of gases. In thermodynamics, probability is employed because tracking the immense number of molecules is impractical, yet their movements and interactions were still thought to be deterministic under Newton's laws. In contrast, the probability in quantum mechanics is intrinsic to nature itself and is not merely a reflection of our inability to calculate accurately.

Although some physicists, such as Einstein, have resisted relinquishing a deterministic view of the universe and have hesitated to fully embrace quantum mechanics as a complete theory, the vast majority of physicists accept the probabilistic interpretation. This widely accepted perspective, known as the Copenhagen interpretation of quantum mechanics, was largely formulated through discussions among Niels Bohr and other prominent physicists in Bohr's home city of Copenhagen.

Because electrons exhibit both wave-like and particle-like properties, they cannot be regarded as following specific paths through space and time. This implies that our understanding of matter within the frameworks of space and time may not be entirely accurate. Such profound conclusions have sparked lively debates among philosophers. Niels Bohr, in particular, has been a significant and influential figure in the philosophy of quantum mechanics, arguing that experiments involving atoms or electrons must be described using concepts familiar to our everyday experiences, such as waves and particles, along with space and time.

We must be careful not to let our interpretations of experiments lead us to believe that atoms or electrons exist in space and time as discrete particles. This distinction between our experimental interpretations and what is "really" occurring in nature is essential.

Up to this point, we have discussed the position and velocity of an electron as if it were merely a particle. However, it is not just a particle; electrons—and matter in general—exhibit both wave and particle properties. The uncertainty principle reveals that if we insist on viewing an electron solely as a particle, we must acknowledge certain limitations of this simplified perspective: specifically, that both the position and velocity cannot be known with precision at the same time, and that energy can be uncertain (or non-conserved) within the relationship *(delta – E) for a time delta-E approx. = (h / delta-E)*. Since Planck's constant *h* is exceedingly small, the uncertainties described by the uncertainty principle are typically negligible at the macroscopic level. However, at the atomic scale, these uncertainties become significant.

Because we understand ordinary "objects" to be composed of atoms containing nuclei and electrons, the uncertainty principle is crucial for comprehending all of nature. It encapsulates the probabilistic nature of quantum mechanics and serves as a foundation for philosophical discussions.

While we still do not fully understand how gravity operates, Einstein's limitation regarding its speed was experimentally validated in the early twenty-first century: gravity travels at the speed of light. Therefore, if the Sun were to vanish suddenly, we would not experience the loss of its gravitational pull until eight minutes later. Locality, it seems, governs our understanding.

However, this notion was challenged by experiments inspired by the work of the lesser-known physicist John Bell from Northern Ireland, which demonstrated the phenomenon of entanglement. Entanglement represents genuine action at a distance—an idea that continues to perplex many scientists today.

Modern theories provide us with a more nuanced understanding of the universe and challenge the clarity of the concept of "distance." A theorist from Duke University has proposed that the quantum realm may possess an additional unseen dimension, enabling spatially separated objects to communicate as though they were adjacent. Some theorists even suggest that spatial separation could be an illusion for entangled particles.

Nevertheless, there remains a strong reluctance to accept that anything, no matter how insubstantial or incapable of conveying information, could travel faster than light.

Although Einstein's critiques of quantum theory, particularly his objections to its reliance on probability, are often quoted—most famously regarding *God does not play dice*—what truly unsettled him was the violation of "locality," which contradicted his understanding of reality.

The phenomenon that challenges locality and reintroduces the possibility of action at a distance is entanglement, which arises from quantum theory, the contemporary science of the very small. To understand entanglement, we must trace the evolution of quantum theory from its origins as a practical workaround for addressing puzzling phenomena to its emergence as a comprehensive framework that ultimately undermines the principles of classical physics.

14.4 Holding on tight to Determinism

Quantum physics was fully developed by 1925, beginning with Einstein's 1905 paper on the photoelectric effect, where he demonstrated that light is quantized. Over the years, quantum theory achieved remarkable success, becoming the most successful physical theory of its time, according to Einstein himself. However, despite its achievements, Einstein never accepted quantum physics as the definitive explanation of reality.

What troubled Einstein about quantum theory was its rejection of determinism. Heisenberg's uncertainty principle, which lies at the core of quantum physics, states that you cannot simultaneously know everything about the behavior of objects. While this principle applies universally—from electrons to stars—the noticeable effects are confined to the atomic realm, particularly when dealing with subatomic particles.

If the Universe follows Heisenberg's principle, it means that you can never know for certain the "past" or "present" of anything, nor can you predict the future with any degree of certainty. Everything becomes probabilistic. As Einstein famously remarked, "God doesn't play dice with the Universe." He was reluctant to abandon the classical deterministic view of physics and engaged in intense debates with Niels Bohr, the primary advocate of the new physics. Although Bohr often convinced Einstein of the shortcomings in specific arguments against quantum physics, he could not address Einstein's concerns regarding the EPR problem, which is discussed in this chapter.

Einstein proposed a way to bypass Heisenberg's uncertainty principle and accurately determine the outcomes of collisions between two particles and their entire future histories. At the time, technology had not advanced enough to conduct the delicate EPR experiment. However, since 1982, several real EPR experiments have been performed, demonstrating that Einstein was mistaken; the fundamental premise of quantum physics is valid—the world is not deterministic.

Quantum theory is widely regarded as resistant to any realist interpretation, marking the onset of a 'postmodern' science characterized by paradox, uncertainty, and the limits of precise measurement. Confusion between the realms of macro-physical objects and subatomic particles was evident early in the history of quantum physics, particularly during the famous debates between Einstein and Bohr. These discussions became most pronounced in the 1932 Einstein-Podolsky-Rosen (EPR) paper, which established criteria for a realist interpretation that aligns with the known laws of physics, including those of relativity.

Probability vs Determinism
In courtesy of The California Institute of Technology,
The University of Jerusalem and The University of Copenhagen

The EPR argument led to J.S. Bell's renowned theorem, which states that any interpretation suggesting the existence of "hidden variables" would result in highly problematic implications, including nonlocal effects of quantum "entanglement" across arbitrary space-time distances—something Einstein was unwilling to accept. In contrast, David Bohm's theory addresses this issue while preserving a realist ontology and aligning with established quantum mechanics (QM) observational results and predictions. Furthermore, Bohm's theory avoids the extravagant conjectures associated with interpretations like the 'many-worlds' hypothesis, currently promoted by David Deutsch, which proposes a radical and difficult-to-conceptualize revision of what constitutes a "realist" worldview based on orthodox QM.

14.5 The Clock in the Box

The famous debate between Bohr and Einstein regarding the interpretation of quantum theory began at the fifth Solvay Congress in 1927 and continued until Einstein's death in 1955. Einstein also engaged in correspondence with Max Born on the topic, offering insights into their discussions through *The Born-Einstein Letters*. Central to their debate were hypothetical tests of the predictions made by the Copenhagen interpretation, rather than real experiments. The premise was that Einstein would devise a thought experiment aimed at measuring two complementary properties of a particle simultaneously, such as its position and mass or its exact energy at a specific time.

In response, Bohr and Born would demonstrate how Einstein's proposed experiments could not be feasibly executed in a way that would undermine the theory. One illustrative example of this back-and-forth is the "clock in the box" thought experiment proposed by Einstein.

In this scenario, imagine a box with a hole in one wall, which is covered by a shutter controlled by a clock inside the box. The interior of the box is filled with radiation. The apparatus is set up so that at a precise, predetermined time, the clock will open the shutter to allow one photon to escape before closing it again.

After the photon escapes, the box is weighed, and then weighed again. Because mass is equivalent to energy, the difference in the two weights reveals the energy of the escaping photon. This seemingly demonstrates that we can know both the exact energy of the photon and the precise moment it passed through the hole, seemingly contradicting the uncertainty principle.

However, Bohr countered by examining the practicalities of how such measurements could be conducted. For example, the box must be suspended, perhaps by a string, in a gravitational field. Before the photon escapes, the experimenter notes the position of a pointer fixed to the box against a scale. After the photon has escaped, the experimenter could theoretically add weights to the box to return the pointer to its original position.

Yet, this process itself introduces uncertainty. The position of the pointer can only be measured within the limits dictated by Heisenberg's uncertainty principle, resulting in an uncertainty in the box's momentum. The more precisely the weight of the box is measured, the greater the uncertainty in its momentum. Even when attempting to restore the original state by adding a small weight to return the spring to its original position, the experimenter can only reduce uncertainty to the limits allowed by Heisenberg's relation, illustrating the inherent challenges in Einstein's thought experiment.

The specifics of this and other thought experiments from the Einstein-Bohr debate are thoroughly examined in Abraham Pais's *Subtle is the Lord...*. Pais emphasizes that Bohr's insistence on a comprehensive and precise description of the hypothetical experiments is not merely fanciful. He highlights essential components such as the heavy bolts that secure the balance framework, the spring that enables mass measurement while allowing the box to move, and the small weight that must be added, among other details.

All experimental results must be interpreted using classical language, which reflects our everyday reality. For instance, if we were to rigidly fix the box in place to eliminate any uncertainty about its position, we would then be unable to measure the change in mass. The dilemma of quantum uncertainty arises when we attempt to articulate quantum concepts using familiar language, which is why Bohr emphasized the practical details of the experiments.

14.6 Einstein Forgets Relativity

As mentioned earlier, over the past three years, Bohr had revisited the hypothetical experiments proposed by Einstein at the Solvay conference in October 1927. Each experiment aimed to demonstrate inconsistencies in quantum mechanics, but Bohr identified flaws in Einstein's reasoning in every instance. Not one to rest on his achievements, Bohr created his own thought experiments, incorporating various slits, shutters, clocks, and other elements, as he scrutinized his interpretation for any weaknesses. However, he found none. Yet, Bohr had not devised anything as simple and clever as the thought experiment that Einstein had just shared with him in Brussels at the sixth Solvay conference.

The theme of the six-day meeting, which began on October 20, 1930, was the magnetic properties of matter. The format remained consistent, featuring a series of commissioned reports on topics related to magnetism, each followed by discussions. Bohr and Einstein both served on the nine-member scientific committee, ensuring their automatic invitation to the conference. Following Lorentz's death, Paul Langevin had taken on the demanding dual roles of presiding over both the committee and the conference. Among the 34 participants were notable figures such as Dirac, Heisenberg, Kramers, Pauli, and Sommerfeld.

In the Newtonian universe, everything is purely deterministic, leaving no room for chance. In this framework, a particle possesses a definite momentum and position at any given moment. The forces acting on the particle dictate how its momentum and position change over time. Physicists like James Clerk Maxwell and Ludwig Boltzmann could only account for the properties of gases made up of many such particles by employing probability and settling for a statistical description. This reliance on statistical analysis arose from the challenges of tracking the motion of vast numbers of particles. In this deterministic universe, probability was seen as a result of human ignorance, where everything unfolded according to the laws of nature.

If the present state of any system and the forces acting upon it are known, the future behavior of that system can be precisely predicted. In classical physics, determinism is intricately linked to causality, the idea that every effect has a specific cause. However, it is impossible to determine the exact location of an electron after a collision. Instead, physics can only calculate the probability of the electron being scattered at a certain angle. This concept forms the basis of Born's "new physical content," which hinges on his interpretation of the wave function.

The wave function itself lacks physical reality; it exists in the enigmatic, ghost-like realm of possibilities. It encompasses abstract possibilities, such as all the angles at which an electron could scatter after colliding with an atom. There is a significant distinction between what is possible and what is probable. Born contended that the square of the wave function—a real number, not a complex one—resides in the realm of probability. Squaring the wave function does not yield the actual position of the electron; rather, it provides the odds of finding the electron in one location compared to another. For instance, if the value of the wave function at point X is twice that at point Y, the probability of finding the electron at X is four times greater than at Y. The electron could ultimately be located at X, Y, or another position entirely.

Niels Bohr would later assert that until an observation or measurement occurs, a microphysical entity like an electron does not exist in any specific location. Between measurements, it lacks existence outside the abstract possibilities represented by the wave function. It is only when a measurement is made that the "wave function collapses," transitioning from one of the many possible states of the electron to a definitive state, rendering all other possibilities irrelevant.

Schrödinger rejected Born's probability interpretation, arguing that the collision of an electron or an alpha particle with an atom cannot be considered "absolutely accidental" or "completely undetermined." He believed that accepting Born's view would imply the inevitability of quantum jumps, once again putting causality at risk.

The upcoming meeting, featuring twelve current and future Nobel laureates, was a significant backdrop for the "second round" of the ongoing debate between Einstein and Bohr regarding the meaning of quantum mechanics and the nature of reality. Einstein arrived in Brussels with a new thought experiment intended to deliver a decisive blow to the uncertainty principle and the Copenhagen interpretation. An unsuspecting Bohr was taken by surprise after one of the formal sessions.

As previously mentioned, Einstein posed a challenge to Bohr, asking him to consider, "Imagine a box full of light. In one of its walls is a hole equipped with a shutter that can be opened and closed by a mechanism connected to a clock inside the box." This clock is synchronized with another in the laboratory. First, weigh the box. Then, set the clock to open the shutter at a specific time, allowing just enough time for a single photon to escape. "Now we know," Einstein explained, "precisely when the photon left the box."

Bohr listened without concern; everything Einstein proposed seemed straightforward and indisputable. He believed that the uncertainty principle applied only to pairs of complementary variables—like position and momentum or energy and time—without restricting the accuracy with which any single variable could be measured. Then, with a hint of a smile, Einstein dropped the bombshell: "Weigh the box again." In an instant, Bohr realized that he and the Copenhagen interpretation were in serious trouble.

To determine how much light was locked in that single photon, Einstein drew upon a remarkable insight he had made while working as a clerk in the Patent Office in Bern: energy is mass, and mass is energy. This profound relationship was succinctly expressed in his famous equation, *$E=mc^2$*, where E represents energy, m stands for mass, and c is the speed of light.

By weighing the box of light before and after the photon escapes, one could easily calculate the difference in mass. Although such an incredibly small change would have been impossible to measure with the technology available in 1930, in the realm of thought experiments, it was a trivial task. Using *$E=mc^2$* to convert the lost mass into an equivalent amount of energy allowed for precise calculation of the energy of the escaped photon. The moment of the photon's departure was known due to the synchronization between the laboratory clock and the one inside the box that controlled the shutter. It seemed that Einstein had devised an experiment capable of measuring both the energy of the photon and the timing of its escape with a degree of accuracy limited only by Heisenberg's uncertainty principle.

"It was quite a shock for Bohr," recalled Belgian physicist Leon Rosenfeld, who had recently begun a long-term collaboration with Bohr. "He did not see the solution immediately." While Bohr was deeply troubled by Einstein's latest challenge, Pauli and Heisenberg dismissed the concerns. "Ah, well, it will be all right, it will be all right," they reassured him. However, Bohr spent the entire evening feeling extremely unhappy, moving from one person to another, trying to convince them that Einstein's assertion couldn't possibly be true and that it would spell the end of physics if he were correct. Yet, Rosenfeld noted, "he couldn't produce any refutation."

Rosenfeld was not invited to the Solvay conference in 1930 but traveled to Brussels to meet Bohr. He vividly recalled the sight of the two quantum rivals walking back to the Hotel Metropole that evening: "Einstein, a tall, majestic figure, walked quietly, wearing a somewhat ironic smile, while Bohr hurried alongside him, visibly agitated and pleading that if Einstein's device were to work, it would spell the end of physics." For Einstein, however, it was neither an ending nor a beginning; it merely demonstrated that quantum mechanics was inconsistent and not the closed and complete theory that Bohr claimed. His latest thought experiment was simply an attempt to salvage a version of physics that sought to understand an observer-independent reality.

Bohr spent a sleepless night scrutinizing every aspect of Einstein's thought experiment. He took apart the imaginary box of light, hoping to identify a flaw. Unlike Bohr, who aimed to grasp the device's intricacies, Einstein had not visualized the details of the light box's inner workings or how to weigh it. Desperate to comprehend the apparatus and the necessary measurements, Bohr sketched what he called a "pseudo-realistic" diagram of the experimental setup.

Focusing on the weighing process, Bohr, filled with mounting anxiety and limited time, opted for the simplest method: he suspended the light box from a spring attached to a supporting frame. To transform it into a weighing scale, Bohr affixed a pointer to the light box, allowing its position to be read against a scale on a vertical arm resembling a gallows. To ensure the pointer rested at zero on the scale, he attached a small weight to the bottom of the box. Every detail was meticulously accounted for, even the nuts and bolts that secured the frame to its base, and he included the clockwork mechanism responsible for opening and closing the shutter through which the photon would escape.

Initially, the light box's configuration was set with the attached weight, ensuring the pointer indicated zero. After the photon escaped, the box became lighter, causing it to rise under the tension of the spring. To return the pointer to zero, the initial weight had to be replaced with a slightly heavier one. Importantly, there was no time limit on how long the experimenter could take to adjust the weights. The difference in the weights indicated the mass lost due to the escaped photon, and from $E=mc^2$, the energy of the photon could be calculated precisely.

Drawing on his arguments from the 1927 Solvay conference, Bohr asserted that any attempt to measure the position of the light box would inherently introduce uncertainty in its momentum, as illuminating the scale for reading would be necessary.

The act of measurement

The act of measuring the light box's weight would inevitably result in an uncontrollable transfer of momentum due to the exchange of photons between the pointer and the observer, causing the box to move. To enhance the accuracy of the position measurement, Bohr concluded that the balancing of the light box and the adjustment of the pointer to zero would need to be performed over an extended period. However, he argued that this approach would lead to a corresponding uncertainty in the box's momentum; the more precisely its position was measured, the greater the uncertainty in the measurement of its momentum.

Unlike at the 1927 Solvay conference, Einstein was now challenging the "energy-time" uncertainty relation instead of the "position-momentum" version. In the early hours of the morning, as Bohr's fatigue began to set in, he suddenly recognized a flaw in Einstein's experiment. Gradually, he reconstructed the analysis until he was convinced that Einstein had made an almost inconceivable mistake. Relieved, Bohr finally went to sleep, knowing that he would wake up to savor his triumph over breakfast.

In his quest to undermine the Copenhagen interpretation of quantum reality, Einstein had overlooked his own theory of general relativity, failing to consider the effects of gravity on the time measurement by the clock inside the light box. General relativity was Einstein's crowning achievement. Max Born described it as "the greatest feat of human thinking about Nature, an astonishing combination of philosophical insight, physical intuition, and mathematical skill." He likened it to "a great work of art, to be enjoyed and admired from a distance." The confirmation of the bending of light predicted by general relativity in 1919 made headlines worldwide, with J.J. Thomson declaring it "a whole new continent of new scientific ideas."

One of these groundbreaking ideas was gravitational "time dilation." Two identical and synchronized clocks placed in a room, with one fixed to the ceiling and the other on the floor, would be out of sync by 300 parts in a *billion billion* because time flows more slowly at the floor than at the ceiling, a consequence of gravity. According to general relativity, the rate at which a clock ticks depends on its position within a gravitational field. Additionally, a clock moving within a gravitational field ticks slower than one that remains stationary. Bohr realized that this meant that weighing the light box would impact the timekeeping of the clock inside it.

The position of the light box within the Earth's gravitational field is altered by measuring the pointer against the scale. This positional change affects the rate of the clock inside the box, resulting in a loss of synchronization with the laboratory clock. Consequently, it becomes impossible to measure with the precision Einstein assumed the exact moment the shutter opened, and the photon escaped. The greater the accuracy in measuring the energy of the photon using $E=mc^2$, the greater the uncertainty in the light box's position within the gravitational field.

This uncertainty in position, influenced by gravity's effect on the flow of time, means that it is impossible to measure both the energy of the photon and the time of its escape simultaneously. Heisenberg's uncertainty principle remained unchallenged, along with the Copenhagen interpretation of quantum mechanics.

When Bohr came down to breakfast, he no longer resembled "a dog who has received a thrashing" the night before. Instead, it was Einstein who sat in stunned silence as Bohr explained why his latest challenge, like those from three years earlier, had failed. However, some would later question Bohr's refutation, arguing that he had treated macroscopic elements—such as the pointer, scale, and light box—as if they were quantum objects subject to the uncertainty principle's limitations. This approach contradicted his insistence that laboratory equipment should be treated classically. Yet, Bohr had never clearly defined where the line between micro and macro should be drawn, as every classical object ultimately consists of atoms.

Despite any reservations, Einstein accepted Bohr's counterarguments, as did the broader physics community. Consequently, he ceased his attempts to demonstrate that quantum mechanics was logically inconsistent and instead began to focus on revealing the theory's incompleteness.

In November 1930, Einstein lectured on the light box in Leiden. Afterward, a member of the audience suggested there was no conflict within quantum mechanics. "I know; this business is free of contradictions," Einstein replied, "yet in my view, it contains a certain unreasonableness." Nonetheless, in September 1931, he nominated Heisenberg and Schrödinger for a Nobel Prize once again. However, after his confrontations with Bohr at the Solvay conferences, a telling sentence appeared in Einstein's nomination letter: "In my opinion, this theory contains, without a doubt, a piece of the ultimate truth." His inner voice persisted in whispering that quantum mechanics was incomplete and not the comprehensive truth that Bohr suggested.

In the years following the 1930 Solvay conference, direct contact between Bohr and Einstein diminished. A valuable communication channel was severed with Paul Ehrenfest's suicide in September 1933. In a poignant tribute, Einstein reflected on his friend's struggle to understand quantum mechanics and the challenges of adapting to new ideas, particularly for someone over fifty.

Many who read Einstein's words misinterpreted them as a lament for his own situation. Now in his mid-fifties, he felt he was viewed as a relic of a bygone era, unable or unwilling to accept quantum mechanics. However, he also recognized what set him and Schrödinger apart from most of their contemporaries: "Almost all the other fellows do not look from the facts to the theory but from the theory to the facts."

14.7 In search of a complete unified theory

There is a fundamental paradox in the quest for a complete unified theory. The idea behind scientific theories assumes that we, as rational beings, can freely observe the universe and draw logical conclusions from our observations. Under this assumption, it seems reasonable to expect that we could progressively approach the ultimate laws governing the universe. However, if such a completely unified theory truly exists, it would presumably "determine" our actions as well meaning the theory would dictate the outcome of our search for itself. But why should it ensure we draw the correct conclusions from the evidence? Could it not just as easily lead us to the wrong conclusion, or no conclusion at all?

The only response to this paradox lies in Darwin's principle of natural selection. In any population of self-reproducing organisms, variations in genetics and upbringing result in some individuals being better equipped to draw accurate conclusions about the world, which improves their chances of survival and reproduction. Over time, their pattern of thought and behavior comes to dominate. Historically, intelligence and scientific discovery have conferred on survival advantages, though it is uncertain whether this is still true. Our discoveries could potentially lead to our destruction, and even if they don't, a completely unified theory might not significantly affect our survival. Nevertheless, assuming the universe evolved in an orderly manner, we can expect that the reasoning abilities shaped by natural selection would remain valid in our search for this theory, preventing us from arriving at false conclusions.

14.8 The EPR paradox

Einstein accepted Bohr's critiques of various thought experiments, but by the early 1930s, he shifted his focus to a new kind of theoretical test of quantum mechanics. This new approach aimed to use experimental information about one particle to infer properties—such as position and momentum—of a second particle. Although this version of the quantum debate was unresolved during Einstein's lifetime, it has since been confirmed, not through thought experiments, but by real laboratory experiments. Once again, Bohr's interpretation prevailed over Einstein's.

In the early 1930s, Einstein's personal life was fraught with turmoil. He fled Germany to escape Nazi persecution and settled in Princeton by 1935. In December 1936, his second wife, Elsa, passed away after a prolonged illness. Despite these challenges, Einstein continued to grapple with the interpretation of quantum theory, defeated by Bohr's arguments but still unconvinced that the Copenhagen interpretation—with its inherent uncertainty and lack of strict causality—could be the final word on the nature of reality. Max Jammer meticulously detailed Einstein's thoughts during this period in *The Philosophy of Quantum Mechanics*.

In 1934 and 1935, Einstein collaborated with Boris Podolsky and Nathan Rosen at Princeton on a paper presenting what is now known as the "EPR Paradox," although it doesn't describe a paradox in the strictest sense. The paper, *Can Quantum-Mechanical Description of Physical Reality Be Considered Complete?*, published in *Physical Review* (vol. 47, pp. 777-780, 1935), argued that the Copenhagen interpretation must be incomplete, suggesting there must be an underlying mechanism that governs the universe, giving the illusion of uncertainty and unpredictability at the quantum level due to statistical variations.

Alternatively, we could have measured the precise position of the first particle and, by extension, deduced the position of the second particle. It is one thing to argue that measuring the momentum of particle A destroys knowledge of its position, preventing us from knowing its exact location, and that, similarly, its momentum remains unknown. However, to Einstein and his colleagues, it seemed an entirely different issue when considering the two measurements we choose to make on particle A. How could particle B "know" whether it should have a precisely defined momentum or a precisely defined position? This appeared to violate causality, suggesting that measurements on a particle here somehow affect its distant partner instantaneously, a phenomenon referred to as "action at a distance."

In the EPR paper, Einstein, Podolsky, and Rosen argued that if one accepts the Copenhagen interpretation, the reality of position and momentum in the second system would depend on the measurement process of the first system, which does not disturb the second system in any way.

According to the authors, "no reasonable definition of reality could be expected to permit this." This is where the team diverged from most of their colleagues, particularly from the Copenhagen school. While no one disputed the logical coherence of the argument, they disagreed on what constituted a "reasonable" definition of reality. Bohr and his colleagues accepted a reality in which the position and momentum of the second particle had no objective meaning until measured, regardless of what was done to the first particle. Ultimately, a choice had to be made between "objective reality" and the quantum world, and Einstein remained in a small minority, favoring objective reality while rejecting the Copenhagen interpretation.

Despite his objections, Einstein was always honest and willing to accept sound experimental evidence. Had he lived to see the recent experimental tests of the EPR effect, he would have been persuaded that he was wrong. Objective reality does not play a role in our fundamental description of the universe, but action at a distance—or acausality—does. The experimental confirmation of this principle is so crucial that it warrants a chapter of its own. But before delving into that, it is necessary to explore other paradoxical possibilities inherent in quantum mechanics, such as particles traveling backward in time and Schrödinger's famous half-dead cat.

14.9 EPR's impact on the Scientific community

Despite these mutual misgivings, young physicists were always eager to work with Einstein. One such individual was Nathan Rosen, a 25-year-old New Yorker who arrived from MIT in 1934 to serve as his assistant. A few months earlier, Boris Podolsky, a 39-year-old Russian-born physicist, had joined the institute. Podolsky had first met Einstein at Caltech in 1931, and they had collaborated on a paper. Now, Einstein had an idea for another paper that would mark a new phase in his ongoing debate with Bohr, launching a fresh critique of the Copenhagen interpretation of quantum mechanics.

At the Solvay Conferences in 1927 and 1930, Einstein had attempted to challenge the uncertainty principle, trying to show that quantum mechanics was inconsistent and therefore incomplete. However, Bohr, along with Heisenberg and Pauli, successfully countered each of Einstein's thought experiments and defended the Copenhagen interpretation. While Einstein eventually conceded that quantum mechanics was logically consistent, he did not accept that it was the definitive theory Bohr claimed it to be. Realizing he needed a new approach to demonstrate that quantum mechanics was incomplete, Einstein devised his most famous thought experiment.

In early 1935, Einstein met with Podolsky and Rosen in his office for several weeks to develop his ideas. Podolsky was tasked with writing the paper, while Rosen performed most of the necessary mathematical calculations. Einstein, according to Rosen's later recollections, "contributed the general point of view and its implications."

The result was the Einstein-Podolsky-Rosen paper, or the EPR paper, completed and mailed by the end of March. Titled "Can Quantum Mechanical Description of Physical Reality Be Considered Complete?"—notably missing the article "the"—it was published on May 15 in *Physical Review*. The paper's answer to the question it posed was a resolute "No!" However, even before it appeared in print, Einstein's name ensured that the EPR paper generated unwanted attention.

On Saturday, May 4, 1935, the *New York Times* ran an article on page eleven with the headline: "Einstein Attacks Quantum Theory." The article stated, "Professor Einstein will attack science's important theory of quantum mechanics, a theory of which he was a sort of grandfather. He concluded that while quantum mechanics is 'correct,' it is not 'complete.'"

Three days later, *The New York Times* published a response from an evidently disgruntled Einstein. While no stranger to the press, he clarified, "It is my invariable practice to discuss scientific matters only in the appropriate forum, and I deprecate advance publication of any announcement in regard to such matters in the secular press."

In their published paper, Einstein, Podolsky, and Rosen began by distinguishing between reality as it exists and a physicist's understanding of it: "Any serious consideration of a physical theory must take into account the distinction between the objective reality, which is independent of any theory, and the physical concepts with which we describe this reality, and by means of these concepts we picture this reality to ourselves."

To evaluate the success of any physical theory, EPR argued, two questions must receive an unequivocal "Yes": Is the theory correct? Is the description it provides complete?

"The correctness of the theory is judged by the degree of agreement between the conclusions of the theory and human experience," EPR stated. This was a principle every physicist would accept, where "experience" in physics is understood through experiment and measurement. To date, there had been no discrepancy between laboratory experiments and the theoretical predictions of quantum mechanics, making it a seemingly correct theory. However, for Einstein, it was not enough for a theory to be correct; it also had to be complete.

Whatever "complete" might mean, EPR established a necessary condition for the completeness of a physical theory: "Every element of the physical reality must have a counterpart in the physical theory." For this argument to hold, EPR needed to define what constituted an "element of reality."

Einstein was wary of getting bogged down in the philosophical quagmire that had swallowed many before him in attempting to define "reality." Few had emerged unscathed from trying to pinpoint the nature of reality. To avoid this, EPR astutely sidestepped offering a "comprehensive definition of reality," deeming it "unnecessary" for their purposes. Instead, they adopted what they considered a "sufficient" and "reasonable" criterion for defining an "element of reality": "If, without in any way disturbing a system, we can predict with certainty (i.e., with probability equal to unity) the value of a physical quantity, then there exists an element of physical reality corresponding to this physical quantity."

Einstein sought to refute Bohr's claim that quantum mechanics was a complete, fundamental theory of nature by demonstrating that there were objective "elements of reality" the theory failed to account for. In doing so, Einstein shifted the focus of the debate with Bohr and his supporters away from the internal consistency of quantum mechanics and toward the nature of reality and the role of theory.

EPR asserted that for a theory to be considered complete, there must be a one-to-one correspondence between an element of the theory and an element of reality. A sufficient condition for the reality of a physical quantity, such as momentum, is the ability to predict it with certainty without disturbing the system. If an element of physical reality exists that the theory does not account for, the theory is incomplete. This can be likened to finding a book in a library, but upon attempting to check it out, being told by the librarian that there is no record of the book in the catalogue. Despite the book bearing all the necessary markings to indicate it belongs to the collection, the only explanation is that the library's catalogue is incomplete.

According to the uncertainty principle, measuring the exact value of a system's momentum excludes the possibility of simultaneously measuring its position. The question Einstein sought to answer was: does the inability to measure a particle's exact position mean the particle doesn't have a definite position? The Copenhagen interpretation claimed that without a measurement determining its position, the electron lacks one.

EPR aimed to show that physical elements, such as an electron having a definite position, exist but quantum mechanics cannot account for them—making the theory incomplete.

EPR attempted to prove their point with a thought experiment involving two particles, A and B, that briefly interact before moving in opposite directions. The uncertainty principle forbids exact measurements of both the position and momentum of either particle simultaneously. However, it allows for an exact measurement of one property while leaving the other unknown.

The crux of the EPR thought experiment is leaving particle B undisturbed by avoiding direct observation. Even if A and B are separated by light-years, quantum mechanics does not prohibit the exact measurement of A's momentum from providing information about B's momentum without disturbing B. When A's momentum is measured precisely, it allows an exact determination of B's momentum through the law of conservation of momentum. According to the EPR criterion of reality, B's momentum must be an element of physical reality. Similarly, measuring A's exact position, knowing the physical distance separating A and B, enables the deduction of B's position without measuring it directly. Thus, EPR argued, the position of B must also be an element of physical reality.

EPR appeared to have devised a way to determine, with certainty, the exact values of either the momentum or position of particle B by performing measurements on particle A, all without any chance of physically disturbing particle B. Based on their reality criterion, EPR argued that they had proven both the momentum and position of particle B to be "elements of reality," implying that B could simultaneously possess exact values for both position and momentum.

**Einstein around 1940
In courtesy of The California Institute of Technology
and The University of Jerusalem**

Since quantum mechanics, through the uncertainty principle, precludes the possibility of a particle possessing both position and momentum simultaneously, these "elements of reality" lack counterparts in the theory. Consequently, EPR concludes that the quantum mechanical description of physical reality is incomplete.

Einstein's thought experiment was not intended to measure the position and momentum of particle B at the same time. He acknowledged that it is impossible to directly measure either property without causing an irreducible physical disturbance. Instead, the two-particle thought experiment was designed to demonstrate that these properties could have a definite simultaneous existence, asserting that both the position and momentum of a particle are indeed "elements of reality."

If the properties of particle B can be determined without directly observing (measuring) it, then those properties exist as elements of physical reality independently of observation. Thus, particle B has a position that is real and a momentum that is real.

EPR recognized a potential counterargument: that multiple physical quantities can only be considered simultaneous elements of reality if they can be measured or predicted at the same time. However, this perspective would imply that the reality of the momentum and position of particle B is contingent on the measurement conducted on particle A, which could be light years away and does not disturb particle B in any way. EPR asserted that "no reasonable definition of reality could be expected to permit this."

Central to the EPR argument was Einstein's assumption of locality—the belief that there is no instantaneous *"action-at-a-distance."* Locality rejects the idea that an event in one region of space could instantaneously influence another event elsewhere, particularly at faster-than-light speeds.

For Einstein, the speed of light represented an unbreakable limit on how fast anything could travel from one location to another. As the pioneer of relativity, he found it inconceivable that a measurement taken on particle A could instantaneously influence the independent elements of physical reality associated with particle B, even at a distance. When the EPR paper was published, it raised alarms among leading quantum physicists across Europe. Pauli, furious in Zurich, wrote to Heisenberg in Leipzig, expressing his dismay: "Einstein has once again made a public statement about quantum mechanics, and even in the issue of *Physical Review* of May 15 (together with Podolsky and Rosen, not good company by the way). As is well known, that is a disaster whenever it happens."

Reflecting on the challenges of interpretation, Paul Dirac remarked fifty years after the 1927 Solvay conference, "This problem of getting the interpretation proved to be rather more difficult than just working out the equations." Nobel laureate Murray Gell-Mann suggested that part of the issue stemmed from Niels Bohr having "brainwashed" a whole generation of physicists into believing the problem had been solved. A poll conducted in July 1999 at a quantum physics conference in Cambridge revealed the perspectives of a new generation on this complex issue: of the 90 physicists surveyed, only four supported the Copenhagen interpretation, while 30 favored a modern version of Everett's many-worlds theory. Notably, 50 respondents chose "none of the above or undecided." The unresolved conceptual difficulties, including the measurement problem and the challenge of defining where the quantum realm ends and the classical world begins, have prompted an increasing number of physicists to seek a deeper understanding beyond quantum mechanics. Dutch Nobel Prize-winning theorist Gerard 't Hooft contends, "A theory that yields 'maybe' as an answer should be recognized as an inaccurate theory." He advocates for a deterministic universe and seeks a more fundamental theory that would explain the peculiar, counterintuitive aspects of quantum mechanics.

Conversely, others like Nicolas Gisin, a leading experimenter in entanglement research, express no hesitation in believing that quantum theory is incomplete. The emergence of alternative interpretations and growing doubts about the completeness of quantum mechanics have led to a reevaluation of the long-standing verdict against Einstein in his debate with Bohr. British mathematician and physicist Sir Roger Penrose questions, "Can it really be true that Einstein, in any significant sense, was as profoundly 'wrong' as the followers of Bohr might maintain?" He expresses his alignment with Einstein's belief in a sub-microscopic reality and concurs with his view that contemporary quantum mechanics is fundamentally incomplete.

Although Einstein never landed a decisive blow in his debates with Bohr, his challenge was both sustained and thought-provoking. It inspired thinkers like David Bohm, John Bell, and Hugh Everett to critically examine Bohr's Copenhagen interpretation. For the last 30 years of his life, Einstein faced marginalization due to his critiques of this interpretation and his attempts to confront the uncertainties of quantum mechanics. However, he has seen partial vindication since then.

The conflict between Einstein and Bohr extended beyond the equations and numbers produced by quantum mechanics; it delved into deeper philosophical questions about the meaning of quantum mechanics and its implications for the nature of reality. Their differing answers to such questions were what ultimately distinguished them. Einstein never proposed his own interpretation; rather, he evaluated quantum mechanics through his belief in an observer-independent reality, concluding that the theory was lacking.

What is the true nature of the electron? After years of contemplation and discussions with Einstein, Bohr concluded that this question lacks meaning. He argued that "it's meaningless to ask what an electron really is. Physics isn't about what is; it's about providing insights into the world."

Heisenberg's uncertainty principle states that it is impossible to determine both the exact position of an electron and its momentum with complete accuracy simultaneously. An electron only occupies a specific location in space when measured. If a subsequent measurement finds it elsewhere, that's all that can be said. The question of how it moved from one place to another holds no significance in the realm of physics. This principle applies not only to electrons and atoms but to all physical entities. However, because it involves Planck's constant, a very small number, its effects go unnoticed when observing larger objects like baseballs, cars, or planets.

14.10 The Proof

Direct experimental evidence for the paradoxical nature of the quantum world emerges from modern adaptations of the EPR thought experiment. Unlike the original, these contemporary experiments focus on measuring spin and polarization—properties of light that share certain similarities with the spin of material particles.

In 1952, David Bohm of Birkbeck College in London proposed the concept of spin measurements in a revised version of the EPR thought experiment. However, it wasn't until the 1960s that researchers began to seriously consider conducting experiments to test the predictions of quantum theory in these contexts.

The conceptual breakthrough emerged from a 1964 paper by John Bell, a physicist at CERN, the European research center near Geneva. Before diving into the experiments, it's important to take a step back and clarify the meanings of "spin" and "polarization."

14.11 The strength of the Copenhagen Interpretation

Each challenge to the Copenhagen interpretation has ultimately reinforced its validity. When influential thinkers like Einstein scrutinize a theory, and its defenders successfully counter all criticisms, the theory emerges stronger from the confrontation. The Copenhagen interpretation is undeniably "correct" in the sense that it works effectively; any superior interpretation of quantum mechanics must incorporate it as a foundational perspective that allows experimenters to reliably predict outcomes—at least statistically—and enables engineers to develop functioning laser systems, computers, and more.

There's no need to revisit the extensive groundwork that has led to the discrediting of various counterproposals to the Copenhagen interpretation, as this has been thoroughly documented by others. However, a crucial observation made by Heisenberg in his 1958 book, *Physics and Philosophy*, is worth noting. Heisenberg emphasized that all counterproposals are "compelled to sacrifice the essential symmetry of quantum theory"—for instance, the symmetry between waves and particles or between position and momentum. He concluded that the Copenhagen interpretation is likely indispensable if these symmetry properties are considered genuine aspects of nature, and every experiment conducted thus far supports this viewpoint.

While there is an enhancement of the Copenhagen interpretation that preserves this essential symmetry—a refined picture of quantum reality that will be discussed in Chapter Eleven—it is not surprising that Heisenberg did not mention it in his 1958 publication, as this new perspective was still being developed by a PhD student in the United States at that time. Before we delve into that, it's important to trace the trajectory of theory and experimentation that, by 1982, had firmly established the Copenhagen interpretation as a reliable framework for understanding quantum reality. This narrative begins with Einstein and culminates in a physics laboratory in Paris over fifty years later; it stands as one of the great tales in the history of science.

14.12 Schrodinger's Cat

The famous cat paradox first appeared in print in *Naturwissenschaften* (vol. 23, p. 812) in 1935, the same year as the EPR paper. Einstein regarded Schrödinger's proposal as the "prettiest way" to illustrate that the wave representation of matter is an incomplete depiction of reality. Along with the EPR argument, the cat paradox remains a topic of discussion in quantum theory today. However, unlike the EPR argument, it has not been satisfactorily resolved.

The concept behind this thought experiment is quite simple. Schrödinger proposed imagining a box that contains a radioactive source, a detector (like a Geiger counter) to sense radioactive particles, a glass bottle filled with a poison such as cyanide, and a live cat. The setup is arranged so that there is a fifty-fifty chance of one of the atoms in the radioactive material decaying, triggering the detector. If the detector registers an event, the glass container shatters, resulting in the cat's death; if it does not, the cat remains alive. The outcome of the experiment remains unknown until we open the box; radioactive decay is entirely random and only predictable in a statistical sense.

According to the strict Copenhagen interpretation, similar to the two-slit experiment where there is an equal probability that an electron passes through either slit, the overlapping possibilities of radioactive decay and no decay create a superposition of states. The entire experiment, including the cat, is governed by the principle that the superposition is "real" until we observe the experiment. Only at the moment of observation does the wave function collapse into one of the two possible states.

Until we peek inside, the radioactive sample exists in a state where it has both decayed and not decayed, the glass vessel is both broken and unbroken, and the cat is both alive and dead, existing in a paradoxical state. While particles like neutrons can briefly change into a proton and a charged pion—as long as they reunite quickly—visualizing a familiar object, like a cat, in such a suspended state is much more challenging. Schrödinger devised this example to highlight a flaw in the strict Copenhagen interpretation, as it seems impossible for the cat to be both alive and dead simultaneously. However, is this any more "obvious" than the assertion that an electron cannot simultaneously be a particle and a wave? As it stands, "common sense has already been tested as a guide to quantum reality and has been found wanting."

The one certainty we have about the quantum world is that we should not rely on common sense and must only trust what we can directly observe or detect clearly with our instruments. We cannot know what occurs inside a box unless we open it and look.

Debates about the cat in the box have persisted for over seventy years. One perspective suggests there is no issue because the cat can determine for itself whether it is alive or dead, and its consciousness is sufficient to trigger the collapse of the wave function. This raises the question of where to draw the line—would an ant or a bacterium possess awareness of what's happening inside the box?

Conversely, since this is merely a thought experiment, we can envision a human volunteer taking the cat's place (this volunteer is often referred to as "Wigner's friend," named after physicist Eugene Wigner, who contemplated variations of the cat-in-the-box experiment and is also Dirac's brother-in-law).

The human occupant of the box is clearly a competent observer capable of collapsing wave functions. When we open the box and, assuming we find him alive, we can be confident he will not recount any mystical experiences but will simply state that the radioactive source failed to emit any particles during the allotted time. Yet, for those of us outside the box, the most accurate way to describe the conditions inside remains as a superposition of states until we look.

In Richard Feynman's space-time diagrams, genuine interactions between several particles are depicted, revealing how a single proton can engage in a network of virtual interactions. Such interactions occur constantly; no particle is as solitary as it may first appear. For instance, a proton, an antineutron, and a pion can spontaneously emerge from nothing due to vacuum fluctuations, existing briefly before annihilating. This same interaction can also be represented as a loop in time, with a proton and neutron pursuing each other around a time eddy linked by a pion. Both interpretations are equally valid.

A proton can similarly chase its own tail through time, creating an endless chain of interactions. Imagine that we have publicly announced the experiment to an eager world, but to avoid media interference, it takes place behind locked doors. Even after we open the box and either greet our friend or retrieve the corpse, the reporters outside remain oblivious to what has transpired. For them, the entire building housing our laboratory exists in a superposition of states, leading to an infinite regression.

Now, suppose we substitute Wigner's friend with a computer. This computer can register information about the radioactive decay—or lack thereof. Can a computer also collapse the wave function, at least within the confines of the box?

From another perspective, what matters is not human awareness of the experiment's outcome or even the awareness of a living creature, but rather that the result of an event at the quantum level has been recorded (or interacted with) and made an impact on the macroscopic world. While the radioactive atom may exist in a superposition of states, once the Geiger counter has "probed" the atom, its properties collapse into one definitive state: either decayed or not decayed.

In contrast to the EPR thought experiment, the cat-in-the-box scenario does possess genuinely paradoxical implications. It is challenging to reconcile with the strict Copenhagen interpretation without accepting the "reality" of a cat that is both dead and alive. This paradox has led physicists like Wigner and John Wheeler to ponder the possibility that, due to the infinite regression of "cause and effect," the entire Universe may owe its "real" existence to being observed by intelligent beings.

The most paradoxical possibilities emerging from quantum theory are direct descendants of Schrödinger's cat experiment, particularly the concept of what Wheeler refers to as a delayed-choice experiment.

14.13 The participatory Universe

American physicist John Wheeler has written extensively on the implications of quantum theory across various publications over four decades. Perhaps his most lucid explanation of the concept of the "participatory Universe" appears in his contribution to *Some Strangeness in the Proportion*, the proceedings of a symposium celebrating Einstein's centenary, edited by Harry Woolf.

In the quantum realm, our understanding is limited to the results of experiments. The cloud of possibilities is, in a sense, shaped by the act of questioning, much like the electron is brought into existence through our experimental probing. This highlights a fundamental principle of quantum theory: no elementary phenomenon can be deemed a phenomenon until it is recorded. The act of recording can lead to unexpected implications for our everyday understanding of reality. To illustrate this, Wheeler proposed a thought experiment that modifies the traditional two-slit experiment. In his variation, the two slits are combined with a lens that focuses the light passing through the system, while the usual screen is replaced by another lens that causes the photons from each slit to diverge. As a result, a photon passing through one slit is directed by the second lens to a detector on the left, while a photon passing through the other slit is sent to a detector on the right.

In this experimental setup, we can determine precisely which slit each photon passes through, just as we would if we directly observed each slit. When we allow one photon at a time to traverse the apparatus, we can clearly identify its path, eliminating any interference due to the lack of superposition of states.

Now, let's modify the apparatus again by covering the second lens with strips of photographic film arranged like a venetian blind. These strips can either be closed to form a complete screen, blocking the photons from passing through the lens, or opened to allow the photons to pass as before. When the strips are closed, the photons reach the screen just as they would in the classic two-slit experiment, resulting in an interference pattern that suggests each photon has passed through both slits simultaneously.

Here's the intriguing part: with this configuration, we don't need to decide whether to open or close the strips until after the photon has already passed through the slits. We can wait until the photon has traveled past both openings and then choose whether to create an experiment that shows it went through one hole alone or through "both at once." In this delayed-choice experiment, our actions now irreversibly influence what we can conclude about the past. The historical narrative for a single photon depends on how we choose to measure it.

Philosophers have long grappled with the idea that history lacks meaning—the past carries no experience—except as it is recorded in the present. Wheeler's delayed-choice experiment makes this abstract notion tangible. He asserts, "We have no more right to say 'what the photon is doing'—until it is registered—than we do to say 'what word is in the room'—until the game of question and response is terminated" (*Some Strangeness*, p. 358).

How far can we extend this idea? Quantum engineers, busy building computers and manipulating genetic material, might dismiss this as mere philosophical speculation, irrelevant to the everyday macroscopic world. Yet, everything in the macroscopic realm is composed of particles governed by quantum rules. Everything we deem real is made of entities that defy the notion of reality; "what choice do we have but to say that in some way, yet to be discovered, they all must be built upon the statistics of billions upon billions of such acts of observer-participation?"

Never hesitant to embrace grand intuitive leaps (such as his vision of a single electron navigating through space and time), Wheeler envisions the entire Universe as a participatory, self-excited circuit. From the Big Bang onward, the Universe expands and cools; after billions of years, it gives rise to beings capable of observing it. These "acts of observer-participation," facilitated by the delayed-choice experiment, confer tangible "reality" upon the Universe not only in the present but also stretching back to its inception.

By observing the photons of cosmic background radiation, the remnant of the Big Bang, we might be actively participating in the creation of the Big Bang and the Universe itself. If Wheeler is correct, Feynman was even more insightful than he realized when he remarked that the two-slit experiment "contains the only mystery." Following Wheeler's line of thought, we find ourselves venturing into the realm of metaphysics, which may lead many readers to believe that, since all this is rooted in hypothetical thought experiments, one can interpret reality in any way without consequence. However, what we truly need is solid evidence from actual experiments to inform our understanding of the most valid interpretations.

The Aspect experiment conducted in the early 1980s provides proof that quantum phenomena are not only real but also observable and measurable. This leads to a fascinating perspective: the entire Universe can be envisioned as a delayed-choice experiment, where the existence of observers—those who perceive what is happening—imparts tangible reality to the origin of everything.

14.14 Quantum Uncertainty

All quantum systems are inherently uncertain. As a typical system evolves, numerous possible outcomes and competing realities are presented. For example, in various laser experiments mentioned earlier, a photon has the choice of which path to take through the apparatus. In a laboratory setting, the observer will always witness a single, specific, concrete reality chosen from among these potential outcomes. Thus, a measurement of the photon's path will yield a result corresponding to one path or the other, but never both.

When considering the Universe as a whole, there is no external observer since the Universe encompasses everything, which leads quantum cosmology to face significant interpretive challenges. One prevalent solution is to assume that all contending quantum realities hold equal status. They are not mere "phantom worlds" or "potential realities," but rather "really real"—each corresponding to its own complete Universe, with its own space and time. These multiple Universes exist in parallel, not connected through space and time, and there may be an infinite number of them.

You might wonder why we perceive only one of these Universes. This can be explained by the idea that when a Universe splits into, say, two alternative worlds, the observers also split, with each copy experiencing only their respective reality. In practice, the quantum processes occurring at the atomic level continually split the Universe, resulting in a vast number of copies of each observer.

Each version of you poignantly believes she or he is unique. Bizarre as it may seem, this belief aligns with experience, provided the various Universes remain distinct. However, complications arise if these Universes begin to overlap or interfere with one another.

This brings us to a second question: is it possible to observe another Universe? The conventional answer is no, but there is no consensus on this matter. David Deutsch, a physicist at Oxford University known for his unconventional views, posits that microscopic experiments could, in principle, be conducted where two or more worlds temporarily merge, allowing physical influences to pass between them. Under normal circumstances, a connection between two quantum worlds would only produce atomic-level effects, not manifesting as "paranormal" phenomena on a macroscopic scale. However, some scientists speculate that certain conditions might allow a mixing of quantum realities to have dramatic effects observable at the human level.

Tachyons—hypothetical particles that would always travel faster than light—introduce the idea that "faster than light" could also imply "backwards in time." While Einstein's special theory of relativity clearly prohibits ordinary matter, and by extension human beings, from traveling into the past, the general theory of relativity raises questions on this topic. The speed of light serves as a barrier to cause and effect. British physicist Stephen Hawking proposed a "chronology protection hypothesis," suggesting that nature will always find a way to prevent mechanisms like "wormholes" from allowing time travel. The most challenging argument against time travel is undoubtedly the grandmother paradox.

During brain surgeries, patients are typically kept conscious. Benjamin Libet took the opportunity to attach electrodes to exposed brains, stimulating the cortex electrically to evoke sensations such as tingling in the patient's hand. Descartes' dualism suggests that these sensations arise from a non-material mind somehow instructing the brain.

Our consciousness of time differs significantly from our awareness of other physical properties, such as spatial size or shape. When we perceive a shape like a square, the electrical activity in our brains does not mirror that shape. There is no "little square" inside our heads displayed on a mental movie screen. Instead, a complex pattern of electrical activity generates the sensation of "squareness." In essence, the square is represented by an electrical pattern. Galileo, Newton, and Einstein all positioned time as the central conceptual pillar around which they constructed their scientific understanding of physical reality.

14.15 Enters David Bohm

One of the most intriguing and unconventional scientists of the postwar era was David Bohm, an American theoretical physicist who primarily worked in London. He grappled with the paradox surrounding the nature of time. We often take for granted that when a radio station transmits a signal, we receive it at home only after it has been sent out by the transmitter. The delay is minimal—just a fraction of a second between points on Earth—so we usually don't notice it. However, satellite phone conversations can introduce a noticeable lag. The key point is that we never hear a radio signal before it is transmitted.

You might wonder why this matters. After all, effects typically do not occur before their causes. The source of my concern traces back to the mid-19th century when James Clerk Maxwell formulated his famous equations describing the propagation of electromagnetic waves, including light and radio. While Maxwell's theory predicts that radio waves travel through empty space at the speed of light, it does not clarify whether these waves arrive before or after they are sent. The equations are indifferent to the distinction between past and future.

According to the equations, it is entirely permissible for radio waves to travel backward in time as well as forward. Given a pattern of electromagnetic activity—like the radio waves emitted from a transmitter—its time-reversed counterpart (in this case, converging waves) is equally valid according to the laws of electromagnetism.

In physics terminology, waves that move forward in time are referred to as "retarded" (as they arrive late), while waves that move backward in time are termed "advanced" (as they arrive early). Despite Bohm's fame for his writings and philosophical explorations—particularly among those with a "mystical" inclination—he remained a somewhat isolated figure in the physics community. He is perhaps best known for his 1950s textbook on quantum mechanics, but early on, he rejected the conventional formulation of quantum mechanics championed by Niels Bohr. This led to a rivalry of sorts: Bohm versus Bohr. Bohm took up the mantle of quantum dissent, continuing where Einstein had left off at his deathbed. With the support of a small group of followers, notably his Birkbeck colleague Basil Hiley, Bohm sought a theory that explained the seemingly random and unpredictable aspects of quantum phenomena as stemming from deeper, deterministic processes.

Bohm proposed a captivating notion: while certain aspects of the world may appear complex or even random, there exists a hidden order beneath it all, intricately "folded up." He later termed this concept "the implicate order." Bohm asserted, "In my opinion, progress in science is usually made by dropping assumptions." Though this statement felt like a stinging rebuke at the time, it has stuck with me ever since. History supports his claim; significant advancements in science often occur when the established paradigm confronts a new set of ideas or experimental evidence that doesn't align with current theories. In those moments, someone might discard a long-cherished assumption—often one that was taken for granted and not explicitly articulated—and suddenly, everything changes.

14.16 Hidden Variables

An interpretation that counters Bohr's positivism in favor of realism—sometimes referred to as "naïve realism" by its critics—leans on the concept of "hidden variables." This suggests that a quantum object possesses inherent attributes, even if they cannot be directly observed. The primary theory in this realm is the de Broglie-Bohm model (DBB), named after Louis de Broglie, who first proposed the idea of matter waves, and David Bohm, who further developed these concepts in the 1950s and 1960s.

In DBB theory, both the particle's position and its wave function are considered real attributes at all times. The wave evolves according to quantum laws, while the particles are influenced by this wave and the classical forces acting on them. Consequently, the path of any specific particle is fully determined, eliminating uncertainty at this level. However, different particles may end up in various locations depending on their initial conditions, ensuring that the distribution of arrivals aligns with the probabilities predicted by quantum mechanics. For example, in the double-slit experiment, the wave's configuration directs most particles to positions with high intensity in the interference pattern, while none reach points where the wave amplitude is zero.

As previously mentioned, the emergence of seemingly random statistical outcomes from deterministic systems is well-known in classical contexts. For instance, when tossing a large number of coins, we typically find that about half land heads up and the other half tails, even though each individual coin's outcome is influenced by the forces acting on it and its initial spin. Similarly, we can analyze the behavior of gas atoms statistically, despite their individual movements and collisions being governed by classical mechanics.

It's important to note that any measurement involves some degree of uncertainty or error. For example, measuring the length of a table cannot yield a perfectly accurate result; even with a ruler marked in millimeters, there could be an inaccuracy of approximately 1–2 mm.

Bohm

More precise instruments lead to more accurate measurements. However, there will always be some uncertainty involved in any measurement, regardless of the accuracy of the measuring device. While we might expect that using more precise instruments could reduce uncertainty indefinitely, quantum mechanics posits that there is actually a limit to the accuracy of certain measurements. This limitation is not due to the capabilities of the instruments but is instead an intrinsic property of nature. It arises from two key factors: wave-particle duality and the unavoidable interaction between the observed object and the measuring instrument.

Let's delve into this further. Measuring an object without disturbing it—even slightly—is impossible. For example, if you try to locate a ping-pong ball in a completely dark room, you might feel around for its position. However, as soon as you touch it, it bounces away.

Whenever we measure the position of an object, whether it's a ping-pong ball or an electron, we must interact with it in some way to gain information about its location. To find a lost ping-pong ball in the dark, you could use your hand or a stick to probe around, or you could shine a light to detect the reflection off the ball. When you feel around with your hand or a stick, you determine the ball's position at the moment you make contact. Yet, when you touch the ball, you inevitably impart some momentum to it, causing it to move.

As a result, you won't be able to predict its future position. A similar principle applies, albeit to a lesser degree, when observing the ping-pong ball with light. To "see" the ball, at least one photon must scatter off it and enter your eye or another detector. When a photon strikes an object of ordinary size, it does not significantly alter the object's motion or position. However, when a photon interacts with a tiny object like an electron, it can transfer momentum, causing a considerable and unpredictable change in the object's motion and position. Thus, the act of measuring the position of an object at one moment renders our knowledge of its future position imprecise.

Now, let's explore how wave-particle duality fits into this scenario. Imagine a thought experiment where we attempt to measure the position of an electron using photons (the same arguments would apply if we used an electron microscope). As discussed in a previous chapter, the best accuracy for observing objects is limited to about the wavelength of the radiation used. For precise position measurements, we need a short wavelength. However, short wavelengths correspond to high frequencies and high energy (as given by the equation $E=hf$). The greater the energy of the photons, the more momentum they can impart to the object upon impact.

Conversely, using photons with longer wavelengths and lower energy will have a lesser effect on the object's motion when they strike it. However, this longer wavelength leads to lower resolution, meaning the object's position will be known less accurately. Thus, the act of observation introduces significant uncertainty regarding either the position or momentum of the electron. This concept is at the heart of the uncertainty principle, first articulated by Heisenberg in 1927.

Heisenberg's uncertainty principle states that we cannot measure both the position and momentum of an object with precision at the same time. The more accurately we measure the position (resulting in a smaller uncertainty Δ-x), the greater the uncertainty in momentum (Δ-p). Conversely, if we attempt to measure momentum very precisely, the uncertainty in position becomes large. The uncertainty principle does not prohibit single exact measurements, though; for instance, we could, in theory, measure an object's position precisely. However, this would render its momentum completely unknown. Therefore, while we might know an object's position exactly at one instant, we would have no idea where it would be just a moment later.

14.17 David Bohm and the hidden variables

This book chapter constructively presents various arguments supporting an alternative to the Copenhagen Interpretation, specifically the "hidden-variables" theory developed by David Bohm since the early 1950s. This theory has often been neglected or marginalized by advocates of the Copenhagen doctrine. Bohm's approach aligns with the pilot-wave hypothesis first proposed by Louis de Broglie, which posits that a particle is "guided" by a wave whose probability amplitudes accurately reflect the predictions of quantum mechanics (QM) and experimental results.

Where Bohm's theory diverges from orthodox views is in its realist premise that a particle possesses precise simultaneous values of both position and momentum. Furthermore, these values relate to the particle's objective state at any given moment, regardless of the limitations imposed on our knowledge by the inherent precision limits of measurement.

David Deutsch argues that the many-worlds theory, or multiverse theory, offers the "sole plausible" solution to various well-known QM paradoxes, including wave-particle dualism, remote simultaneous interaction, and the observer-induced collapse of the wave packet. According to this hypothesis, all possible outcomes must be realized in each instance of wave packet collapse, as the observer splits into numerous parallel, coexisting, but epistemically non-interaccessible "worlds." Each of these worlds constitutes a unique lifeline or series of experiences for its respective centers of consciousness.

From the current literature, interested laypersons can glean several key points:

1. Quantum mechanics has generated numerous problems and paradoxes, including wave-particle dualism, regarding our understanding of physical "reality" and the levels of precision we can achieve in scientific knowledge about it.
2. These issues stem from inherent limits on detecting or measuring quantum phenomena, such as the impossibility of assigning precise simultaneous values of location and momentum. Heisenberg's uncertainty principle famously illustrates that any observation of subatomic particles, such as through an electron microscope, exposes them to other energy-bearing particles, which in turn affects what can be "actually" observed.
3. Quantum paradoxes arise from the apparent necessity of abandoning local realism—the idea that no causal influence can propagate faster than the speed of light—in favor of remote superluminal interactions between particles, regardless of the distance separating them. A substantial body of experimental evidence now supports the existence of such nonlocal effects, necessitating that any realist interpretation account for them, thereby introducing additional challenges.

The notion that a body remains in a state of uniform motion unless acted upon by a resultant force may be counterintuitive to Aristotle but feels natural to Galileo. These were among Einstein's chief objections in his famous debates with Niels Bohr, where he argued that the orthodox (Copenhagen) theory of quantum mechanics was inherently "incomplete," as it implied the existence of perplexing phenomena like instantaneous remote correlations, or "spooky action at a distance." Although he played a significant role in the early development of quantum mechanics, Einstein grew increasingly dissatisfied with its inability to provide a coherent realist or causal explanation for QM phenomena.

The debate between Einstein and Bohr centered on whether the orthodox interpretation could be considered a "Complete Theory." Adherents of the orthodox view—especially Heisenberg—firmly believed that quantum mechanics was indeed "complete" in the sense that no alternative account, such as Bohm's hidden-variables theory, could uphold a realist interpretation while also conforming to the established QM formalisms and predictions.

Bohr strongly disagreed with Einstein on the question of whether the orthodox theory might ultimately prove "incomplete" or allow for future advancements that could reconcile quantum mechanics with classical physics, particularly the special and general theories of relativity. A significant issue with orthodox QM was that it seemed to necessitate nonlocal simultaneous (faster-than-light) "communication" between particles that had previously interacted, regardless of the distance separating them.

Preface: Until recently, the prevailing belief in physics, especially among quantum mechanics textbook authors, was that Niels Bohr had definitively solved the problem of interpreting quantum mechanics. Many writers preferred to sidestep philosophical discussions and focus directly on the physics.

Niels Bohr, incidentally, would have disagreed with that perspective. Currently, it's common for physicists to claim that quantum mechanics is so strange that no one truly understands it—not even themselves. This shift in attitude may have been influenced by Bell's theorem in the 1960s, or perhaps it was merely a matter of prevailing fashion. However, to assert that quantum mechanics is incomprehensible does not imply that it is false; rather, it suggests that the human mind may not be attuned to the fundamental nature of reality. The philosophy of quantum mechanics grapples with the question: What is the nature of reality if quantum mechanics is accurate? In this philosophical context, realism posits that quantum systems behave similarly to classical particles.

14.18 The Bell test

According to Einstein, no influence can travel faster than the speed of light, a principle known as "locality." The initial tests of Bell's inequality were conducted at the University of California, Berkeley, using photons, and the results were published in 1972. In the mid-1970s, researchers performed the first measurements using a different approach, employing gamma rays produced by the annihilation of an electron and a positron. In these experiments, the polarizations of the two photons had to be correlated, and the evidence suggested that when measuring those polarizations, the results violated Bell's inequality. Out of the first seven tests, five supported the predictions of quantum mechanics.

In his article for *Scientific American*, d'Espagnat emphasizes that this constitutes stronger evidence for quantum theory than it might initially appear. The complexity of these experiments and the challenges involved mean that "a variety of systematic flaws in the experimental design could undermine the evidence of a real correlation... On the other hand, it's hard to conceive of an experimental error that could create a false correlation in five independent experiments." Not only did these experiments violate Bell's inequality, but they did so in precise agreement with quantum mechanics' predictions.

Since the mid-1970s, even more tests have been conducted to address any remaining gaps in the experimental design. The apparatus components are now positioned far enough apart that any potential "signal" between the detectors, which could lead to a spurious correlation, would have to travel faster than light. Despite these precautions, Bell's inequality was still violated. Alternatively, one might speculate that the photons "know," even at their creation, what experimental setup has been established to capture them. This could occur without requiring faster-than-light signals if the apparatus is preconfigured to establish an overall wave function that influences the photons from the moment they are born.

The ultimate test of Bell's inequality so far has involved altering the experimental setup while the photons are in flight, reminiscent of John Wheeler's thought experiment involving the double-slit experiment. In 1982, Alain Aspect's team at the University of Paris-South successfully closed the last major loophole for local realistic theories.

They had previously tested Bell's inequality using photons from a cascade process, finding it was violated. Their enhancement involved a switch that redirected a light beam through either of two polarizing filters, each connected to its own photon detector, measuring different polarization directions. This switch could change the beam's direction every 10 nanoseconds (10 x 10^-9 seconds) using an automatic device that generated a pseudorandom signal. Given that it takes 20 nanoseconds for a photon to travel from its source to the detector, no information about the experimental setup could influence the measurement outcome—unless any influence were to travel faster than light.

Even at this stage, we encounter the kinds of issues that perplexed Bohr for so long. The only realities we can truly grasp are the outcomes of our experiments, and the manner in which we measure directly impacts what we observe. In the 1980s, physicists employed a laser beam simply to excite atoms, a tool they could only utilize because of their understanding of excited states and the quantum principles that guided their work. Ironically, the objective of the experiment was to verify the accuracy of quantum mechanics, the very theory that informed their experimental methods.

I'm not suggesting that these experiments are flawed; alternative methods to excite atoms could yield consistent results. However, just as earlier generations of physicists were influenced by the tools of their time, today's researchers are similarly affected by the quantum technologies they utilize. Philosophers may want to explore the implications of Bell's experiment results when quantum processes are involved in setting up the experiments. In line with Bohr's perspective, "what we see is what we get; nothing else is real." By 1975, six tests had been conducted, and four of them produced results that violated Bell's inequality.

Despite uncertainties surrounding the interpretation of locality in relation to photons, the findings provide compelling evidence in favor of quantum mechanics, particularly given that the experiments employed two distinct techniques. In the earliest photon version, the photons originated from atoms of calcium or mercury, which were excited into a specific energy state using laser light. The transition back to the ground state involves two steps: first, the electron moves to a lower excited state, and then it returns to the ground state, producing a photon with each transition. For the chosen transitions, the resulting two photons exhibit correlated polarizations.

These cascade-produced photons can then be analyzed using photon counters placed behind polarizing filters. The Bell test is rooted in a "local" realistic perspective of the world. For instance, in a proton spin experiment, while an experimenter cannot determine all three components of spin for a single particle simultaneously, they can measure any one of them. If the components are designated as X, Y, and Z, every time a +1 value is recorded for the X spin of one proton, a corresponding -1 value is found for its counterpart's X spin. However, the experimenter can measure the X spin of one proton and either the Y or Z spin of its partner (but not both), potentially revealing information about both the X and Y spins of each pair.

Even in principle, this task is complex and requires measuring the spins of many proton pairs at random, discarding those that yield the same spin vector for both members of the pair. Particles with half-integer spins can only align parallel or anti-parallel to a magnetic field, while particles with integer spins can also align across the field.

Nevertheless, this method can yield sets of results where pairs of spins are identified for proton pairs in combinations labeled as XY, XZ, and YZ. In his seminal 1964 paper, Bell demonstrated that if such an experiment is conducted, the realistic local viewpoint suggests that the number of pairs with both X and Y components exhibiting positive spins (X+Y+) must always be less than the total number of pairs in which the XZ and YZ measurements yield positive spin values (X+Z+ + Y+Z+). This conclusion stems from the straightforward logic that if a measurement reveals a particular proton with spin X+ and Y-, its total spin state must be either X+ Y- Z+ or X+ Y- Z-.

The rest of the argument follows from a mathematically simple application of set theory. However, in quantum mechanics, mathematical principles differ. When these are applied correctly, they predict the opposite outcome: the number of X+ Y+ pairs exceed, rather than falls short of, the combined number of X+Z+ and Y+Z+ pairs.

Since the calculation was initially framed from a local realistic perspective, it is commonly referred to as "Bell's inequality." If this inequality is violated, it indicates that the realistic local view of the world is incorrect, while simultaneously confirming the validity of quantum theory in yet another test.

14.19 The measurement problem

While this may be challenging to accept, it is effective: if we apply the rules and utilize the mapping accurately, we can reliably calculate predictable measurement outcomes, such as the energy levels of the hydrogen atom, the electrical properties of a semiconductor, and the results produced by a quantum computer.

However, this reliance on measurement implies that we fully understand what "measurement" entails, which proves to be one of the most complex and contentious issues in the interpretation of quantum physics. When a 45-degree photon encounters a horizontal-vertical (HV) polarizer, it may not be detected; instead, the two possible paths are combined, allowing them to interfere similarly to the two-slit experiment. Just like in that experiment, the uncertainty of which path the photon took prevents us from assigning reality to either.

Consequently, the 45-degree polarization is reconstructed through the combination of the horizontal (H) and vertical (V) components. This can be demonstrated by passing the photon through another ±45-degree polarizer, where all photons exit in the ±45-degree channel. However, if we place a detector in one of the paths between the two polarizers, we will either detect the photon or not, thus confirming its polarization as either H or V.

In this scenario, reconstructing the original state becomes practically impossible, and the emerging photons will definitively be either H or V.

This leads us to conclude that detection is a fundamental aspect of the measuring process, determining whether the photon assumes an H or V state. This aligns with the earlier outlined positivist approach, as we cannot assume the photon possesses polarization in the absence of detection. Hence, we seem to delineate the quantum world from the classical world based on the presence or absence of a detector within the experimental setup.

However, this raises the question: what is it about a detector that distinguishes it, and why can't we consider it a quantum object? What if it were governed by the principles of quantum physics?

To maintain consistency, we would have to assert that a detector does not possess the attribute of having detected a photon—or not—until its state is recorded by another piece of equipment, such as a camera aimed at the detector's output.

However, if we consider the camera as a quantum object, we encounter the same dilemma. At some point, we must draw a distinction between the quantum and classical realms, ultimately relinquishing our ambition of a single fundamental theory that unifies both. This impasse leads to a well-known thought experiment in quantum physics known as Schrödinger's cat.

Erwin Schrödinger, a pioneer of quantum physics whose equation was mentioned earlier, proposed a scenario where a 45-degree photon passes through a horizontal-vertical (HV) apparatus and interacts with a detector connected to a lethal device. In this setup, the cat is killed if the photon is detected. This situation leads to the argument that if the HV attribute cannot be assigned to the photon, then the detection attribute cannot be assigned to the detector, nor can the life-or-death attribute be assigned to the cat. Consequently, the cat exists in a state that is simultaneously alive and dead.

When light is split into two components by an HV polarizer, it can be reunited by a second polarizer oriented in the opposite direction (designated as VH). If the crystals are configured such that both paths through the apparatus are identical, the light emerging from the right retains the same polarization as that incident from the left. This holds true for individual photons, which complicates the notion that measurement alters a photon's polarization state.

The quantum measurement problem highlighted here lies at the heart of the conceptual challenges of quantum physics and is a source of ongoing controversy. Detectors and cats appear to be fundamentally different types of objects governed by distinct physical laws compared to polarized photons. The former exist in definite states (detector: particle detected or not; cat: dead or alive), while the latter inhabit superpositions until measured by something that exhibits the properties of the former.

It seems increasingly unlikely that we will achieve a comprehensive single theory, as quantum objects differ from classical ones not only in degree but also in kind. We may conclude that a significant and essential distinction exists between experimental apparatus, particularly the detector, and quantum objects like photons or electrons. In practice, this difference appears quite apparent: the measuring apparatus is large, composed of countless particles, and bears no resemblance to an electron.

Yet, defining this difference objectively and in principle proves challenging: just how large must an object be to be classified as classical?

What about a water molecule made up of two hydrogen atoms and one oxygen atom, containing ten electrons? Or consider a speck of dust composed of millions of atoms. Some suggest that the laws of quantum physics fundamentally differ when applied to objects composed of a large number of particles.

From a philosophical standpoint, it would be appealing to have a single overarching theory for the physical world, rather than distinct theories for quantum and classical regimes. Is it possible that a universal fundamental theory exists, one that reduces to quantum physics when applied to a single particle or a small number of particles, while aligning with classical physics for larger objects? Physics has witnessed something similar in the development of the theory of relativity, which predicts that objects moving near the speed of light follow different laws than those of Newton. These new principles apply to all objects; even at slow speeds, they adhere to the rules of relativity, though the effects are negligible and often unnoticed. Perhaps a similar scenario exists here, where the role of speed is replaced by the number of fundamental particles in an object.

To test this hypothesis, we could demonstrate the wave properties of a large-scale object through an interference experiment. If the laws of quantum physics differ at larger scales, we should be unable to detect interference in situations where it would typically be expected. The largest object to exhibit wave properties in an interference experiment akin to Young's double-slit setup is the buckminsterfullerene molecule, which consists of sixty carbon atoms arranged in a football-like structure. However, this does not imply that larger objects lack wave properties; rather, no experiments have yet been devised to demonstrate them. As the size of the object increases, the practical challenges of such experiments become significantly greater. Nevertheless, no experiment has been conducted in which predicted quantum properties were not observed when they should have been.

Now, let's consider how the Copenhagen interpretation addresses this issue. Niels Bohr stated: "Every atomic phenomenon is closed in the sense that its observation is based on registrations obtained by means of suitable amplification devices with irreversible functions, such as, for example, permanent marks on a photographic plate caused by the penetration of the electrons into the emulsion."

From this, we can conclude that Bohr was comfortable making a distinction between quantum systems and classical apparatus. As we noted earlier, this distinction is not particularly difficult for practical purposes, and our ability to consistently make it underpins the remarkable success of quantum physics.

However, the Copenhagen interpretation takes this a step further by denying the reality of anything beyond the changes observed in classical apparatuses. According to this view, only the cat's life or death or the "permanent marks on a photographic plate" are truly real. The polarization state of a photon is seen as an idealistic concept derived from our observations, with no greater reality attributed to it. From this perspective, the role of quantum physics is to make statistical predictions about experimental outcomes, and we should refrain from ascribing any truth value to conclusions drawn about the nature of the quantum system itself.

Not all physicists and philosophers accept this *positivist* stance, leading to considerable efforts to develop alternative interpretations that aim to address these issues.   Positivism is a philosophical school of thought that holds that knowledge is based on sensory experience and scientific methods. Each of these interpretations has its supporters, but none has garnered enough backing to displace the Copenhagen interpretation as the prevailing consensus in the scientific community. We will now explore some of these alternative interpretations.

14.20 Some more evidence

The test should be applicable to both the spin measurements of material particles, which are quite challenging to conduct, and the polarization measurements of photons, which, while easier, still present difficulties. However, some physicists express discomfort with photon experiments because photons have zero rest mass, travel at the speed of light, and lack a clear understanding of time, making the concept of locality ambiguous for them. Consequently, although most tests of Bell's inequality have focused on photon polarization measurements, it is essential to note that the only test conducted so far using proton spin measurements also produced results that violate Bell's inequality, thereby supporting the quantum perspective.

This particular test was not the first instance of Bell's inequality being tested; it was reported in 1976 by a team at the Saclay Nuclear Research Centre in France. The experiment closely followed the original thought experiment, involving low-energy protons directed at a target filled with hydrogen atoms. When a proton collides with a hydrogen nucleus—another proton—the two particles interact via a singlet state, allowing for the measurement of their spin components. However, the challenges in obtaining these measurements are significant. Only a fraction of the protons are detected, and unlike the idealized scenario of the thought experiment, it is not always feasible to record the spin components with complete clarity.

Nonetheless, the results of this French experiment provide clear evidence that local realistic perspectives of the world are incorrect.

14.21 What does it all mean

The experiment is nearly flawless; even though the switching of the light beam isn't entirely random, it changes independently for each of the two photon beams. The only significant loophole that remains is that many photons produced go undetected due to the inefficiency of the detectors. One could argue that only the photons violating Bell's inequality are being detected, and that those not detected would conform to the inequality if they were. However, no experiments have been planned to test this unlikely scenario, making such an argument seem quite desperate.

After the announcement of the results from Aspect's team just before Christmas in 1982, there was little doubt that the Bell test confirmed the predictions of quantum theory. In fact, the results of this experiment—representing the best achievable outcomes with current technology—violated the inequalities more significantly than any previous tests and aligned closely with the predictions of quantum mechanics. As d'Espagnat remarked, "Experiments have recently been carried out that would have forced Einstein to change his conception of nature on a point he always considered essential… we may safely say that non-separability is now one of the most certain general concepts in physics."

However, this does not suggest that faster-than-light communication is possible. There is no way to transmit useful information through this process, as there's no method to connect one event to the event it causes. This effect only applies to events sharing a common cause—such as the annihilation of a positron/electron pair, the return of an electron to the ground state, or the separation of a pair of protons from the singlet state.

You might envision two detectors positioned far apart in space, with photons from a central source traveling to each. You could imagine a technique for altering the polarization of one photon beam, causing an observer at the second detector to notice changes in the polarization of the other beam. However, what kind of signal is being altered? The original polarizations or spins of the particles in the beam arise from random quantum processes and carry no inherent information.

Ultimately, the observer will only perceive a different random pattern compared to the pattern they would have observed without the manipulations of the first polarizer.

Since a random pattern contains no information, it would be entirely useless. The meaningful information lies in the differences between the two random patterns, but the first pattern never existed in reality, making it impossible to extract that information. However, don't be too disheartened, as the Aspect experiment and its predecessors present a fundamentally different worldview than our everyday common sense. They suggest that particles that were once part of the same interaction remain, in some sense, components of a unified system, responding collectively to subsequent interactions.

**Einstein around 1950
In courtesy of The California Institute of Technology
and The University of Jerusalem**

Everything we see, touch, and feel is made of particles that have interacted with other particles throughout the history of the universe, tracing back to the Big Bang. The atoms in my body contain particles that once moved in close proximity with those in distant stars, or perhaps with particles forming life on an unknown planet. In fact, the particles making up my body once interacted with those that now make up your body. In this sense, we are part of a connected system, much like the two photons flying apart in the Aspect experiment.

Theorists like d'Espagnat and Bohm suggest that everything in the universe is interconnected, and that only a "holistic" approach can explain complex phenomena like human consciousness. Although physicists and philosophers are still working towards understanding this holistic view of the universe, it's too early to offer a clear picture of what it will look like. Instead of speculating on the possibilities, I can draw from my own background in physics and astronomy. One enduring mystery in physics is the property of inertia—the resistance of an object, not to motion, but to changes in motion.

In free space, an object continues to move in a straight line at a constant speed unless acted upon by an external force—one of Newton's great discoveries. The amount of force needed to change the object's motion depends on its mass. But how does the object "know" it's moving in a straight line at a constant speed? What does it measure its velocity against? Since Newton's time, philosophers have recognized that inertia seems to be measured against the frame of reference provided by distant stars, or, as we now say, distant galaxies. The Earth spinning in space, a Foucault pendulum in a museum, or even an atom, all seem to "know" the average distribution of matter in the universe. How or why this happens is still unknown and has led to fascinating, if unresolved, questions.

If there were only one particle in an empty universe, it wouldn't have any inertia, since there would be nothing to measure its motion or resistance against. But what if there were two particles in an otherwise empty universe—would they have the same inertia as they do in our universe? If half of the matter in our universe disappeared, would the remaining particles still have the same inertia? These questions remain as puzzling today as they were three hundred years ago.

However, the death of local realistic views of the world may offer a clue. If everything that interacted during the Big Bang maintains some connection, then every particle in every star and galaxy "knows" about the existence of all others. Inertia, then, becomes less of a cosmological or relativistic puzzle and more a question for quantum mechanics.

While this may seem paradoxical, as Richard Feynman said in his *Lectures*, "The paradox is only a conflict between reality and your feelings of what reality ought to be."

This idea is not just abstract. Early in 1983, just weeks after the Aspect team published their results, scientists at the University of Sussex in England confirmed the "connectedness" of quantum-level phenomena. Their experiments not only supported earlier findings but hinted at practical applications, such as the development of quantum computers—devices that could surpass modern solid-state technology as dramatically as the transistor radio replaced semaphore flags for communication.

14.22 Confirmation and applications

The team at Sussex, led by Terry Clark, approached the challenge of measuring quantum reality from a different perspective. Instead of conducting experiments at the scale of typical quantum particles—such as atoms—they aimed to create "quantum particles" closer in size to conventional measuring devices. Their method relies on superconductivity and involves a superconducting ring about half a centimeter wide. This ring features a "weak link," a constriction at one point that narrows the cross section to just one ten-millionth of a square centimeter. This weak link, inspired by Brian Josephson's development of the Josephson junction, makes the ring function like an open-ended cylinder, much like an organ pipe or a tin can with both ends removed.

The Schrödinger waves that describe the superconducting electrons within the ring behave similarly to standing sound waves in an organ pipe. By applying an electromagnetic field at varying radio frequencies, the team can "tune" these waves. The electron wave traveling around the entire ring mimics the behavior of a single quantum particle, and with a sensitive radio-frequency detector, the team can observe quantum transitions in the electron wave. Essentially, it's as if they created a single quantum particle half a centimeter in size—an even more striking example than the earlier experiment with superfluid helium.

This experiment allows for direct measurements of individual quantum transitions and offers further evidence of quantum non-locality. Because the electrons in the superconducting ring act collectively as a single boson, the Schrödinger wave that undergoes the quantum transition spreads across the entire ring. Remarkably, the whole "pseudo-boson" transitions simultaneously, without one part of the ring changing first and the other catching up after a light-speed signal travels around the ring.

In some respects, this experiment is even more compelling than the Aspect test of Bell's inequality. While the Bell test, though mathematically clear, can be difficult for non-specialists to grasp, the concept of a half-centimeter "particle" that behaves like a single quantum entity is much easier to visualize. This quantum particle responds instantaneously as a whole to any external influence, making the non-local behavior even more tangible.

Clark and his team are already planning their next step in this groundbreaking work. They aim to create a large "macro-atom," possibly in the form of a six-meter-long cylinder. If this new device responds to external stimuli as predicted, it could potentially pave the way for faster-than-light communication. A detector at one end of the cylinder would instantly register changes in the quantum state caused by a disturbance at the other end. While this wouldn't be immediately useful for conventional communication, it's not feasible to build a macro-atom stretching from Earth to the Moon to eliminate the lag in communications between lunar explorers and ground control—it would have significant practical applications.

In today's most advanced computers, performance is often limited by how quickly electrons can travel between components. Even though these time delays are minuscule, measured in nanoseconds, they are still crucial. While the Sussex experiments don't suggest the possibility of instantaneous communication over long distances, they do hint at the potential to build computers where all components respond simultaneously to changes in any part of the system. This possibility has led Terry Clark to suggest that such technology could revolutionize electronics, making today's advanced systems seem as primitive as semaphore signals.

Thus, while the Copenhagen interpretation has been largely confirmed for practical purposes by these experiments, even more exciting developments could lie ahead, pushing beyond classical devices into realms once thought impossible. However, the Copenhagen interpretation remains intellectually unsatisfying to many. What happens to the "ghostly" quantum worlds that collapse when we measure a subatomic system? How can a parallel reality, just as real as the one we observe, vanish upon measurement? The most compelling explanation is that these alternate realities don't disappear—they persist, and Schrödinger's cat is indeed both alive and dead, but in different worlds. This leads us to the "many worlds" interpretation, which encompasses the practical success of the Copenhagen view within a more complete understanding of reality.

14.23 Many Worlds Interpretation

Most physicists who engage with these concepts are generally content with the collapsing wave functions of the Copenhagen interpretation. However, a respectable minority holds an alternative view, one that has the advantage of incorporating the Copenhagen interpretation within its framework.

The reason this more expansive interpretation hasn't swept through the physics community is its unsettling implication: the existence of many other worlds—perhaps even an infinite number—running parallel to our own, existing "sideways" across time. These worlds remain forever isolated from our reality, unable to interact with our Universe.

14.24 Who observes the observers

The many-worlds interpretation of quantum mechanics was first proposed by Hugh Everett, a graduate student at Princeton in the 1950s. Troubled by the Copenhagen interpretation's idea that wave functions collapse mysteriously upon observation, Everett explored alternatives, with encouragement from his mentor John Wheeler, who urged him to develop his ideas for his PhD thesis.

Everett's approach stemmed from a fundamental question. When we carry out an experiment in isolation, then share the results with others—eventually, spreading the information across the world—the wave function describing the system grows more complex, encompassing more of reality at each step. Yet, all possible outcomes remain valid until each observer "receives" the information. According to this view, the entire Universe could be described by a web of overlapping wave functions, with different realities coexisting until an observation forces a collapse. But, since the Universe is a closed system, there is no external observer to collapse its wave function.

One attempt to address this paradox is Wheeler's notion that consciousness itself, through reverse causality, could serve as the ultimate observer. However, this solution introduces its own circular reasoning, leaving the core mystery unresolved.

I would even find the solipsist argument more appealing—the idea that I am the sole observer in the Universe and that my observations alone crystallize reality from the web of quantum possibilities. However, such extreme solipsism is ultimately unsatisfying, especially for someone like me, whose contribution to the world involves writing books intended for others to read. In contrast, Everett's many-worlds interpretation offers a more complete and fulfilling explanation, one that goes beyond individual perception and embraces a broader, interconnected view of reality.

Everett

Everett's interpretation suggests that the overlapping wave functions of the entire Universe—the alternative realities that interact to produce measurable quantum interference—do not collapse. Instead, all these realities are equally real, existing in their own realms of "super-space" (and "super-time"). When we make a quantum measurement, we are essentially selecting one of these alternatives, which then becomes part of our perceived "real" world. The act of observation severs the connections between these parallel realities, allowing each to diverge into its own path through super-space. In each of these realities, an observer believes they have collapsed the wave function and arrived at a single quantum outcome, unaware of the others.

14.25 More on Schrodinger's cats

It's challenging to fully comprehend what it means when we talk about the collapse of the wave function of the entire Universe. However, Everett's interpretation becomes clearer when we consider a more relatable example. The paradox of Schrödinger's cat provides the perfect illustration of how the many-worlds interpretation offers a significant step forward. The surprising conclusion is that the search for the "real" cat doesn't end with finding just one—but two.

Quantum mechanics tells us that inside Schrödinger's famous box, both the "live cat" and "dead cat" wave functions exist as equally real possibilities. In the traditional Copenhagen interpretation, both of these outcomes are considered unreal until we observe the cat, at which point one outcome "crystallizes" into reality. But Everett's interpretation takes the quantum equations literally, asserting that both cats are real—each in its own separate world.

It's not that the radioactive atom inside the box either decayed or didn't decay; rather, it did both. At the point of decision, the entire Universe split into two versions: in one, the atom decayed, and the cat died; in the other, the atom remained intact and the cat lived. While this might sound like science fiction, it's actually a profound consequence of faithfully interpreting quantum mechanics—and it's based on precise mathematical principles.

14.26 Everett's work

The significance of Everett's work, published in 1957, lies in how he took this seemingly outrageous idea of multiple realities and grounded it in solid mathematical principles, all within the established framework of quantum theory. While it's one thing to speculate about the nature of the Universe, Everett managed to turn those speculations into a coherent, self-consistent theory of reality. He wasn't the first to consider such possibilities, but his ideas appear to have developed independently from earlier musings on parallel worlds and multiple realities.

Earlier versions of such concepts—and many more since 1957—mostly found their place in the realm of science fiction. One of the earliest examples can be traced to Jack Williamson's *The Legion of Time*, first serialized in 1938. In science fiction, alternate realities often feature "what if" scenarios, such as a world where the South won the Civil War or the Spanish Armada conquered England. These stories sometimes include characters who jump between realities or offer convoluted explanations for how alternative worlds split off from our own. Williamson's tale describes a "ghost reality" that only becomes concrete when a crucial decision in the past causes two realities to diverge. While Williamson's world collapses one reality in favor of another, echoing the Copenhagen interpretation's collapse of a wave function, Everett's worlds are all equally real, and sadly, no hero can move between them. His interpretation, however, is grounded in science fact, not fiction.

To better understand Everett's interpretation, consider one of the foundational experiments in quantum physics: the two-slit experiment. According to the Copenhagen interpretation, the interference pattern seen when a single particle travels through the apparatus is explained as interference from two alternative realities—one where the particle passes through slit A and another where it goes through slit B. But when we observe the particle, we only find it at one of the slits, seemingly chosen at random, governed by quantum probabilities. On Everett's many-worlds interpretation, no such choice is made. Instead, at the quantum level, the entire Universe splits into two versions: in one, the particle goes through slit A; in the other, it goes through slit B. In each Universe, an observer sees the particle pass through just one slit, and from that point on, the two Universes are completely separate, which is why we don't observe interference when we detect the particle at the slits.

The phrase "parallel worlds" often suggests multiple realities lying side by side in some kind of "superspace-time," but this is a misleading image. A more accurate analogy is that of the Universe constantly branching like a tree. However, even this picture falls short when you consider the sheer number of quantum events happening at every moment in every region of the Universe. This overwhelming complexity is part of why many physicists are hesitant to embrace the idea. Still, as Everett demonstrated, the many-worlds interpretation offers a logical, self-consistent description of quantum reality that doesn't conflict with experimental evidence.

Despite the sound mathematical foundation of Everett's interpretation, it barely caused a ripple in the scientific community when it was first introduced in 1957. His work was published in *Reviews of Modern Physics*, alongside a paper by John Wheeler that highlighted its significance, but the ideas were largely overlooked. It wasn't until Bryce DeWitt of the University of North Carolina revisited the concept more than a decade later that the many-worlds interpretation began to gain traction.

It's unclear why the many-worlds interpretation took so long to gain traction, even in the limited way it did in the 1970s. Beyond the complex mathematics involved, Everett addressed this issue in his paper for *Reviews of Modern Physics*, asserting that the argument against the reality of the Universe's splitting—based on our lack of direct experience with it—doesn't hold up.

He pointed out that all components of a superposition of states follow the wave equation without regard to the existence of the other components. This lack of interaction between branches suggests that no observer can ever detect the splitting process. To argue otherwise is akin to claiming that the Earth cannot possibly orbit the Sun because we don't feel that motion. As Everett stated, "In both cases, the theory itself predicts that our experience will be what in fact it is."

14.27 Beyond Einstein

The many-worlds interpretation offers a conceptually simple, causal framework that aligns well with observed experiences. Wheeler made concerted efforts to highlight this new idea, but it can be challenging to convey how radically the relative state formulation departs from classical concepts. This shift in thinking has few historical parallels: Newton's notion of gravity as action at a distance, Maxwell's field theory, and Einstein's rejection of any privileged coordinate system are all similarly disruptive. The only comparable principle in physics might be general relativity's assertion that all coordinate systems are equally valid.

Wheeler emphasized that apart from Everett's concept, no other self-consistent framework exists to clarify what it means to quantize a closed system, such as the Universe in general relativity. However, the Everett interpretation faces a significant challenge in displacing the Copenhagen interpretation. Both interpretations yield identical predictions regarding the outcomes of experiments or observations, which is both a strength and a weakness. Since the Copenhagen interpretation has consistently proven effective, any new framework must produce the same results in testable scenarios, and the Everett interpretation does pass this initial evaluation.

Nonetheless, it only enhances the Copenhagen perspective by resolving the seemingly paradoxical elements found in double-slit experiments and those related to the Einstein-Podolsky-Rosen (EPR) thought experiments. From the standpoint of various quantum experiments, the differences between the two interpretations are subtle, leading many to prefer the familiar Copenhagen approach. However, for those who have engaged with the EPR experiments and the tests of Bell's inequality, the appeal of the Everett interpretation lies in its assertion that our choice of which spin component to measure determines the branch of reality in which we find ourselves. In that specific branch of superspace, the spin of the distant particle is always complementary to the one we measure.

It is our choice that dictates which quantum world we inhabit during experiments, rather than chance. Since all possible outcomes of an experiment occur, with each outcome observed by its own set of observers, it is unsurprising that what we observe aligns with one of the potential results.

14.28 A deeper look at the many worlds interpretation

The many-worlds interpretation of quantum mechanics remained largely overlooked by the physics community until Bryce DeWitt revived interest in the idea in the late 1960s. He not only wrote about the concept himself but also encouraged his student Neill Graham to expand upon Everett's work for his PhD thesis. In a 1970 article for *Physics Today*, DeWitt highlighted the immediate appeal of the Everett interpretation in resolving the paradox of Schrödinger's cat. No longer did we have to grapple with the conundrum of a cat being simultaneously dead and alive. Instead, we understand that in our world, the box contains a cat that is either alive or dead, while in a parallel universe, another observer has an identical box with a cat that is also either dead or alive.

DeWitt posited that if the Universe is "constantly splitting into a stupendous number of branches," then "every quantum transition taking place on every star, in every galaxy, in every remote corner of the Universe is splitting our local world on Earth into myriads of copies of itself." He recalls the initial shock he felt upon encountering this idea: the notion of 10^{100} slightly imperfect copies of oneself perpetually splitting into further duplicates. Yet, he found himself convinced by his own work, Everett's thesis, and Graham's renewed exploration of the phenomenon. He even contemplated the extent to which this splitting could continue.

In a finite universe—which is likely the case if general relativity accurately describes reality—there can only be a limited number of "branches" on the quantum tree. This means that super-space may not have enough capacity to accommodate more bizarre possibilities, including what DeWitt refers to as "maverick worlds," which exhibit unusually distorted patterns of behavior. While the strict Everett interpretation suggests that anything possible does occur somewhere in reality within super-space, it does not imply that anything imaginable can manifest. We can conceive of impossible scenarios, but the real worlds that exist could not support them. For example, in a universe otherwise identical to ours, even if pigs had wings, they still wouldn't be able to fly; likewise, no matter how extraordinary, heroes cannot navigate sideways through cracks in time to visit alternative realities, despite what science fiction writers might speculate.

DeWitt's conclusion echoes the earlier sentiments of Wheeler: from the perspectives of Everett, Wheeler, and Graham, the view is genuinely remarkable. However, it remains a rather casual perspective—one that even Einstein might have found acceptable. It arguably stands as a natural culmination of the interpretative program initiated by Heisenberg in 1925.

However, it is worth noting that Wheeler later expressed some doubts about the entire concept. At a symposium marking the centennial of Einstein's birth, he stated regarding the many-worlds theory, "I confess that I have reluctantly had to give up my support of that point of view in the end—much as I advocated it in the beginning—because I fear it carries too great a load of metaphysical baggage."

General relativity describes "closed systems," and Einstein initially viewed the Universe as a closed, finite system. Although discussions of open, infinite universes abound, such descriptions do not align strictly with relativity theory. For the Universe to be closed, it would need enough matter for gravity to curve space-time back on itself, much like the curvature around a black hole. This requires more matter than we can observe in visible galaxies. Nevertheless, most observations of the Universe's dynamics suggest that it is near a state of closure, either "just closed" or "just open." Consequently, there is no observational basis for dismissing the fundamental relativistic implications of a closed and finite Universe, and every reason to pursue the dark matter that maintains its gravitational cohesion.

This should not be interpreted as undermining the Everett interpretation; Einstein's shift in perspective regarding the statistical foundations of quantum mechanics did not negate the validity of that interpretation either.

Wheeler's statement from 1957 remains valid today; as of 2011, no other self-consistent system of ideas exists to clarify what it means to quantize the Universe beyond Everett's theory. However, Wheeler's change of perspective illustrates the difficulty many individuals have in accepting the "many-worlds" theory.

The metaphysical implications of the many-worlds interpretation can be seen as less troubling than those posed by the Copenhagen interpretation or Schrödinger's cat experiment, as well as the necessity for three times as many dimensions of "phase space" as there are particles in the Universe. The concepts involved are no stranger than those that seem familiar simply because they are frequently discussed. Moreover, the many-worlds interpretation provides valuable insights into the nature of our Universe. The theory is far from exhausted and continues to warrant serious consideration.

14.29 Beyond Everett

Cosmologists today readily discuss events that took place just after the Universe's birth in the Big Bang, calculating reactions that occurred at just 10^{-35} seconds or even earlier. These reactions involve a chaotic mix of particles and radiation, including pair production and annihilation. The assumptions underlying these processes are derived from a combination of theoretical models and observations of particle interactions in massive accelerators like CERN in Geneva.

According to these calculations, the laws of physics, as derived from our limited experiments here on Earth, can logically and consistently explain how the Universe transitioned from a state of nearly infinite density to the state we observe today. These theories even attempt to predict the balance between matter and antimatter, as well as the relationship between matter and radiation.

Everyone with a passing interest in science has likely heard of the Big Bang theory regarding the Universe's origin.

Theorists eagerly manipulate numbers to describe events that supposedly unfolded in mere split seconds over 15 billion years ago. Yet, who today takes the time to ponder the true implications of these concepts? It's absolutely mind-boggling to grasp what a timeframe like 10^{-35} seconds actually signifies, let alone to comprehend the nature of the Universe at that age. Scientists who grapple with such extraordinary extremes of nature should find it quite reasonable to stretch their minds to encompass the idea of parallel worlds.

The appealing phrase often used in science fiction to describe alternative realities is actually misleading. A more accurate depiction is that these alternative realities resemble branches radiating from a central stem, running parallel through "superspace," much like the intricate lines of a railway junction.

Sci-fi writers envision these worlds as existing side by side in time, with our nearest neighbors being nearly identical to our own world. However, the differences become more pronounced the further we move "sideways in time." This imagery naturally invites speculation about changing lanes on this superhighway of existence, allowing for a seamless transition into a neighboring world. Unfortunately, the mathematics behind these concepts doesn't align neatly with this picture.

Mathematicians easily navigate dimensions beyond the three spatial dimensions we experience daily. In fact, the entirety of our world—one branch of Everett's many-worlds reality—is described mathematically in four dimensions: three of space and one of time, all oriented at right angles to each other. The math for describing additional dimensions, all perpendicular to our four, is standard number manipulation. This is where alternative realities actually exist—not parallel to our own world, but at right angles to it, forming perpendicular branches that extend through superspace.

While this perspective is challenging to visualize, it clarifies why shifting sideways into an alternative reality is impossible. Moving at right angles to our world would generate a new reality entirely.

In the many-worlds theory, this branching occurs every time the Universe encounters a quantum choice. The only way to access an alternative reality created by such a split—like the outcomes of the Schrödinger's cat thought experiment or the double-slit experiment—would involve traveling back in time to the moment of the experiment in our four-dimensional reality, and then progressing forward along the alternative branch, at right angles to our own. This scenario may prove impossible. Conventional wisdom suggests that true time travel cannot occur due to the paradoxes it introduces, such as the infamous grandfather paradox, where one could potentially prevent their own existence.

However, at the quantum level, particles appear to engage in a form of "time travel." Frank Tipler has demonstrated that the equations of general relativity allow for time travel under specific conditions. It's conceivable to envision a type of genuine time travel—both forward and backward—that avoids paradoxes and relies on the existence of alternative universes.

If you're struggling to accept this concept, you might find yourself yearning for the comfort of the familiar Schrödinger equation. However, it's not as straightforward as it seems. While the wave interpretation of quantum mechanics begins with a simple wave equation familiar in other areas of physics, the correct quantum description for a single particle involves a wave in three dimensions—not in our everyday spatial dimensions but in what is known as "configuration space."

The challenge arises because each particle requires three distinct dimensions in this wave function. For two interacting particles, you need six dimensions; for three particles, nine dimensions; and so forth. Thus, the wave function for the entire Universe—which is an ambiguous concept—requires three times as many dimensions as there are particles within it. Physicists who dismiss the Everett interpretation as too laden with metaphysical baggage conveniently overlook that the wave equations they utilize also rely on an equally staggering amount of extra-dimensional complexity.

David Gerrold explores these themes in his engaging science fiction novel, *The Man Who Folded Himself*, which serves as a compelling guide to the intricacies of a many-worlds reality. To illustrate, consider the classic scenario where if you travel back in time and kill your grandfather, you create or enter an alternative world branching off at "right angles" to your original timeline. In this new reality, both your father and you would never be born. However, there is no paradox because you were still born in the "original" reality, which allowed you to make the journey back in time and into an alternative branch. If you then attempt to return to undo the damage, you merely re-enter the original branch of reality—or one that closely resembles it.

However, even Gerrold doesn't frame the bizarre events his main character experiences in terms of perpendicular realities. The physical explanation of the mathematics behind the Everett interpretation is indeed a novel perspective on time travel—one that hasn't yet been fully embraced by science fiction writers. It's essential to emphasize that, in this model, alternative realities do not simply exist "alongside" ours, easily accessible with minimal effort. While there may be a world where Bonaparte's given name was Pierre instead of Napoleon, and where history flowed similarly to our own, there could also be a reality where that Bonaparte never existed at all. Both scenarios are equally distant and unreachable from our world. Accessing them would require traveling backward in time to the appropriate branching point and then moving forward through time at right angles (in one of the many possible orientations) to our own reality.

This concept can be expanded to eliminate the paradoxes commonly associated with time travel, which are often the delight of science fiction writers and the subject of philosophical debate. In the many-worlds interpretation, all possible events do occur in some branch of reality. The key to accessing these possible realities lies not in traveling sideways through time but in moving backward and then forward into another branch.

One of the finest science fiction novels that consciously employs the many-worlds interpretation is Gregory Benford's *Timescape*. In this story, the fate of a world is fundamentally altered by messages sent back to the 1960s from the 1900s. The narrative is expertly crafted and gripping, standing strong even outside its sci-fi context. However, a crucial point to note is that because the world is transformed by the actions of those receiving messages from the future, the future from which those messages originate no longer exists for them. So, where did these messages come from? While one might argue from the perspective of the old Copenhagen interpretation that these are ghost worlds sending back ghost messages that influence the collapse of the wave function, such an argument is tenuous at best.

In contrast, the many-worlds interpretation allows for a more straightforward visualization: messages from one reality can travel back in time to a branching point, where they are received by individuals who then move forward into their own distinct branch of reality. Both alternative worlds coexist, and communication between them is severed once critical decisions that shape the future are made. *Timescape* not only serves as an enjoyable read but also presents a "thought experiment" that is as intriguing and pertinent to the quantum mechanics debate as the EPR experiment or Schrödinger's cat. While Everett may not have fully recognized it, a many-worlds reality is precisely the kind of framework that facilitates time travel. It also provides an explanation for why we find ourselves engaged in discussions about these profound issues.

Additionally, it's essential to highlight that even if time travel is theoretically feasible, there may be insurmountable practical challenges that prevent us from sending physical objects through time. However, transmitting messages through time might be more achievable if we can leverage the particles that travel backward in time, as suggested by Feynman's interpretation of reality.

14.30 Our place in the multiverse

According to the many-worlds theory, the future is not predetermined from our conscious perspective, while the past is. Through the act of observation, we select a "real" history from a multitude of realities. Once someone sees a tree in our world, it remains there, even when no one is watching. This principle extends back to the Big Bang; at every junction in the quantum continuum, many new realities may have emerged, but the path leading to our current existence is clear and distinct. There are numerous routes into the future, with some version of "us" pursuing each one.

Each version of ourselves believes it is following a unique path and reflects on a distinct past, yet it is impossible to foresee the future due to the vast number of possibilities. We may even receive messages from the future, whether through mechanical means, as depicted in *Timescape*, or through dreams and extrasensory perception. However, such messages are unlikely to be beneficial. Because multiple future worlds exist, any messages we receive are expected to be confusing and contradictory. If we act on them, we are more likely to veer into a reality branch different from the one from which the "messages" originated, making it improbable that they will ever "come true." Those who suggest that quantum theory provides a key to practical ESP, telepathy, and similar phenomena are likely deluding themselves.

The notion of the Universe as a straightforward Feynman diagram with the present moment moving steadily along it is an oversimplification. The more accurate picture is a multidimensional Feynman diagram encompassing all possible worlds, with the present unfolding across every branch and detour. The most significant question within this framework is why our perception of reality is as it is. Why has the path through the quantum maze that began with the Big Bang led to the emergence of intelligence in the Universe?

The answer lies in the concept known as the "anthropic principle." This principle posits that the conditions present in our Universe are the only ones, aside from minor variations, that could have allowed life as we know it to evolve. Consequently, it is inevitable that any intelligent species, like us, would perceive a Universe that appears as it does. If the Universe were different, we would not be here to observe it. We can envision the Universe taking various quantum paths from the Big Bang.

In some of those worlds, due to differences in quantum choices made during the early expansion, stars and planets never form, and life as we know it does not exist. For instance, our Universe appears to have a predominance of matter particles with little or no antimatter. This phenomenon may not have a fundamental reason; it could simply be an accident of the reactions during the fireball phase of the Big Bang.

It's equally possible for the Universe to be empty or primarily composed of what we call antimatter, with little to no matter present. In an empty Universe, life as we know it would not exist; in an antimatter Universe, life could resemble ours—a kind of mirror world made real. The real puzzle is why a Universe ideally suited for life emerged from the Big Bang. The anthropic principle suggests that many possible worlds could exist and that we are an inevitable outcome of our specific Universe.

But where are these other worlds? Are they mere ghosts, akin to the interacting worlds of the Copenhagen interpretation? Do they correspond to different life cycles of the entire Universe, existing before the Big Bang that initiated our conception of time and space? Or could they be Everett's many worlds, existing at right angles to our own?

This idea appears to be the most compelling explanation available today, and the resolution of the fundamental question of why we perceive the Universe as it is justifies the complexities associated with the Everett interpretation. Most alternative quantum realities are inhospitable to life and are empty. The conditions ideal for life are unique, so when living beings reflect on the quantum path that led to their existence, they observe significant events—branches in the quantum road that may not be statistically the most likely but are the ones that culminate in intelligent life. The various worlds resembling our own but with different histories—such as one where Britain still governs all its North American colonies or one where Native Americans colonized Europe—constitute just a small fraction of a far larger reality. It's not mere chance that has favored the special conditions conducive to life among the multitude of quantum possibilities; it is a matter of choice. All worlds are equally real, but only the worlds that are suitable harbor observers.

The success of the Aspect team's experiments testing Bell's inequality has narrowed down the interpretations of quantum mechanics to just two: the Copenhagen interpretation, with its ghostly realities and half-alive cats, or the Everett interpretation, with its many worlds. It's conceivable that neither of these two leading explanations is correct and that an alternative interpretation of quantum reality exists, one that resolves all the dilemmas that both the Copenhagen and Everett interpretations address, including the Bell test, while surpassing our current understanding. This new interpretation would need to account for everything we've learned since Planck's groundbreaking insights and must do so at least as effectively as, if not better than, the current theories.

That's a formidable challenge, and science doesn't typically allow itself to simply wait for someone to propose a "better" solution to our dilemmas. In the absence of a superior explanation, we must confront the implications of the best one we have.

After over half a century of dedicated inquiry into the nature of quantum reality by some of the most brilliant minds of the 20th century, we must acknowledge that science currently provides us with only two alternative interpretations of how the world is structured. Neither option is particularly comforting at first glance: either nothing is real, or everything is real.

This issue may never reach a resolution, as it might be impossible to create an experiment capable of distinguishing between the two interpretations, barring time travel. However, it is clear that Max Jammer, a leading quantum philosopher, was not exaggerating when he claimed that "the multi-Universe theory is undoubtedly one of the most daring and ambitious theories ever constructed in the history of science." It literally explains everything, including the life and death of cats. As an incurable optimist, I find the Everett interpretation of quantum mechanics most appealing. In this view, all possibilities exist, and our actions determine our paths through the many worlds of quantum reality. In the world we inhabit, what you see is what you get; there are no hidden variables; God doesn't play dice; and everything is real.

One of the well-known anecdotes about Niels Bohr involves someone presenting him with a wild idea aimed at solving one of the puzzles of quantum theory in the 1920s. Bohr responded, "Your theory is crazy, but it's not crazy enough to be true." In this perspective, Everett's theory is crazy enough to be true.

14.31 Anti-worlds

Schwarzschild black holes are not physical barriers but rather gateways to a bizarre region of space-time beyond. The strangeness of this area became apparent upon closer examination of the algebra that describes it. According to this idealized mathematical framework, the "other universe" mirrors our own and extends infinitely. However, there is a crucial difference: the direction of time in this other universe is reversed compared to ours. The concept of a parallel universe with time flowing in the opposite direction—an anti-world, if you will—holds a certain allure. We have previously encountered such speculation in discussions surrounding the kaon.

All quantum systems are inherently uncertain. As a typical system evolves, multiple potential outcomes and competing realities emerge. For example, in the laser experiments mentioned earlier, a photon can choose which path to take through a device. In a lab setting, the observer will always perceive a single, concrete reality chosen from among these possibilities. Thus, when measuring the photon's path, the result will reveal either one path or the other, but never both.

In the context of the Universe as a whole, there is no external observer because the Universe encompasses everything. This creates a significant interpretive challenge for quantum cosmology. A common solution is to assume that all contending quantum realities hold equal status. They are not merely "phantom worlds" or "potential realities," but rather "really real"—each corresponding to its own Universe, complete with distinct space and time. These multiple Universes exist parallel to one another, though they are not interconnected, leading to an infinite number of them.

This philosophical perspective was largely shaped by Niels Bohr and his contemporaries in Denmark during the 1920s and 1930s, giving rise to what is known as the "Copenhagen interpretation" of quantum physics. While Albert Einstein and others have heavily criticized it, it remains the orthodox interpretation accepted by most working physicists. We will explore this further before addressing its perceived weaknesses and discussing alternative viewpoints.

At the core of Bohr's philosophy is a form of "positivism," encapsulated in a phrase from philosopher Ludwig Wittgenstein: "whereof we cannot speak, thereof we should remain silent." In this context, it implies that if something is unobservable (like simultaneous knowledge of the HV and ±45-degree polarization of a photon), we should not assume it has any reality. While this notion may be a matter of choice in some philosophical discussions—such as imagining angels dancing on a pinhead—within the Copenhagen framework, it is a matter of necessity. This line of thinking challenges those accustomed to classical physics, which assumes objects must always exist "somewhere."

Upon learning about wave-particle duality, one might accept the notion that "when the particle is not being observed, it is actually a wave." However, this statement still attributes reality to something unobservable. Earlier chapters demonstrated that wave properties emerge when experiments involve a large number of particles. This is evident, for instance, when an electromagnetic wave is detected by a television or when interference patterns are observed after many photons have reached a screen (even if they passed through the apparatus one at a time). Yet, if we assert that the wave is "real" when considering an individual object, we again suggest that something unobservable truly exists. The wave function should not be viewed as a physical wave; instead, it is a mathematical construct used to predict the probabilities of possible experimental outcomes.

Philosophers often describe an object's characteristics as its "attributes," which for a classical object include permanent quantities like mass, charge, and volume, alongside variables such as position and speed that may change during motion. In the Copenhagen interpretation, an object's attributes depend on the context of observation.

For instance, a photon emerging from the H channel of an HV polarizer possesses the attribute of horizontal polarization, but if it passes through a ±45-degree polarizer, it loses that attribute and gains either +45-degree or -45-degree polarization. It may be more challenging to accept that an attribute like position has a similarly restricted application, but quantum physics compels us to adopt this "counterintuitive" perspective.

To further develop these ideas, let's consider the meaning and purpose of a scientific theory. A useful analogy is a map used to navigate a new city. While the map is typically much smaller than the actual area it represents, it aims to accurately depict the terrain: the relationships between streets and buildings on the map mirror those in reality. However, the map is not the same as the terrain it models; it differs in significant ways, such as size and the materials it is made from.

A scientific theory also seeks to model reality. Setting aside quantum physics for a moment, a classical theory constructs a "map" of physical events. For example, consider an apple released to fall under gravity: the apple starts at rest, is released, accelerates, and stops upon reaching the ground. At each stage of its motion, all relevant attributes (such as time, height, and speed) are represented on our map through algebraic variables, and we use mathematical calculations to depict the apple's motion. Yet, the real apple falls to the ground in the predicted time, even though it cannot perform mathematical calculations itself.

Creating this map often involves extensive mathematics, but it is crucial to remember that the map is not reality itself. We must also select an appropriate map for the physical situation at hand. For instance, a map based on Maxwell's electromagnetism theory would be ineffective if we sought to understand the apple's fall under gravity. Even when using a map based on Newton's laws, it must incorporate relevant parameters; for example, it should account for air resistance unless the apple falls in a vacuum.

In quantum physics, we must acknowledge that there is no single map of the quantum world. Instead, quantum theory provides various maps; the choice of which one to use depends on the experimental context or even the experimental outcome, and it may change as the system evolves over time. Continuing our analogy, quantum physics allows us to create a "map book," where we must refer to the appropriate page for the specific situation we are considering. For instance, when a 45-degree photon passes through an HV polarizer, the relevant map before it reaches the polarizer depicts a +45-degree polarized photon moving from left to right. Once it exits the polarizer, the appropriate map represents the two possibilities of the photon being horizontally or vertically polarized; finally, when it is detected, the relevant map focuses solely on the actual outcome.

14.32 Alternative Interpretations (Subjectivism)

One response to the "quantum measurement problem" is to embrace **subjective idealism**. This perspective suggests that quantum physics indicates the impossibility of providing an objective description of physical reality. The only certainty we have is our personal subjective experience: the counter may both register a hit and not register a hit, the cat may be both alive and dead, but once that information reaches our mind through our brain, we know what has truly happened. While quantum physics applies to photons, counters, and cats, it seemingly does not pertain to you or me! However, we also can't be entirely sure that the states of our mind are real, putting us at risk of slipping into **solipsism**, where only our minds have any semblance of reality.

Philosophers have long debated whether they can prove the existence of an external physical world, but the goal of science is not to resolve this philosophical dilemma; rather, it seeks to provide a coherent explanation of any objective reality that may exist.

It would be ironic if quantum physics ultimately undermined this quest. Most of us would prefer to explore alternative paths forward.

14.33 The Unfinished Business of Physics

The narrative of quantum mechanics presented here appears clear-cut, aside from the semi-philosophical debate over whether one favors the Copenhagen interpretation or the many-worlds theory. While this may be the most straightforward way to convey the story in a book, it doesn't encompass the entire truth. The tale of quantum physics is still unfolding, and contemporary theorists are wrestling with challenges that could lead to breakthroughs as significant as Bohr's quantization of the atom. Writing about this ongoing exploration can be messy and unsatisfying; the consensus on what is crucial and what can be overlooked may shift dramatically by the time this account is published. However, to provide a sense of how the field might be evolving, this epilogue includes insights into the unresolved elements of the quantum narrative and hints about what to anticipate in the future.

Ch 15—Extra Dimensions

15.1 The Kaluza-Klein Theory—A look at Extra Dimensions

Extra dimensions have transformed the way physicists conceptualize the Universe. Their potential connections to well-established principles in physics provide a novel framework for revisiting and exploring foundational truths about the cosmos.

One of the earliest efforts to unify gravity and electromagnetism was the Kaluza-Klein theory, which proposed the existence of an additional spatial dimension. In this model, the extra dimension was compactified to an incredibly small, "microscopic" scale. Although the theory ultimately fell short, many of its ideas later influenced the development of string theory, where extra dimensions play a central role.

Einstein's theory of gravity, General Relativity, was so remarkably successful in explaining gravity that physicists began to wonder if it could also describe the only other known force at the time—the electromagnetic force. Could electromagnetism, like gravity, be a result of the geometry of space-time? Even before Einstein finalized his general relativity field equations in 1915, British mathematician David Hilbert noted that research by Nordström and others suggested "gravitation and electrodynamics are not really different." Einstein himself admitted, "I have often tortured my mind in order to bridge the gap between gravitation and electromagnetism."

A significant attempt at unification came in 1919 when German mathematician Theodor Kaluza presented Einstein with a bold idea. Drawing inspiration from earlier work by Nordström—who in 1914 had reformulated Maxwell's equations in five dimensions to derive gravity equations—Kaluza extended the field equations of general relativity into a fifth dimension. Remarkably, his approach produced not only Einstein's gravity equations but also Maxwell's equations of electromagnetism.

When Kaluza wrote to Einstein about his findings, Einstein replied that the concept of adding dimensions "never dawned on me," hinting that he may have been unaware of Nordström's earlier attempts to unify electromagnetism and gravity, despite being familiar with Nordström's gravity theory. Kaluza envisioned the universe as a 5-dimensional cylinder, with our familiar 4-dimensional world existing as a projection on its surface. However, without experimental evidence for this extra dimension, Einstein hesitated to fully embrace the idea.

Einstein incorporated aspects of Kaluza's ideas into a unified field theory he published in 1925, though he quickly withdrew it. In 1926, Swedish physicist Oskar Klein revisited Kaluza's theory and redefined it into what is now known as the Kaluza-Klein theory. Klein proposed that the fourth spatial dimension was compactified into an extremely small circle, rendering it effectively undetectable by direct observation.

In Kaluza-Klein theory, the geometry of the extra, hidden spatial dimension determined the properties of the electromagnetic force, the size of the compactified circle, and a particle's motion within this extra dimension, which was tied to its electric charge. However, the theory encountered significant issues. Predictions of the electron's charge and mass failed to align with observed values. Additionally, as quantum mechanics—supported by experimental evidence—rapidly as many physicists shifted their focus away from Kaluza-Klein theory.

Another major flaw in the theory was its prediction of a particle with zero mass, zero spin, and zero charge. Despite being a low-energy particle that should have been easily detectable, such a particle was never observed. Even more problematic, this particle was associated with the radius of the extra dimension, leading to a paradox: the theory introduced extra dimensions but implied their effects were effectively nonexistent.

An alternative, less conventional explanation for the failure of Kaluza-Klein theory points to a fundamental theoretical limitation: for electromagnetism to function within the framework, the geometry of the extra dimension had to remain completely fixed. However, in a theory of dynamic space-time, introducing an extra dimension should ideally result in a framework where all dimensions, including the extra one, remain dynamic. Having a fifth dimension that is static while the other four are flexible undermines the consistency of the theory. This issue, known as background dependence, resurfaces as a significant critique of string theory.

Regardless of the precise reasons for its shortcomings, Kaluza-Klein theory had a relatively short-lived prominence. However, there are indications that Einstein continued revisiting its concepts sporadically into the early 1940s, incorporating elements into his unsuccessful attempts at a unified field theory. In the 1970s, when physicists recognized that string theory also required extra dimensions, the original Kaluza-Klein theory provided a historical precedent. Once again, extra dimensions were compactified in the manner Klein had proposed, rendering them effectively undetectable. These modern adaptations became known collectively as Kaluza-Klein theories.

15.2 Defining Dimensionality

No physical theory requires that space be limited to just three dimensions. To dismiss the possibility of extra dimensions without fully exploring their existence would be premature. Just as "up-down" differs from "left-right" or "forward-backward," entirely new dimensions could exist within our Universe. While we cannot see or directly experience these additional dimensions, their existence remains a logical and viable possibility.

At the time, these hypothetical, unseen dimensions lacked a proper name. If they do exist, they would represent new directions through which something could move. These dimensions could be flat, like the familiar ones we know, or warped, resembling the distorted reflections in a funhouse mirror. They might be incredibly small—far smaller than an atom—which was the prevailing assumption among proponents of extra dimensions until recently. However, new research suggests that these dimensions could also be large, or even infinite, while still remaining elusive to detection.

While our senses perceive only three expansive dimensions, the idea of an infinite extra dimension, though seemingly incredible, is one of many possibilities for what might exist in the Universe. Exploration into extra dimensions has also inspired extraordinary concepts that verge on science fiction, including parallel universes, warped geometries, and three-dimensional sinkholes.

15.3 How Unseen dimensions can help our understanding

Recent advances suggest that extra dimensions—unseen and not yet fully understood—could hold the key to solving some of the most profound mysteries of our Universe. These dimensions may have far-reaching implications for the reality we perceive, offering new insights and revealing hidden connections that remain elusive in our three-dimensional understanding of space.

Extra dimensions could also provide clarity on long-standing puzzles in particle physics. Decades-old mysteries about the relationships between particle properties and forces, which appear disjointed in three-dimensional space, seem to align more elegantly in a higher-dimensional framework.

For years, physicists have relied on the Standard Model of particle physics to describe the fundamental nature of matter and the forces governing its interactions. This theory has been rigorously tested through experiments that recreate particles not present since the Universe's earliest moments. While the Standard Model has been remarkably successful in describing many observed properties, it leaves some critical questions unanswered. Investigating extra dimensions may help resolve these lingering mysteries and deepen our understanding of the Universe.

15.4 Why is gravity so weak

One of the greatest mysteries in physics is why gravity is so much weaker than the other fundamental forces. While gravity may feel significant when hiking up a mountain, that's only because the entire Earth is pulling on you. In contrast, a small magnet can lift a paper clip against the pull of the planet's mass. Why is gravity so easily overpowered by such a tiny force? Within the framework of standard three-dimensional particle physics, this weakness of gravity remains a perplexing puzzle. However, the existence of extra dimensions could offer an explanation.

In 1998, physicists Lisa Randall and Raman Sundrum proposed a groundbreaking model of gravity based on the concept of "warped geometry," a notion rooted in Einstein's General Theory of Relativity. According to Einstein, space and time are woven into a single spacetime fabric that can be distorted, or warped, by the presence of matter and energy. Randall and Sundrum identified a configuration in which spacetime is warped so drastically that gravity, while strong in one region of space, becomes extraordinarily weak everywhere else.

Their model revealed something even more extraordinary: while physicists had long assumed that extra dimensions must be tiny to remain hidden, Randall and Sundrum showed that warped space could explain gravity's weakness even if an extra dimension were infinite in size. An invisible extra dimension could stretch to infinity, provided it is sufficiently distorted in a curved spacetime. In this scenario, gravity's apparent weakness in our world arises because it is concentrated in another region of this higher-dimensional space.

The Randall-Sundrum model also suggests that we may exist in a three-dimensional "pocket" of space, while the rest of the Universe behaves as though it is higher-dimensional. This opens up new possibilities for the structure of spacetime, which could consist of distinct regions with varying numbers of dimensions. Just as Copernicus revolutionized our understanding of the Universe 500 years ago by showing that we are not at its center, the Randall-Sundrum model suggests we may be living in an isolated neighborhood of three spatial dimensions within a higher-dimensional cosmos.

Central to these ideas are membrane-like objects called branes, which play a vital role in higher-dimensional theories. Branes provide a rich and intricate framework, with some models proposing universes in which we live on a single brane, while others suggest multiple branes, each potentially housing unseen worlds. These "braneworlds" create a multi-layered, multi-faceted playground for physicists, encompassing possibilities such as curled-up, warped, large, or infinite dimensions.

In 2007, the Large Hadron Collider (LHC) began operations, producing particles at energies high enough to explore these ideas. If theories about extra dimensions are correct, the LHC could provide visible evidence. One such clue would be the discovery of particles known as Kaluza-Klein modes, which travel through extra dimensions but leave detectable traces in our familiar three-dimensional space. These particles would serve as "fingerprints" of extra dimensions. With even greater luck, experiments at the LHC might uncover other phenomena, such as higher-dimensional black holes, offering deeper insights into the fabric of our Universe.

15.5 What are dimensions?

Working with spaces of many dimensions is something we all do daily, even if we don't think of it that way. The number of dimensions corresponds to the number of quantities needed to uniquely define a point in space. This multidimensional space might be abstract—such as the list of features you're considering when searching for a house—or concrete, like the physical space we experience. For example, when buying a house, you can think of the dimensions as the number of characteristics you'd record for each listing in a database, representing the features you deem important to investigate.

A space of a particular dimension is one that requires a specific number of quantities to locate a point within it. In one dimension, this would be a single value, like a point on a line represented by an x-axis. In two dimensions, it would be a point on a plane, requiring both x and y coordinates. In three dimensions, it would be a point in space, determined by x, y, and z coordinates.

String theory proposes the existence of six or seven additional spatial dimensions, meaning six or seven extra coordinates are required to define a point in space. More recent developments in string theory suggest the possibility of even more dimensions. The principles related to extra dimensions discussed here apply regardless of the number of additional dimensions.

To understand these dimensions, we rely on a *metric*, which acts as a measuring tool, defining the scale of space. For instance, on a map, a half-inch might represent one mile, just as the metric system provides a universally accepted meter stick. However, a metric does more than establish scale—it also describes the shape of space. It reveals whether space is flat or curved, like the surface of a balloon expanding into a sphere. Metrics for curved spaces provide information about both distances and angles, where an inch or degree can correspond to different shapes depending on the curvature.

As physics has progressed, the metric's significance has grown. Einstein's theory of relativity introduced the concept of time as the fourth dimension, inseparable from the three spatial dimensions. Time also requires a scale, so Einstein formulated gravity using a metric for four-dimensional spacetime. Recent advances suggest the potential existence of additional spatial dimensions, implying that the true spacetime metric could involve more than three spatial dimensions. The number of dimensions and their corresponding metric are key to describing these multidimensional spaces.

To visualize this, consider a garden hose stretched across a football field. From a distance, the hose appears one-dimensional, like a line. But up close, you see the hose's surface is two-dimensional, and its enclosed volume is three-dimensional.

Now imagine the entire universe as a garden hose—a long sheet of rubber rolled into a tube with a small circular cross-section. In this scenario, the universe has one very long dimension and one small, rolled-up dimension. For a tiny creature, a flat bug, for instance—living on the hose's surface, the universe would appear two-dimensional. The bug could move in two directions: along the length of the hose or around its circumference. If the bug crawled around the hose, it would eventually return to its starting point. Because the second dimension is small, the bug wouldn't need to travel far to complete the loop. This analogy provides a glimpse into how additional dimensions might exist in our universe.

If a population of bugs lived on the surface of a garden hose, they would experience forces—like gravity or an electric force—that could attract or repel them in any direction on the hose. These forces would act along both dimensions of the hose: the long axis and the circular circumference. When the bugs could distinguish distances as small as the diameter of the hose, both dimensions would become apparent in the way forces and objects behaved.

However, a bug observing its universe would notice that the two dimensions were very different. The dimension along the hose's length might be vast, possibly even infinite. In contrast, the circular dimension would be extremely small. Bugs separated along the circumference could never get very far from one another before looping back to their starting points. A perceptive bug, eager to explore its universe, would realize it was two-dimensional—one expansive dimension stretching along the hose and another compact, rolled-up dimension forming a circle.

Yet, the bug's perspective would be vastly different from our own in a universe like Klein's, where the extra dimension is minuscule—on the order of *10^{-33} cm*. Unlike the bug, we are far too large to perceive, let alone travel within such a tiny dimension. Physical forces and effects in our scale of reality would not reveal the existence of this extra dimension. In the garden-hose analogy, beings large enough to span the entire small dimension would fail to detect its presence. Without perceivable structures, variations, or oscillations of matter and energy along the extra dimension, the small rolled-up dimension would remain hidden.

Similarly, in a four-dimensional Kaluza-Klein universe with one tiny, rolled-up spatial dimension, we would perceive the universe as having only three dimensions. The extra dimension would go unnoticed unless we could detect structures at their incredibly small scale. Rolled-up, or compactified, dimensions effectively vanish from perception when they are small enough. As we'll see in later chapters, such dimensions remain undetectable at scales below the Planck length—an unimaginably tiny threshold. For now, it's enough to recognize that these compactified dimensions may exist, even if their size keeps them hidden from view.

Ch 16—Parallel Worlds

16.1 The Cosmos composed of Parallel Worlds

Cosmology is the scientific study of the Universe in its entirety, encompassing its origins, development, and potential fate. Over time, this field has undergone significant transformations, often hindered by the influence of religious doctrines and superstitions that shaped humanity's understanding of the cosmos.

The first major revolution in cosmology began with the invention of the telescope in the 1600s. Using this groundbreaking tool, Galileo Galilei expanded on the foundational work of astronomers Nicolaus Copernicus and Johannes Kepler, allowing the splendor of the heavens to be explored through rigorous scientific investigation for the first time. This transformative period culminated in the contributions of Isaac Newton, who established the fundamental laws governing the motion of celestial bodies. No longer attributed to magic or mysticism, the behavior of heavenly objects was now understood to be governed by forces that could be calculated and consistently reproduced.

The second great revolution in cosmology came with the advent of the powerful telescopes of the 20th century, such as the 100-inch reflecting telescope at Mount Wilson Observatory. In the 1920s, astronomer Edwin Hubble utilized this massive instrument to challenge centuries-old beliefs that the universe was static and eternal. Hubble revealed that galaxies are racing away from Earth at extraordinary speeds, providing evidence that the universe is expanding. This groundbreaking discovery aligned with Einstein's theory of general relativity, which redefined the structure of space-time as dynamic and curved rather than static and linear.

Hubble's findings offered the first credible explanation for the universe's origin: a dramatic event now known as the "big bang," a cosmic explosion that set stars and galaxies into motion. Building on this foundation, George Gamow and his colleagues advanced the big bang theory, while Fred Hoyle explored the origins of the elements, creating a framework that outlined the universe's evolution and paved the way for modern cosmology.

A third revolution in cosmology is currently unfolding, driven by cutting-edge technologies such as space-based satellites, lasers, gravity wave detectors, X-ray telescopes, and high-speed supercomputers. These advanced tools have provided the most precise insights yet into the universe's nature, including its age, composition, and potential future.

Astronomers have discovered that the universe is not only expanding but doing so at an accelerating pace. This runaway expansion suggests a grim future: the "big freeze." Over time, the universe will grow increasingly cold and dark, eventually becoming inhospitable to intelligent life as it succumbs to eternal darkness and frigid isolation.

16.2 Trapped on Branes

It's highly unlikely that you'll ever explore all the space available to you. There may be places you dream of visiting, like outer space, but currently cannot—though, in principle, no physical law makes such travel impossible. However, if you were inside a black hole, you would face an entirely different scenario. A black hole's immense gravitational pull would trap you within its interior, making escape physically impossible. You would remain confined until the black hole eventually decayed away.

There are many familiar examples of objects with restricted movement, which highlight regions of space that are genuinely inaccessible. For instance, a charge on a wire exists in a three-dimensional world but only moves along one of those dimensions. Similarly, some objects are confined to two-dimensional surfaces, like water droplets on a shower curtain, which can only move along the curtain's two-dimensional surface.

Branes, similar to shower curtains, constrain objects to lower-dimensional surfaces. They introduce the idea that, in a world with additional dimensions, not everything can move freely in all directions. Just as the droplets on a curtain are confined to a two-dimensional brane within a higher-dimensional space, some particles may be similarly bound. However, unlike the droplets, these particles are truly trapped.

Particles that are confined to branes are restricted to those branes by physical laws, unable to move into the extra dimensions that extend beyond them. While not all particles are trapped on branes—some may be free to travel through the extra dimensions—what sets multidimensional theories apart is the distinction between particles that are confined to branes and those that can move throughout the entire higher-dimensional space.

In theory, both branes and the bulk can have any number of dimensions, as long as a brane does not have more dimensions than the bulk. The dimensionality of a brane refers to the number of dimensions in which particles confined to the brane can move. While there are various possibilities, the most interesting branes for our purposes will be the three-dimensional ones. Although we don't fully understand why three dimensions seem so special, branes with three spatial dimensions could be significant to our universe because they could span the same three spatial dimensions that we experience. These three-dimensional branes could exist within a bulk space that has more than three dimensions—such as four, five, or even more dimensions.

Even if the universe contains many dimensions, if particles and forces are confined to a three-dimensional brane, they would behave as though they exist in only three dimensions. Particles restricted to the brane would only move along the brane itself. Similarly, if light were also confined to the brane, light rays would spread only along the brane. In this case, light would behave exactly as it does in a three-dimensional universe.

Moreover, forces that are confined to a brane would only affect particles within that brane. The materials we are made of, such as nuclei and electrons, and the forces governing their interactions, like the electric force, could be confined to a three-dimensional brane. These brane-bound forces would extend only along the brane, and particles within the brane would interact and move only along its dimensions.

So, if you lived on such a three-dimensional brane, you would be able to move freely within its dimensions, just as you do in our current three-dimensional world. Anything confined to this three-dimensional brane would appear exactly as it would in a truly three-dimensional universe. While other dimensions would exist adjacent to the brane, anything bound to the brane would never be able to penetrate the higher-dimensional bulk.

While forces and matter can be confined to a brane, brane-worlds are particularly fascinating because not everything is restricted to a single brane. Gravity, for instance, is never confined to a brane. According to general relativity, gravity is an intrinsic part of the fabric of space-time, meaning it must be present throughout space and across all dimensions. If gravity could be confined to a single brane, we would have to abandon general relativity.

Fortunately, this is not the case. Even if branes exist, gravity will be felt everywhere—on and off branes. This is significant because it means braneworlds must interact with the bulk, at least through gravity. Since gravity extends into the bulk and everything interacts via gravity, braneworlds are always connected to the extra dimensions. They do not exist in isolation but are part of a larger, interconnected whole.

In addition to gravity, there could be other particles and forces in the bulk. If such particles exist, they could interact with those confined to a brane, linking brane-bound particles to the higher-dimensional bulk. The string theory branes we'll explore later have unique characteristics beyond what we've discussed. They can carry specific charges and respond in particular ways when influenced by external forces. Branes are lower-dimensional surfaces that can house forces and particles and may act as the boundaries of higher-dimensional spaces.

Since branes have the ability to trap most particles and forces, it's possible that the universe we experience could exist on a three-dimensional brane, suspended in a higher-dimensional space. While gravity would extend into extra dimensions, everything we perceive—stars, planets, people, and all matter—could be confined to this three-dimensional brane. In this scenario, we would be living on a brane, with it serving as our habitat.

16.3 Branes Worlds

If one brane can exist within a higher-dimensional spacetime, it's certainly possible that many more could be present. Brane-world models often involve multiple branes, though we don't yet know how many types of branes could exist in the cosmos. The term "multiverse" is sometimes used to describe theories that involve more than one brane. This term often refers to a cosmos made up of separate, non-interacting, or weakly interacting regions. Therefore, the universe could contain multiple branes that either interact only through gravity or don't interact at all.

Different branes could be so far apart that they can never communicate, or they might only interact weakly through mediating particles traveling between them. As a result, particles on distant branes would experience different forces, and particles confined to one brane would never have direct contact with those on another.

When multiple branes exist without sharing any forces other than gravity, the universe containing them can be described as a multiverse. Some of these branes might be parallel to ours, each hosting its own parallel world. However, there are many other possible types of brane-worlds. Branes could intersect, with particles becoming trapped at the intersections. They might have different dimensionalities, curves, or even movement. Some could wrap around hidden, invisible dimensions. It is entirely possible that such geometries exist within the Cosmos.

In a universe where branes are embedded within a higher-dimensional bulk, some particles may explore the higher dimensions, while others remain confined to the branes. If the bulk separates different branes, particles could be distributed across the branes or within the bulk itself, with some on one brane, others on another, and some in between. Theories suggest various ways in which particles and forces could be spread across the branes and the bulk. Even within string theory, we still don't fully understand why it would favor any particular distribution of particles and forces. Brane-worlds introduce new physical possibilities that could explain not only the world we observe but also other worlds on branes we have yet to discover, separated from our own by unseen dimensions.

There may be new forces confined to distant branes, and new particles that we will never directly interact with could propagate on these other branes. The mysterious substances we associate with dark matter and dark energy—whose existence we infer from their gravitational effects but whose nature remains unknown—could be spread across different branes, or even within both the bulk and other branes. Gravity might also affect particles in different ways as one moves from one brane to another.

If life exists on another brane, those beings will likely experience entirely different forces, detected by senses adapted to their unique environment. Our senses are attuned to the chemistry, light, and sound around us, but since fundamental forces and particles are likely different on other branes, life forms on those branes, if they exist, would probably bear little resemblance to life on our own brane. The other branes would likely be vastly different from ours, with gravity being the only force necessarily shared—though even gravity's influence may differ.

The implications of a brane-world depend on the number and types of branes, and their locations. Unfortunately, particles and forces confined to distant branes might have little effect on us. They could only influence what moves through the bulk and emit weak signals that may never reach our brane. As a result, many possible braneworlds would be incredibly difficult to detect, even if they exist. Since gravity is the only force known to be shared between our brane and others, and gravity is extremely weak, without direct evidence, the existence of other branes will remain speculative.

Ch 17—The Fabric of Space-Time

17.1 Quantum Reality

The second anomaly mentioned by Lord Kelvin played a pivotal role in sparking the quantum revolution, a transformative event that reshaped modern human understanding. Once the turmoil settled, classical physics had effectively given way to the emerging paradigm of quantum reality.

A fundamental aspect of classical physics is that if you know the positions and velocities of all objects at a specific time, Newton's equations, along with their Maxwellian updates, can predict their positions and velocities at any other moment, whether in the past or future. Classical physics firmly asserts that the present moment holds the key to both the past and future. This idea is also true in both special and general relativity. While the concepts of past and future in relativity are more nuanced than in classical physics, the equations of relativity, combined with a thorough understanding of the present, determine them with equal certainty.

By the 1930s, however, physicists had to adopt an entirely new framework known as quantum mechanics. To their surprise, they discovered that only quantum principles could resolve numerous puzzles and account for various pieces of new data from the atomic and subatomic world. Yet, under quantum laws, even with the most precise measurements of the present, the best you can achieve is predicting the probability of how things might be at a specific time in the future or how they might have been at a specific time in the past. According to quantum mechanics, the Universe is not fixed in the present; instead, it engages in a form of chance.

While there remains debate about the exact interpretation of these developments, most physicists concur that probability is inherently embedded in the structure of quantum reality. Unlike classical physics, which aligns with human intuition by presenting a world where things are always clearly one way or another, quantum mechanics portrays a reality where things can exist in a state of uncertainty, partly one way and partly another. Only when an appropriate observation is made do things become definite, as the act of measurement causes them to abandon their quantum possibilities and settle into a specific result.

The outcome that occurs, however, cannot be predicted—what we can determine is only the likelihood of things turning out one way or another. Simply put, this is strange. We are not accustomed to a reality that remains uncertain until it is observed. But the strangeness of quantum mechanics doesn't end there. Equally remarkable is a concept introduced in a 1935 paper by Einstein, along with his colleagues Nathan Rosen and Boris Podolsky, which was meant to challenge quantum theory. Over time, this paper has come to be seen as one of the first to highlight that quantum mechanics, if taken literally, suggests that an action on one side of the universe can instantaneously affect something on the other side, regardless of distance. Einstein found this instantaneous connection absurd, interpreting it as a flaw in quantum theory that needed further development to become credible. However, by the 1980s, both theoretical advancements and experimental breakthroughs confirmed that such instantaneous connections between distant locations are indeed possible. Under controlled laboratory conditions, what Einstein dismissed as absurd was proven to be true.

The consequences of these aspects of quantum mechanics for our understanding of reality are still being explored. Many scientists consider them as part of a profound rethinking of the nature and properties of space. Typically, spatial separation suggests physical independence. For instance, if you want to influence something happening on the other side of a football field, you must go there or, at the very least, send something—whether it's a person, sound waves, or light signals—to transmit your influence. If you stay in one place, without any means of communication, you can't affect what happens on the other side, since the intervening space prevents any physical connection. Quantum mechanics challenges this idea by showing that, under certain conditions, it's possible to transcend space. Long-range quantum connections can overcome spatial separation. Even if two objects are far apart, quantum mechanics treats them as if they were a single entity. Furthermore, due to the close relationship between space and time identified by Einstein, these quantum connections also extend across time.

Soon, we will explore some remarkable and ingenious experiments that have recently investigated several of the surprising spatio-temporal connections revealed by quantum mechanics. These experiments, as we will see, present a powerful challenge to the classical, intuitive worldview that many of us maintain.

Despite the many profound discoveries, there remains one fundamental aspect of time—the sense that it moves from the past to the future—for which neither relativity nor quantum mechanics offers an explanation. Instead, the most promising advancements in understanding this phenomenon have come from the field of cosmology.

17.2 Entangling Space—What does it mean to be separate in a quantum universe

Accepting the principles of special and general relativity means abandoning the notion of absolute space and time as proposed by Newton. Although it's challenging, you can train your mind to adapt to this new way of thinking. When you move, for instance, imagine that your "now" is shifting away from the "nows" experienced by others who aren't moving with you. While driving on a highway, think of your watch ticking at a different rate compared to the clocks in the homes you pass by. When standing atop a mountain, imagine that due to the warping of spacetime, time flows faster for you than for those on the ground below, where gravity is stronger. I say "imagine" because in normal circumstances, the effects of relativity are so minuscule that they go unnoticed. Our everyday experience doesn't reveal the true workings of the Universe, which is why, a century after Einstein, few people, even professional physicists, feel relativity in a tangible way. This isn't surprising; there's little evolutionary benefit to having a deep understanding of relativity. Newton's ideas about absolute space and time work perfectly well at the relatively low speeds and moderate gravitational forces we encounter in everyday life, so there's no evolutionary pressure for our senses to detect relativistic effects. Achieving true understanding, therefore, requires us to use our intellect to bridge the gap left by our senses.

While relativity marked a groundbreaking shift in traditional views of the Universe, between 1900 and 1930, another revolution was reshaping physics. It began at the start of the twentieth century with two papers on the properties of radiation—one by Max Planck and the other by Einstein. After three decades of intense research, these works culminated in the development of quantum mechanics. Similar to relativity, whose effects become noticeable under extreme conditions of speed or gravity, the principles of quantum mechanics manifest most clearly in another extreme: the realm of the incredibly small. However, the upheavals of relativity and quantum mechanics are fundamentally different. The strangeness of relativity stems from the fact that our personal experience of space and time can differ from that of others. This weirdness arises from comparison, forcing us to recognize that our perception of reality is just one among countless others, all of which fit together within the unified fabric of spacetime.

17.3 The World According to the Quantum

Each era creates its own stories or metaphors to explain how the Universe is conceived and structured. In an ancient Indian creation myth, the Universe came into being when the gods dismembered the primordial giant Purusa. His head became the sky, his feet the earth, and his breath the wind. For Aristotle, the Universe consisted of fifty-five concentric crystalline spheres, with heaven at the outermost layer, followed by the spheres of the planets, Earth, its elements, and the seven circles of hell. With Newton's precise, deterministic mathematical formulation of motion, the conception of the Universe shifted once again. It was compared to a vast, grand clockwork: once wound and set into motion, the Universe would proceed from one moment to the next with perfect regularity and predictability.

Special and general relativity highlighted key nuances in the clockwork metaphor: there is no universal, preferred clock, and no agreed-upon definition of what constitutes a moment or a "now." Nonetheless, it is still possible to tell a clockwork-like story about the Universe's evolution. The clock in this narrative is yours, and the story is uniquely yours, but the Universe continues to unfold with the same regularity and predictability as it did within the Newtonian framework.

If, through some means, you know the current state of the Universe—if you can pinpoint the position of every particle and determine their speed and direction—then, according to both Newton and Einstein, you can, in principle, use the laws of physics to predict everything about the Universe far into the future or determine what it was like at any point in the past.

Quantum mechanics challenges this long-standing tradition. It is impossible to know both the exact position and velocity of a particle simultaneously. We cannot predict the outcome of even the simplest experiment with absolute certainty, much less the future of the entire Universe. Quantum mechanics reveals that the best we can achieve is predicting the likelihood of a particular outcome. And as quantum mechanics has been validated through decades of highly accurate experiments, the Newtonian view of the cosmic clock, even with Einstein's refinements, proves to be an untenable metaphor—it is clearly not how the world operates.

The break with the past is even more profound. While Newton's and Einstein's theories sharply differ in their understanding of space and time, they do share agreement on certain fundamental principles—truths that seem self-evident. For instance, if there is space between two objects, like two birds in the sky, one to your right and the other to your left, we treat them as independent. We perceive them as separate and distinct entities. Space, regardless of its ultimate nature, serves as the medium that separates and differentiates one object from another. That is its role.

Objects in different locations in space are considered separate from one another. For one object to influence another, it must somehow engage with the space that separates them, either by direct contact or through some form of interaction with the space between them. For instance, a person can affect another by using a slingshot to send a pebble across the gap between them, or by yelling, which sets off a chain reaction of air molecules that carry the sound until it reaches the listener's eardrum. More complex interactions may involve a laser, where an electromagnetic wave travels across space separating two points. On a larger scale, one could even move a massive object, such as the moon, creating a gravitational wave that propagates through the intervening space.

It is true that if we are here, we can influence someone over there, but no matter how it's done, the process always involves something traveling from one place to another, and only once it arrives can the influence take effect. Physicists refer to this characteristic of the universe as locality, stressing that we can only directly affect things that are close to us, or local. Voodoo contradicts locality, as it involves influencing something far away without anything physically traveling between the two locations. However, everyday experience leads us to believe that verifiable, repeatable experiments would confirm locality—and for the most part, they do. Yet, a series of experiments conducted in the last few decades has shown that actions taken here (such as measuring the property of a particle) can be subtly connected to events happening there (such as measuring the properties of a distant particle), without any physical transmission between the two. Although this phenomenon seems counterintuitive, it fully aligns with the principles of quantum mechanics, which predicted it long before the technology was available to experimentally confirm it. This phenomenon might sound like voodoo; in fact, Einstein, one of the first to recognize—and criticize—this feature of quantum mechanics, famously called it "spooky." However, as we'll explore, these long-distance connections revealed by the experiments are very delicate and, in a precise sense, fundamentally beyond our ability to control.

Nevertheless, both theoretical and experimental evidence strongly suggest that the universe allows for interconnections that are non-local. Events that occur here can be linked to those happening elsewhere, even if nothing physically travels between the two locations—and even if there isn't enough time for anything, including light, to move from one to the other. This indicates that space can no longer be viewed in the traditional sense, as simply a separator between objects. Quantum mechanics permits a kind of connection, or entanglement, between two objects, meaning that space itself doesn't guarantee separation. A particle, like one of the countless that make up our bodies, can try to escape, but it cannot fully avoid being connected. According to quantum theory and the experiments that validate its predictions, this quantum connection can endure even across vast distances, such as particles being on opposite ends of the universe. Despite the trillions of miles separating them, from the perspective of their entanglement, it's as if they are right next to each other. Modern physics continues to challenge our understanding of reality, but among the experimental revelations, none are as astonishing as the realization that our universe is not local.

17.4 Probability and the Laws of Physics

If an electron behaves like a wave, what exactly is oscillating? Erwin Schrodinger offered the first hypothesis: perhaps the substance making up the electron is spread out in space, and it is this spread-out essence of the electron that is oscillating. In this view, the electron particle would be a sharp peak in a cloud of electron-like matter. However, it was soon realized that this idea couldn't be correct, because even a sharply peaked wave—like a massive tidal wave—ultimately spreads out. If the electron's wave were spreading, we would expect to detect portions of its mass or charge scattered across space. Yet, when we measure an electron, we always find its entire mass and charge concentrated in a small, point-like region. In 1927, Max Born introduced a different idea that would become a crucial turning point in physics, propelling it into a completely new domain. He proposed that the wave is not a smeared-out electron or anything familiar from prior scientific understanding. Instead, he suggested that the wave represents a probability wave.

To grasp this concept, imagine a snapshot of a water wave, showing areas of high intensity (near the peaks and troughs) and areas of low intensity (in the flatter regions between peaks and troughs). The higher the intensity, the greater the force the water wave can exert on nearby ships or coastal structures. Similarly, the probability waves that Born described have regions of varying intensity, but their significance was quite different from that of water waves. Born proposed that the size of the wave at a particular point in space corresponds to the likelihood of finding the electron at that location. In areas where the probability wave is large, the electron is more likely to be found. In areas where the wave is small, the electron is less likely to be located there. And in regions where the probability wave is zero, the electron cannot be found at all.

Figure 17-1 presents a "snapshot" of the probabilistic interpretation. However, unlike a photo of water waves, this image could not have been captured with a camera. A probability wave has never been directly observed, and traditional quantum mechanical theory suggests that it never will be. Instead, we rely on mathematical equations—formulated by Schrodinger, Niels Bohr, Werner Heisenberg, Paul Dirac, and others—to determine what the probability wave should look like in any given scenario.

We test theoretical calculations by comparing them with experimental data in the following manner. After determining the predicted probability wave for an electron in a specific experimental setup, we repeatedly perform identical experiments, starting from scratch each time, and record the electron's measured position. Unlike Newton's expectations, repeating the experiment with the same initial conditions does not always result in the same measurements. Instead, the positions where the electron is found vary. Occasionally, we observe the electron at one location, other times at a different point, and sometimes even far from both. According to quantum mechanics, the frequency with which we detect the electron at a particular location should correspond to the square of the size of the probability wave at that point.

For over eighty years, experiments have demonstrated that the predictions of quantum mechanics are accurate to an extraordinary degree. The portion of an electron's probability wave shown in represents only a small fraction of the entire wave: quantum mechanics suggests that the probability wave extends across all of space, spanning the entire Universe. However, in many situations, the probability wave rapidly diminishes to nearly zero outside a small region, indicating a very high likelihood that the particle is within that area. In these cases, the portion of the probability wave not depicted (which stretches across the rest of the Universe) resembles the edges of the figure: nearly flat and close to zero. Nevertheless, as long as the probability wave has a nonzero value anywhere—no matter how small—in the Andromeda galaxy, there remains a tiny but real chance that the electron could be found there.

The success of quantum mechanics compels us to accept that the electron, typically thought of as a tiny, point-like particle, also exhibits a wave-like nature that extends throughout the entire Universe. Furthermore, quantum mechanics tells us that this particle-wave duality applies to all fundamental particles, not just electrons. Protons exhibit both particle-like and wavelike properties, as do neutrons. Early 20th-century experiments even showed that light, which is clearly wave-like as seen in Figure 9-1, can also be described as consisting of particles called photons. For instance, the electromagnetic waves emitted by a 100-watt light bulb can be described in terms of the bulb emitting approximately 100 billion billion photons per second. In the quantum realm, we have come to understand that everything exhibits both particle-like and wavelike characteristics.

17.5 Unification in Higher Dimensions

In 1910, Einstein received a paper that, at first glance, could have easily been dismissed as the work of a crackpot. The paper was written by an obscure German mathematician named Theodor Kaluza, and in just a few pages, it presented an idea for unifying the two known forces of the time—gravity and electromagnetism. To achieve this, Kaluza proposed a bold idea that challenged something so fundamental, so universally accepted, that it seemed unquestionable.

Kaluza suggested that the Universe doesn't just have three spatial dimensions. Instead, he proposed that the Universe might have four spatial dimensions, making a total of five spacetime dimensions when combined with time.

So, what does this mean? When we refer to the three space dimensions, we mean three "independent" directions or axes along which movement is possible. These directions can be thought of as left/right, back/forth, and up/down. In a universe with three spatial dimensions, any movement you make is a combination of movement along these three axes. Similarly, in a three-dimensional universe, you need three pieces of information to specify a location. For example, to locate a dinner party in a city, you need the street, the cross street, and the floor number of the building. But to ensure people arrive on time, you also need to include a fourth piece of information: time. That's what we mean when we say spacetime is four-dimensional.

Kaluza proposed that, in addition to the familiar left/right, back/forth, and up/down dimensions, the universe actually contains one more spatial dimension that, for some unknown reason, has never been observed. If this theory is accurate, it would imply the existence of an additional independent direction in which movement is possible. As a result, to pinpoint a precise location in space, we would need four pieces of information, and a total of five if we also account for time.

Kaluza realized that it would be relatively straightforward to extend Einstein's general theory of relativity to a universe with an additional spatial dimension. He undertook this extension and, unsurprisingly, discovered that the higher-dimensional version of general relativity not only encompassed Einstein's original equations for gravity but also introduced additional equations due to the extra spatial dimension. Upon examining these new equations, Kaluza made an astonishing discovery: they were identical to the equations Maxwell had formulated in the 19th century to describe the electromagnetic field. By proposing a universe with one more spatial dimension, Kaluza had essentially solved one of the key challenges Einstein had grappled with in physics. Kaluza had found a way to unify Einstein's general relativity with Maxwell's equations for electromagnetism. This is why Einstein didn't dismiss Kaluza's paper.

To understand Kaluza's proposal, think of it this way. In General Relativity, Einstein revolutionized our understanding of space and time, showing how they can bend and stretch, and in doing so, he revealed the geometric nature of gravity. Kaluza's paper suggested that the geometry of space and time could extend even further. While Einstein demonstrated that gravitational fields are represented by distortions in the familiar three spatial dimensions and one-time dimension, Kaluza proposed that with an extra spatial dimension, additional distortions would occur. His analysis showed that these extra distortions could perfectly describe electromagnetic fields. In this way, Kaluza showed that Einstein's geometric approach could be powerful enough to unify gravity and electromagnetism.

The problem, of course, remained. Although the mathematics worked out, there was—and still is—no direct evidence of a spatial dimension beyond the three we are familiar with. So, the question was whether Kaluza's discovery was simply an interesting mathematical curiosity or if it had some real relevance to our universe. Kaluza had great confidence in theory—he had, for example, learned to swim by studying a treatise on the subject and then diving into the sea—but the concept of an unseen spatial dimension, no matter how convincing the theory, still seemed implausible. Then, in 1926, the Swedish physicist Oskar Klein introduced a new idea that added an intriguing possibility: he suggested a potential hiding place for the extra dimension.

Klein's contribution was to propose that what applies to objects within the Universe might also apply to the Universe's very structure. He suggested that, just like a tightrope has both large and small dimensions, one long and noticeable, the other small and curled up—the fabric of space could be similar. The three familiar dimensions of space—left/right, back/forth, and up/down—could be seen as analogous to the long, visible extent of the tightrope. However, just as the tightrope's surface includes an additional small, curled-up dimension, perhaps space itself contains a tiny, curled-up dimension, too. Klein argued that this dimension would be so minuscule that no current technology could detect it, making it effectively hidden.

How small would this extra dimension be? By incorporating elements of quantum mechanics into Kaluza's original idea, Klein's mathematical analysis suggested that the radius of this additional curled-up spatial dimension would be around the Planck length, which is incredibly tiny—far too small for any current experiments to detect. Modern technology can only resolve objects down to about one-thousandth the size of an atomic nucleus, which is still more than a million billion times larger than the Planck length. However, for a hypothetical creature the size of the Planck length, this minuscule, curled-up dimension would offer a new direction to move in, much like how a regular-sized creature navigates the circular dimensions of a tightrope. Of course, just as a regular creature finds little room to move in a tightrope's circular direction before returning to its starting point, a Planck-sized creature would continually circle back to where it began in the curled-up dimension. Still, despite the limited journey, the curled-up dimension would provide direction for movement just as the three familiar dimensions do for larger creatures.

By modifying Kaluza's original idea, Klein offered an explanation for how the universe could contain more than the three familiar spatial dimensions while keeping these extra dimensions hidden. This concept became known as Kaluza-Klein theory. Since adding just one extra spatial dimension was all Kaluza needed to unify general relativity and electromagnetism, Kaluza-Klein theory seemed to be the answer Einstein had been searching for. Excited by the prospect of unification through a hidden space dimension, Einstein and many others enthusiastically pursued this idea, launching significant efforts to explore its potential. However, it wasn't long before Kaluza-Klein theory ran into problems. One of the most significant issues was that attempts to fit the electron into this extra-dimensional framework proved impossible. Einstein continued to explore the Kaluza-Klein theory into the early 1940s, but the initial excitement waned as the theory's potential failed to materialize. Interest in it gradually faded, but within a few decades, Kaluza-Klein theory would be revived in a dramatic way—through String Theory.

17.6 The Shape of Hidden Dimensions

String theory's equations do more than just determine the number of spatial dimensions; they also define the specific shapes that the extra dimensions can take. In the previous figures, we focused on simple shapes like circles, hollow spheres, and solid balls, but string theory's equations point to much more complex shapes. These shapes are part of a six-dimensional class called *Calabi-Yau* spaces, named after mathematicians Eugenio Calabi and Shing-Tung Yau, who discovered them long before their connection to string theory was understood. It's important to remember that the figures shown represent a two-dimensional depiction of a six-dimensional object, which leads to various distortions. Nonetheless, these images provide a rough idea of what these shapes might look like.

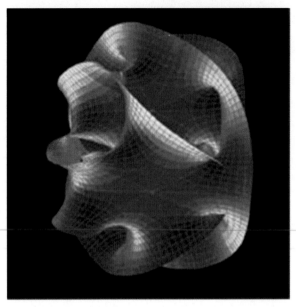

Figure 17-1: An example of a Calabi-Yau shape. A highly magnified portion of space with additional dimensions in the form of a tiny Calabi-Yau shape.

If the Calabi-Yau shape shown in Figure 9-2 represents the extra six dimensions in string theory, then on an ultra-microscopic scale, space would take the form depicted in Figure 12.b. The Calabi-Yau shape would be attached to every point in the familiar three dimensions, meaning that you, I, and everyone else would be constantly surrounded by and filled with these tiny shapes. Figure 15-2 shows: (a) an example of a Calabi-Yau shape, and (b) a highly magnified view of space with additional dimensions manifested as tiny Calabi-Yau shapes.

As you move from one location to another, your body would traverse all nine dimensions, quickly and repeatedly circling around the entire shape, making it appear as if you weren't moving through the extra six dimensions at all. If these concepts are correct, the ultra-microscopic structure of the universe is intricately woven with the most complex textures.

17.7 String Physics and Extra Dimensions

One of the remarkable aspects of general relativity is that the physics of gravity is governed by the geometry of space. With the additional spatial dimensions suggested by string theory, it's reasonable to expect that geometry's influence on physics would grow even more significant—and indeed, it does.

Why does string theory require ten spacetime dimensions? This is a challenging question to address without delving into mathematics, but I'll provide enough explanation to show how it ultimately involves a balance between geometry and physics.

Imagine a string that is limited to vibrating on the two-dimensional surface of a flat tabletop. It can produce various vibrational patterns, but these patterns are restricted to movement along the left/right and back/forth directions of the table's surface. If the string is allowed to vibrate in the third dimension, meaning it can move up and down, new vibrational patterns become possible that extend off the surface of the table. Although it's challenging to visualize in higher dimensions, the key takeaway is that more dimensions allow for more vibrational patterns. If a string can vibrate in a fourth spatial dimension, it can produce more patterns than in three dimensions; if it can vibrate in a fifth spatial dimension, it can generate even more patterns, and so on.

This is a crucial insight because string theory contains an equation that requires the number of independent vibrational patterns to satisfy a very specific constraint. If this constraint is not met, the mathematics of string theory breaks down, and its equations lose meaning.

In a universe with three spatial dimensions, the number of vibrational patterns is insufficient to meet the constraint. The same is true for four dimensions, and even for five, six, seven, or eight dimensions—the number of vibrational patterns remains too small. However, when there are nine spatial dimensions, the constraint is perfectly fulfilled. This is how string theory determines the required number of spatial dimensions.

While this example highlights the relationship between geometry and physics, their connection in string theory extends even further, offering a potential solution to a significant issue we encountered earlier. Recall that when physicists attempted to link string vibrational patterns to known particle types, they faced difficulties. They discovered there were far too many massless string vibrational patterns, and more importantly, the properties of these patterns did not align with those of the known matter and force particles. What I didn't mention earlier—since we hadn't yet discussed the concept of extra dimensions—is that although those calculations accounted for the number of extra dimensions (which explained, in part, the abundance of string vibrational patterns), they overlooked the small size and intricate shape of the extra dimensions. These calculations assumed all space dimensions were flat and fully extended, which makes a significant difference.

Strings are so small that even when the additional six dimensions are compacted into a Calabi-Yau shape, the strings still vibrate in those directions. This is important for two reasons. First, it ensures that the strings vibrate across all nine space dimensions, thereby keeping the constraint on the number of vibrational patterns intact, even when the extra dimensions are tightly curled. Second, just like the vibrations of air moving through a tuba are shaped by the instrument's twists and curves, the vibrational patterns of strings are influenced by the geometry of the extra six dimensions. For instance, altering the shape of a tuba—by narrowing a passage or elongating a chamber—would change the air's vibrational patterns and the resulting sound. Similarly, changing the size or shape of the extra dimensions would significantly affect the vibrational patterns of strings. Since a string's vibrational pattern determines its mass and charge, this means that the extra dimensions play a crucial role in shaping the properties of particles.

This is a crucial insight. The exact size and shape of the extra dimensions have a profound effect on string vibrational patterns, which in turn influence particle properties. Given that the fundamental structure of the universe—ranging from the formation of galaxies and stars to the existence of life as we know it—depends on these particle properties, the very fabric of the Cosmos might be encoded in the geometry of a Calabi-Yau shape.

We saw one example of a Calabi-Yau shape in Figure 15-2, but there are likely hundreds of thousands of other possibilities. The critical question, then, is which Calabi-Yau shape, if any, forms the extra-dimensional component of spacetime. This is one of the central challenges of string theory, as the specific choice of Calabi-Yau shape determines the detailed characteristics of string vibrational patterns. To date, this question remains unresolved. The problem arises because the current understanding of string theory's equations offers no guidance on how to select one shape over the others; from the perspective of the equations, each Calabi-Yau shape is equally valid. The equations do not even provide a means to determine the size of the extra dimensions. While we know they must be small since we cannot observe them, the precise scale remains an open question.

Is this a critical flaw in string theory? Perhaps, but I don't think so. The full equations of string theory have eluded theorists for many years, with much of the work relying on approximations. While these approximations have provided significant insights into string theory, certain questions—such as the precise size and shape of the extra dimensions—remain unsolved. Determining the form of the extra dimensions is a primary goal, one that I believe is attainable, though it remains out of reach for now.

Even so, we can still explore whether any Calabi-Yau shape produces string vibrational patterns that align with known particles, and the answer here is promising. While we haven't yet explored every possibility, examples of Calabi-Yau shapes have been found that generate string vibrational patterns roughly matching those of known particles, as illustrated in Figure **17**-1.

For example, in the mid-1980s, physicists Philip Candelas, Gary Horowitz, Andrew Strominger, and Edward Witten, who recognized the relevance of Calabi-Yau shapes for string theory, discovered that each "hole" (used in a specific mathematical sense) within a Calabi-Yau shape leads to a set of lowest-energy string vibrational patterns. A Calabi-Yau shape with three holes could therefore explain the existence of three families of elementary particles, as shown in Figure 9-2. Indeed, several three-holed Calabi-Yau shapes have been identified, and among them are those that also produce the correct number of messenger particles, electric charges, and more, as seen in Figure **17**-2.

This is an incredibly encouraging finding, one that was far from guaranteed. String theory, by unifying general relativity and quantum mechanics, could have achieved one goal while failing to explain the properties of known matter and force particles. Thankfully, this disappointing outcome has been avoided. However, calculating the precise masses of these particles remains a much more difficult challenge. As previously discussed, some particle masses deviate from the lowest-energy string vibrations by less than one part in a quadrillion. Accurately calculating such tiny deviations requires a level of precision that goes beyond our current understanding of string theory's equations.

Many string theorists believe that the tiny masses in string theory arise in a manner similar to the standard model. In the standard model, a Higgs field takes on a nonzero value throughout space, with a particle's mass determined by how much drag it experiences as it moves through this field. A similar process likely occurs in string theory, where a large network of strings vibrating in a coordinated way across space could create a uniform background resembling the Higgs field. In this scenario, string vibrations that would initially have zero mass could acquire small nonzero masses as they move through this string-theory version of the Higgs field.

In contrast to the standard model, where the drag force—and thus the mass of a particle—is determined experimentally and input into the theory, the drag force in string theory (and thus the masses of the vibrational patterns) would emerge from interactions between strings themselves. Since the Higgs field in string theory would consist of strings, the masses should, in theory, be calculable from these interactions. String theory, in principle, allows all particle properties to be derived directly from the theory.

While no one has yet achieved this, it's important to remember that string theory is still in development. With time and continued effort, researchers hope to unlock the full potential of this unified approach. The motivation is strong, as the rewards could be monumental. With a combination of hard work and some luck, string theory may eventually provide an explanation for fundamental particle properties, and ultimately, for the very nature of the universe itself.

17.8 The Fabric of the Cosmos According to String Theory

Although much of string theory remains beyond our full understanding, it has already opened up remarkable new perspectives. Most notably, by bridging the gap between general relativity and quantum mechanics, string theory has suggested that the fabric of the universe may include many more dimensions than we directly observe—dimensions that could hold the answers to some of the universe's most profound mysteries. Additionally, the theory hints that our current concepts of space and time may not apply at the sub-Planckian level, implying that what we understand as space and time may be simplified approximations of deeper, yet-to-be-discovered fundamental concepts.

In the early moments of the Universe, the aspects of spacetime that we can now only describe mathematically would have been directly observable. At that time, when the three familiar spatial dimensions were also compact, there would likely have been little to no distinction between the large and the curled-up dimensions of string theory. The current difference in their sizes would be the result of cosmological evolution, which, in ways we don't yet fully understand, selected three of the spatial dimensions as special, subjecting only them to the 14 billion years of expansion. As the observable Universe expanded, it would have shrunk into the sub-Planckian domain, which we can now recognize as the "fuzzy patch" where space and time as we know them had yet to emerge from more fundamental entities—still elusive and being explored by current research. To make further strides in understanding the primordial Universe, and to address the origins of space, time, and the direction of time, we will need to significantly refine the theoretical tools used to comprehend string theory—something that, not long ago, seemed like a distant but worthy goal. As we will soon observe, the advancement of M-theory has surpassed even the most hopeful expectations of its proponents.

17.9 The Second Superstring Revolution

Over the past thirty years, five distinct versions of string theory have emerged. While the specific names—Type I, Type IIA, Type IIB, Heterotic-O, and Heterotic-E—aren't crucial, they all share the fundamental concept of vibrating energy strands. As calculations from the 1970s and 1980s demonstrated, each of these theories requires six additional spatial dimensions. However, upon closer examination, notable differences emerge between them. For instance, Type I theory involves both closed strings (vibrating loops) and open strings, which are vibrating segments with two free ends—unlike the other versions. Additionally, the vibrational patterns of the strings and how they interact vary across the different theories.

In the early days of string theory, the most optimistic theorists hoped that the differences between the five versions would eventually narrow down to one theory, which could be tested against experimental data. However, the fact that five distinct versions existed was uncomfortable for many in the field. The idea of unification implies that scientists should arrive at a single theory that explains the entire Universe. If one framework could successfully combine quantum mechanics and general relativity, it would provide a compelling argument for its validity, even without direct experimental evidence.

After all, both quantum mechanics and general relativity already have extensive experimental support, and it seems clear that the laws of nature should be compatible. If a theory could seamlessly bridge the two well-established pillars of 20th-century physics, it would offer strong, though indirect, evidence of its correctness.

However, the existence of five distinct yet similar string theories seemed to challenge the idea of a unique solution. Even if experimental evidence were to eventually support only one of the five versions, questions would remain: why did the others exist? Were they simply mathematical curiosities, or might they have some deeper significance? Could there be even more versions waiting to be discovered?

In the late 1980s and early 1990s, as physicists were intensely focused on understanding one of these string theories, the issue of having five different versions wasn't a major concern on a day-to-day basis. It was seen as one of those long-term questions that would be tackled once the understanding of each theory was more advanced.

Then, in the spring of 1995, this issue was unexpectedly resolved. Edward Witten, with the help of string theorists like Chris Hull, Paul Townsend, and John Schwarz, discovered a hidden unity that linked all five versions together. Witten showed that these five theories were not separate at all but were simply different mathematical representations of a single, unified theory. Much like translations of a book in multiple languages appear distinct but are essentially the same story, the five string theories seemed different only because no one had yet figured out how to translate between them. Once the "dictionary" was created, it became clear that all five versions were just different perspectives on a single, underlying framework.

The master theory that unifies all five string theories has been tentatively named M-theory. The "M" in M-theory remains a placeholder, with its exact meaning—whether Master, Majestic, Mother, Magic, Mystery, or Matrix—left open to interpretation as researchers continue their global effort to further develop the vision Witten's groundbreaking insight has unveiled.

This discovery marked a significant advancement. Witten's work, along with subsequent contributions from Petr Horava, revealed that string theory is actually a single, cohesive framework. No longer would string theorists need to apologize for the fact that the theory they were proposing as a potential unified theory came in five different versions. Instead, it was fitting that the boldest unification proposal itself underwent a process of "meta-unification." Witten's breakthrough demonstrated that the unity within each individual string theory extended to the entire string theory framework.

This chapter provides an overview of the state of the five string theories both before and after Witten's discovery. It serves as a useful reference for understanding M-theory not as a completely new approach but as an advancement that clarifies and refines the existing physical laws. M-theory unifies the five string theories, showing that each one is an integral part of a larger, more complete theoretical framework.

17.10 Eleven Dimensions

With the new ability to analyze string theory more precisely, many important insights have emerged, particularly those that have significantly impacted our understanding of space and time.

A major breakthrough from Witten's work was the realization that the string theory equations used in the 1970s and 1980s, which suggested that the Universe must have nine space dimensions, were off by one. His analysis showed that, according to M-theory, the true number of dimensions is ten spatial dimensions—eleven in total, when including time. Just as Kaluza's five-dimensional universe provided a unified framework for electromagnetism and gravity, and string theory's ten dimensions unified quantum mechanics and general relativity, Witten's eleven dimensions unified all the various string theories. This additional dimension was key in revealing connections between the five string theories, much like discovering a hidden path between five seemingly separate villages when seen from above.

Witten's announcement at the 1995 international string theory conference was a revelation, shaking the foundations of the field. Researchers had been confident that the existing analyses, based on approximate equations, had settled the issue of the number of dimensions. But Witten's discovery was a game changer.

He showed that previous analyses had simplified the mathematics by assuming that an unrecognized tenth spatial dimension would be so small—far smaller than the others—that it wasn't visible in the equations being used. This led everyone to the conclusion that string theory had only nine space dimensions. However, Witten's insights within the M-theoretic framework allowed him to go beyond those approximations and uncover the overlooked dimension. As a result, the five ten-dimensional string theories were shown to be approximate descriptions of a unified, underlying eleven-dimensional theory.

One might wonder if this discovery invalidated the previous work in string theory. Fortunately, it did not. The addition of the tenth spatial dimension introduced a new feature to the theory, but if string/M-theory is correct, and if the tenth dimension is as small as previously assumed, much of the earlier work remains valid.

However, the inability of current equations to definitively determine the sizes or shapes of extra dimensions has led string theorists to explore the possibility of a tenth dimension that is not as small as originally thought. This ongoing research has strengthened the mathematical foundation of M-theory, solidifying its potential as a unifying framework.

17.11 Testing String theory

Testing string theory presents a significant challenge because strings are incredibly small. To understand this, we need to consider the physics behind the string's size. The graviton, which is the messenger particle for gravity, corresponds to one of the lowest-energy string vibrational patterns. The strength of the gravitational force it conveys is directly tied to the string's length. Since gravity is a weak force, the string must be extremely small calculations indicate it must be only about a hundred times the Planck length for its graviton vibrations to produce a gravitational force of the observed strength.

From this, we can understand that a highly energetic string isn't necessarily small. When it gains energy, it detaches from its connection to the graviton, which is a low-energy pattern. Initially, as energy increases, the string vibrates more intensely. However, once it reaches a certain energy level, something different happens: the string's length begins to increase. There is no limit to how large the string can grow. With sufficient energy, a string could even grow to macroscopic sizes. Although current technology can't achieve this, it's possible that in the extreme conditions following the Big Bang, such long strings were created. If any of these strings have survived, they might span across the sky. While unlikely, these long strings could potentially leave observable imprints in the data we gather from space, offering a chance to confirm string theory through astronomical observations.

Additionally, higher-dimensional objects known as p-branes don't need to be tiny either. Because they possess more dimensions than strings, a new possibility arises. For example, imagine a long string that could extend infinitely. Unlike a string in three-dimensional space, which is one-dimensional, a p-brane exists in higher dimensions. If we picture a long, infinitely long, one-dimensional object like a power line, we get a sense of how strings might look. A two-brane, which has two dimensions, could be visualized as an enormous movie screen, thin but extending infinitely in height and width. A three-brane, however, introduces a whole new scenario. A three-brane has three dimensions, so if it were sufficiently large, it could occupy all the three dimensions of space we are familiar with.

This idea leads to an intriguing possibility: could we, in fact, be living on a three-brane? Just as Snow White exists within a two-dimensional world (the movie screen), which itself resides within a higher-dimensional universe, it's possible that we exist within a three-dimensional brane that exists within a higher-dimensional space of string/M-theory.

Could the three-dimensional space we experience, which Newton, Leibniz, Mach, and Einstein described, actually be a specific three-dimensional entity within string/M-theory? Or, in relativistic terms, could the four-dimensional spacetime described by Minkowski and Einstein be the result of a three-brane evolving through time? In essence, could the Universe we know be a brane?

The idea that we might live on a three-brane, known as the brane-world scenario, is a fascinating and novel development in string/M-theory. It offers a fresh perspective on the theory, with profound implications. In this framework, branes behave like cosmic "Velcro," with properties that make them exceptionally sticky in certain ways.

17.12 String Theory Confronts Experiment

The idea that we might be living on a large three-brane is, of course, speculative. Likewise, within the brane-world scenario, the possibility that the extra dimensions and strings could be much larger than previously assumed remains just that: a possibility. However, these are incredibly exciting prospects. While it's possible that even if the brane-world scenario is correct, the extra dimensions and strings could still be at Planck size, there's also a chance they could be much larger—just beyond the reach of current technology. This opens up the thrilling possibility that, in the near future, string/M-theory could intersect with observable physics and become an experimental science.

How likely is this? It's hard to say. My intuition suggests it's unlikely, but that intuition comes from years of working with the conventional model of Planck-sized strings and extra dimensions. Perhaps my instincts are outdated. Fortunately, the outcome will be determined without regard to anyone's intuition. If strings or extra dimensions are indeed large, the experimental implications would be extraordinary.

Currently, several experiments aim to test the possibility of relatively large strings and extra dimensions. If strings are as large as a billionth of a billionth of a meter (10^8 meters), the particles associated with higher harmonic vibrations would not have enormous masses, as predicted by traditional theory. Instead, their masses would be only a few thousand times that of a proton, which is within the reach of the Large Hadron Collider (LHC) at CERN. If these string vibrations are excited by high-energy collisions, the accelerator's detectors would register massive energy releases. A host of never-before-seen particles would emerge, and their masses would be related to each other in patterns similar to the harmonics of a cello. This would leave a distinct signature in the data that would be impossible to miss.

Moreover, in the brane-world scenario, high-energy collisions could even produce miniature black holes. While we usually think of black holes as massive objects in space, it's been known since the early days of general relativity that a sufficiently large concentration of matter could create a tiny black hole. This doesn't happen in everyday circumstances because the necessary compression forces are beyond the capabilities of any known technology. Instead, black holes are typically formed when the immense gravitational pull of a massive star overcomes the outward pressure from nuclear fusion, causing the star to collapse. However, if gravity behaves more strongly on small scales than previously thought, it might take far less force to create microscopic black holes. Some calculations suggest that the LHC could have just enough energy to produce microscopic black holes in high-energy proton collisions. The idea of creating these black holes in a lab setting is truly remarkable.

While these microscopic black holes would be too small and short-lived to pose any danger to us—thanks to Stephen Hawking's theory that black holes evaporate via quantum processes—creating them would provide experimental confirmation of some of the most mind-boggling ideas ever proposed in physics.

17.13 Braneworld Cosmology

A central focus of current research, which is being intensely pursued by scientists globally, is to develop a cosmological understanding that integrates the new insights from string/M-theory. The motivation is obvious: cosmology tackles fundamental questions, and we have come to realize that familiar phenomena—like the arrow of time—are deeply connected to the conditions at the Universe's inception. Additionally, cosmology offers a valuable testing ground for theories. If a theory can hold up under the extreme conditions of the Universe's earliest moments, it is likely to be robust enough for broader application.

Currently, cosmology within the framework of string/M-theory is still evolving, with researchers exploring two main approaches. The first, more traditional approach suggests that, similar to how inflation provided a profound early phase in the standard Big Bang theory, string/M-theory may provide an even earlier, and potentially more significant, phase preceding inflation. The goal is for string/M-theory to clarify the "fuzzy patch"—the unresolved period marking the earliest moments of the Universe—and allow the cosmological story to then proceed according to the well-established narrative of inflationary theory.

While progress has been made on some of the specifics needed to support this vision, such as understanding why only three of the Universe's spatial dimensions underwent expansion and developing mathematical techniques for analyzing the pre-inflationary space-time realm, the major breakthroughs have yet to materialize. The underlying idea is that while inflationary cosmology describes the observable Universe shrinking to smaller and hotter states at earlier times—becoming denser and more energetic—string/M-theory addresses this singular behavior by introducing a minimal size, at which new, less singular physical quantities begin to emerge.

This reasoning is key to string/M-theory's successful unification of general relativity and quantum mechanics, and many researchers are hopeful that the same approach will soon be applied to cosmology. However, at this point, the "fuzzy patch" remains unresolved, and it is uncertain when a clear understanding will emerge.

The second approach involves the brane-world scenario and, in its most radical form, proposes a completely new cosmological framework. While it remains uncertain whether this model will withstand rigorous mathematical examination, it offers an example of how breakthroughs in fundamental theory can lead to innovative ways of revisiting established ideas. This proposal is known as the *cyclic* model.

17.14 Cyclic Cosmology

From the perspective of time, ordinary experiences present us with two types of phenomena: those that unfold in a clear sequence—beginning, middle, and end—and those that repeat cyclically, such as the changing seasons or the daily rising and setting of the Sun. However, upon closer inspection, even these cyclical phenomena have their own beginnings and endings. For instance, while the Sun has risen and set for about 5 billion years, the Sun and the solar system didn't exist at the beginning of time, and they will eventually end when the Sun transforms into a red giant and engulfs the inner planets, including Earth, around 5 billion years from now.

These observations are relatively modern scientific realizations. For ancient cultures, cyclical phenomena, endlessly repeating, were considered the core nature of time. The recurring cycles of day and night, and the changing of the seasons, influenced the rhythm of life. It's not surprising, then, that many of the earliest cosmological models imagined the Universe's evolution as a cyclical process. Rather than having a distinct beginning, middle, and end, these cosmologies viewed the world's changes over time as a series of repeated cycles, similar to the lunar phases, where after one cycle ends, another begins.

Since the advent of general relativity, various cyclic cosmological models have been proposed. One of the earliest and most well-known models was developed by Richard Tolman in the 1930s. Tolman suggested that the Universe's expansion might eventually slow, stop, and then reverse into contraction. However, instead of collapsing to a fiery end, he proposed that the Universe could experience a "bounce," where it shrinks to a small size, rebounds, and begins another expansion phase. This cyclical model avoided the issue of a single origin, as the Universe would be in a continuous loop of expansion, contraction, and rebounding, making the concept of a beginning irrelevant.

Yet, Tolman realized that even in this cyclical model, the second law of thermodynamics would prevent an infinite number of cycles. Over time, entropies would increase in each cycle, meaning each cycle would last progressively longer. The cumulative effect of increasing entropy would imply that the cycles couldn't extend infinitely into the past, and therefore, the Universe must have had a beginning.

Tolman's model relied on a spherical Universe, a concept that has been disproven by modern observations. However, a more modern version of cyclic cosmology, based on a flat Universe and incorporating string/M-theory, has recently been proposed by Paul Steinhardt and Neil Turok, with input from other theorists like Burt Ovrut, Nathan Seiberg, and Justin Khoury. This new model suggests that our Universe exists on a "three-brane" that periodically collides with another parallel three-brane. These collisions, or "bangs," trigger new cycles of cosmic evolution.

The basic idea of this model was inspired by earlier work by Horava and Witten, who found a way to connect different string theories using a unique shape for one of the seven extra dimensions in M-theory. This configuration, involving end-of-the-world branes, naturally led to the concept of parallel branes in a cosmological context. Steinhardt and Turok extended this idea to propose their cyclic model.

In their model, each three-brane represents a universe with three spatial dimensions, while the fourth dimension lies between the branes. The remaining six dimensions are compactified into a Calabi-Yau space, which supports the string vibrational patterns that give rise to known particle species. Our Universe corresponds to one of these three-branes, and a second three-brane exists parallel to it, separated by a tiny fraction in the fourth dimension. Despite their proximity, the weak gravitational force we experience prevents us from detecting the other brane or its potential inhabitants.

In the cyclic cosmology model, the two branes are tethered to each other, constantly colliding and rebounding, with each collision initiating a new cycle of cosmic evolution. The first stage involves the two branes rushing toward each other and colliding, with the energy from the collision generating high-temperature radiation and matter. Steinhardt and Turok argue that the conditions produced by this collision resemble the initial conditions in the inflationary model of the Universe.

While some debate exists over this point, the cyclic model suggests that after the collision, the Universe rebounds, expanding and cooling, with structures like stars and galaxies forming from primordial plasma. As the cycle progresses, dark energy eventually becomes the dominant force, driving accelerated expansion about 7 billion years into the cycle. This matches observations in the current Universe, with accelerated expansion continuing for roughly a trillion years, eventually leading to an almost empty and uniform Universe (Stage 4).

As the branes approach each other again, quantum fluctuations in the strings on our brane cause tiny ripples in the uniformity of the Universe (Stage 5). These ripples grow as the branes collide again, starting a new cycle with conditions similar to those observed in the inflationary model. The cyclic model and the inflationary model both rely on quantum fluctuations to generate the nonuniformities that lead to structure formation in the Universe, though the cyclic model accomplishes this over a much longer period.

In essence, the cyclic model offers a slower, more gradual approach to resolving cosmological issues like flatness and horizon problems. By stretching the Universe over long periods of accelerated expansion, it gradually smooths out imperfections and makes the Universe appear nearly uniform, much like the rapid inflationary expansion, but over a vastly longer timeframe. Thus, the cyclic model presents a fascinating alternative to the standard cosmological model, with its own unique insights and challenges.

17.15 A brief Assessment—Criticisms of the models

Both the brane-world scenario and the cyclic cosmology model, alongside the inflationary model, remain highly speculative at their current stage of development. While both the inflationary and cyclic models offer valuable cosmological frameworks, neither provides a complete theory. Due to a lack of understanding about the conditions at the Universe's earliest moments, proponents of inflationary cosmology must assume, without theoretical justification, that the necessary conditions for inflation to occur were present. If this assumption holds, the theory addresses several significant cosmological puzzles and sets time's arrow in motion. However, these successes are contingent on inflation actually taking place.

Moreover, inflationary cosmology has yet to be fully integrated into string theory, meaning it is not yet part of a consistent unification of quantum mechanics and general relativity.

The cyclic model faces its own set of challenges. Similar to Tolman's model, considerations of entropy buildup and quantum mechanics ensure that the cycles in the cyclic model cannot continue indefinitely.

Each cycle must begin at a specific point in the past, raising the same question as inflation: how did the first cycle begin? If it did, the model resolves key cosmological issues and sets time's arrow in motion from each low entropy "splat" through the following stages. However, the cyclic model, as currently proposed, lacks an explanation for why or how the Universe is in the necessary configuration to begin with. For instance, why do six dimensions compactify into a specific Calabi-Yau shape, and how does one extra dimension take the form of a spatial segment separating two three-branes? What causes the two end-of-the-world branes to align so precisely and exert the right amount of attraction to initiate the stages described? Furthermore, what occurs when the two three-branes collide in the model's "bang"?

On the latter point, there is hope that the cyclic model's "splat" avoids the singularity issue found at time zero in inflationary cosmology. In the cyclic approach, only the dimension between the branes is compressed, while the branes themselves undergo expansion during each cycle, not contraction. This suggests finite temperatures and densities on the branes. However, this conclusion is still tentative, as no one has yet been able to solve the equations to predict what would occur when the branes collide. Early analyses indicate that the "splat" faces a similar problem to the singularity at time zero in inflationary theory: the mathematics break down. Therefore, cosmology still lacks a definitive solution to its singular starting point, whether that refers to the Universe's true origin or the beginning of our current cycle.

One of the most compelling aspects of the cyclic model is how it naturally incorporates dark energy and the observed accelerated expansion. In 1998, the discovery that the Universe is expanding at an accelerating rate surprised many in the scientific community.

While the inflationary model can incorporate this by assuming the Universe contains the correct amount of dark energy, the role of accelerated expansion feels somewhat like an afterthought. In contrast, the cyclic model treats dark energy's role as intrinsic and central. The trillion-year phase of gradual accelerated expansion in this model is vital for "clearing the slate," diluting the observable Universe to near nothingness, and resetting conditions for the next cycle. From this perspective, both the inflationary and cyclic models involve accelerated expansion—early in the inflationary model and at the end of each cycle in the cyclic model—but only the latter has direct observational support. This is significant since we are just entering the trillion-year phase of accelerated expansion in the cyclic model, and this expansion has been observed. This provides a boost for the cyclic model, but it also means that if accelerated expansion is not confirmed by future observations, the inflationary model could still be valid (though the issue of missing 70 percent of the Universe's energy would need to be reconsidered), whereas the cyclic model would not survive.

Ch 18—In Search of the Multiverse

18.1 The Multiverse

In certain experiments, light behaves similarly to ripples on a pond, while in others, it acts like a stream of tiny particles, much like billiard balls. However, this does not mean that light is strictly a wave or a particle, or even a combination of both. Light, and quantum entities in general, are not something we can fully conceptualize. Depending on how we ask, they may behave like a wave in one scenario and like a particle in another. This applies not only to light but also to electrons and other quantum particles. Perhaps we are asking the wrong questions, limited by our human experience, but the questions we do ask and the answers we receive are still striking.

Arthur Eddington, a physicist, captured this dilemma back in 1929 in his book *The Nature of the Physical World*. He remarked, "No familiar concepts can be woven around the electron; something unknown is doing we don't know what."

Even after eighty years, nothing has changed. We still don't understand what electrons, or any other quantum entities, truly are, nor how they perform the behaviors they exhibit.

Instead of clinging to familiar concepts like waves and particles, it may be more productive to reframe quantum physics in an entirely new way. This new perspective could be just as effective, even though we may never fully grasp what quantum entities are or how they function. Remarkably, physicists can still make use of quantum entities by understanding how they behave under certain conditions, even without fully comprehending the underlying mechanisms—much like learning to drive a car by mastering the controls without understanding its internal workings.

For instance, quantum physics plays a crucial role in the development of computer chips, which power everything from smartphones to supercomputers used in weather predictions. It also helps explain the functioning of large molecules like DNA and RNA, which are fundamental to life. The study of quantum physics, then, is not just an abstract pursuit; it has tangible, real-world applications.

Moreover, quantum theory raises fascinating philosophical and esoteric questions. One of the most perplexing aspects is the famous thought experiment involving Schrödinger's cat, which illustrates some of the counterintuitive aspects of quantum mechanics.

The fact that quantum entities are neither purely waves nor particles, or a simple mixture of both, highlights how ill-equipped our everyday experiences are to understand the quantum realm. However, the most effective way to start making sense of the subatomic world is to examine how quantum entities can exhibit behaviors resembling waves or particles. This offers some insight into one of the most fundamental and non-intuitive characteristics of the quantum world: uncertainty.

18.2 A Quantum of Uncertainty

In quantum physics, uncertainty is a precise concept. For quantum entities, there are pairs of quantities called conjugate variables, where it is impossible to determine both values with absolute precision at the same time. The more accurately one property is measured, the less precisely the other can be known, and vice versa. This isn't due to limitations in our measuring devices, but rather a fundamental law of nature discovered by physicist Werner Heisenberg in 1927, known as Heisenberg's Uncertainty Principle. Two of the most significant conjugate pairs are position/momentum and energy/time.

Heisenberg's original example focuses on position and momentum, where momentum is related to velocity, which describes both speed and direction. Heisenberg found that the uncertainty in the position of a quantum entity, like an electron, multiplied by the uncertainty in its momentum, is always greater than a constant value: Planck's constant divided by 2π. This limit can be approached but never fully surpassed. The more accurately the position of an electron is determined, for instance, the less we can know about its momentum and vice versa.

This uncertainty is inherent to the electron (or any quantum entity) itself. An electron, in a sense, cannot "know" both its position and its momentum at the same time.

This is where the wave-particle analogy comes in, though it's important to note these are just analogies. A wave, being spread out, may travel in a certain direction at a specific speed, but it cannot be pinpointed to a single location. A particle, on the other hand, can almost be localized to a specific point if it isn't moving with a well-defined momentum. But once it begins to move, its position becomes less certain. The more a quantum entity behaves like a wave, the less certain its position, and the more it behaves like a particle, the less certain its momentum.

The typical way to represent this in quantum mechanics is through probabilities. For example, if an electron is shot from a gun toward a phosphorescent screen, its position is uncertain, and its wave function begins to spread out as it travels through space. The electron could theoretically end up anywhere, but there is a high probability it will hit the screen and create a spot of light. Upon striking the screen, the uncertainty about its position collapses drastically to the size of the spot created. This is known as the collapse of the wave function. Afterward, the wave function spreads out again from this new position.

If the electron remains free, its position becomes increasingly uncertain as time goes on. If it is captured by an atom, the uncertainty still applies, but this does not have direct implications for theories like the Multiverse.

18.3 The only mystery

The quantum world is challenging to grasp, but its essence can be captured through a simple experiment involving light or electrons passing through two holes in a barrier. Richard Feynman, a Nobel Prize-winning physicist, stated that this experiment "contains the heart of quantum mechanics" and holds "the only mystery." He further noted, "I think I can safely say that nobody understands quantum mechanics... Nobody knows how it can be like that," so you're not alone if you're struggling to comprehend it.

This experiment is often called the double-slit experiment. When light is used, two parallel slits are cut in a piece of paper or card, and light is shone through them in a darkened room. The light spreads out and forms a pattern of alternating light and dark stripes on a second piece of card. In the 19th century, this pattern was interpreted as evidence of light behaving as waves, with the light waves from the two slits interfering with each other—similar to ripples overlapping when two stones are dropped into a pond. Where the waves align, they create a bright stripe; where they are out of phase, they cancel out, producing a dark stripe. This seemed to confirm that light was a wave.

However, at the start of the 20th century, Albert Einstein demonstrated that light could also behave as particles. In the photoelectric effect, light striking a metal surface ejected electrons. The energy of the ejected electrons only took certain values, which led Einstein to conclude that light consisted of discrete packets of energy, now known as photons. For this work, rather than his theories of relativity, Einstein was awarded the Nobel Prize.

So, experiments reveal two different behaviors of light: one showing it as a wave and another showing it as a particle. A similar situation arose with the study of electrons. In the late 19th century, J.J. Thomson's experiments confirmed that electrons were particles. But in the 1920s, further experiments, including those by J.J.'s son George, showed that electrons could behave like waves. Ironically, J.J. Thomson won the Nobel Prize for demonstrating that electrons were particles, while George Thomson received the same honor for showing they were waves. This encapsulates the puzzling nature of the quantum world.

Today, the double-slit experiment has been refined to such an extent that it can be performed with individual photons or electrons. Instead of a card, a detector screen like a computer monitor records the impact of each particle. Initially, each electron hits the screen as a single spot, behaving like a particle. However, after many electrons are fired through the slits, an interference pattern forms on the screen—something typical of waves.

What's strange is that each electron seems to pass through both slits simultaneously and interfere with itself, ultimately contributing to the overall interference pattern alongside past and future electrons. Quantum entities appear to "know" about the entire experiment, both spatially (with the two slits) and temporally (as electrons influence the past and future).

Experimenters can also place detectors to observe which slit each electron passes through. When this happens, the electrons are no longer seen as passing through both slits. Instead, they behave like particles, and the interference pattern disappears, replaced by two blobs on the screen—just as particles would behave. This suggests that the electron "knows" if it's being observed, and this is true for all quantum entities, including photons.

This is why Feynman said the two-slit experiment is at the heart of quantum mechanics and why no one understands how it works. Despite our inability to fully comprehend the behavior of quantum entities, quantum mechanics provides precise equations that allow us to describe what happens. By understanding the conditions under which electrons act like waves or particles, for instance, we can create technologies like computer chips. It may seem bizarre, but it works.

18.4 Interpreting the unimaginable

In an effort to make quantum physics more understandable, people have created various interpretations that offer ways to imagine how the quantum world works. These interpretations aim to put the complex concepts of quantum theory into terms that humans can grasp. One of the earliest and most widely known interpretations is the Copenhagen Interpretation, named after the city where many of its key ideas were developed. This interpretation dominated thinking about quantum physics from the 1930s until the 1980s and continues to be taught today, though it raises as many questions as it answers.

According to the Copenhagen Interpretation, it's meaningless to ask what quantum entities like atoms and electrons are doing when they aren't being observed. We can never know with certainty the precise outcome of a quantum experiment. Instead, we can only calculate the probabilities of different outcomes. This approach is similar to rolling dice: while you can't predict exactly what total will come up, you can determine the probabilities for each possible outcome. Similarly, in quantum experiments, some results are more likely than others, and some outcomes are impossible.

In the case of dice, you know the dice exist and are behaving predictably, even if you're not looking at them. But according to the Copenhagen Interpretation, when quantum entities aren't being observed, they exist in a state where all possible outcomes are mixed together, represented by something called a wave function. This mixture, known as a superposition of states, includes all the probabilities of where the particle might be or what it might be doing. When a measurement is made, the act of measuring forces the quantum entity to "choose" one of these outcomes, and the wave function collapses to reflect the measurement. But as soon as the measurement is complete, the quantum entity dissolves back into a new superposition of states.

For instance, in the double-slit experiment, once an electron is fired through the slits, the Copenhagen Interpretation suggests that it dissolves into a superposition of waves passing through both slits. These waves interfere with each other, and when they hit the detector, the wave function collapses to a single point, and the electron behaves like a particle again—at least momentarily.

However, if the experiment is altered to observe which slit the electron goes through, the act of observation forces the wave function to collapse before the electron reaches the detector. This results in the electron behaving like a particle, with no interference pattern forming. Instead, two blobs form on the screen, each corresponding to one of the slits, as if the electron traveled through just one slit.

Heinz Pagels, in his book *The Cosmic Code*, sums up the Copenhagen Interpretation by stating that the electron does not have an independent, objective existence at a specific point in space, such as at one of the slits, unless it is observed. According to this interpretation, reality is partly created by the observer.

Erwin Schrödinger, a Nobel-winning physicist who contributed significantly to quantum theory, disliked the Copenhagen Interpretation. In fact, he once said of the quantum theory he helped develop, "I don't like it, and I wish I'd never had anything to do with it." To illustrate the absurdity of the interpretation, he devised his famous thought experiment of a "cat in a box."

The complexity of the Copenhagen Interpretation raises further concerns. Some physicists pointed out that this leads to an infinite regression. If you're the only one observing the experiment, does the wave function collapse when you look? Or do you become part of the superposition of states? If someone else asks about the outcome, does the wave function collapse then, or does that person become part of the superposition? This argument even extends to cosmology, raising the question of who or what could observe the entire Universe to make its wave function collapse into a definite state. If no one observes everything, why isn't the entire universe in a superposition of states?

If a better interpretation than the Copenhagen Interpretation existed, it would have replaced it long ago. However, no interpretation surpasses it. Other interpretations offer alternative views that are just as good at predicting quantum experiments, but they don't predict anything the Copenhagen Interpretation doesn't also predict.

These alternative interpretations still involve elements of quantum weirdness, such as time-reversed signals or instantaneous communication between quantum entities across vast distances. Choosing an interpretation comes down to personal preference, based on which aspects of quantum weirdness one is most comfortable with. One of the most significant alternative interpretations, particularly in discussions about the Multiverse, is the Many Worlds Interpretation, proposed by Hugh Everett in the 1950s. This interpretation is considered just as valid as the Copenhagen Interpretation in terms of making accurate predictions about quantum experiments.

18.5 The Branching Tree of History

According to the equations that describe Everett's Multiverse, communication between different branches, often referred to as "parallel worlds," is impossible. However, there is one intriguing exception to this rule, which, though beyond the scope of this book, is too fascinating not to mention briefly: time travel.

The concept of parallel worlds was popularized in science fiction long before it gained any credibility in the scientific community. Time travel is another idea that, like parallel worlds, appeared in fiction before being considered seriously by physicists. In many stories involving parallel universes, characters either shift "sideways in time" into an alternate universe or find themselves in a reality where history has branched off at some pivotal moment—such as an alternate timeline where the Axis powers won World War II.

This leads to an image of history as a tree with many branches, each representing different universes that diverge at various points based on different outcomes. Though this analogy works to some extent, it's not perfect. If Everett's theory is correct, there is no single "main trunk" of history; rather, the branching is far more complex. Many time-travel stories explore the idea of travelers going back in time and, either intentionally or accidentally, altering events so that they return to a present that is drastically different from the one they left.

If you combine these two concepts, you can imagine a traveler going back in time along one branch of history (in one universe), only to move forward in time along another branch (another universe). The key point here is that this doesn't mean the traveler has altered history; both versions of history always existed.

A well-known example of this is the "grandmother paradox," where a time traveler goes back and accidentally causes the death of her grandmother before her mother is born. In a single timeline scenario, this creates a paradox: if the grandmother dies, the traveler is never born, meaning she can't go back in time to cause the death of her grandmother, and so on. But in the many-worlds version, the traveler goes back and causes the death of her grandmother, which leads to the creation of a new universe branching from that moment. The traveler could then move forward in time into this new branch where she was never born, or she might return to her original timeline to find that her grandmother survived. Either way, no paradox occurs.

This makes for engaging science fiction, but the fascinating thing is that there is no physical law preventing time travel, even though creating a time machine would be an immense challenge. The equations of general relativity, our best theory of space and time, do allow for time travel. However, these equations also prevent time travel into the past before the time machine is built, which is why we don't have visitors from the future just yet.

Returning to the Multiverse, the grandmother paradox brings us back to Schrödinger's cat thought experiment. This is exactly the kind of puzzle that Everett's Many Worlds Interpretation solves.

In Schrödinger's original setup, there are only two possible outcomes: the cat is either dead or alive. According to Everett's theory, this means that two equally real worlds are created, one where the cat is dead and one where the cat is alive. While this example has just two outcomes, it's easy to imagine a more complex version where the outcomes depend on random events, such as the roll of dice. In this case, there could be dozens of different cats, each in their own compartment, with different outcomes depending on the dice roll. This would lead to a corresponding number of parallel universes. However, this number would be vast, but not infinite. Though there are many worlds, the Many Worlds Interpretation does not suggest an infinite number of them.

In this immense collection of parallel universes, the closest one to ours would be nearly identical, with universes further away becoming progressively more different, having branched off at earlier times. Universes across the Multiverse would diverge further and further, becoming increasingly distinct.

The main reason for taking the Many Worlds Interpretation seriously is that no alternative has yet been found to fully explain the entire universe in quantum terms. John Wheeler, an influential physicist, recognized this from the start. In a 1957 paper in *Reviews of Modern Physics*, he concluded with the remark: "Apart from Everett's concept of relative states, no self-consistent system of ideas is at hand to explain what one shall mean by quantizing a closed system like the universe of general relativity."

At the time, this statement didn't have as much impact as it does today. In the 1950s, our understanding of the universe was far more limited. There was still significant debate between cosmologists who believed the universe began with the Big Bang and those who thought it had existed forever in a Steady State. One of the major scientific breakthroughs of the late 20th and early 21st centuries was confirming that the universe did, in fact, originate from the Big Bang around 13.7 billion years ago and has been expanding ever since, in accordance with the theory of general relativity. As our understanding of the universe has deepened, it has become increasingly clear that the Many Worlds Interpretation is the best way to reconcile quantum physics with our understanding of space-time, which is why Everett's ideas are now seen as more credible than ever.

18.6 Three Dimensions work well

Most people never question the fact that the Universe exists in three spatial dimensions, plus one dimension of time. It's simply a given. Yet, some of the most groundbreaking scientific discoveries arise from questioning why things we take for granted are the way they are. A classic example is why an apple falls from a tree or why the Moon orbits the Earth. These kinds of questions led Isaac Newton to his profound insights into gravity.

The question of why space has three dimensions is particularly relevant to the understanding of one of the key principles that allows planets like Earth to maintain stable orbits around stars like the Sun—the inverse-square law of gravity, formulated by Newton in the 1680s.

According to the inverse-square law, the force of gravity between two objects is inversely proportional to the square of the distance between them. While Einstein's general theory of relativity further explains this relationship by describing it in terms of curved space, it doesn't invalidate Newton's law. In fact, Einstein's theory incorporates Newton's description within a broader framework. The behavior of falling apples and the Moon's orbit remained unchanged when Einstein's theory emerged.

What is particularly interesting is that the inverse-square law is the only type of law that allows stable orbits to exist. In our universe, if the Earth's orbit were to shift slightly in either direction—speeding up or slowing down—the inverse-square law would restore the Earth to its original orbit, acting as a stabilizing force. This is a result of the balance between the Earth's orbital speed and the gravitational pull it experiences from the Sun—essentially, a balance between centrifugal force and gravity. But in a universe governed by a different kind of law, such as an inverse-cube law, planetary orbits would be unstable. A slight reduction in speed or a shift closer to the Sun would increase gravitational pull, causing the planet to spiral inward, while an increase in speed or a slight outward shift would result in a weaker force, allowing the planet to drift away. Even small disturbances, like a meteorite impact, would cause the orbit to destabilize due to positive feedback. This delicate balance only works with an inverse-square law.

The general theory of relativity further explains this by showing that the dimensionality of gravitational law is always one less than the spatial dimensions. In a two-dimensional space, gravity's strength would depend on the inverse of the distance; in four-dimensional space, gravity would follow an inverse cube law, and so on. Therefore, stable planetary orbits are only possible in a three-dimensional space.

At the same time, in the early 20th century, scientists discovered that the equations of electromagnetism, developed by the Scottish physicist James Clerk Maxwell in the 19th century, also only function in a universe with three spatial dimensions and one time dimension. Gravity keeps planets in orbit, and electromagnetism holds atoms and molecules together to form the matter we're made of. In 1955, just before Hugh Everett proposed his Many Worlds Interpretation, British cosmologist Gerald Whitrow suggested that the reason we observe the Universe as having three spatial dimensions is that only in a universe with three spatial dimensions can observers—like us—exist. If life is only possible in a three-dimensional space, it's no surprise that we find ourselves in such a universe.

18.7 In an Infinite Existence, where Anything Is Possible

Five centuries ago, the Universe appeared to be a small, Earth-centered space. People believed Earth was the most significant entity in existence, sitting at the very center of the cosmos. The Sun and the five known planets (Mercury, Venus, Mars, Jupiter, and Saturn) were seen as small objects revolving around Earth, while stars were thought to be fixed points of light on a celestial sphere surrounding the Earth. Aside from the daily and seasonal cycles, this view seemed eternal and unchanging. The idea of other worlds was considered heresy, as evidenced by the execution of Giordano Bruno at the end of the 16th century for suggesting that stars were other suns with their own planets and life.

The shift began with Nicolaus Copernicus, whose groundbreaking work, *De Revolutionibus Orbium Coelestium*, was published in 1543. Though ancient philosophers had entertained the idea of the Earth orbiting the Sun, it was Copernicus who sparked a revolution in thought. His theory not only proposed that the Earth moved to orbit the Sun but also implied that the Earth was just one of many planets around the Sun.

This realization led to another upheaval: the Sun, once considered the center of the Universe, was revealed to be just an ordinary star. In 1576, English astronomer Thomas Digges, after observing the Milky Way, claimed it was made of countless stars stretching infinitely in all directions. Giordano Bruno later picked up on these ideas. In the 17th century, astronomers such as Galileo and Johannes Kepler built upon Copernicus's work, estimating star distances by assuming that stars were as bright as the Sun, but far more distant. Isaac Newton in 1728 estimated that Sirius, a star, was about a million times farther away than the Sun, a distance not far off from modern measurements. Over the course of two centuries, the once Earth-centered Universe became a small part of an expansive and potentially infinite cosmos.

The next 200 years saw advancements in telescopic technology and new tools like astronomical photography and spectroscopy. The discovery of planets beyond Saturn—Uranus and Neptune—was less significant than the advancements in measuring star distances and analyzing the composition of stars. By the 1920s, these advances expanded our understanding of the Universe's structure and history.

When Digges first used a telescope to explore the Milky Way, he saw countless stars, a discovery later independently made by Galileo. Digges believed the stars extended infinitely, a view that Durham astronomer Thomas Wright challenged in 1750. Wright proposed that the Milky Way had a finite size, like a disc, with the Sun not at its center. He also speculated that nebulae, the fuzzy light patches seen through telescopes, were outside the Milky Way—a view that was later confirmed with 20th-century observations.

From these observations, we now know that the Milky Way is a disc-shaped galaxy, containing hundreds of billions of stars, including our Sun. The galaxy spans 100,000 light-years, with the Sun located about two-thirds of the way out from the center. But in the grand scheme, the Milky Way is just one of many galaxies, all part of an enormous Universe, with no special significance attached to our position in it.

Estimates suggest there are hundreds of billions of galaxies observable by modern telescopes, though only a small fraction has been studied in depth. These galaxies form clusters spread across the Universe, with the most distant galaxies emitting light that took over ten billion years to reach Earth. However, due to the expanding nature of the Universe, these galaxies are no longer at the same distance as when the light first began its journey.

The expansion of the Universe was an accidental discovery, though it was predicted by Albert Einstein's general theory of relativity. In the late 1920s, Edwin Hubble observed that galaxies were moving away from each other, a phenomenon explained by the stretching of space itself. This cosmological redshift, which shifts light towards the red end of the spectrum, is not due to galaxies moving through space but rather because the space between galaxies is expanding. It is not a movement toward a specific center but rather a universal expansion.

This expansion is best understood through the analogy of a balloon with dots on its surface. As the balloon inflates, each dot moves away from every other dot, illustrating that no spot is at the center. The Earth is not at the center of the Universe, and no single point can be considered its center.

The idea of a cosmological redshift suggests that, in the distant past, everything in the visible Universe was compressed into a much smaller space. This leads to the conclusion that the Universe emerged from a hot, dense state about 13.7 billion years ago, an idea that has been refined over time. The precision of this age estimate has become much more accurate, with the current consensus placing the age of the Universe between 13.6 and 13.8 billion years.

This event, known as the Big Bang, marked the beginning of the Universe. Although the term was initially coined mockingly by Fred Hoyle, it stuck. The finite age of the Universe, combined with its evolving nature, places us in a specific moment in cosmic history. The Sun and Earth formed about 4.5 billion years ago, shortly after the Big Bang, when conditions allowed for the creation of planets like Earth, rich in the elements necessary for life.

Bruno's vision of an infinite Universe filled with stars and planets, many hosting life, foreshadowed modern theories of a "many worlds" hypothesis. Though he couldn't know about the finite speed of light or the Universe's age, his ideas were strikingly prescient. The concept of countless Earth-like planets and the possibility of life beyond our world remains a key part of modern cosmology.

The concept of a finite Universe, limited by the speed of light, only became fully understood in the 17th century, with discoveries like Ole Romer's measurement of light's speed. The concept of the multiverse, where other "worlds" could exist in parallel with our own, has evolved with time. In the 20th century, the term "multiverse" gained new significance in the context of quantum theory and the idea that other universes might exist, possibly with different physical laws.

The modern understanding of the multiverse was shaped by figures like Hugh Everett, who proposed the "many-worlds" interpretation of quantum mechanics. The term "multiverse" was later popularized by figures such as David Deutsch, who used it to describe the totality of reality. The notion of a multiverse is important in addressing fundamental questions about why the laws of physics are the way they are and why the Universe is capable of supporting life.

This concept challenges our understanding of the Universe, suggesting that there may be an infinite number of "worlds" or universes, each with its own set of properties. While these universes may be inaccessible to one another, their existence raises profound questions about the nature of reality. The idea of the multiverse opens new avenues of inquiry and offers a deeper understanding of our place in the cosmos.

18.8 The Quantum Cat

Quantum physics governs the behavior of matter and energy at tiny scales, specifically at the level of atoms and smaller particles. To give you an idea of how small that is, imagine lining up around ten million atoms to span the distance between two points on the jagged edge of a postage stamp.

In contrast, Newtonian physics explains how everyday objects behave—like how billiard balls roll and collide, waves ripple on a pond, or a rocket heads to Mars. But what's truly remarkable is how quantum physics differs so drastically from Newtonian physics. The difference isn't just in small details; it's a fundamental shift in how things behave. After all, billiard balls, water in a pond, and rockets are all made of atoms. So, why does the whole behave so differently from its individual components?

There's no single answer to this question, and the multiple possible answers are all valid. However, this creates an unsatisfying situation because none of these answers align with our daily experiences of the world. This is crucial to understand when studying quantum physics: it's entirely outside of our ordinary experience. Our minds can't fully grasp what quantum entities like light or electrons truly "are." Instead, we can only conduct experiments and interpret the results, often relying on analogies from the familiar world around us.

18.9 Neither wave nor particle

In certain experiments, light behaves like ripples on a pond, while in others, it acts like a stream of tiny billiard balls. However, this doesn't mean light is actually a wave or a particle, nor is it a simple combination of the two. Light is something beyond our full understanding, responding like a wave when one type of question is asked and like a particle when another is posed. The same applies to electrons and all quantum entities. Perhaps, constrained by our human perspective, we are simply asking the wrong questions. But we are limited to the questions and answers we currently have.

18.10 Worlds Colliding

Branes exist in higher dimensions, much like a two-dimensional membrane can move in three-dimensional space. Instead of being stacked like sheets of paper, branes—essentially entire universes—can move, collide, or orbit each other, much like how the Moon orbits Earth or planets orbit the Sun. A more fitting analogy than paper stacked quietly on a desk would be sheets blown around in a gust of wind.

While we can't directly perceive the extra dimensions of space, there are fields that could be influenced by the geometry of higher-dimensional space and the nearby branes. These fields might eventually be detected, and physicists have proposed that these fields could drive inflation.

Pushing two branes together requires energy, similar to how it takes energy to collide two positively charged atomic nuclei. If branes collide at high speeds, their motion could release energy, much like how nuclear energy is released when an atomic nucleus splits after being struck by a fast particle. In the late 1990s, Georgi Dvali and Henry Tye suggested that when branes collide, some of the energy could trigger inflation. However, the energy involved was too small to be effective, but this wasn't the end of the story.

A breakthrough came in 2001, when a team of researchers considered another way to extract energy from brane collisions. They speculated that far more energy could be released when a brane collides with its counterpart, or "antibrane," much like the energy released when matter and antimatter collide. This could trigger inflation, since the energy from a brane-antibrane collision would spill over into nearby branes, providing enough energy for inflation. Moreover, this process might naturally lead to the formation of universes with fewer dimensions, similar to our own. For instance, when a 7-dimensional brane and its counterpart collide, they don't convert entirely into energy at once. Instead, energy is released along with fragments like 5-branes, which eventually lead to 3-branes and then 1-branes, until everything is finally converted into energy. Larger branes quickly collide in higher-dimensional space, but smaller ones like our universe are more spread out and can remain intact for long periods before being destroyed. This could explain why three-dimensional universes like ours are relatively common.

However, this idea suggests a one-time evolution of the Multiverse, shifting from a small number of high-dimensional universes to many lower-dimensional ones. This is similar to the discomfort of the unique Big Bang theory that was once thought to perfectly produce conditions suitable for life. A more satisfying scenario would involve an ongoing process, akin to the concept of eternal inflation.

Paul Steinhardt and Neil Turok, who were not aware of Dvali and Tye's work when they met in 1999, discussed the idea of multiple brane worlds separated by a minuscule distance along the eleventh dimension. This conversation led to the idea that the Big Bang could be the result of a collision between two of these branes. In their cyclic universe model, two smooth, flat, and empty 3-branes would begin far apart in the eleventh dimension. As they move closer, drawn together by a spring-like force, they would collide, heating both branes to extreme temperatures.

Crucially, this model doesn't rely on an enormous amount of energy to trigger inflation. Instead, the temperature reaches around 10^{20} K, enough to form the particles of our world and cool into atoms and radiation. The energy involved is far smaller than what would occur at the Planck scale, avoiding singularities.

Steinhardt and Turok's model also explains how the irregularities that led to galaxies formed. Like inflation, quantum fluctuations meant that spacetime could never be perfectly flat, causing slight variations in the branes before they collided. After the branes collided, they bounced apart, and while the temperature was high everywhere, it was a little higher in some areas. Over time, the irregularities caused by these fluctuations formed the same patterns observed by the WMAP satellite, linking regions of matter across universes. The gravitational influence of branes on each other also explains dark matter—matter from adjacent universes, whose proximity is much closer than the size of an atom.

The bouncing brane model suggests a universe much like our own, born from a Big Bang. The force holding the branes together is related to dark energy, so unlike the inflationary model, dark energy is essential. As the Universe continues to expand, dark energy accelerates its expansion, spreading matter thin until it's almost entirely empty, and the Universe reaches a state of de Sitter space.

Meanwhile, the other brane world, after bouncing away, also undergoes expansion. Despite being pulled apart, the branes remain connected by the spring-like force, eventually drawing back together. This process has happened over trillions of years. As the branes collide again, a new universe or two is born from the old one. This cyclical process offers an infinite series of universes, each with potentially different constants of nature. The energy for each new Big Bang comes from gravity, and this continual rebirth of the Universe could explain its self-sustaining cycle, creating and re-creating life-friendly conditions.

18.11 The Bottomless Pit

Each time the two branes collide, some of their motion energy along the "eleventh dimension" is transformed into radiation and matter. This energy originates from the gravitational pull between the two branes, which assists the spring-like force that draws them together. Based on our everyday experiences, we might expect that the strength of each collision should decrease over time, similar to how a ball bounces less with each drop onto a hard surface. This suggests that the distance between the two branes should be reduced with each cycle. However, this view does not account for the unique nature of gravity.

Gravity is not simply a force with negative effects; it represents an infinite void. Unlike other measurable quantities like temperature, gravity has no ultimate lower limit. For temperature, there's a defined lowest point (absolute zero), which can be measured in different units, such as -273.15°C or 0 K. However, gravity doesn't have such a limit. There's no definitive "zero" for gravitational energy, so we cannot measure a drop in energy with each cycle, even in theory. Instead, only observable properties like temperature, matter density, and expansion rate during the closest approach of the branes can be detected. According to Steinhardt and Turok, their calculations show that these properties remain the same with each cycle, making the process truly cyclic.

Their model also addresses the entropy issues that arise in other bouncing universe models, which typically involve the Universe contracting before expanding again. In the Phoenix model, however, there is no contraction of three-dimensional space. The Universe never experiences a "big crunch" where matter concentrates, and the accumulated entropy prevents the bounce from being identical to the prior Big Bang. As the Universe expands after each Big Bang, entropy increases as usual, but the vast stretching of space due to dark energy creates enough room for the entropy to remain low. Even as the branes approach each other along the "eleventh" dimension, space continues to expand, avoiding the contraction phases seen in older cyclic models. By the time the branes collide, the entropy density is so low that the cycle can repeat without issue.

The infinite negative nature of gravity, along with the endless expansion of space, allows for an infinite series of Big Bangs. Rather than producing a limitless variety of universes scattered throughout space, this model suggests an infinite variety of universes emerging from each Big Bang over time, driven by the unique conditions that each new cycle presents.

18.12 There's lots of places like home

While the Phoenix model generates a universe that closely resembles one produced by inflation, the two concepts have some important distinctions, which might allow us to determine which type of universe we inhabit. The ideas are conceptually quite different. Inflation proposes that although our universe seems infinite, it is a rare bubble isolated from similar bubbles by vast expanses of inflating space. There could be infinitely many of these bubbles, each with different variations of inflationary outcomes, but they are not in proximity, and it's unlikely that any randomly selected bubble will resemble our own. Most of the Multiverse would be very different from our universe.

In contrast, the Phoenix model suggests that in each cycle, conditions everywhere are similar, with the local universe being representative of the Universe as a whole. The entire infinite Universe contains stars and galaxies arranged in much the same way as those in our neighborhood, and any creatures on other planets would see the same general cosmic features we observe. Essentially, everywhere is like home, even if not the same.

This distinction cannot be directly tested through observations, as the maximum observable region of space, even with the best telescopes, would likely appear similar to what we've already seen. However, there is a way the two models make different, observable predictions.

Inflation suggests that the extreme conditions at the Universe's birth should have created significant gravitational radiation, or ripples in space, which would have left an imprint on the cosmic microwave background (CMB) radiation. While the imprint is expected to be tiny, it could potentially be detected by instruments on the European Planck satellite, launched in 2009, which is currently analyzing the CMB. On the other hand, the Phoenix model, being less extreme, predicts that there should be no gravitational wave imprint in the background radiation.

If Planck fails to detect these waves, it wouldn't necessarily validate the Phoenix model, as the waves could still be present at a lower level. However, if Planck detects traces of gravitational waves, it will contradict the Phoenix model. Steinhardt and Turok appreciate the simplicity of their model, emphasizing that the underlying mechanism is gentle and self-regulating. The brane collisions happen at moderate speeds, the dark energy density remains low, and there is no runaway expansion like inflation. Dark energy acts as a cushion, controlling the cycles and minimizing the impact of random fluctuations, thus maintaining a regular, periodic evolution.

While their arguments are compelling, I remain unconvinced. I expect that Planck, or future missions will detect traces of gravitational waves in the early Universe. The Phoenix model has one significant flaw: why limit the explanation to just two branes? What else is happening in the 11-dimensional space while these two branes are locked in their endless cycle? The most exciting aspect of M-theory—and the most persuasive reason to consider the Multiverse seriously—is that it offers a vast array of possible worlds, not just a repetitive interaction between two branes. Leonard Susskind refers to this variety as 'the cosmic landscape,' which has become one of the most exciting and talked-about ideas in cosmology. The landscape may even accommodate cyclic universes as one possibility among many, rather than as the sole solution to the mysteries of our existence.

Ch 19— The latest discoveries of the Cosmos

19.1 The latest discoveries of the Cosmos that researchers have found

Recent Discoveries in Cosmology: Unveiling the Mysteries of the Universe

The cosmos has long held its secrets close, but recent advancements in observational technology and innovative research techniques have led to a flurry of discoveries that deepen our understanding of the universe. These findings encompass various phenomena, including primordial galaxies, the nature of dark matter and dark energy, the intricacies of black holes, and the hunt for exoplanets that could harbor life. This essay explores these latest discoveries, their implications, and how they reshape our view of cosmic evolution.

Discovery of Primordial Galaxies

One of the most significant recent discoveries is the identification of primordial galaxies, which are believed to be some of the first structures formed after the Big Bang. Research utilizing the James Webb Space Telescope (JWST) has provided unprecedented insights into these ancient systems. Observations have revealed that these galaxies, existing over 13 billion years ago, played a crucial role in the cosmic history by contributing to star formation and the assembly of larger galactic structures (Cosmology News - ScienceDaily, 2024). These primordial galaxies are typically smaller and less massive than the present-day galaxies, highlighting the evolution of cosmic structures over time.

Furthermore, studies indicate that many of these early galaxies were highly efficient at forming stars, contrary to previous assumptions that low mass would correlate with low star formation rates (Cosmology News - ScienceDaily, 2024). Discovering the properties of primordial galaxies potentially sheds light on the conditions of the early universe, informing models that describe galaxy formation and the later evolution of the cosmos.

Dark Matter and Dark Energy

The study of dark matter and dark energy continues to be a cornerstone of contemporary cosmology. Dark matter, which makes up approximately 27% of the universe, is inferred from its gravitational effects on visible matter. Recent findings related to galaxy rotation curves and gravitational lensing have provided robust evidence for its existence. These studies show that galaxies rotate at speeds that cannot be explained solely by the gravitational pull of the visible matter they contain, indicating a significant presence of dark matter in the universe.

Equally intriguing is the role of dark energy, which constitutes about 68% of the universe and drives its accelerated expansion. Recent analyses of distant supernovae and cosmic background radiation have provided insight into the behavior of dark energy. For instance, data from the Dark Energy Survey revealed unexpected results about how this mysterious force might vary over time and its implications on the fate of the universe. Understanding dark energy is not only essential for cosmology but also has implications for future cosmic events and the ultimate fate of galaxies.

Black Holes and Gravitational Waves

The realm of black holes has continued to captivate astronomers, especially with the advancements in gravitational wave detection technologies. The LIGO (Laser Interferometer Gravitational-Wave Observatory) has made significant strides in detecting gravitational waves generated by black hole mergers. Since the first detection in 2015, LIGO has recorded numerous events, confirming that black holes not only exist but can collide and merge, releasing colossal amounts of energy in the process. These gravitational waves provide a unique method to study black holes, allowing researchers to explore their properties and formation mechanisms in ways that optical telescopes cannot.

Recent observations have identified "black hole triples," where a single black hole is in the process of consuming a star while being orbited by another distant star. This discovery offers new insights into the formation and interaction of black holes in various environments. This ongoing research helps refine our understanding of black hole evolution and their role in the larger context of galaxy formation.

Exoplanet Research

The exploration of exoplanets—planets outside our solar system—has expanded dramatically with technological advancements. The transit method and radial velocity method are principal techniques used to discover and confirm the existence of exoplanets. The JWST's capabilities allow for more detailed studies of these worlds, including their atmospheres, providing crucial data for determining their habitability.

Recent discoveries include Earth-like planets found in the habitable zones of their stars, which raises exciting possibilities regarding the existence of extraterrestrial life. For example, findings from the TOI-700 system have reported multiple planets that exhibit conditions conducive to liquid water. These discoveries emphasize the diverse characteristics of exoplanets, some of which challenge our existing models of planetary formation and behavior.

Additionally, the identification of unique exoplanets, such as those with extreme atmospheric conditions or unusual compositions, provides insight into the processes that govern planetary system development. Such findings could eventually inform the search for life beyond Earth.

The Role of Space Missions

Recent space missions have significantly impacted our understanding of the cosmos. Missions such as NASA's Perseverance rover on Mars have not only searched for signs of past life but also collected samples for future return to Earth, promising rich insights that could answer fundamental questions about life's existence beyond our planet.

Missions exploring outer solar system bodies, such as the upcoming Europa Clipper, aim to investigate the icy moon of Jupiter, which is believed to harbor a subsurface ocean. These missions highlight the importance of international collaboration in space exploration, pooling resources and expertise to tackle the challenges of deep-space research.

Conclusion

The discoveries of primordial galaxies, the nature of dark matter and dark energy, and advancements in black hole and exoplanet research have reshaped our understanding of the universe. Each finding serves as a testament to the power of modern observational technology and collaborative scientific efforts, fostering a richer comprehension of the cosmos and its myriad mysteries. Future research endeavors will undoubtedly continue to unravel the complexities of the universe, prompting new questions and challenges in the field of cosmology. As humanity strives towards understanding our place in the cosmos, the pursuit of knowledge remains an enduring adventure.

19.2 Latest Discoveries of the Cosmos

The cosmos, often defined as the universe viewed as an orderly system, encompasses everything from the smallest particles to the largest galaxies. Astrophysics, as the branch of astronomy that applies the laws of physics and chemistry to understand celestial bodies and phenomena, plays a crucial role in exploring this vast universe. The study of the cosmos is not merely an academic endeavor; it aids humanity in understanding its origin and evolution and impacts technology and society in profound ways. This essay will explore some of the latest discoveries in cosmology, examining their implications and the ongoing quest to understand the universe.

19.3 Definition of Key Concepts

The term "cosmos" refers to the universe at large, incorporating all matter and energy, galaxies, stars, planets, and beyond (Cosmos Definition & Meaning - Merriam-Webster, 2024). It signifies an orderly or harmonious system, which aligns with how ancient civilizations perceived the universe (Cosmos - Definition, Meaning & Synonyms - Vocabulary.Com, 2024). Astrophysics, in contrast, is the scientific field dedicated to understanding the natural phenomena occurring in the cosmos, employing principles from physics and chemistry to investigate the behavior and interactions of cosmic entities, from the behavior of black holes to the life cycles of stars (Astrophysics Definition & Meaning - Merriam-Webster, 2024).

19.4 Importance of Studying the Cosmos

Studying the cosmos is integral to our understanding of fundamental questions regarding existence. Foremost, it provides insights into the universe's origin and evolution, revealing how galaxies formed, how stars are born and die, and the fate of cosmic structures over time (History of Cosmology - University of Oregon, n.d.). For instance, recent research has illuminated the history of our universe from the delicate aftermath of the Big Bang to the present-day clusters of galaxies, offering a narrative of cosmic evolution that stretches back 13.8 billion years (The Past Decade and the Future of Cosmology and Astrophysics, 2019).

The impact of these studies extends beyond theoretical knowledge, influencing technology and society at large. Technologies developed for space exploration, such as advanced imaging systems and data processing techniques, have found applications in various Earthly sectors, including healthcare, telecommunications, and environmental management (Earthly Innovations From Outer Space: The Cosmic Impact On ..., 2023). Hence, the pursuit of knowledge about the cosmos not only feeds scientific curiosity but also drives innovation that affects everyday life.

19.5 Overview

This chapter will discuss several landmark findings in cosmic research, including the mapping of cosmic structures, the exploration of exoplanets, and advancements in understanding dark matter and dark energy. Each section will address recent breakthroughs and their implications for both astrophysics and broader scientific endeavors. By understanding these discoveries, we gain critical insights into not only our place in the universe but also the underlying principles that govern cosmic phenomena.

19.6 Mapping Cosmic Structures

In the last decade, one of the most compelling developments in cosmology has been the ability to map the large-scale structure of the universe. Utilizing advanced telescopes and observational techniques, astronomers have created three-dimensional maps detailing the distribution of galaxies and galaxy clusters across vast distances. Projects like the Sloan Digital Sky Survey have revolutionized our understanding of these structures, showing how galaxies are interconnected through filaments of dark matter, sculpting the cosmic web that underlies the universe (The Past Decade and the Future of Cosmology and Astrophysics, 2019). This mapping has significant implications for cosmological models, helping to refine our understanding of the universe's expansion and age, alongside providing insights into the presence of dark matter.

19.7 The Search for Exoplanets

The search for exoplanets, or planets outside our solar system, has accelerated in recent years, yielding an incredible array of discoveries. The Kepler Space Telescope and more recent instruments like the Transiting Exoplanet Survey Satellite (TESS) have identified thousands of exoplanets, many of which exist within their star's habitable zone—the region where conditions may support life. Notably, the discovery of Earth-like planets in habitable zones offers tantalizing possibilities regarding extraterrestrial life and the conditions that foster it (The Past Decade and the Future of Cosmology and Astrophysics, 2019). In addition, techniques such as transit photometry and radial velocity measurements have become highly sophisticated, enabling researchers to glean information about an exoplanet's atmosphere and composition, further enriching our understanding of planetary systems beyond our own.

19.8 Understanding Dark Matter and Dark Energy

Perhaps one of the most profound areas of study in contemporary astrophysics concerns dark matter and dark energy, which together constitute about 95% of the universe's total energy density. Despite their significance, both phenomena remain enigmatic. Recent studies utilizing gravitational lensing—where massive objects bend light from objects behind them—have provided indirect evidence supporting the existence of dark matter. Similarly, observations of supernovae and the cosmic microwave background have led to the conclusion that dark energy is driving the accelerated expansion of the universe.

One of the latest discoveries in this field is the potential for dark matter to interact in ways beyond gravitational influence, suggesting that future investigations could redefine our understanding of fundamental physics and potentially lead to new physics beyond the Standard Model. Efforts to detect dark matter directly through various underground experiments and astronomical observations continue to be a priority, as researchers develop new technologies aimed at resolving these cosmic mysteries.

19.9 Summary

The exploration of the cosmos is an ongoing journey that brings with it exciting discoveries and profound implications for science and society. By venturing into the depths of space, we not only seek answers about our own origins but also develop technologies that can enhance life on Earth. From mapping the intricate web of galaxies to unveiling the secrets of dark matter, recent advancements in cosmology reflect humanity's quest for knowledge and understanding of its place in the universe. As we confront the unknown, the future of astrophysics will likely be marked by new discoveries that challenge our perceptions and inspire generations to come. In navigating these cosmic inquiries, we stand at the brink of profound discoveries that could reshape our understanding of the universe and our existence within it.

19.10 Recent Advancements in Observational Technology

The field of astronomy has witnessed significant growth and sophistication over the past few decades, particularly in the development and implementation of advanced observational technologies. These innovations not only enhance our ability to explore the cosmos but also facilitate deeper insights into fundamental questions regarding the universe's origins, composition, and ultimate fate. Recent advancements include the deployment of state-of-the-art telescopes, advanced imaging and data analysis techniques, and the utilization of artificial intelligence, culminating in transformative discoveries in cosmology. This manuscript delves into these advancements, with a particular focus on the James Webb Space Telescope (JWST) and the Extremely Large Telescope (ELT), enhanced imaging techniques, and key discoveries related to the early universe through observations of cosmic microwave background radiation.

Development of Advanced Telescopes

The design and construction of new telescopes have revolutionized our ability to observe celestial phenomena. Two of the most noteworthy advancements in this area are embodied in the James Webb Space Telescope (JWST) and the Extremely Large Telescope (ELT).

James Webb Space Telescope (JWST)

Launched on December 25, 2021, the JWST represents a monumental leap in observational capabilities. Equipped with a large primary mirror measuring 6.5 meters in diameter and several advanced instruments, the JWST is designed to observe infrared wavelengths, allowing astronomers to see through cosmic dust and capture images of distant, cold objects in the universe (James Webb Space Telescope - NASA Science, 2024). The telescope's location at the second Lagrange point (L2) provides a stable environment and minimizes interference from Earth's atmosphere, significantly improving the clarity of its observations (James Webb Space Telescope - NASA Science, 2024). The JWST aims to study a wide range of celestial phenomena, from the formation of stars and galaxies to the atmospheres of exoplanets and the conditions for potential life (James Webb Space Telescope - NASA Science, 2024). Its first images, released in July 2022, showcased the telescope's ability to reveal the previously unseen intricacies of the universe, providing deeper insights into cosmic evolution and structure (James Webb Space Telescope - NASA Science, 2024).

Extremely Large Telescope (ELT)

The ELT, currently under construction in the Atacama Desert in Chile, aims to be the largest optical/near-infrared telescope in existence, with a primary mirror measuring 39 meters in diameter. Scheduled to see its first light in 2028, the ELT is expected to revolutionize observational astronomy through its exceptional light-gathering power and angular resolution. The configuration of the ELT's segmented mirror will enable it to gather 13 times more light than current leading optical telescopes. This ambitious project includes advanced adaptive optics capabilities that will mitigate atmospheric distortions, allowing astronomers to capture images 16 times sharper than those from the Hubble Space Telescope. The ELT's capacity to unveil the hidden universe and its potential for groundbreaking discoveries in exoplanet characterization and galaxy formation make it a landmark advancement in astronomical technology.

19.11 Enhanced Imaging and Data Analysis Techniques

In addition to the development of sophisticated telescopes, advancements in imaging and data analysis techniques are reshaping the landscape of observational astronomy. Enhanced methods for processing large datasets and the integration of artificial intelligence have opened new avenues for research.

Use of Artificial Intelligence in Astronomy

Artificial intelligence (AI) technologies are being increasingly employed to analyze and interpret the vast amounts of data generated by modern telescopes. For example, AI algorithms are used to classify astronomical images, detect anomalies, and identify celestial objects with unprecedented accuracy (NCSA News Staff, 2024). Machine learning techniques, such as convolutional neural networks, have shown promise in reducing noise and enhancing image quality, significantly improving the effectiveness of observational data (astroimager, 2023). Additionally, AI facilitates automated analysis, enabling astronomers to sift through data efficiently and focus on significant discoveries.

One notable application of AI in astronomy includes the DeepDISC project, which utilizes machine learning algorithms to identify and classify astronomical objects from telescope images. This initiative demonstrates the transformative potential of AI to enhance observational astronomy by accurately distinguishing between stars and galaxies, thereby streamlining data analysis (NCSA News Staff, 2024). Moreover, AI-driven tools have the capacity to model complex physical processes, aiding in our understanding of cosmic phenomena and contributing to the rapid advancement of astronomical research (astroimager, 2023).

Improvements in Spectroscopy

Advancements in spectroscopy, the study of how light interacts with matter, play a critical role in modern astronomical observations. Sophisticated spectroscopic techniques enable scientists to extract valuable information about celestial objects, including their composition, temperature, motion, and distance. Recent developments in spectroscopy techniques, such as multiplexing, have improved the efficiency and resolution of spectral data collection, allowing for a more comprehensive understanding of cosmic phenomena.

The burgeoning field of precision spectroscopy also enables astronomers to detect faint signals from distant objects, revealing insights into elements present in the atmosphere of exoplanets and determining the physical properties of stars (Spectroscopy | Center for Astrophysics | Harvard & Smithsonian, 2023). Enhanced spectroscopic instruments, like those being developed for the ELT, will further push the boundaries of our knowledge about the cosmos, providing more detailed analyses of chemical compositions and the dynamics of various celestial bodies.

19.12 Key Discoveries in Cosmology

The advancements in observational technology discussed above have facilitated groundbreaking discoveries within the field of cosmology, particularly in understanding the early universe.

Insights into the Early Universe

Recent observations of the cosmic microwave background (CMB) radiation have transformed our understanding of the universe's infancy and provided crucial evidence for the Big Bang theory. The CMB represents the afterglow of the Big Bang, consisting of nearly uniform radiation that permeates the universe, revealing the conditions of the early cosmos approximately 380,000 years after the event (Latest Results from Cosmic Microwave Background Measurements, 2021). The remarkable uniformity of the CMB, with slight temperature fluctuations, offers critical clues to the underlying structure of the universe and the processes that shaped its evolution.

Recent studies utilizing telescopes such as the South Pole Telescope (SPT) and BICEP have focused on measuring the polarization of CMB radiation. This polarization serves as a vital tracer of the universe's early expansion, illuminating the process of cosmic inflation that took place in the first moments of the universe's existence (ZACHARY BAHAR, 2024). Understanding the CMB polarization and the correlations of its temperature fluctuations allows astronomers to infer the content, geometry, and dynamics of the cosmos, paving the way for testing various cosmological models and informing our knowledge of dark energy and dark matter (Latest Results from Cosmic Microwave Background Measurements, 2021).

Overall, advancements in observational technologies are not just enhancing our ability to view celestial objects; they are ushering in a new era of astronomical discovery. Each development, from advanced telescopes like the JWST and ELT to cutting-edge techniques in data analysis and spectroscopy, contributes to our understanding of the cosmos and uncovers the mysteries of the universe's origins. The ongoing pursuit of knowledge continues to shape our understanding of cosmology, revealing more about the universe and our place within it.

19.13 Conclusion

In summary, the recent advancements in observational technology have propelled humanity into an exciting new realm of understanding about the universe. The deployment of sophisticated telescopes like the JWST and upcoming ELT signifies a monumental leap in our capacity to explore the cosmos. Enhanced imaging and data analysis techniques, particularly the application of artificial intelligence and innovative spectroscopic methods, are transforming how we detect, analyze, and interpret celestial phenomena. These developments culminate in significant discoveries within cosmology, particularly the insights gained from observing cosmic microwave background radiation, which illuminate our understanding of the early universe. As we continue to push the boundaries of astronomical research, we can only anticipate the rich discoveries that lie ahead in our quest to comprehend the cosmos.

19.19 Discovery of Primordial Galaxies and their Cosmological Significance

The discovery of primordial galaxies has significantly expanded our understanding of the early universe and the processes that governed galaxy formation. These ancient systems are thought to be among the first structures to emerge after the Big Bang, and their study offers valuable insights into the conditions of the early cosmos. This discussion will be articulated in three parts: the nature of primordial galaxies, the role of dark matter and dark energy, and the implications of these discoveries for our understanding of the universe's evolution.

Nature of Primordial Galaxies

Primordial galaxies are the earliest galaxies formed in the universe, existing just a few hundred million years after the Big Bang. Their characteristics are inferred from recent observations, primarily based on advanced telescopes like the James Webb Space Telescope (JWST). Evidence suggests that these early galaxies were significantly smaller than contemporary galaxies and were likely composed mainly of hydrogen and helium, the primordial elements forged in the Big Bang.

Studies have revealed that many primordial galaxies possess an active star formation process. This rapid star formation likely played a critical role in how these galaxies evolved into the larger structures observed in the current universe. For instance, the analysis of very distant light from these early galaxies has provided tantalizing glimpses into the mechanisms of star creation during an epoch known as Cosmic Dawn.

Dark Matter and Dark Energy

Dark matter and dark energy are two fundamental concepts in cosmology that help explain the behavior and distribution of matter throughout the universe. Dark matter, which does not emit, absorb, or reflect light, is believed to constitute about 27% of the universe's total mass-energy content.

Evidence Supporting Dark Matter's Existence

The evidence for dark matter comes from several sources, particularly the gravitational effects it has on visible matter. One of the most notable pieces of evidence comes from observations of galaxy rotation curves, which show that stars at the outer edges of galaxies rotate at velocities that cannot be explained by the gravitational pull of visible matter alone. This discrepancy suggests that there is a significant amount of unseen mass, or dark matter, exerting a gravitational influence.

Observations from the Bullet Cluster, which consists of two colliding galaxy clusters, further bolster the existence of dark matter. In this case, gravitational lensing studies indicated that most of the mass resides in the regions where dark matter is concentrated rather than in the hot gas that collided during the event, highlighting a clear separation between visible and dark matter.

Implications of Dark Energy on Cosmic Expansion

Dark energy, making up approximately 68% of the universe's energy content, is an even more mysterious force believed to drive the accelerated expansion of the universe. This accelerative phenomenon was shocking to astronomers who expected gravity to slow cosmic expansion over time. Instead, observations revealed that distant supernovae appear dimmer than predicted, indicating they are farther away than expected, providing strong evidence for dark energy's prevalence.

The implications of dark energy are profound, as they suggest a universe that will continue to expand indefinitely, leading to a future where galaxies drift apart beyond detection—a scenario often articulated in the concept of the "Big Freeze".

19.14 Black Holes and Gravitational Waves

The study of black holes has become a focal point for astrophysical research, especially with the advent of gravitational wave astronomy.

Recent Black Hole Observations

Recent observations have confirmed the existence of black holes of varying masses, from stellar mass black holes to supermassive black holes residing at the centers of galaxies. The detection of black holes has been facilitated by programs like LIGO (Laser Interferometer Gravitational-Wave Observatory), which has captured gravitational waves from black hole mergers, validating previous theoretical models of these enigmatic celestial objects.

Notably, research suggests that black holes might not only form from collapsing stars but could also result from the merging of smaller black holes, contributing to a growing population of massive black holes in the universe. This discovery shapes our understanding of galaxy evolution, as the growth of supermassive black holes is thought to correlate with the development of their host galaxies.

LIGO's Ongoing Discoveries of Gravitational Waves

Since its inception, LIGO has transformed our understanding of the cosmos. From its first detection of gravitational waves in 2015 to the recent identification of events involving multiple black holes merging, LIGO has revealed a dynamic universe rich with information. The ability to observe these waves provides new insights into the population of black holes and their evolutionary paths.

Gravitational waves carry information about their origins, potentially shedding light on the formation processes of black holes and other extreme cosmic phenomena. Future observatories like the space-based LISA (Laser Interferometer Space Antenna) aim to broaden this field, detecting gravitational waves in different frequency bands and providing insights into events that LIGO cannot capture.

Exoplanet Research

The discovery of exoplanets—planets outside our solar system—has also received considerable attention, with astronomers developing a variety of techniques to uncover their secrets.

Techniques for Discovering Exoplanets

The two primary techniques for exoplanet discovery are the transit method and the radial velocity method.

Transit Method: This method involves monitoring the brightness of stars for periodic dips in their luminosity, which indicate a planet passing in front. The Kepler Space Telescope and TESS have successfully identified thousands of exoplanets using this method, allowing scientists to gather significant data on their sizes and orbits.

Radial Velocity Method: The radial velocity method detects the wobble of a star caused by the gravitational pull of an orbiting planet. As the star moves towards Earth, its light shifts to short wavelengths (blue), and as it moves away, it shifts to longer wavelengths (red). This technique has confirmed many exoplanets, particularly those that may be similar in size to Earth.

Notable Exoplanet Discoveries

Recent discoveries have highlighted several intriguing exoplanets, particularly those in habitable zones—regions around stars where conditions might support liquid water. The TRAPPIST-1 system, which includes seven Earth-sized planets, has garnered significant interest due to its proximity and potential for habitability. Another notable discovery is Proxima Centauri b, which resides within the habitable zone of our nearest stellar neighbor, sparking discussions about the possibilities of extraterrestrial life nearby.

Unique characteristics of discovered exoplanets contribute to the ongoing excitement in this field; planets like TOI 849 b, which have an unusual hybrid nature, challenge our understanding of planetary formation and classification.

19.15 Implications for the Search for Extraterrestrial Life

The detection of exoplanets, particularly Earth-like ones, raises essential questions about the potential for life beyond our planet. Determining the habitability of these worlds, especially those within their star's habitable zones, underscores the significance of ongoing research efforts. Technologies being developed for atmospheric analysis, such as those employed by the JWST, hope to identify biosignatures—indicators of life—in exoplanet atmospheres, further bridging the gap between astrophysics and the search for extraterrestrial organisms.

The Role of Space Missions

Recent space missions have played a crucial role in expanding our understanding of the cosmos and facilitating international cooperation in exploration efforts.

Highlights from Recent Space Missions

Mars Explorations (e.g., Perseverance Rover): Launched in July 2020, NASA's Perseverance rover aims to seek signs of ancient life on Mars and collect soil samples for future return to Earth. Its discoveries, particularly in the Jezero Crater—a site believed to have hosted an ancient lake—are poised to enhance our understanding of Martian history and habitability.

Missions to Outer Solar System Bodies (e.g., Europa Clipper): The Europa Clipper mission, scheduled to launch in 2024, aims to investigate the potential habitability of Jupiter's moon Europa, believed to harbor a subsurface ocean. By conducting detailed reconnaissance of Europa, this mission could provide insights into conditions suitable for life in our solar neighborhood.

Impact of International Collaboration in Space Exploration

International collaboration has been integral to the success of various space missions. Initiatives like the International Space Station (ISS) and joint exploratory missions between agencies like NASA, ESA, and JAXA underline the power of shared resources, knowledge, and goals in advancing our understanding of space.

As nations continue to collaborate on ambitious space explorations, the efficacy of data sharing and joint scientific endeavors is increasingly recognized, paving the way for future discoveries that extend purposefully into previously inaccessible boundaries of the cosmos.

19.16 Conclusion

In summary, the discovery of primordial galaxies, dark matter, dark energy, black holes, gravitational waves, and exoplanets encapsulates the spirit of modern astronomical exploration. The advancements in observational technologies, including the JWST and ELT, enhance our capabilities to investigate the universe and uncover its myriad mysteries. The implications of these discoveries stretch beyond theoretical knowledge, influencing technology, society, and our understanding of life itself.

Ongoing research in cosmology remains critical, as each advancement deepens our comprehension of the universe's structure and evolution. The future directions of cosmological research, including upcoming missions and unanswered questions, promise to yield further discoveries and insights that challenge our understanding and expand our horizons.

In the realm of scientific exploration, the endeavor is far more than mere observation; it is a quest stitched into the very fabric of humanity, bridging the gap between where we are and an expansive universe filled with wonders yet to be discovered.

Ch 20— Recent Advancements in Observational Technology

20.1 Recent Advancements in Observational Technology on the Universe

The study of the universe has been revolutionized by rapid advancements in observational technology. From improved ground-based telescopes to innovative space observatories, modern astronomy leverages cutting-edge tools to answer some of humanity's deepest questions. These technologies provide unprecedented insight into the structure, formation, and evolution of cosmic phenomena, allowing researchers to probe the universe across a range of wavelengths and at an unparalleled level of precision.

Ground-Based Telescope Advancements

Ground-based observatories have undergone significant improvements, particularly through adaptive optics and interferometry. Adaptive optics correct for atmospheric distortions, enabling telescopes to achieve near-space quality resolution. Instruments like the Very Large Telescope (VLT) in Chile and the upcoming Extremely Large Telescope (ELT) are pushing the boundaries of what can be observed from Earth. Interferometry, as employed by the Atacama Large Millimeter/submillimeter Array (ALMA), combines data from multiple telescopes to simulate a much larger aperture, providing highly detailed views of distant star-forming regions and protoplanetary disks.

Space-Based Observatories

Space-based telescopes bypass atmospheric limitations altogether, observing the universe in wavelengths that cannot penetrate Earth's atmosphere, such as ultraviolet, X-rays, and infrared. The James Webb Space Telescope (JWST), launched in 2021, is among the most significant recent advancements. Its highly sensitive instruments and large infrared mirror allow it to study the formation of the first galaxies, peer through dense clouds of dust to observe star formation, and analyze the atmospheres of exoplanets for potential signs of life. JWST has already provided groundbreaking images and data that deepen our understanding of cosmic evolution.

High-Resolution Imaging

The development of more sensitive imaging technologies, such as charge-coupled devices (CCDs) and advanced spectrographs, has drastically improved the quality of data collected. For example, the European Space Agency's Gaia mission has mapped the positions and motions of over a billion stars in the Milky Way with unprecedented accuracy, transforming our understanding of galactic dynamics and structure.

Radio Astronomy

Advances in radio astronomy have expanded our ability to detect and study phenomena like pulsars, black holes, and cosmic microwave background radiation. The Square Kilometre Array (SKA), currently under construction, will be the world's largest radio telescope, capable of studying the early universe and detecting faint signals from distant galaxies and extraterrestrial technologies, if they exist.

20.2 Gravitational Wave Detectors

Gravitational wave astronomy has emerged as a groundbreaking field, made possible by observatories like LIGO (Laser Interferometer Gravitational-Wave Observatory) and Virgo. These detectors measure ripples in spacetime caused by events like black hole mergers and neutron star collisions. Their discoveries provide a new way to observe the universe, complementing traditional electromagnetic observations.

20.3 Machine Learning and Big Data

Modern telescopes generate vast amounts of data that require sophisticated analysis. Machine learning algorithms and big data analytics are now essential in processing these datasets, identifying patterns, and discovering new phenomena. Projects like the Vera C. Rubin Observatory's Legacy Survey of Space and Time (LSST) rely heavily on such computational techniques to catalog billions of objects and detect transient events.

The Interplay Between Machine Learning and Big Data: Transforming Insights into Action

The advent of big data and machine learning (ML) represents a paradigm shift in how businesses, researchers, and policymakers gather, analyze, and extract insights from vast amounts of information. The merging of these two powerful domains has opened up new avenues for automating tasks, generating predictive models, and driving decision-making processes. This essay delves into the relationship between machine learning and big data, discussing their definitions, the benefits of their integration, key methodologies, and real-world applications.

Understanding Big Data

Big data refers to the massive volume of structured and unstructured data generated every day, characterized by its volume, variety, velocity, and veracity. The ability to collect and process such large datasets presents both opportunities and challenges. As organizations produce terabytes of logs, customer transactions, and sensor data, traditional data processing methods fall short, prompting the need for advanced analytical techniques capable of handling this expansion. Hence, big data is not simply about volume; it also involves the capacity to derive meaningful insights from complex datasets that can fuel strategic decision-making.

The Role of Machine Learning

Machine learning is a subset of artificial intelligence that enables systems to learn from and make predictions based on data without being programmed explicitly. The synergy between machine learning and big data lies in ML's ability to analyze large datasets proficiently and uncover patterns that would otherwise go unnoticed. Machine learning models evolve by continuously improving their performance based on training data, thus providing significant advantages in handling big data challenges.

Benefits of Integrating Machine Learning with Big Data

The integration of machine learning with big data analytics allows organizations to automate processes, enhance efficiency, and foster innovation. Specific benefits include:

Improved Decision-Making: By leveraging machine learning algorithms to analyze enormous datasets, organizations can uncover actionable insights that facilitate informed decision-making processes. For instance, predictive analytics can forecast trends, enabling organizations to adjust their strategies accordingly.

Enhanced Customer Experience: Machine learning models enable personalized customer interactions by analyzing customer data to predict preferences, behaviors, and needs. This capability enhances customer satisfaction and loyalty through tailored recommendations and targeted marketing campaigns.

Operational Efficiency: By automating routine tasks and processes through ML algorithms, organizations can reduce operational costs, minimize human error, and optimize resource allocation. For example, businesses can use machine learning to automate inventory management and predict demand fluctuations, leading to more efficient supply chain operations.

Speed of Analysis: Machine learning algorithms can process and analyze big data much faster than traditional methods, allowing for real-time insights. This speed is crucial in industries like finance and healthcare, where timely information can significantly impact outcomes.

Key Methodologies

Several methodologies are at the forefront of applying machine learning to big data analytics:

Supervised Learning: This approach involves training machine learning models on labeled datasets, enabling the system to make predictions based on input data (GeeksforGeeks, 2023). Supervised learning techniques are prevalent in applications such as fraud detection, where historical transaction data is labeled as legitimate or fraudulent.

Unsupervised Learning: Unsupervised learning algorithms work with unlabeled data, identifying hidden patterns and relationships. This technique is valuable for clustering similar items in marketing segmentation or identifying anomalies in network security (GeeksforGeeks, 2023), contributing to discoveries driven by big data's complex nature.

Reinforcement Learning: Reinforcement learning applies algorithms that learn to make decisions by interacting with an environment, receiving feedback through rewards or penalties (GeeksforGeeks, 2023). This approach is increasingly utilized in dynamic sectors such as robotics and gaming, where the need for adaptive decision-making processes is paramount.

20.4 Real-World Applications including cosmology research

The collaboration between machine learning and big data has yielded transformative applications across industries:

Healthcare: In healthcare, machine learning models analyze massive datasets, including electronic health records and genomic data, to predict patient outcomes, assist in diagnosing ailments, and personalize treatment plans. For instance, machine learning algorithms can identify patterns in patient data, leading to earlier intervention strategies and improved healthcare delivery.

Finance: The financial sector utilizes machine learning for risk assessment, fraud detection, and algorithmic trading. By analyzing large volumes of transactional data, ML models help identify suspicious activities and optimize trading strategies, enhancing profitability and security.

Retail: Retailers leverage machine learning to analyze consumer behavior, optimize inventory levels, and enhance customer experience through personalized marketing campaigns. Big data analytics enables retailers to make informed decisions regarding product placement and pricing strategies, driving sales.

Manufacturing: In manufacturing, predictive maintenance powered by machine learning algorithms helps companies anticipate equipment failures before they occur, reducing downtime and maintenance costs. By analyzing historical sensor data, companies can optimize operational workflows for greater efficiency (What's the difference between Artificial Intelligence and Machine Learning, 2021).

Summary

The convergence of machine learning and big data offers unprecedented opportunities for organizations seeking to harness the power of information for strategic advantage. By employing advanced analytical techniques, businesses can uncover hidden insights, improve decision-making, and foster innovation across sectors. As technology continues to evolve, the interplay between these two domains will likely lead to even deeper insights, shaping the future of industries and society as a whole. The journey of exploration unraveling the complexities of big data through machine learning has only just begun, and its implications are destined to be transformative for generations to come.

The critical challenge moving forward will be addressing the ethical implications of data use and ensuring that the benefits of these technologies are accessible to all without compromising individual privacy or security. The societal impact of big data and machine learning will demand continued vigilance and a commitment to ethical standards as stakeholders navigate this complex and rapidly evolving landscape.

Interdisciplinary Collaborations

Recent advancements also stem from interdisciplinary collaborations, integrating physics, engineering, and computer science. These collaborations have led to the development of instruments like the Event Horizon Telescope (EHT), which captured the first-ever image of a black hole, and the upcoming Laser Interferometer Space Antenna (LISA), designed to detect low-frequency gravitational waves.

Summary

Recent advancements in observational technology have ushered in a golden age of discovery, allowing scientists to explore the universe in unprecedented detail and across multiple dimensions. From detecting the faint signals of the early universe to analyzing the atmospheres of distant exoplanets, these innovations continue to expand our understanding of the cosmos. The future promises even greater breakthroughs as technology evolves, offering humanity an ever-clearer window into the mysteries of the universe

Ch 21—Development of advanced telescopes - James Webb Space Telescope (JWST) Extremely Large Telescope (ELT) B. Enhanced imaging and data analysis techniques

21.1 Advancements in Astronomy: A Detailed Analysis of Modern Observational Tools and Techniques

Astronomy has witnessed unprecedented progress due to the development of advanced telescopes, enhanced imaging and data analysis techniques, and breakthroughs in spectroscopy. These advancements enable scientists to study celestial phenomena with greater precision, paving the way for transformative discoveries about the universe's origins, structure, and evolution. This paper explores the contributions of advanced telescopes such as the James Webb Space Telescope (JWST) and the Extremely Large Telescope (ELT), the role of artificial intelligence in astronomy, and improvements in spectroscopy.

21.2 Development of Advanced Telescopes

James Webb Space Telescope (JWST)

The James Webb Space Telescope, launched in December 2021, represents a monumental leap in space-based observation. Developed as a successor to the Hubble Space Telescope, JWST is equipped with a 6.5-meter primary mirror and operates primarily in the infrared spectrum. This capability allows it to peer through cosmic dust clouds and observe phenomena that are billions of light-years away.

Key Features and Innovations

- **Infrared Capabilities:** JWST's infrared sensors can detect light from the first galaxies formed after the Big Bang, revealing information about the early universe.
- **Near Infrared Camera (NIRCam):** This instrument identifies exoplanets and examines their atmospheres, searching for potential biosignatures such as water, methane, and carbon dioxide.
- **Sunshield Technology:** A multi-layered sunshield protects the telescope from solar radiation, enabling it to operate at cryogenic temperatures essential for infrared observations.

Scientific Contributions

- **Galactic Formation and Evolution:** JWST has already provided insights into the structure of early galaxies, contributing to models of galactic evolution.
- **Exoplanet Exploration:** Its precision allows the study of atmospheric compositions of distant exoplanets, aiding in the search for habitable worlds.

21.3 Extremely Large Telescope (ELT)

The Extremely Large Telescope, under construction in Chile, represents the forefront of ground-based observational technology. When completed, its 39-meter primary mirror will make it the world's largest optical/near-infrared telescope.

Innovative Technologies

- **Adaptive Optics:** The ELT employs advanced adaptive optics to correct for atmospheric distortions, ensuring sharp and clear images comparable to those from space-based telescopes.
- **Modular Mirror Design:** Its segmented mirror system consists of nearly 800 individual mirrors, working together to collect unprecedented levels of light.
- **Multi-Object Spectroscopy:** The ELT can observe multiple targets simultaneously, significantly enhancing efficiency for large-scale surveys.

Expected Impact

- **Star and Planet Formation:** The ELT will provide detailed views of protoplanetary disks and nascent planetary systems.
- **Dark Matter and Dark Energy Studies:** By mapping the distribution of galaxies and measuring their redshifts, the ELT will contribute to understanding dark matter and dark energy.

Exoplanetary Atmospheres: Its sensitivity will allow direct imaging and spectroscopic analysis of Earth-like exoplanets.

Enhanced Imaging and Data Analysis Techniques: Use of artificial intelligence in astronomy Improvements in spectroscopy.

21.4 Artificial Intelligence in Astronomy

The integration of artificial intelligence (AI) in astronomy has revolutionized data analysis, enabling researchers to process and interpret vast amounts of information generated by modern telescopes.

Applications of AI

- **Pattern Recognition:** AI algorithms excel at identifying celestial objects such as galaxies, supernovae, and exoplanets from large datasets.
- **Predictive Modeling:** Machine learning tools predict stellar behaviors and simulate cosmic events based on historical data.
- **Automated Surveys:** Projects like the Vera C. Rubin Observatory rely on AI to detect transient phenomena, such as gamma-ray bursts and gravitational wave events, in real time.

Impact on Astronomy

- AI reduces human error and increases the speed of data analysis, allowing researchers to focus on interpretation and theoretical modeling.
- The discovery of unusual or rare events, such as rogue planets or atypical star systems, has been significantly enhanced by AI-driven algorithms.

21.5 Improvements in Spectroscopy

Spectroscopy, the study of light spectra, remains a cornerstone of modern astronomy. Advances in spectroscopic technology have greatly improved the resolution and accuracy of observations, enabling detailed analysis of celestial objects.

Key Innovations

- **High-Resolution Spectrographs:** Instruments like HARPS (High Accuracy Radial Velocity Planet Searcher) detect minute shifts in spectral lines caused by exoplanetary orbits.
- **Multi-Channel Detectors:** New spectrographs can simultaneously observe multiple wavelengths, providing a comprehensive view of celestial compositions and dynamics.
- **Ultraviolet and Infrared Spectroscopy:** Enhancements in non-visible spectroscopy allow astronomers to study objects obscured by dust or emitting in non-optical wavelengths.

Applications

- **Chemical Composition Analysis:** Spectroscopy reveals the elemental makeup of stars, nebulae, and galaxies, offering insights into their formation and evolution.
- **Doppler Effect Studies:** Measuring shifts in spectral lines enables the determination of object velocities, contributing to studies of cosmic expansion and orbital mechanics.
- **Atmospheric Characterization:** Spectroscopic analysis of exoplanet atmospheres helps identify potential biosignatures, advancing the search for extraterrestrial life.

Summary

The development of advanced telescopes like JWST and ELT, coupled with enhanced imaging, AI-driven data analysis, and improved spectroscopic techniques, has transformed our understanding of the universe. These innovations empower astronomers to explore phenomena from the first galaxies to exoplanetary atmospheres, bridging the gap between theoretical predictions and observational evidence. As technology continues to evolve, the potential for groundbreaking discoveries in cosmology and astrophysics remains boundless.

Ch 22— Discovery of primordial galaxies

22.1 The Discovery of Primordial Galaxies: Unveiling the Universe's Origins

The study of primordial galaxies has become a cornerstone in understanding the early universe. These ancient structures, formed less than a billion years after the Big Bang, provide crucial insights into cosmic evolution, star formation, and galaxy assembly. Recent advancements in observational technology, particularly in infrared astronomy and high-resolution imaging, have revolutionized the discovery and analysis of these distant galaxies. This paper explores the significance of primordial galaxies, the methods used to detect them, and the profound implications of their study on cosmology.

Primordial Galaxies: A Window into the Early Universe

Primordial galaxies are the first galactic structures to emerge after the Big Bang, approximately 13.7 billion years ago. These galaxies are key to understanding several fundamental processes:

- **Cosmic Reionization:** The first stars and galaxies reionized the universe's neutral hydrogen, making it transparent to light.
- **Star Formation:** Studying these galaxies reveals the initial stages of star formation and the production of heavy elements.
- **Galactic Evolution:** They provide insights into how small protogalaxies merged and evolved into the massive galaxies observed today.

Technological Advancements in Detecting Primordial Galaxies

Role of Advanced Telescopes

The discovery of primordial galaxies relies heavily on cutting-edge telescopes capable of observing faint, distant objects.

22.2 James Webb Space Telescope (JWST)

The JWST has been instrumental in uncovering primordial galaxies due to its infrared capabilities, which penetrate cosmic dust and detect the redshifted light from early galaxies. Key contributions include:

- **High-Redshift Observations:** JWST can identify galaxies with redshifts greater than 10, corresponding to ages less than 500 million years after the Big Bang.
- **Spectroscopic Analysis:** The telescope's spectrometers analyze the chemical composition and star formation rates of these galaxies.

Hubble Space Telescope (HST)

Before JWST, the HST played a pioneering role in identifying distant galaxies through deep-field imaging campaigns such as the Hubble Ultra-Deep Field (HUDF). Although limited to ultraviolet and optical wavelengths, the HST provided the first glimpses of galaxies at redshifts of 6 to 8.

Ground-Based Telescopes

Large ground-based telescopes, like the Atacama Large Millimeter/submillimeter Array (ALMA) and the Extremely Large Telescope (ELT), complement space telescopes by providing high-resolution imaging and spectroscopic data. Adaptive optics systems mitigate atmospheric distortions, enhancing observations of distant galaxies.

22.3 Techniques for Identifying Primordial Galaxies

Redshift Measurement

Redshift is the primary tool for determining the distance and age of primordial galaxies. The further the galaxy, the more its light has been stretched to longer, redder wavelengths due to cosmic expansion. Techniques include:

- **Spectroscopic Redshift:** Precise measurements using spectral lines.
- **Photometric Redshift:** Estimations based on broadband photometry across multiple filters.

Gravitational Lensing

Massive galaxy clusters act as natural lenses, magnifying the light from primordial galaxies behind them. This technique, used in conjunction with space telescopes, has uncovered some of the faintest galaxies at extreme distances.

Lyman-Break Technique

This method identifies galaxies based on the characteristic dropout of light at specific wavelengths caused by absorption from intergalactic hydrogen. It is especially effective for high-redshift galaxy surveys.

22.4 Key Discoveries and Observations

Ancient Star-Forming Galaxies

Many primordial galaxies are small but exhibit intense star formation. Observations reveal starburst activity due to the abundance of gas and a lack of heavy elements that dampen cooling processes. These galaxies help trace the early stages of stellar nucleosynthesis.

Formation of Supermassive Black Holes

Some early galaxies host supermassive black holes, which likely played a role in their evolution. Understanding the co-evolution of galaxies and black holes at these early epochs is a key area of study.

Clumpy and Irregular Structures

Unlike the organized spirals and ellipticals observed in the modern universe, primordial galaxies are often irregular and clumpy, reflecting ongoing mergers and turbulent environments

22.5 Implications for Cosmology

Understanding Reionization

Primordial galaxies are believed to be the primary sources of ionizing photons responsible for cosmic reionization. Studying their properties allows astronomers to reconstruct the timeline and mechanisms of this critical phase.

Insights into Dark Matter

The formation and clustering of primordial galaxies are influenced by dark matter halos. Observations of these galaxies help constrain the properties and distribution of dark matter in the early universe.

Evolutionary Pathways

By studying primordial galaxies at various redshifts, astronomers can piece together the history of galaxy formation, including how early structures evolved into the diverse galaxy populations seen today.

22.5 Challenges and Future Prospects

Despite significant progress, studying primordial galaxies presents challenges due to their faintness and extreme distances. Future advancements aim to overcome these limitations:

- **Next-Generation Telescopes:** Instruments like the Square Kilometre Array (SKA) and the Nancy Grace Roman Space Telescope promise deeper and more comprehensive surveys of the early universe.
- **Artificial Intelligence:** Machine learning algorithms are increasingly used to analyze large datasets, identify faint objects, and model galaxy formation processes.
- **Improved Spectroscopy:** High-resolution spectrographs will allow detailed chemical and kinematic studies of individual galaxies.

Conclusion

The discovery of primordial galaxies marks a significant achievement in astronomy, offering a glimpse into the universe's infancy. Advanced telescopes, innovative detection techniques, and improved analytical tools have made these discoveries possible, shedding light on the origins of stars, galaxies, and cosmic structure. As technology continues to evolve, future observations will further unravel the mysteries of these ancient galaxies, enriching our understanding of the cosmos.

Ch 23—Dark matter and dark energy

23.1 Dark Matter and Dark Energy: Mysteries of the Universe

The universe, as observed through astronomical data, is composed of more than just the matter and energy we can see and measure. Dark matter and dark energy, which together constitute approximately 95% of the cosmos, remain among the greatest mysteries in physics. These enigmatic components play critical roles in the formation and expansion of the universe, yet their true nature continues to elude scientists. This paper delves into the concepts of dark matter and dark energy, the evidence supporting their existence, and the ongoing efforts to uncover their properties.

Dark Matter: The Invisible Mass

23.1 Defining Dark Matter

Dark matter refers to a form of matter that does not emit, absorb, or reflect electromagnetic radiation, making it invisible to conventional observational instruments. It interacts gravitationally with visible matter, shaping the structure and dynamics of galaxies and galaxy clusters.

23.2 Evidence for Dark Matter

The existence of dark matter is supported by several lines of observational evidence:

23.2.1 Galaxy Rotation Curves

Measurements of the rotational velocities of stars and gas in galaxies reveal discrepancies between the observed speeds and the mass inferred from visible matter. The constant velocities at large distances from galactic centers imply the presence of an unseen mass halo.

23.2.2 Gravitational Lensing

Gravitational lensing, predicted by Einstein's General Theory of Relativity, occurs when massive objects bend the light of background sources. Observed lensing effects often exceed what can be attributed to visible matter, suggesting the presence of dark matter.

23.2.3 Cosmic Microwave Background (CMB)

The CMB, the afterglow of the Big Bang, carries imprints of the early universe's matter distribution. Analysis of CMB anisotropies indicates that dark matter constitutes a significant portion of the universe's mass-energy content shortly after its formation.

23.2.4 Large-Scale Structure

Simulations of cosmic structure formation, including galaxy clusters and filaments, align with observations only when dark matter is included in the models.

23.3 Candidates for Dark Matter

The precise nature of dark matter remains speculative, but several theoretical candidates have been proposed:

- **WIMPs (Weakly Interacting Massive Particles):** Hypothetical particles that interact only through gravity and the weak nuclear force.
- **Axions:** Light particles with properties predicted by certain extensions of the Standard Model of particle physics.
- **MACHOs (Massive Compact Halo Objects):** Stellar remnants like black holes, brown dwarfs, or neutron stars.

Dark Energy: The Driving Force of Cosmic Expansion

23.4.1 Defining Dark Energy

Dark energy is a hypothetical form of energy that permeates space and counteracts the attractive force of gravity, driving the accelerated expansion of the universe.

23.4.2 Evidence for Dark Energy

The concept of dark energy arises from several key observations:

23.4.2.1 Accelerated Expansion

In the late 1990s, studies of Type Ia supernovae revealed that the universe's expansion rate is increasing rather than slowing down. This unexpected finding necessitated the introduction of dark energy to explain the observations.

23.4.2.2 CMB and Large-Scale Structure

The distribution of galaxies and the CMB's power spectrum support the existence of a repulsive force influencing the universe's dynamics.

23.4.2.3 Baryon Acoustic Oscillations (BAOs)

BAOs, periodic fluctuations in the density of visible matter, act as standard rulers for measuring cosmic expansion. Their patterns indicate the influence of dark energy.

23.4.3 Theories of Dark Energy

Various theories attempt to explain the nature of dark energy:

- **Cosmological Constant (Λ):** Einstein's proposed constant represents a uniform energy density inherent in space.
- **Quintessence:** A dynamic scalar field that evolves over time, potentially varying in density and influence.
- **Modified Gravity Theories:** Alternative frameworks, such as f(R) gravity, propose modifications to General Relativity to account for accelerated expansion.

Interplay Between Dark Matter and Dark Energy

Dark matter and dark energy are distinct but interrelated phenomena. While dark matter governs the formation of structures by attracting mass, dark energy drives the large-scale dynamics of the universe by accelerating its expansion. Together, they shape the observable universe:

- **Structure Formation:** Dark matter provides the gravitational framework for galaxy and cluster formation, countering dark energy's expansion.
- **Cosmic Fate:** The ultimate destiny of the universe, whether a "Big Freeze," "Big Rip," or another scenario, depends on the balance between dark matter and dark energy.

Experimental Efforts and Future Prospects

23.5.1 Direct Detection of Dark Matter

Efforts to detect dark matter particles include:

- **Cryogenic Detectors:** Instruments like LUX-ZEPLIN (LZ) aim to observe interactions between WIMPs and ordinary matter.
- **Particle Colliders:** Experiments at CERN's Large Hadron Collider (LHC) seek to produce dark matter particles through high-energy collisions.

23.5.2 Observing Dark Energy

Dark energy studies focus on refining cosmic expansion measurements:

- **Surveys:** Projects like the Dark Energy Survey (DES) and Euclid mission map galaxy distributions to study dark energy's influence.
- **Gravitational Waves:** Observations of cosmic events may reveal indirect effects of dark energy.

23.5.3 Theoretical Developments

Advances in theoretical physics and computational modeling aim to integrate dark matter and dark energy into a unified framework, potentially revealing connections to quantum mechanics and string theory.

Implications for Cosmology and Fundamental Physics

Dark matter and dark energy challenge our understanding of the universe and suggest the need for new physics beyond the Standard Model. Their study has implications for:

- **Cosmology:** Providing insights into the universe's origin, evolution, and fate.
- **Astrophysics:** Enhancing our understanding of galaxy formation and dynamics.
- **Fundamental Physics:** Pushing the boundaries of particle physics and quantum field theories.

Conclusion

Dark matter and dark energy represent profound mysteries that lie at the heart of modern cosmology. Together, they shape the universe's structure and expansion, yet their true nature remains elusive. Advances in observational technology, experimental physics, and theoretical modeling continue to drive the quest for answers, promising a deeper understanding of the cosmos and our place within it. These discoveries may ultimately redefine our knowledge of the fundamental forces and components that govern reality.

Ch 24—Evidence supporting the existence of dark matter & Implications of dark energy on cosmic expansion

24.1 Evidence Supporting the Existence of Dark Matter and the Implications of Dark Energy on Cosmic Expansion

The universe is governed by forces and constituents that are not fully understood, yet they profoundly shape its structure and evolution. Among these mysteries are dark matter, an unseen mass that binds galaxies together, and dark energy, an enigmatic force driving the accelerated expansion of the universe. This paper explores the evidence supporting the existence of dark matter and examines the implications of dark energy on cosmic expansion, highlighting their importance in modern cosmology.

24.1.0 Evidence Supporting the Existence of Dark Matter

Dark matter constitutes about 27% of the universe's total mass-energy content, yet its direct detection remains elusive. The existence of dark matter is inferred from its gravitational effects on visible matter, radiation, and the large-scale structure of the universe. Several lines of evidence substantiate its presence:

24.1.1 Galaxy Rotation Curves

In the 1970s, Vera Rubin and Kent Ford observed that the rotational velocities of stars in spiral galaxies did not decrease with distance from the galactic center as expected if only visible matter were present. Instead, the velocities remained nearly constant, suggesting the presence of an unseen "halo" of dark matter providing additional gravitational pull.

24.1.2 Gravitational Lensing

Gravitational lensing occurs when massive objects bend the light of background sources, as predicted by Einstein's General Theory of Relativity. Observations show that lensing effects often exceed what can be accounted for by visible matter alone. For instance, studies of galaxy clusters reveal that dark matter contributes significantly to their total mass.

24.1.3 Cosmic Microwave Background (CMB)

The CMB, the afterglow of the Big Bang, provides a snapshot of the universe approximately 380,000 years after its inception. The fluctuations in the CMB's temperature and polarization patterns indicate the presence of dark matter, which influenced the early universe's structure and density distribution.

24.1.4 Large-Scale Structure

The large-scale distribution of galaxies and cosmic filaments matches simulations that include dark matter. These structures could not have formed within the universe's age without the gravitational influence of dark matter to accelerate clumping and growth.

24.1.5 Bullet Cluster

The Bullet Cluster, a collision of two galaxy clusters, provides compelling evidence for dark matter. Observations show that the distribution of mass (inferred from gravitational lensing) is offset from the hot gas observed in X-rays. This separation suggests the presence of a non-colliding component—dark matter—distinct from ordinary matter.

Implications of Dark Energy on Cosmic Expansion

Dark energy accounts for approximately 68% of the universe's total mass-energy content. It is the driving force behind the accelerated expansion of the universe, a discovery that revolutionized cosmology.

24.2.1 Discovery of Cosmic Acceleration

The late 1990s marked a turning point in cosmology when independent teams studying Type Ia supernovae found that the universe's expansion was accelerating. These supernovae, used as "standard candles" due to their consistent brightness, appeared dimmer than expected, indicating a faster-than-anticipated expansion.

24.2.2 The Role of Dark Energy

Dark energy's nature remains speculative, but its effects are evident in several phenomena:

- **Repulsive Gravity:** Unlike ordinary matter, dark energy exerts a repulsive gravitational force, counteracting the pull of gravity from matter.
- **Expansion Rate:** Observations of the Hubble parameter (which measures the universe's expansion rate) reveal its dependence on dark energy's density and properties.

24.2.3 Models of Dark Energy

Several theoretical models attempt to explain dark energy:

- **Cosmological Constant (Λ):** Proposed by Einstein, this constant represents a uniform energy density inherent in space itself. It is consistent with current observations but raises questions about fine-tuning.

- **Quintessence:** A dynamic scalar field that evolves over time, potentially varying in density and impact.
- **Modified Gravity Theories:** Alternative approaches suggest changes to General Relativity to account for the observed acceleration.

Interplay Between Dark Matter and Dark Energy

While dark matter governs the formation of structures like galaxies and clusters by providing gravitational scaffolding, dark energy influences the universe's overall dynamics. Their interplay shapes the observable cosmos:

- **Structure Growth:** Dark matter facilitates the growth of cosmic structures, but dark energy slows this growth by accelerating expansion.
- **Cosmic Fate:** The balance between dark matter and dark energy determines the universe's ultimate destiny—continued acceleration, a deceleration leading to collapse, or a steady state.

Implications for Cosmology and Physics

The study of dark matter and dark energy has profound implications for our understanding of the universe:

- **Unifying Physics:** Dark matter and dark energy challenge the Standard Model of particle physics and General Relativity, pointing toward new frameworks that incorporate quantum mechanics and gravity.
- **Cosmic Evolution:** Understanding dark energy is crucial for predicting the universe's fate, whether it will expand forever, reach equilibrium, or end in a "Big Rip."
- **Technological Advances:** Observing these phenomena drives the development of cutting-edge technologies, including advanced telescopes, detectors, and computational models.

Summary

Dark matter and dark energy remain central to our understanding of the universe, yet their true nature continues to elude scientists. The evidence supporting dark matter, from galaxy rotation curves to the Bullet Cluster, underscores its gravitational significance. Meanwhile, dark energy's role in cosmic expansion raises profound questions about the universe's fate and the fundamental laws of physics. Continued observational, experimental, and theoretical efforts promise to illuminate these cosmic enigmas, potentially reshaping our understanding of reality.

Ch 25—Black holes and gravitational waves recent black hole observations

25.1 Black Holes and Gravitational Waves: Recent Observations and Their Significance

Black holes and gravitational waves are two of the most fascinating phenomena in modern astrophysics. Black holes, regions of spacetime with gravitational forces so intense that nothing—not even light—can escape, have long been a subject of theoretical and observational research. Gravitational waves, ripples in spacetime caused by accelerating massive objects, provide a new way of observing the universe. This paper explores the nature of black holes, the discovery of gravitational waves, and the recent advancements in black hole observations that have revolutionized our understanding of the cosmos.

25.1.1 Black Holes: Fundamental Properties

Black holes are solutions to Einstein's field equations in General Relativity and are categorized based on mass and spin. They arise from the collapse of massive stars or through the merging of smaller black holes.

Types of Black Holes

- **Stellar Black Holes:** Formed by the gravitational collapse of massive stars, these have masses ranging from a few to tens of solar masses.
- **Intermediate Black Holes:** A rare class with masses between 100 and 10,000 solar masses, likely formed in dense star clusters.
- **Supermassive Black Holes:** Found at the centers of galaxies, these black holes have masses ranging from millions to billions of solar masses. Their formation mechanisms remain a topic of active research.
- **Primordial Black Holes:** Hypothetical black holes formed shortly after the Big Bang due to density fluctuations.

25.1.2 Event Horizon and Singularity

The event horizon is the boundary beyond which nothing can escape a black hole. At the core lies the singularity, where density becomes infinite, and the laws of physics as we know them break down.

Gravitational Waves: A New Window to the Universe

Predicted by Einstein in 1916, gravitational waves are distortions in spacetime caused by accelerating massive objects, such as merging black holes or neutron stars.

25.2.1 Discovery of Gravitational Waves

In 2015, the Laser Interferometer Gravitational-Wave Observatory (LIGO) detected gravitational waves for the first time, originating from the merger of two stellar-mass black holes. This discovery confirmed a major prediction of General Relativity and earned the 2017 Nobel Prize in Physics.

25.2.2 Properties of Gravitational Waves

- **Amplitude and Frequency:** Gravitational waves vary in amplitude and frequency based on the mass and velocity of the source.
- **Detection Techniques:** Interferometers like LIGO and Virgo measure tiny spacetime distortions caused by gravitational waves. Future observatories, like LISA, aim to detect lower-frequency waves from supermassive black hole mergers.

Recent Black Hole Observations

Technological advancements have enabled unprecedented observations of black holes, deepening our understanding of their nature and behavior.

25.3.1 Event Horizon Telescope (EHT)

In 2019, the EHT collaboration released the first-ever image of a black hole's event horizon in the galaxy M87. This groundbreaking observation:

- Confirmed the presence of a supermassive black hole.
- Provided direct evidence of the black hole's shadow, a ring of light formed by matter spiraling into the event horizon.

In 2022, the EHT unveiled an image of Sagittarius A* (Sgr A*), the supermassive black hole at the center of the Milky Way. This achievement marked another milestone in black hole imaging and helped refine measurements of its mass and spin.

25.3.2 Black Hole Mergers

Gravitational wave observatories have detected dozens of black hole mergers since 2015. These events reveal:

- **Massive Stellar Black Holes:** Observations of black holes with masses exceeding 30 solar masses challenge traditional models of stellar evolution.
- **Spin Alignments:** Analyzing the spins of merging black holes provides insights into their formation, such as whether they formed in isolation or dense stellar environments.

Observations by LIGO and Virgo in 2020 revealed the first-ever intermediate-mass black hole formed from the merger of two smaller black holes. This detection provides evidence for the existence of this elusive class of black holes.

25.3.4 Accretion Disks and Jet Formation

Observations with X-ray and radio telescopes have studied accretion disks and relativistic jets, the energetic outflows produced by black holes. These studies explore:

- The mechanisms driving jet formation.
- The role of magnetic fields near the event horizon.

Implications of Recent Discoveries

The study of black holes and gravitational waves has significant implications for astrophysics and fundamental physics.

25.4.1 Testing General Relativity

Black hole and gravitational wave observations test General Relativity in extreme conditions, such as strong gravitational fields and high velocities. So far, these studies have validated Einstein's theory.

25.4.2 Understanding Cosmic Evolution

Supermassive black holes influence galaxy formation and evolution through feedback processes like jet outflows and energy injection into the interstellar medium. Observations provide insights into how galaxies grow and evolve.

25.4.3 Dark Matter and Quantum Gravity

- **Dark Matter Candidates:** Primordial black holes are potential dark matter candidates.
- **Quantum Gravity:** Studying black hole singularities and event horizons could bridge the gap between General Relativity and quantum mechanics.

Future Prospects

The next generation of observatories promises even more groundbreaking discoveries:

- **James Webb Space Telescope (JWST):** Will study the environments of black holes and their impact on surrounding galaxies.
- **LISA:** A space-based gravitational wave detector will observe supermassive black hole mergers and cosmic gravitational wave backgrounds.
- **Einstein Telescope:** A proposed ground-based detector will increase sensitivity to gravitational waves from a wider range of sources.

Summary

Black holes and gravitational waves are at the forefront of modern astrophysics, revealing the universe's most extreme phenomena. Recent observations, from imaging black hole shadows to detecting gravitational waves, have validated theoretical predictions and opened new research avenues. As technology advances, these enigmatic objects will continue to provide profound insights into the universe's origins, structure, and ultimate fate.

Ch 26—LIGO's ongoing discoveries of gravitational waves and Exoplanet research

26.1 Techniques for discovering exoplanets 1. Transit method 2. Radial velocity method B. Notable exoplanet discoveries Earth-like planets in habitable zones Unique characteristics of other discovered exoplanets

LIGO's Ongoing Discoveries of Gravitational Waves and Advancements in Exoplanet Research

The study of the universe has entered a golden era of discovery, fueled by groundbreaking technologies and methodologies. The Laser Interferometer Gravitational-Wave Observatory (LIGO) has revolutionized our understanding of the cosmos by directly detecting gravitational waves, ripples in spacetime caused by massive cosmic events. Concurrently, advances in exoplanet research have unveiled thousands of planets beyond our solar system, some of which may harbor the conditions necessary for life. This paper explores LIGO's ongoing contributions to astrophysics, techniques used to discover exoplanets, and notable exoplanet discoveries that reshape our understanding of planetary systems.

LIGO's Ongoing Discoveries of Gravitational Waves

LIGO, along with its European counterpart Virgo, continues to provide a new lens for observing the universe. Since its first detection of gravitational waves in 2015, LIGO has revealed numerous astrophysical phenomena.

Detection of Compact Binary Mergers

Gravitational waves are primarily detected from merging black holes and neutron stars.

- **Black Hole Mergers:** LIGO has identified dozens of binary black hole collisions, including some involving unusually massive black holes, challenging existing models of stellar evolution.
- **Neutron Star Mergers:** In 2017, LIGO detected gravitational waves from the merger of two neutron stars (GW170817). This event was accompanied by electromagnetic radiation, marking the first multi-messenger astronomy observation.

Unique Discoveries

- **Intermediate-Mass Black Holes:** Observations in 2020 provided evidence of an intermediate-mass black hole formed from the merger of two smaller black holes.

- **Asymmetric Mergers:** LIGO has detected mergers with significant mass imbalances, providing insights into binary formation scenarios.
- **Stellar Population Insights:** Gravitational wave detections contribute to understanding stellar populations and the environments in which compact objects form.

Future Discoveries

As LIGO's sensitivity improves, it is expected to detect new types of gravitational wave sources, such as:

- **Continuous Waves:** From isolated spinning neutron stars.
- **Stochastic Background:** Gravitational waves from the early universe.

Exoplanet Research

The discovery of exoplanets has transformed our view of the cosmos, revealing a vast diversity of planetary systems.

26.1 Techniques for Discovering Exoplanets

The transit method detects periodic dips in a star's brightness as a planet crosses in front of it.

- **Strengths:** Highly effective for identifying planets in large surveys, especially those orbiting close to their host stars.
- **Notable Missions:** NASA's Kepler and TESS (Transiting Exoplanet Survey Satellite) have identified thousands of exoplanets using this technique.

26.1.2 Radial Velocity Method

The radial velocity method measures shift in a star's spectral lines due to the gravitational pull of an orbiting planet.

- **Strengths:** Sensitive to planets with large masses and those farther from their host stars.
- **Notable Instruments:** Ground-based observatories like HARPS (High Accuracy Radial Velocity Planet Searcher) have been instrumental in confirming exoplanet candidates.

26.2 Emerging Techniques

- **Direct Imaging:** Advances in adaptive optics allow astronomers to directly image exoplanets by blocking out starlight.
- **Gravitational Microlensing:** Detects planets by observing the bending of light from a background star as a planet passes in front of it.

26.3 Notable Exoplanet Discoveries

26.3.1 Earth-like Planets in Habitable Zones

- **Kepler-452b:** A "super-Earth" located in its habitable star zone, with conditions potentially suitable for liquid water.
- **TRAPPIST-1 System:** Comprising seven Earth-sized planets, three of which lie within the habitable zone, making it one of the most intriguing systems for astrobiology.
- **Proxima Centauri b:** An Earth-like planet orbiting the nearest star to the Sun, Proxima Centauri, in the habitable zone.

26.3.2 Unique Characteristics of Other Discovered Exoplanets

- **Hot Jupiters:** Gas giants like 51 Pegasi b orbit very close to their stars, with extreme surface temperatures.
- **Rogue Planets:** Planets like OGLE-2016-BLG-1928 that float freely in space, unbound to any star.
- **Ultra-Dense Planets:** Some exoplanets, such as Kepler-10b, have densities far exceeding that of Earth, indicating unusual compositions.

Implications of LIGO and Exoplanet Discoveries

26.4 Expanding Cosmological Frontiers

Gravitational wave astronomy and exoplanet research complement traditional observational methods, providing a more comprehensive understanding of the universe. While LIGO uncovers cosmic-scale phenomena like black hole mergers, exoplanet studies explore the microcosm of planetary systems.

26.4.1 Search for Extraterrestrial Life

The discovery of potentially habitable exoplanets fuels the search for extraterrestrial life. Advanced telescopes like the James Webb Space Telescope (JWST) are now capable of analyzing the atmospheres of these planets for biosignatures, such as water vapor, methane, and oxygen.

26.4.2 Technological Advancements

Both fields benefit from rapid technological progress:

- **Machine Learning:** Improves data analysis in detecting gravitational waves and identifying exoplanets.
- **Interferometry and Adaptive Optics:** Enhance observational precision, enabling discoveries at unprecedented scales.

Summary

The ongoing work of LIGO in detecting gravitational waves and advancements in exoplanet research underscores the dynamic nature of modern astrophysics. From unveiling the violent mergers of black holes to discovering planets that may resemble Earth, these fields continue to challenge and expand our understanding of the cosmos. As technology advances, the synergy between gravitational wave astronomy and exoplanet studies promises even more profound insights into the universe's origins, structure, and the potential for life beyond our solar system.

Epilogue

Gravitational waves and Exoplanet Research seems to be a natural ending point in our long discussion of the Cosmos, and that perhaps is a sign that the unification of quantum theory and gravity may be on the horizon. But there are no certain endings yet, and many hope there never will be. In the last few decades, physics has come a long way. Quantum field theories such as *QCD* and the electroweak theory, once considered esoteric and treated with suspicion, have now become the Standard Model, routinely tested and confirmed by precision experiments to many decimal places. But precision experiments, both on the ground and in the sky, have also raised questions.

As regards to *"pure"* particle physics, the discovery that neutrinos are not quite massless has taught us that there must be physics beyond the Standard Model and many theories have been put forward in an attempt to describe it. The evidence that will hopefully confirm these theories may be dramatic such as the possible discovery of super-symmetric particles or large extra dimensions at the *LHC*. Either way, the next few years promise to be exciting ones, both theoretically and experimentally.

As regards to cosmology, things are perhaps even more exciting. Experiments have entered a new era of precision and the puzzles they raise are truly profound. What is the real nature of the dark matter and energy which make up 96 per cent of the Universe? What caused the fantastic inflation that we believe occurred at the beginning of time? Is there a real possibility of parallel universes? Will grand unification, super-symmetry or string theory answer any or all of these questions? Nobody knows what the future may bring, but the only real surprise would be if there were no surprises.

If the business of physics is ever finished, the world will be a much less interesting place in which to live, which is why many physicists and other scientists are happy to leave us with loose ends, and with the prospect of more interesting stories yet to be told.

Physics Formulas – with explanations

1. **Explain in simple terms what the following formula in physics means: $P^2 = a^3$**

The formula $P^2 = a^3$ comes from Kepler's Third Law of Planetary Motion and describes the relationship between a planet's orbital period and the size of its orbit around the Sun.

- **P** represents the **orbital period** of a planet, or the time it takes for the planet to complete one orbit around the Sun. It is measured in **Earth years**.
- **a** represents the **semi-major axis** of the planet's orbit, or the average distance from the planet to the Sun. This is measured in **astronomical units (AU)**, where 1 AU is the distance from the Earth to the Sun.

The formula $P^2 = a^3$ means that the square of the planet's orbital period (P^2) is equal to the cube of its average distance from the Sun (a^3). In simple terms, it shows that planets farther from the Sun take longer to complete an orbit, and the relationship between time and distance follows this mathematical pattern.

For example:

- If a planet is twice as far from the Sun as Earth, a=2a, then its orbital period **P** would be around 2.83 Earth years because = approx. 2.838≈**2.83**.

2. **Explain in simple terms what the following formula in physics means: $E = mc^2$**

The formula $E = mc^2$ is one of the most famous equations in physics, discovered by Albert Einstein as part of his theory of relativity. It explains the relationship between **energy (E)** and **mass (m)**. Here's what it means in simple terms:

- **E** represents **energy**, which is the ability to do work or cause change.
- **m** represents **mass**, which is the amount of matter or stuff in an object.
- **c** is the **speed of light** in a vacuum, which is about 300,000 kilometers per second (or 3×10^8 meters per second). It's a very large number.

The equation tells us that **mass and energy are equivalent**—they can be converted into each other. A small amount of mass can be turned into a huge amount of energy because the speed of light (**c**) is such a large number, and when you square it c^2 it becomes even bigger.

In simple terms, this equation shows that matter can be transformed into energy, and that a tiny amount of mass can release an enormous amount of energy. This principle is what makes nuclear reactions, like those in the Sun or nuclear power plants, so powerful. For example, when atoms split (in nuclear fission) or fuse together (in nuclear fusion), a tiny bit of their mass is converted into a lot of energy.

3. **Explain in simple terms what the following formula in physics means: G/m = g or G = mg**

The formula **G=mg** is related to the **force of gravity** acting on an object near the Earth's surface. Let's break it down:

- **G** represents the **gravitational force** or the weight of the object. It's the force with which gravity pulls an object toward the Earth.
- **m** represents the **mass** of the object (the amount of matter in the object).
- **g** represents the **acceleration due to gravity**, which on Earth is approximately **9.8 m/s²**

This formula tells us that the gravitational force (or weight) acting on an object is equal to the mass of the object multiplied by the acceleration due to gravity.

In simple terms:

- The **heavier** an object is (the greater the mass), the **greater** the gravitational force pulling it down.
- The acceleration due to gravity, **g**, is a constant near Earth's surface, so you just multiply the object's mass by **g** to get its weight.

For example, if an object has a mass of 10 kg, its weight (gravitational force) on Earth would be G=10 kg×**9.8 m/s²** =98 N (newton) is the unit of force.

4. **Explain in simple terms what the following formula in physics means: mb = G = mg**

The formula **mb = G = mg** can be broken down as follows:

1. **G** represents the **gravitational force** (or weight) acting on an object.
2. **m** represents the **mass** of the object.
3. **g** is the **acceleration due to gravity** (approximately **9.8 m/s²** on Earth).
4. **mb** may be an unfamiliar term. If you meant **mb** (perhaps for "body mass" or something specific), it's likely another way to refer to the object's mass, like **m**.

Putting it all together:

- The **gravitational force G** acting on an object with mass mmm is equal to the product of the object's mass and the acceleration due to **gravity, mg**.
- If **mb** also represents mass, then **mb = G** could be another way to express that the **gravitational force G** is related to an object's mass times gravity.

In essence, it's reiterating the idea that the gravitational force (or weight) on an object depends on its mass and the gravitational pull of Earth (or any other planet).

5. **Explain in simple terms what the following formula in physics means: b = v² /r**

The formula **b = v² /r**, describes the **centripetal acceleration** (denoted by b here) of an object moving in a circular path. Here's a simple breakdown:

- **b** represents the **centripetal acceleration**, which is the acceleration that keeps an object moving in a circle, always directed towards the center of the circle.
- **v** is the **velocity** or speed of the object as it moves around the circle.
- **r** is the **radius** of the circular path.

In simple terms:

- This formula tells us that an object moving in a circle at speed **v** experiences centripetal acceleration **b**, which depends on its speed squared divided by the radius of the circle.
- The faster the object moves (**v** gets larger), the greater the centripetal acceleration.
- A smaller **radius r** (a tighter circle) also increases the centripetal acceleration.

This is the acceleration required to keep the object in circular motion, and without it, the object would move in a straight line instead of following the circle.

6. **Explain in simple terms what the following formula in physics means: $r^3/T^2 = C$ or $r/T^2 = C/r^2$.**

The formulas you're asking about, $r^3/T^2 = C$ or $r/T^2 = C/r^2$ are related to **Kepler's Third Law of Planetary Motion**, which describes the (body) and its orbital period. Let's break it down:

1. **r** is the **average distance** from the planet to the Sun (or another central object). This distance is the semi-major axis of the planet's elliptical orbit.
2. **T** is the **orbital period**, the time it takes for the planet to complete one orbit around the Sun.
3. **C** is a constant that depends on the mass of the central object (like the Sun).

Formula: $r^3/T^2 = C$ or $r/T^2 = C/r^2$

- This tells us that for any object orbiting a central body, the **cube of the distance** (how far it is from the central body) divided by the **square of the orbital period** (how long it takes to go around) is always the same constant, **C**.
- This constant **C** is the same for all planets orbiting the same central body, meaning that planets farther away from the Sun have longer orbital periods, but there is a predictable relationship between their distance and time.

Formula: $r^3/T^2 = C$ or $r/T^2 = C/r^2$. This is a rearranged version of the first formula, but it essentially conveys the same idea: the orbital properties of a planet (distance and time) are related in a specific mathematical way.

- It emphasizes how the ratio of the distance to the square of the orbital period changes with the distance to the central body.

In Simple Terms:

- These formulas describe how the farther a planet is from the Sun, the longer it takes to complete one orbit, and the relationship between distance and time is very specific and predictable.
- It's a mathematical way to explain why planets like Jupiter (which are far from the Sun) take much longer to orbit the Sun compared to planets like Earth or Venus, which are closer.

7. **Explain in simple terms what the following formula in physics means: b = 4(pi)2 C/r^2.**

The formula **b = 4(pi)2** appears to describe a relationship in orbital mechanics or circular motion, where **b** could represent a force or acceleration related to the orbit of a body around a central object, such as a planet orbiting the Sun. Let's break it down:

Components:

1. **4(pi)2**: This is a constant, which comes from the geometry of circular motion (since $2\pi r^2$ is the circumference of a circle).
2. **C**: This could represent a **constant** related to the system, often depending on properties like the gravitational force between two bodies. In many orbital equations, **C** might be tied to factors like the gravitational constant and the mass of the central body.
3. **R^2**: This represents the **square of the distance** from the object to the central body (like the distance from a planet to the Sun in orbital motion).

Formula Explanation:

This formula suggests that **b**, which could be an acceleration or force, is inversely proportional to r^2, meaning that as the distance **r** increases, **b** decreases. This is common in gravitational and centripetal force formulas, where the force or acceleration decreases as the distance from the central body increases.

In Simple Terms:

- This formula describes how some quantity **b** (likely a force or acceleration) behaves in a system involving circular or orbital motion.
- The farther an object is from the central point (as **r** increases), the smaller **b** becomes, following the inverse-square law, which is a common feature in forces like **gravity** or **centripetal acceleration**.

For example, if this relates to gravity, it could describe how gravitational force weakens as you move farther from the source of gravity.

8. **Explain in simple terms what the following formula in physics means: K = k Mm/r^2,**

The formula **K = k Mm/r^2**, represents **Newton's law of universal gravitation**, which describes the gravitational force between two masses. Let's break it down:

Components:

1. **K**: This represents the **gravitational force** between two objects.
2. **k**: This is the **gravitational constant (G)** = **(6.67430 * 10^{-11} * m³ kg^{-1} * s^{-2})** which is a constant value that makes the units work out correctly and describes the strength of gravity in our universe.
3. **M**: This is the **mass** of the first object, such as a planet or a star.
4. **m**: This is the **mass** of the second object, such as a moon, satellite, or another planet.
5. **r²**: This is the **square of the distance** between the two objects' centers. As the distance increases, the gravitational force decreases following the inverse-square law.

Formula Explanation:

This formula tells you that the gravitational force **K** between two objects depends on:

- The product of their masses **M** and **m**: Larger masses mean a stronger gravitational pull.
- The square of the distance between them **r²**: As the distance increases, the force decreases significantly because it's divided by the square of the distance.

In Simple Terms:

- **Gravitational force** increases if the objects are more massive and decreases quickly as they move farther apart.
- This formula is a way to calculate how much two objects, like the Earth and the Moon, pull on each other due to gravity. For example, doubling the distance between two objects makes the gravitational force four times weaker.

This equation describes why larger objects (like planets) exert stronger gravitational forces and why gravity gets weaker the farther you are from the source.

9. **Explain in simple terms what the following formula in physics means: $mv^2 = k\, Mm/r^2$.**

The formula **$mv^2 = k\, Mm/r^2$** represents a balance between **kinetic energy** (or more precisely, centripetal force) and **gravitational force**. This type of formula often shows up when describing the motion of an object in orbit, such as a planet orbiting a star or a satellite orbiting Earth.

Let's break it down:

Components:

1. **mv²**: This represents the **centripetal force** acting on an object with mass mmm that is moving in a circular path with velocity **v**. This force is necessary to keep the object moving in a curved orbit rather than flying off in a straight line.
2. **k Mm/r²**: This represents **Newton's gravitational force** between two objects, where:
    - **k** is the **gravitational constant**.

- M is the **mass of a larger object** (like a planet or star).
          - m is the **mass of a smaller object** (like a satellite or moon).
          - r is the **distance between the two objects** (from the center of the large object to the center of the small object).

Explanation

This formula expresses that, for an object to stay in a stable orbit, the **centripetal force** (which depends on the object's mass **m**, velocity **v**, and radius of the orbit **r**) must be equal to the **gravitational force** pulling the two objects together.

- The **left-hand side** of the equation, **mv²**, shows the force needed to keep the smaller object in a circular orbit due to its velocity.
- The **right-hand side**, **k·M·mr²**, shows the gravitational pull exerted by the larger object on the smaller one.

In Simple Terms:

- If an object, like a satellite, is orbiting a planet, the **centripetal force** keeping it in orbit must be exactly balanced by the **gravitational pull** of the planet.
- If the object moves too fast, it will escape orbit (because the left side would be larger). If it moves too slowly, it will fall toward the planet (because the right side would dominate).

This equation is key in understanding why objects in space stay in orbit and how their velocity and distance from the larger object determine the stability of that orbit.

10. Explain in simple terms what the following formula in physics means: K = mb = m(4π²C/r²).

The formula K = mb = m(4π²C/r²) appears to describe a relationship between several physical quantities involving mass (**m**), a constant (**C**), and some other variables. Let's break this down:

Components:

1. **K**: This likely represents a force or energy in the system. It could represent **kinetic energy**, **force**, or another physical quantity, depending on the context.
2. **m**: This is the **mass** of the object in question.
3. **b**: This could represent an **acceleration** or some other factor that depends on distance and time (like the gravitational constant or a centripetal acceleration, depending on the system).
4. **(4π²C/r²)**: This is a mathematical constant, often seen in equations that involve circular or periodic motion (like orbits or waves). It arises from geometrical and mathematical properties of circles.
5. **C**: This represents a constant specific to the system, such as a gravitational constant or some property of the object, like a characteristic length or energy scale. Without additional context, we can think of it as a constant that makes the units work out correctly.
6. **r**: This is typically the **radius** or **distance** from a center point in a circular or orbital system.

Explanation:

This formula seems to describe a system where an object's motion or acceleration (**b**) is related to its mass (**m**) and a constant factor **4π²C/r²**, which depends on its distance **r** from a central point.

In simple terms:

- The object's **mass** (m) is multiplied by an **acceleration term** (**b**).
- The acceleration term includes the distance from the center, **r**, and a constant C, with **4π²** being a geometrical factor often found in circular motion.

This might represent a case where an object is moving in a circular path, and its **acceleration** or the force acting on it depends on its distance from the center (like a satellite orbiting a planet or a mass on a spring in circular motion).

Simplified Concept:

- **K** is some measure of the **force or energy** acting on a mass mmm.
- The force or acceleration acting on this mass depends on its **distance** from the center (**r**) and involves circular or orbital motion (hence the **4π²**).
- **As the distance r decreases, the force or acceleration increases**, because **r** is in the denominator and squared.

This formula likely applies to situations involving **rotational** or **orbital dynamics**, such as planets orbiting the Sun or an object moving in a circular path.

In the context of **centripetal acceleration**, the "geometric factor" that relates the acceleration term to the distance from the center (R) is simply "1/R".

Explanation:

Centripetal acceleration formula:

The formula for centripetal acceleration is $a = v^2/r$ where,

a is the centripetal acceleration
v is the tangential velocity
r is the radius (distance from the center)

Geometric factor Interpretation:

In this formula "1/r" acts as the geometric factor, directly influencing the magnitude of the acceleration based on the distance from the center. A small radius (**r**) results in a larger acceleration for the same velocity.

Key Point:

When discussing the relationship between acceleration and distance in circular motion, the geometric factor is simply the reciprocal of the radius (1/R).

11. **Explain in simple terms what the following formula in physics means: G = mg.**

The formula **G = mg** is a simple expression in physics that relates to **weight** and **gravitational force**. Here's what each part means:

Components:

1. **G**: This represents the **gravitational force** or **weight** of an object. It tells us how strongly gravity is pulling on that object.
2. **m**: This stands for the **mass** of the object. Mass is a measure of how much matter is in the object, and it doesn't change based on where you are in the universe.
3. **g**: This represents the **acceleration due to gravity**. On Earth, this value is approximately **9.81 m/s**. It tells us how quickly an object will accelerate towards the ground when it is dropped.

Explanation:

- The formula **G = mg** shows that the gravitational force (or weight) **G** of an object is equal to its mass mmm multiplied by the acceleration due to gravity **g**.
- Essentially, it means that **the heavier an object is (more mass), the stronger the gravitational force acting on it.**

Simplified Concept:

- If you have a **5 kg** object on Earth, you can calculate its weight using this formula:
    - **G**=5 kg×9.81 m/s^2.
    - This gives you approximately **49.05 N** (Newtons), which is the weight of the object.

In summary, **G = mg** is a fundamental relationship that helps us understand how gravity affects objects based on their mass.

12. **Explain in simple terms what the following formula in physics means: K = k (mM/r2)**

The formula **K = k (mM/r2)** describes the relationship between gravitational force and mass in a gravitational field. Here's a breakdown of the components:

Components:

1. **K**: This represents the **gravitational force** between two objects. It tells you how strongly these objects attract each other due to gravity.
2. **k**: This is a constant that helps define the strength of the gravitational force in the context of the equation. In the case of gravitational interactions, this is often replaced by the

gravitational constant G, which has a value of approximately 6.674×10^{-11} N m²/kg².

3. **m**: This is the **mass** of the first object. The more massive the object, the stronger its gravitational pull.
4. **M**: This is the **mass** of the second object. Like mmm, a larger mass means a stronger gravitational pull.
5. **r**: This is the **distance** between the centers of the two objects. The farther apart they are, the weaker the gravitational force between them.

Explanation:

- The formula states that the gravitational force **K** is proportional to the product of the two masses mmm and **M** and inversely proportional to the square of the distance **r** between them.
- This means that:
    - If either mass increases, the gravitational force also increases.
    - If the distance **r** increases, the gravitational force decreases. Specifically, if you double the distance, the force becomes four times weaker because you divide by **r²**.

Simplified Concept:

- Imagine two objects, like the Earth and the Moon. The force of gravity between them depends on how heavy they are and how far apart they are. If the Earth were much more massive or the Moon were much less massive, the gravitational pull between them would be stronger. Conversely, if they were farther apart, that pull would be weaker.

In summary, **K = k (mM/r2)** describes how gravity works between two masses: the more massive they are and the closer they are to each other, the stronger the gravitational force.

13. **Explain in simple terms what the following formula in physics means: E = T + U + C + W...**

The formula **E = T + U + C + W...** represents the total energy **E** of a system in terms of different types of energy. Here's a breakdown of the components:

Components:

1. **E**: This stands for **total energy**. It is the sum of all the different forms of energy present in a system.
2. **T**: This represents **kinetic energy**, which is the energy of motion. An object in motion, like a rolling ball or a speeding car, has kinetic energy. The faster it moves, the more kinetic energy it has.
3. **U**: This is **potential energy**, which is the energy stored in an object due to its position or condition. For example, a ball held high above the ground has gravitational potential energy because of its height. If it falls, that potential energy converts to kinetic energy.

4. **C**: This often stands for **chemical energy**, which is the energy stored in the bonds of chemical compounds. This energy is released or absorbed during a chemical reaction. For instance, when you eat food, your body uses the chemical energy stored in it for various functions.
5. **W**: This usually denotes **work**, which is the energy transferred to or from an object via a force causing it to move. For example, pushing a box across the floor does work on the box.

Explanation:

- The formula states that the total energy of a system is equal to the sum of its kinetic energy, potential energy, chemical energy, and work done.
- This means that all these forms of energy can transform into one another. For instance:
    - When you throw a ball, its potential energy at the highest point turns into kinetic energy as it falls.
    - In a chemical reaction, like burning wood, chemical energy is converted into heat and light energy.

Simplified Concept:

Think of a roller coaster:

- At the top of the ride, the coaster has a lot of potential energy because it's high up.
- As it descends, that potential energy converts into kinetic energy, making it go faster.
- The total energy remains constant, but it changes forms between kinetic and potential as the coaster moves.

In summary, **E=T+U+C+WE = T + U + C + WE = T+U+C+W** tells us that the total energy of a system is made up of various forms of energy, and these forms can change into one another while the total amount of energy remains the same.

14. **Explain in simple terms what the following formula in physics means: vT = v = 1/T or T = 1 / v.**

The formula **vT = v = 1/T or T = 1 / v** relates the **velocity v** of an object to its **time period T**, which is the time it takes for one complete cycle of motion, like one complete rotation in a circular path. Here's a breakdown of the components:

Components:

1. **v**: This stands for **velocity**. It measures how fast something is moving. Velocity is usually expressed in units like meters per second **(m/s).**
2. **T**: This represents the **time period**. It is the time it takes for an object to complete one full cycle of motion, such as one complete rotation around a circle. The time period is typically measured in seconds **(s)**.

Explanation:

- The formula can be understood in two ways depending on how it's arranged:
  - vT = v = 1/T or T = 1 / v
    - This implies that the velocity multiplied by the time period gives the same velocity, indicating a consistent speed throughout the cycle.
  - vT = v = 1/T or T = 1 / v.
    - This shows that the time period is the reciprocal of velocity, meaning that as an object moves faster (higher **v**), the time period (**T**) becomes shorter. Conversely, if the object moves slower (lower **v**), the time period increases.

Simplified Concept:

- Think of a carousel at an amusement park:
  - If the carousel is spinning quickly (high velocity), it takes less time to complete one full rotation (short time period).
  - If the carousel is spinning slowly (low velocity), it takes more time to complete one full rotation (long time period).

Summary:

In simple terms, this formula connects how fast something is moving (velocity) to how long it takes to complete a motion (time period). If you know the speed of something, you can easily find out how long it takes to go around once, and if you know how long it takes, you can find out how fast it's going.

15. Explain in simple terms what the following formula in physics means: c=λfc,

The formula **c=λfc,** relates to three important concepts in physics: **speed, wavelength,** and **frequency**. Here's what each term means:

Components:

1. **c**: This is the **speed of light** in a vacuum, which is approximately *299,792,458* meters per second (often rounded to 3×10^8 m/s). It represents how fast light travels.
2. **Λ** (lambda): This symbolizes the **wavelength**, which is the distance between two consecutive peaks (or troughs) of a wave. It is usually measured in meters (m). A longer wavelength means the peaks are farther apart.
3. **f**: This represents the **frequency**, which is the number of times a wave's peak passes a fixed point in one second. It is measured in hertz (**Hz**), where **1** Hz means one cycle (or wave) per second.

Explanation:

- The formula can be understood as stating that the speed of light (**c**) is equal to the product of the wavelength (**λ or lambda**) and the frequency (**f**).

Simplified Concept:

- Imagine waves on a beach:
    - If you have waves that are very close together (high frequency), they move quickly (higher frequency results in shorter wavelengths).
    - If the waves are farther apart (low frequency), they move more slowly (longer wavelength).

Summary:

In simple terms, this formula shows how the speed of light is connected to how long each wave is (wavelength) and how often the waves pass by (frequency). If you know two of these values, you can always calculate the third one. For example:

- If you know the wavelength of a light wave, you can find its frequency by rearranging the formula to *f=cλf*.
- Similarly, if you know the frequency, you can find the wavelength using *f=cλf*. So, the formula helps us understand the relationship between these three properties of light waves!

16. **Explain in simple terms what the following formula in physics means: τ + l/c.**

The formula **τ + l/c** combines two concepts related to time and the speed of light. Here's what each part means:

Components:

1. **τ** This symbol typically represents a **time duration** or a **time constant** in physics, depending on the context. It could refer to any amount of time measured in seconds.
2. **l**: This represents a **distance** or **length**. It's often measured in meters (m).
3. **c**: This is the **speed of light**, which is approximately *299,792,458* meters per second (or about **3×10⁸ m/s**). It indicates how fast light travels in a vacuum.

Explanation:

- The term **l/c** calculates the **time it takes for light to travel a distance l**. Since ccc is the speed of light, dividing a distance by this speed gives you the time in seconds.

Simplified Concept:

- Think of **τ** as a waiting time (like how long you wait for something to happen), and **l/c** as how long it takes for light to reach you from a distance **l**.

Summary:

In simple terms, this formula gives you a total time:

- It combines your waiting time τ with the time it takes for light to travel a certain distance **l**.

So, if you want to know how much time has passed after waiting for τ seconds and then waiting for light to travel a distance **l**, you would add those two times together. This is often relevant in physics when considering how signals or information travel through space!

17. **Explain in simple terms what the following formula in physics means: Einstein's equation, 9.2×107 J or (9.2 x 1014 ergs) of energy corresponds to a mass loss of just 0.0000010g (or 1.0 μg):**

Einstein's equation is often expressed as **E=mc²**, which relates energy (**E**) to mass (**m**) and the speed of light **c**. Here's a breakdown of what this means in simple terms, especially in the context you provided:

Key Components:

1. **Energy (E)**: In this case, the energy value is given as **9.2×107 J or (9.2 x 1014 ergs)** This is a measure of energy, which can be thought of as the ability to do work or produce change.
2. **Mass (m)**: The mass loss mentioned is 0.000001 grams (or 1.0 μg). This is an extremely small amount of mass, equivalent to a millionth of a gram.
3. **Speed of Light (c)**: The speed of light is approximately 299,792,458 meters per second (or about 3×10^8 **m/s**). In Einstein's equation, this value is squared (**c²**), which makes it a very large number.

Explanation:

- **Mass-Energy Equivalence**: The key idea behind **E=mc²** is that mass can be converted into energy and vice versa. This means that even a tiny amount of mass can be converted into a significant amount of energy because of the large value of **c²**.
- **Context of the Values**:
    - In your example, when a mass of just 0.000001 grams is lost, it corresponds to a huge energy release of **9.2×10^7 joules**.
    - This shows that when very small amounts of mass are converted into energy (like in nuclear reactions), the energy produced can be enormous.

Summary:

In simple terms, this equation demonstrates the powerful relationship between mass and energy. A tiny loss of mass can result in a vast amount of energy, which is a fundamental principle of physics that underpins many processes in the universe, including nuclear reactions and the energy produced by stars. So, even though 1.0 μg is extremely small, it can produce a lot of energy when converted according to Einstein's equation!

18. **Explain in simple terms what the following formula in physics means: $V^2=a^2+b^2+c^2$**

The formula **$V^2=a^2+b^2+c^2$** is a mathematical expression used in physics, often related to the concept of vector magnitudes in three-dimensional space. Here's what it means in simple terms:

Key Components:

1. **V**: This represents the magnitude (or length) of a vector in three-dimensional space. It could be the total distance or a measure of how strong a particular quantity is, like velocity or force.
2. **a, b, and c**: These are the components of the vector along the three different axes of a coordinate system (usually labeled as **x, y,** and **z**). Each component tells you how far the vector extends in each direction:
   - **a** could represent the distance along the x-axis,
   - **b** could represent the distance along the y-axis,
   - **c** could represent the distance along the z-axis.

Explanation:

- **Pythagorean Theorem**: This formula is a three-dimensional version of the Pythagorean theorem, which relates the sides of a right triangle. In two dimensions, the theorem states that **$V^2=a^2+b^2+c^2$**. In three dimensions, you add another component **c^2** for the third direction.
- **Finding the Length**: The formula shows how to find the total length (or magnitude) of a vector when you know its components. For example, if you know how far something moves in each direction (x, y, and z), you can calculate how far it has moved overall using this formula.

Summary:

In simple terms, **$V^2=a^2+b^2+c^2$** helps you find the overall distance (or magnitude) of a vector in three-dimensional space by squaring the distances in each direction, adding them together, and then taking the square root. It's a way to combine movements or forces acting in different directions into a single measure of size or strength!

19. **Explain in simple terms what the following formula in physics means: $d=t \cdot c$**

The formula **$d=t \cdot c$** is a straightforward equation used in physics to relate distance, time, and the speed of light. Here's what each component means in simple terms:

Key Components:

1. **d**: This represents **distance**. It tells you how far an object has traveled or how far apart two points are.
2. **t**: This represents **time**. It measures how long something takes, such as the duration of a trip or the time it takes for light to travel from one point to another.

3. **c**: This represents the **speed of light** in a vacuum, which is approximately 299,792,458 meters per second (often rounded to 3×10^8 **m/s**). It is the fastest speed at which information or matter can travel in the universe.

Explanation:

- **Understanding the Relationship**: The formula tells us that the distance (**d**) an object travels (like light) is equal to how long it takes (**t**) to travel that distance multiplied by the speed of light (**c**).
- **Example**: If you know that light takes 2 seconds to reach you from a star, you can calculate how far away that star is by multiplying the time (2 seconds) by the speed of light. So, **d=2 s·3×10⁸ m/s**, which equals about 600,000 kilometers.

Summary:

In simple terms, **d=t·c** helps you find out how far light (or any object moving at the speed of light) has traveled based on how long it has been traveling. It shows the relationship between distance, time, and the speed at which light moves!

20. **Explain in simple terms what the following formula in physics means: f = ma = mk/r²**

The formula **f = ma = mk/r²** combines concepts from Newton's second law of motion and gravitational force. Here's a breakdown of what each part means in simple terms:

Key Components:

1. **f**: This represents **force**. It's the push or pull on an object, measured in newtons (N).
2. **m**: This stands for **mass**. It measures how much matter is in an object, usually measured in kilograms (kg).
3. **a**: This represents **acceleration**. It tells you how quickly an object is speeding up or slowing down, measured in meters per second squared (m/s²).
4. **k**: This is a constant that depends on the specific situation, often related to the force constant in springs or gravitational force.
5. **r**: This represents the **distance** or **radius**. It's the distance from the center of an object (like a planet) to another point, measured in meters (m).

Explanation:

1. **Understanding f = ma = mk/r²**:
    - This part of the equation tells us that the force acting on an object (**f**) is equal to its mass (**m**) multiplied by its acceleration (**a**). If you push a heavier object, it requires more force to accelerate it than a lighter one.
2. **Understanding f = ma = mk/r²**:
    - This part relates to gravitational force or a similar force, where the force (**f**) is also equal to a mass (**m**) multiplied by a constant (**k**) divided by the square of

the distance (**r²**). This indicates that the force decreases rapidly as the distance increases. For example, if you double the distance, the force becomes one-fourth as strong.

Summary:

In simple terms, **f = ma = mk/r²** describes how force, mass, and acceleration are related, as well as how they relate to distance in situations like gravity. It tells us that the force needed to move an object depends on how heavy it is and how quickly we want to change its speed, while also showing how that force changes based on how far apart the objects are.

21. **Explain in simple terms what the following formula in physics means: V=Hr**

The formula **V=Hr** is a simple equation used in physics to relate the volume **V** of a cylinder (or similar shape) to its height **H** and radius **r**. Here's a breakdown of what each component means:

Key Components:

1. **V**: This represents the **volume** of the object, usually measured in cubic units (like cubic meters, **m³**). Volume tells you how much space an object occupies.
2. **H**: This is the **height** of the object. It's the distance from the base to the top, measured in linear units (like meters, **m**).
3. **r**: This is the **radius** of the base of the object. The radius is the distance from the center of the base to its edge, also measured in linear units (like meters, **m**).

Explanation:

- **Understanding the Formula**: The formula states that the volume **V** is equal to the height **H** multiplied by the radius **r**.
- This is often used in contexts where the base area of a shape is circular.

Visualizing it:

- Imagine a cylinder, like a can. The base of the can is a circle. The volume of the cylinder depends on how tall it is (height **H**) and how wide its base is (radius **r**).
- In more complex terms, if you think about the area of the base (which is a circle, calculated as **2πr²**), the full volume would be calculated as **V=πr²H**. However, the simplified equation **V=Hr** is a straightforward way to express the relationship for certain shapes.

Summary:

In simple terms, **V=Hr** tells us that the volume of a cylinder (or similar object) can be found by multiplying its height by its radius. This shows how the size of the object in three-dimensional space is connected to its height and how wide it is at the base.

22. **Explain in simple terms what the following formula in physics means: c=f×λ**

The formula **c=f×λ** is a fundamental equation in physics that relates the speed of a wave **c,** its frequency (**f**), and its wavelength (**λ** or lambda). Here's a breakdown of what each part means:

Key Components:

1. **c**: This represents the **speed of the wave (light)**. For light waves in a vacuum, this speed is approximately 299,792,458 meters per second (often rounded to **3×10^8 m/s).** It indicates how fast the wave travels through space.
2. **f**: This stands for the **frequency** of the wave. Frequency measures how many times the wave oscillates or cycles per second, and it is usually expressed in hertz (Hz). For example, a frequency of 1 Hz means one complete wave cycle occurs every second.
3. **λ (lambda)**: This symbolizes the **wavelength** of the wave. Wavelength is the distance between two consecutive points that are in phase on the wave (like the distance from one crest to the next crest). It is typically measured in meters (**m**).

Explanation:

- **Understanding the Formula**: The formula tells us that the speed of a wave (**c**) is equal to its frequency (**f**) multiplied by its wavelength (**λ**).
- In simpler terms, if you know how fast a wave is moving and either its frequency or wavelength, you can find the other.

Example:

- If you have a sound wave with a frequency of **4186.009 Hz** (which is the pitch of the musical note **A**), and you know that the speed of sound in air is about **343** meters per second, you can rearrange the formula to find the wavelength: **λ=cf=343 m/s 440 Hz≈0.78 meters**.
- This means the wavelength of the sound wave is about 17mm (meters).

Summary:

In simple terms, **c=f×λ** tells us that the speed of a wave is determined by how often it oscillates (frequency) and the distance between its peaks (wavelength). This relationship is crucial in understanding all types of waves, including sound and light.

23. **Explain in simple terms what the following formula in physics means: v= √(GM/r)**

The formula **v= √(GM/r)** is used in physics to describe the **orbital speed** of an object (like a planet or a satellite) moving around another body (like a star or a planet). Here's a breakdown of what each component means:

Key Components:

1. **v**: This represents the **orbital speed** of the object. It is how fast the object needs to move to maintain a stable orbit around the larger body.
2. **G**: This is the **gravitational constant**, a number that tells us how strong the force of gravity is. Its value is approximately **6.674×10^{-11} N m²/kg².**
3. **M**: This stands for the **mass** of the larger body that the object is orbiting. For example, if a satellite is orbiting Earth, **M** would be the mass of Earth.
4. **r**: This is the **distance** from the center of the larger body to the object in orbit. It's important to note that this distance includes both the radius of the larger body and the altitude of the object above the surface.

Explanation:

- **Understanding the Formula**: The formula tells us that the orbital speed (**v**) is equal to the square root of the gravitational force created by the larger body, divided by the distance from that body to the orbiting object.
- In simpler terms, the faster you want to orbit (the higher **v** is), the more massive the body you're orbiting (higher **M**) or the closer you need to be to it (smaller **r**).

Example:

- Let's say you want to calculate the orbital speed of a satellite orbiting Earth.
    - **Earth's mass (M)** is about **5.97×10^{24} kg**
    - **Distance (r)**: If the satellite is orbiting at an altitude of 300km above Earth's surface, the total distance from Earth's center is about 6,371 km+300 km=6,671 km

Using the formula:

v=$(6.674 \times 10^{-11}$ N m2/kg2$)(5.97 \times 10^{24}$ kg$)$ -> 6,671,000 mv

This will give you the speed required for the satellite to maintain its orbit.

Summary:

In simple terms, **v= √(GM/r)** tells us how fast an object needs to travel to stay in orbit around a larger body, depending on the mass of that body and the distance from it. The larger the mass or the closer the object is to the mass, the faster it must move to avoid falling towards the body due to gravity.

24. **Explain in simple terms what the following formula in physics means: $E^2=m^2c^4+p^2c^2$**

The formula $E^2=m^2c^4+p^2c^2$ is a fundamental equation in physics that relates energy, mass, and momentum, particularly in the context of special relativity. Here's a simple breakdown of what each part of the equation means:

Key Components:

1. **E**: This represents the **total energy** of an object. It includes all forms of energy, such as kinetic (energy of motion) and potential energy.
2. **m**: This is the **rest mass** of the object, which is the mass it has when it is not moving. It is a measure of how much matter is in the object.
3. **c**: This is the **speed of light in a vacuum**, approximately 3.00×10^8 m/s. It's a very important constant in physics, especially in relativity.
4. **p**: This stands for the **momentum** of the object, which is a measure of how much motion it has. Momentum depends on both the mass of the object and its velocity (speed and direction).

Explanation:

- **Understanding the Formula**: The equation tells us that the total energy **E** of an object depends not only on its mass mmm but also on its momentum **p**.
- The terms $m^2c^4 + p^2c^2$ represent different contributions to the total energy:
    - $m^2c^4 + p^2c^2$: This part shows how the rest mass contributes to the energy, with a large multiplier because c^4 is a very large number. This illustrates how even a small mass can be associated with a large amount of energy (as seen in nuclear reactions).
    - p^2c^2: This part accounts for the energy due to the object's motion (momentum). The faster the object moves, the more momentum it has, contributing to its total energy.

Example:

- **Imagine a moving car**: If you have a car that weighs 1,000 kg (its rest mass mmm), it has energy due to its mass alone. If the car is also moving at a high speed, it has momentum (**p**). Both the **rest-mass** and the **momentum** contribute to the total energy of the car.

Summary:

In simple terms, the formula $E^2 = m^2c^4 + p^2c^2$ shows that the total energy of an object is determined by both its mass and its momentum. It highlights the relationship between mass, motion, and energy in the framework of special relativity, demonstrating that mass can be converted into energy and that moving objects carry additional energy due to their motion.

25. **Explain in simple terms what the following formula in physics means: Since a light particle has no rest mass, this equation simplifies quickly to: $E^2 = p^2c^2$ which can be expressed as: $p = E/c$**

The formulas you're asking about deal with the relationship between energy and momentum for light particles (photons), which are massless. Let's break it down into simple terms:

Key Components:

1. **E**: This represents the **energy** of a light particle (like a photon).
2. **p**: This is the **momentum** of the light particle. Momentum is a measure of how much motion an object has.
3. **c**: This is the **speed of light in a vacuum**, approximately **3.00×10⁸ m/s**. It's a fundamental constant in physics.

The Formulas:

1. **Starting Equation**: The original equation is **E²=p²c²**. This equation shows how energy and momentum are related for particles, including massless particles like photons.
2. **Simplified Equation for Light Particles**: Because light particles (photons) have **no rest mass**, we can simplify the equation. When we set the rest mass mmm to zero, the equation reduces to: **E²=m²c⁴+p²c²**
3. **Rearranged Formula**: From this, we can express momentum as: **p=E/c**

This means that the momentum of a light particle is equal to its energy divided by the speed of light.

Explanation:

- **Energy and Momentum of Light**: For light particles, their energy is directly related to their momentum. This means that as the energy of a photon increases (for example, a higher frequency of light), its momentum also increases.
- **No Rest Mass**: Since photons don't have mass, we can't use the usual formulas that apply to objects with mass. Instead, we use this special relationship between energy and momentum.

Example:

- **Imagine a photon of light**: If you have a photon with a certain amount of energy (like from a bright flashlight), you can calculate its **momentum** using the formula **p=mv**. So if the photon has an energy of 1 joule, its momentum would be:

$$p=1 J 3.00×10^8 m/s ≈ 3.33×10-9 kg\ m/s$$

Summary:

In simple terms, the equations **E²=p²c² which can be expressed as: p=E/c,** show the relationship between energy and momentum specifically for light particles, which have no mass. They tell us that light, energy and momentum are directly connected, and as energy increases, so does momentum.

26. **Explain in simple terms what the following formula in physics means: (ab−ba)/iℏ**

The formula **(ab−ba)/iℏ** is related to quantum mechanics, specifically to the concept of **commutation** of operators. Let's break it down in simple terms:

Key Components:

1. **Operators**: In quantum mechanics, physical quantities (like position, momentum, energy, etc.) are represented by mathematical entities called **operators**. Here, **a** and **b** are two operators.
2. **Commutator**: The expression **(ab−ba)** is called the **commutator** of the two operators **a** and **b**. It measures how much the two operators do not "commute," meaning that applying them in different orders will yield different results. If **(ab−ba)**, the operators commute.
3. This is the imaginary unit, where **i2=−1i² = -1i²=−1**. It's used in quantum mechanics for various mathematical reasons.
4. **ℏ**: This is **h-bar**, which is a constant that is a version of Planck's constant **(h) divided by 2π**. It plays a crucial role in quantum mechanics.

Meaning of the Formula:

- The formula **(ab−ba)/iℏ** quantifies the **commutation relation** between two operators. This relation is fundamental in understanding the behavior of quantum systems.

- **Physical Implication**:
    - If the commutator is **zero** or **(ab−ba=0)**, it means the two operators can be measured simultaneously without any issues (they commute).
    - If the commutator is **non-zero**, it indicates that there is a fundamental limit to how precisely you can know the values of the two quantities represented by **a** and **b** at the same time. This is related to the **Heisenberg uncertainty principle**.

Example in Quantum Mechanics:

- **Position and Momentum**: A famous example in quantum mechanics is the commutation relation between the position operator **x** and the momentum operator. This means that position and momentum do not commute, leading to the uncertainty principle: you cannot know both the exact position and exact momentum of a particle simultaneously.

Summary:

In simple terms, the formula **(ab−ba)/iℏ** tells us about the relationship between two quantum operators. It helps us understand whether two physical quantities can be measured at the same time without uncertainty. If they don't commute (the result is non-zero), it signifies a limit to our knowledge about those quantities, a central theme in quantum mechanics.

27. **Explain in simple terms what the following formula in physics means: E² = m²*c⁴ + p-²*c²**

The formula **E² = m²*c⁴ + p-²*c²** is a key equation in physics that relates energy, mass, and momentum, particularly in the context of special relativity. Here's a breakdown of what each part means in simple terms:

Key Components:

1. **E**: This represents the **total energy** of an object. In relativity, energy isn't just about motion; it's also related to the mass of the object.
2. **m**: This is the **rest mass** of the object, which is the mass it has when it's not moving. It's an intrinsic property of the object.
3. **c**: This is the **speed of light in a vacuum**, approximately **3.00×10⁸** meters per second or **299,792,458 m/s**. It's a very important constant in physics, especially in relativity.
4. **p**: This symbolizes the **momentum** of the object. Momentum is a measure of how much motion an object has and is calculated as the product of mass and velocity.

What the Formula Means:

- **Total Energy**: The left side of the equation, **E²**, represents the total energy of an object when considering both its mass and its momentum.
- **Right Side Components**:

The term **m²*c⁴ + p-²*c²** shows how much energy is associated with the object's mass. It indicates that even a stationary object has energy due to its mass.

  - The term **p²c²** relates to the energy due to the object's momentum. When an object is moving, its momentum contributes to its total energy.

Interpretation:

1. **Energy-Mass Equivalence**: This formula demonstrates that mass and energy are interchangeable. An object with mass has inherent energy, and the faster it moves (greater momentum), the more energy it has.
2. **Importance in Relativity**: This relationship is fundamental in Einstein's theory of relativity. It shows how energy increases not just with mass but also with velocity (momentum).
3. **Massless Particles**: For particles that have no rest mass (like photons), the formula simplifies. In such cases, mmm is zero, and the energy is purely dependent on momentum:
E² = m²*c⁴ + p-²*c²

This means massless particles carry energy based solely on their momentum.

Summary:

In simple terms, the formula $E^2 = m^2 \cdot c^4 + p^{-2} \cdot c^2$ tells us that the total energy of an object comes from both its mass and how fast it is moving. It illustrates the deep connection between mass and energy, emphasizing that both play crucial roles in determining an object's overall energy, especially in the realm of high speeds and relativistic physics.

28. Explain in simple terms what the following formula in physics means: $pq - qp = \hbar/i$

The formula **pq − qp = ℏ/I** is an important equation in quantum mechanics, specifically related to the concepts of momentum and position operators. Let's break down what each part means in simple terms:

Key Components:

1. **p**: This represents the **momentum operator** in quantum mechanics. Momentum is a measure of how much motion an object has, and in quantum mechanics, it's treated as an operator that acts on the wave function of a particle.
2. **q**: This stands for the **position operator**. It describes the position of a particle in space. Like momentum, position is also treated as an operator in quantum mechanics.
3. **pq**: This notation means to first apply the momentum operator **p** and then the position operator **q**.
4. **qp**: Similarly, this means to first apply the position operator **q** and then the momentum operator **p**.
5. **h-bar**, is a fundamental constant in quantum mechanics. It represents the reduced Planck constant and is equal to Planck's constant divided by **2π**. It has a value of approximately **1.055×10⁻³⁴** Joule-seconds.
6. **i**: This is the imaginary unit, defined as the square root of **−1**. It's often used in physics and mathematics to deal with complex numbers.

What the Formula Means:

- **Commutator**: The expression **pq−qp** is known as the **commutator** of the momentum and position operators. In quantum mechanics, the commutator measures how much two operators "fail to commute," meaning that the order in which you apply them affects the result. If the commutator is zero, it means the operators can be applied in any order without affecting the outcome.
- **Uncertainty Principle**: This formula is related to the **Heisenberg Uncertainty Principle**, which states that you cannot simultaneously know the exact position and momentum of a particle. The non-zero commutator **(pq−qp) − (qp− pq)** indicates that there is an inherent uncertainty in the measurements of these two quantities.

Physical Interpretation: The equation shows that the difference between applying the momentum operator followed by the position operator and vice versa is a constant (**ℏ**). This constant signifies a fundamental limit to our ability to know both the position and momentum of a particle at the same time.

Summary:

In simple terms, the formula **pq – qp = ℏ/i** illustrates a core principle of quantum mechanics: that position and momentum are intertwined in such a way that measuring one affects the other. The non-commutativity of these operators leads to fundamental limits on our knowledge of a particle's state, encapsulated in the uncertainty principle.

29. Explain in simple terms what the following formula in physics means: e- + e- -> 2 gamma.

The formula **e- + e- -> 2 gamma or** (2γ) represents a particle interaction in physics, specifically in the context of particle physics and quantum electrodynamics. Here's a breakdown of what it means in simple terms:

Key Components:

1. **e–**: This symbol represents an **electron**, which is a negatively charged subatomic particle. Electrons are fundamental particles that orbit the nucleus of an atom and are responsible for electricity and chemical reactions.
2. **→**: This arrow means "produces" or "transforms into." It indicates that the particles on the left side of the arrow interact to form the particles on the right side.
3. **2γ**: This notation represents **two gamma rays** (denoted by (**γ**). Gamma rays are a type of high-energy electromagnetic radiation, similar to X-rays but with even higher energy. They are emitted during radioactive decay or from other high-energy processes in the universe.

What the Formula Means:

- **Electron-Electron Interaction**: The formula describes a situation where two electrons collide or interact with each other. Instead of continuing as electrons, they transform into gamma rays.
- **Energy Conversion**: The interaction suggests that the energy carried by the two electrons is converted into the energy of the two gamma rays. According to Einstein's equation **E=mc²**, mass can be converted into energy. Here, the mass (or energy) of the electrons is transformed into the energy of the emitted gamma rays.
- **Conservation Laws**: This interaction respects several conservation laws in physics:
    - **Conservation of Energy**: The total energy before and after the interaction remains the same.
    - **Conservation of Momentum**: The total momentum of the system is conserved during the interaction.
    - **Charge Conservation**: Since the initial and final states are electrically neutral (the electrons are negatively charged, and gamma rays have no charge), charge is conserved.

Summary:

In simple terms, the formula **e- + e- -> 2 gamma** describes a process in which two electrons collide and annihilate each other, resulting in the production of two gamma rays. This illustrates how mass can be converted into energy, a fundamental concept in physics, and highlights the fascinating interactions that occur at the subatomic level.

30. **Explain in simple terms what the following formula in physics means: λ=h/p.**

The formula **λ=h/p** is a key equation in quantum physics that relates to the wave-particle duality of matter. Here's a breakdown of what it means in simple terms:

Key Components:

1. **λ** (lambda): This symbol represents **wavelength**, which is the distance between two consecutive peaks of a wave. In this context, it describes the wavelength of a particle's wave-like behavior.
2. **h**: This is **Planck's constant**, a fundamental constant in physics with a value of about **Planck's Constant = 6.62607015 x $10^{-34}m^2$ kg/s** (the smallest physical size/distance in nature).
3. **joule-seconds**. It helps to describe the sizes of quantum effects.
4. **p**: This symbol represents **momentum**, which is a measure of how much motion an object has. It can be calculated as the product of an object's mass and its velocity (**p=mv**).

What the Formula Means:

- **Wave-Particle Duality**: This formula highlights the concept that particles, like electrons or photons (light particles), exhibit both wave-like and particle-like properties. In quantum mechanics, everything can behave like a wave under certain conditions.
- **Relationship Between Wavelength and Momentum**:
    o The formula shows that the wavelength (λ-lambda) of a particle is inversely proportional to its momentum (p). This means:
        - If the momentum of a particle increases, its wavelength decreases.
        - Conversely, if the momentum decreases, the wavelength increases.
- **Planck's Constant**: The presence of Planck's constant (h) in the formula indicates that these effects are significant at very small scales, such as those of atoms and subatomic particles.

Summary:

In simple terms, **λ=h/p** tells us that the wavelength of a particle is determined by its momentum. This relationship is a key aspect of quantum mechanics, illustrating how particles can behave like waves. It emphasizes that at tiny scales, the properties of matter cannot be fully understood without considering both their wave-like and particle-like behaviors.

31. **Explain in simple terms what the following formula in physics means: density = | omega (x, t)|²**

The formula **density = | omega (x, t)|²** relates to wave physics and represents how the density of a wave (like a sound wave or a light wave) can be described in terms of a mathematical function **omega (x, t):** Here's a breakdown of what this means:

Key Components:

1. **Density**: In this context, density refers to the intensity or strength of the wave at a particular point in space and time. For example, in a sound wave, it could refer to how much energy the sound wave carries at a specific location.
2. **ω omega (x, t)**: This is a complex function that describes the wave.
   - **x** usually represents the position in space, while **t** represents time.
   - The function **ω omega (x, t)** contains information about the wave's amplitude and phase, which are crucial for understanding how the wave behaves.
3. **Absolute Value Squared**: The notation **| omega (x, t)|²** means you take the absolute value of the function **ω omega (x, t)** (which removes any negative signs) and then square it.
   - This squaring is important because it gives a non-negative value that corresponds to the "strength" or "intensity" of the wave at that point.

What the Formula Means:

- **Wave Intensity**: The formula tells us that the intensity or density of the wave at any point xxx and time **t** is determined by taking the square of the magnitude of the wave function **ω omega (x, t)**
  - A larger value of **| omega (x, t)|2** indicates a higher intensity or energy density of the wave at that point, while a smaller value indicates lower intensity.

Summary:

In simple terms, **density = | omega (x, t)|²** tells us that the intensity or strength of a wave at any point in space and time can be calculated by squaring the magnitude of a mathematical function that describes the wave. This is a common way to quantify the energy carried by waves in physics.

32. **Explain in simple terms what the following formula in physics means: p·λ=h**

The formula **p·λ=h** relates momentum (p), wavelength (λ-lambda), and Planck's constant (h). Here's what each part means in simple terms:

Key Components:

1. **p**: This represents the **momentum** of a particle. Momentum is a measure of how much motion an object has and depends on its mass and velocity. In simpler terms, it's how "fast and heavy" something is moving.

2. **Λ-lambda**: This symbol represents the **wavelength** of a wave. Wavelength is the distance between two consecutive peaks (or troughs) of a wave. It tells us how "long" each wave is.
3. **h**: This is **Planck's constant**, a fundamental number in quantum mechanics that relates energy and frequency of waves. Its value is approximately **6.626×10^{-34}** joule-seconds.

What the Formula Means:

- **Relationship Between Momentum and Wavelength**: The formula **p·λ=h** states that the momentum of a particle (like an electron or photon) multiplied by its wavelength equals a constant value, Planck's constant.
- **Quantum Mechanics Context**: In the realm of quantum mechanics, this formula indicates that particles like electrons or photons exhibit both particle-like and wave-like behavior.
    - When you know the wavelength of a particle, you can determine its momentum, and vice versa.

Summary:

In simple terms, **p·λ=h** shows that the momentum of a particle is directly related to its wavelength through Planck's constant. This relationship is fundamental in understanding the wave-particle duality of matter in quantum mechanics.

33. **Explain in simple terms what the following formula in physics means: e √ℏc**

The expression **e √ℏc** combines three important concepts from physics, particularly in quantum mechanics and relativity. Let's break it down:

Key Components:

1. **e**: This symbol typically represents the base of the natural logarithm, approximately equal to **2.71828**. In some physics contexts, it can also represent the charge of an electron (about **1.6×10^{-19}** coulombs), but in this expression, it likely refers to the mathematical constant.
2. **ℏ** (h-bar): This is **Planck's constant divided by 2π(pi)**, approximately equal to **1.055×10^{-34}** joule-seconds. It plays a crucial role in quantum mechanics, particularly in the quantization of energy levels.
3. **c**: This symbol stands for the **speed of light in a vacuum**, which is approximately **3×10^8** meters per second. It's a fundamental constant in physics that represents how fast light travels.

What the Expression Means:

- **Combination of Constants**: The expression **e √ℏc** represents a combination of fundamental constants from quantum mechanics (ℏ) and relativity (c). This combination is often encountered in the context of theoretical physics, particularly in quantum field theory and particle physics.

- **Physical Significance**:
  - The square root of **ℏc** often appears in calculations involving the scale of phenomena at the quantum level. It has dimensions of length (specifically, it's related to the Compton wavelength of a particle).
  - The factor of **e** can indicate the mathematical nature of certain equations, such as those involving exponential growth or decay, or it can relate to statistical mechanics and thermodynamics in quantum contexts.

Summary:

In simple terms, **e √ℏc** combines fundamental constants from quantum mechanics and relativity to describe physical phenomena at the quantum level. It highlights the interconnectedness of different areas of physics and often appears in theoretical discussions about the properties of particles and waves.

34. Explain in simple terms what the following formula in physics means: $E = \hbar\omega$

The formula **E = ℏω** relates the energy (**E**) of a particle (usually a photon or another quantum particle) to its angular frequency (**ω** or **omega**). Here's a simple breakdown of the components and what the formula means:

Key Components:

1. **E**: This represents the energy of the particle. Energy is a measure of the ability to do work or cause change, and in this context, it typically refers to the energy of a photon or quantum particle.
2. **ℏ (h-bar)**: This is **Planck's constant** divided by **2π**, which is approximately 1.055×10^{-34} joule-seconds. It is a fundamental constant in quantum mechanics that helps describe the behavior of particles at the quantum level.
3. **Ω or omega**: This represents the **angular frequency** of the particle. It is related to how quickly the particle oscillates or waves, and it is measured in radians per second. Angular frequency is connected to the regular frequency (**f**) by the formula **E = ℏω** where **f** is the number of cycles per second.
4. **Planck's Constant = $6.62607015 \times 10^{-34} m^2$ kg/s** (the smallest physical size/distance in nature).

What the Formula Means:

- **Direct Relationship**: The formula states that the energy of a quantum particle (like a photon) is directly proportional to its angular frequency. This means that the higher the frequency (or oscillation rate) of the particle, the greater its energy.
- **Photon Example**: For photons (light particles), this formula illustrates that different colors of light correspond to different energies. For instance, blue light has a higher frequency (and thus a higher angular frequency) than red light, which means blue light carries more energy.

Summary:

In simple terms, **E = ℏω** tells us that the energy of a quantum particle is determined by how fast it oscillates. Higher oscillation frequencies mean higher energy, which is a fundamental concept in quantum mechanics, particularly when studying light and other wave-like particles.

35. **Explain in simple terms what the following formula in physics means: F = G M1*m²/r²**

The formula **F = G M1*m²/r²** describes the gravitational force (**F**) between two objects. Here's a breakdown of each part of the formula in simple terms:

Key Components:

1. **F**: This is the gravitational force between the two objects. It's the attraction that pulls them towards each other.
2. **G**: This is the **gravitational constant**; a number that helps quantify the strength of gravity in the universe. Its approximate value is **G** = 6.674×10^{-11} N *m2/kg².
3. **M1** and **m2**: These are the masses of the two objects. **M1** is the mass of the first object (like a planet or a star), and **m2** is the mass of the second object (like a smaller object, such as a satellite or a person).
4. **r**: This is the distance between the centers of the two masses. It tells us how far apart the two objects are.

What the Formula Means:

- **Attraction**: The formula shows that every two objects with mass attract each other with a force. The more massive the objects (**M1** and **m2**), the stronger the gravitational force between them.
- **Distance Matters**: The force decreases as the distance rrr between the objects increases. Specifically, if you double the distance between the objects, the gravitational force becomes four times weaker (because **r²** means that the distance is squared in the denominator).

Summary:

In simple terms, **F = G M1*m2/r2** tells us how strong the gravitational pull is between two objects based on their masses and how far apart they are. More massive objects exert a stronger gravitational force, and the force weakens as the distance between them increases. This formula is fundamental in understanding how gravity works in our universe.

36. **Explain in simple terms what the following formula in physics means: F=ma.**

The formula **F=ma** is one of the most fundamental equations in physics, describing the relationship between force, mass, and acceleration. Here's what each part of the formula means in simple terms:

Key Components:

1. **F**: This represents **force**. Force is a push or pull that can cause an object to move, stop, or change direction. It is measured in **newtons (N)**.
2. **m**: This represents **mass**. Mass is a measure of how much matter is in an object, and it is typically measured in **kilograms (kg)**. More mass means more matter and more resistance to being moved.
3. **a**: This represents **acceleration**. Acceleration is how quickly an object speeds up or slows down. It is measured in **meters per second squared (m/s²)**.

What the Formula Means:

- **Direct Relationship**: The formula tells us that the force acting on an object is equal to the mass of that object multiplied by its acceleration. This means:
    - If you apply a greater force to an object, it will accelerate more.
    - If the mass of the object is larger, it will require more force to achieve the same acceleration.

Summary:

In simple terms, **F=ma** means that the force you apply to an object depends on two things: how heavy (**mass**) the object is and how quickly you want it to speed up (acceleration). For example, pushing a car (which has a lot of mass) requires more force than pushing a bicycle (which has less mass) to get them to move at the same speed.

37. **Explain in simple terms what the following formula in physics means: Gμv=8 * π * GTμv**

The formula **Gμv=8 * π * GTμv** is an equation from Einstein's General Theory of Relativity that describes how matter and energy in the universe affect the curvature of space-time. Here's a breakdown of what each part means in simple terms:

Key Components:

1. **Gμv=8 * π * GTμv:** This represents the **Einstein tensor**, which describes the curvature of space-time due to gravity. In simpler terms, it tells us how space-time is bent or warped by the presence of mass and energy.
2. **Tμv:** This represents the **stress-energy tensor**, which describes the distribution of matter and energy in space. It includes things like mass density, pressure, and energy flow. Essentially, it tells us how much mass and energy are present and how they are distributed in space.
3. **G**: This is the **gravitational constant**, a number that helps measure the strength of gravity in the universe. Its value is approximately G = 6.674×10⁻¹¹N *m2/kg²
4. **8π(pi)**: This is a numerical factor that comes from the mathematics of how gravity works in the context of space-time. It's a constant that helps balance the equation. 5.
5. **Newton meter** squared per kilogram squared (Nm²/kg²).

6. **Value:** 6.674 10⁻¹¹

What the Formula Means:

- **Relationship Between Mass/Energy and Curvature**: The formula states that the curvature of space-time (on the left side) is directly related to the amount of mass and energy present (on the right side). In simple terms, the more mass and energy you have in a region of space, the more it warps the space around it.
- **Gravity as Curvature**: This formula is essentially a way of saying that gravity is not just a force but is a result of the curvature of space-time caused by mass and energy. For example, the Earth bends space-time around it, and this curvature is what keeps the Moon in orbit.

Summary:

In simple terms, **Gµv=8 * π * GTµv** explains how matter and energy influence the shape of the universe. More mass and energy lead to greater curvature of space-time, which is what we perceive as gravity. This equation shows the deep connection between gravity and the structure of the universe.

38. **Explain in simple terms what the following formula in physics means: a = e2/ř* c = 1/137**

The formula **a = e2/ř* c = 1/137** relates to a fundamental constant in physics known as the **fine-structure constant**. Here's what it means in simple terms:

Key Components:

1. **a**: This represents the **fine-structure constant**, a dimensionless number that characterizes the strength of electromagnetic interactions between charged particles (like electrons and protons).
2. **e**: This is the **elementary charge**, which represents the charge of a single electron or proton. It's a fundamental property of particles.
3. **ℏ (h-bar)**: This is the **reduced Planck's constant**, a very small number that is essential in quantum mechanics. It relates to the scale of quantum effects.
4. **c**: This is the **speed of light in a vacuum**, approximately **299,792,458 m/s**. It's a constant that describes how fast light travels.
5. **1/137**: This number is approximately the value of the fine-structure constant. It is significant because it indicates the strength of the electromagnetic force compared to other forces in nature.

What the Formula Means:

- **Characterizing Electromagnetic Interactions**: The fine-structure constant **a** tells us how strong the electromagnetic interactions are between charged particles. A value of **1/137** suggests that these interactions are relatively weak compared to the strong force (which holds atomic nuclei together).

- **Connection to Fundamental Forces**: This constant is a key factor in determining the behavior of electrons in atoms and influences the structure of atoms, the properties of light, and the nature of chemical bonds.

Summary:

In simple terms, the formula **a = e2/ř* c = 1/137** tells us about the strength of electromagnetic forces between charged particles. The value of **1/137** shows that these forces are fundamental to understanding how atoms behave and interact in the universe. This constant is a bridge between quantum mechanics and electromagnetic theory, and it plays a crucial role in the fundamental laws of physics.

39. **Explain in simple terms what the following formula in physics means: The Lagrangian (L) for any system is defined as the difference between its kinetic energy (KE) and potential energy: Formula: (PE), L = KE – PE**

The formula for the **Lagrangian L** is:

KE – PE

Key Terms:

1. **Lagrangian (L)**: This is a function that summarizes the dynamics of a system. It's a central concept in a field of physics called **Lagrangian mechanics**, which helps us understand how systems move and change over time.
2. **Kinetic Energy (KE)**: This is the energy that an object has because of its motion. For example, a rolling ball has kinetic energy due to its speed.
3. **Potential Energy (PE)**: This is the stored energy in an object due to its position or condition. For instance, a book on a shelf has potential energy because of its height; if it falls, that energy can be converted into kinetic energy.

What the Formula Means:

- **Difference Between Energies**: The Lagrangian is calculated by subtracting the potential energy from the kinetic energy of the system.
- **Insight into Motion**: By using this difference, physicists can derive equations of motion for the system. Essentially, the Lagrangian provides a way to describe how an object moves by looking at how much energy it has in motion compared to the energy it could have due to its position.
- **Finding the Best Path**: The Lagrangian helps to determine the "best" path that a system can take through its motion. This path minimizes action, which is a measure of the overall energy used in the system over time.

Summary:

In simple terms, the formula **(PE), L = KE − PE** helps us understand how a physical system behaves by looking at the balance between its motion (kinetic energy) and its position (potential energy). This balance is crucial for analyzing the motion of objects and is a powerful tool in physics to predict how systems evolve over time.

40. **Explain in simple terms what the following formula in physics means: Mu → E+y.**

The formula **Mu → E+y** can be interpreted in a few different contexts, but here's a general explanation of its components in simple terms:

Key Terms:

1. **Mu**: This typically represents a variable or quantity in physics, often used to denote a mass, a coefficient (like friction), or a specific particle type in particle physics.
2. **E**: This usually stands for energy. In physics, energy can come in many forms, such as kinetic energy (energy of motion), potential energy (stored energy), or other forms depending on the context.
3. **y**: This could represent another quantity, which might be a variable related to energy, such as a force, potential, or another contributing factor to the overall system.

What the Formula Means:

- **Transition or Transformation**: The arrow in **Mu → E+y** typically signifies a transformation or transition from one state to another. In this case, it suggests that **mu** is transforming into or leading to the sum of **E** (energy) and **y** (another quantity).
- **Combining Quantities**: The equation suggests that whatever **mu** represents can be expressed as the combination of energy (**E**) and another variable (**y**). This can imply that **mu** is derived from or influenced by these two quantities.

Summary:

In simple terms, the formula **Mu → E+y** indicates that the quantity represented by **mu** is related to or can be expressed as a combination of energy and another variable. This suggests a relationship between these quantities and might be used to describe a physical process or transition in a system.

41. **Explain in simple terms what the following formula in physics means: the 1/r2 problem**

The "**$1/r^2$" problem**" typically refers to how certain forces in physics, like gravity and electrostatics, decrease in strength as the distance from the source increases. Here's a breakdown of what this means in simple terms:

Key Concepts:

1. **Inverse Square Law**: The term "**1/r²**" indicates an inverse relationship between the force and the square of the distance (**r**) from the source. This means that if you double the distance from the source, the force will be reduced to one-fourth of its original strength. If you triple the distance, the force will be reduced to one-ninth, and so on.
2. **Forces Involved**:
    - **Gravity**: The force of gravity between two objects (like the Earth and an apple) follows this rule. The farther away you are from the Earth, the weaker the gravitational pull you feel.
    - **Electric Force**: The same principle applies to electric charges. If you have two charged objects, the force between them decreases rapidly as you move them farther apart.

Implications of the 1/r² Problem:

- **Distance Matters**: The farther you are from the source of a force (like a planet or an electric charge), the weaker that force becomes.
- **Understanding Motion**: This principle helps explain how objects move in space. For example, planets orbiting the Sun experience a gravitational force that weakens as they move farther away, affecting their speed and path.
- **Real-World Examples**: Think about how you feel gravity on Earth. If you were to travel to the Moon, which is farther away from Earth, you would weigh less because the gravitational pull is weaker there.

Summary:

The "1/r²" problem" highlights that certain forces diminish in strength quickly as the distance from their source increases. This principle is fundamental in understanding how gravity and electrostatic forces work in the universe, impacting everything from falling objects to the orbits of planets.

42. Explain in simple terms what the following formula in physics means: $G \cdot \hbar / c^5$

The formula $G \cdot \hbar / c^5$ combines three important constants in physics:

1. **G**: This is the gravitational constant, which measures the strength of gravity between two objects. It helps determine how much force is exerted by gravity based on the mass of the objects and the distance between them.
2. **$\hbar$ (h-bar)**: This is a reduced Planck constant, which is fundamental in quantum mechanics. It relates to the scale of quantum effects and plays a key role in the behavior of particles at very small sizes.
3. **c**: This is the speed of light in a vacuum, a fundamental constant that represents how fast light travels. It is incredibly fast, about **300,000** kilometers per second (or roughly **186,000** miles per second).

What the Formula Means:

- **Combination of Forces**: The formula combines gravitational interactions (through **G**), quantum effects (through **ℏ**), and the speed of light (through **c**). This combination often appears in theoretical physics, especially in discussions of gravity at the quantum level.
- **Dimensional Analysis**: The specific arrangement of these constants shows how they can relate different physical quantities in a way that maintains consistency in the units of measurement. In essence, it helps physicists understand how gravity behaves in the framework of quantum mechanics and relativity.

Applications:

- **Black Hole Physics**: This formula can appear in theories about black holes, where both quantum mechanics and general relativity play significant roles. It can help in deriving units related to energy, mass, and space-time geometry in extreme conditions.
- **Theoretical Investigations**: Physicists use this kind of formulation in research areas such as quantum gravity, where they seek to unify the laws of gravity (general relativity) with quantum mechanics.

Summary:

In simple terms, $G \cdot \hbar / c^5$ is a formula that brings together key aspects of gravity, quantum physics, and the speed of light to help understand complex interactions in the universe, especially in extreme environments like black holes.

43. **Explain in simple terms what the following formula in physics means: N=8" supergravity**

The formula **N=8"** in the context of "supergravity" refers to a specific type of theoretical physics that combines principles of **supersymmetry** and **gravity**.

Key Concepts:

1. **Supergravity**: This is a field of theoretical physics that attempts to unify gravity (described by general relativity) with supersymmetry, a theoretical framework that suggests every particle has a partner particle with different spin properties. In supergravity, the force of gravity is incorporated into the framework of quantum mechanics.
2. **N**: The N=8 typically denotes the number of supersymmetry generators in a particular supergravity theory. In simple terms, these generators are mathematical entities that relate the various particles in the theory and help describe their interactions.
3. **8 Supergravity**: When we say **8** super-gravities, we are specifically referring to a version of supergravity that has 8 supersymmetry generators. This is one of the most studied versions and is known for its rich structure and mathematical properties.

What It Means:

- **Strong Theoretical Framework**: supergravity is significant because it is one of the most promising candidates for a unified theory of physics that might explain all fundamental forces and particles.
- **Mathematical and Physical Properties**: The **N=8"** supergravity theory exhibits certain mathematical properties that make it a candidate for a more complete theory of quantum gravity. This theory has been extensively analyzed for its potential to resolve various issues in physics.
- **Connections to String Theory**: **N=8"** supergravity has connections to string theory, a broader framework that aims to explain all fundamental forces and particles in the universe. In some contexts, studying **N=8"** supergravity can lead to insights about string theory.

Summary:

In simple terms, **N=8** in the context of supergravity refers to a specific theoretical framework that combines gravity with the concept of supersymmetry, featuring 8 generators that describe how particles and forces interact. This framework is important for efforts to unify the fundamental forces of nature and explore the connections between quantum mechanics and gravity.

44. **Explain in simple terms what the following formula in physics means: string coordinates, $X\mu(\tau,\rho)$**

The formula $X\mu(\tau,\rho)$ is used in the context of **string theory**, which is a theoretical framework in physics that describes particles as one-dimensional "strings" rather than point-like objects. Here's a breakdown of the terms in the formula:

Key Concepts:

1. **String Theory**: This theory posits that the fundamental building blocks of the universe are not just particles (like electrons and quarks) but tiny, vibrating strings. These strings can stretch and vibrate in different ways, and their vibrations determine the properties of the particles we observe.
2. **Coordinates**: In this formula, $X\mu(\tau,\rho)$ represents the coordinates of a point in spacetime. The μ index typically runs from 0 to 3, which means it describes a four-dimensional spacetime (1 time dimension and 3 spatial dimensions). So, X0 would be the time coordinate, and X1, X2, and X3 would be the spatial coordinates (like length, width, and height).
3. **Parameters of $X\mu(\tau,\rho)$**
    - **τ (tau)**: This is often referred to as the **world-sheet time** parameter. It represents how time passes along the string as it moves through spacetime. Essentially, it's a way of tracking the evolution of the string over time.
    - **ρ (rho)**: This parameter typically represents a position along the string itself. It describes where you are on the string, similar to how we might describe a point along a line.

What It Means:

- **Describing String Motion**: The formula **Xμ(τ,ρ)** gives a mathematical way to describe how a string moves and vibrates through spacetime. It tells us the position of points on the string at different times and different points along its length.
- **Visualizing the String**: You can think of **Xμ(τ,ρ)** as a way to visualize the entire history of a string as it moves and vibrates. By **Xμ(τ,ρ)** position along the string, you can trace out the path the string takes through the four dimensions of spacetime.

Summary:

In simple terms, **Xμ(τ,ρ)** is a mathematical expression in string theory that describes the position of a vibrating string in spacetime over time. It shows how the string moves and evolves, allowing physicists to study its properties and interactions.

45. Explain in simple terms what the following formula in physics means: compactified dimension of radius α'/R

The phrase **"compactified dimension of radius α›/R"** comes from theories like string theory, where additional dimensions beyond the familiar three dimensions of space and one dimension of time are proposed. Here's a simple explanation:

Key Concepts:

1. **Compactified Dimensions**: In string theory, it is suggested that there are more than the usual four dimensions (three space + one time). Some of these extra dimensions are "compactified," meaning they are curled up into very small shapes. Imagine a circle: while it is one-dimensional, you can think of it as having a second dimension in how it curves.
2. **Radius (R)**: The term **R** refers to the size of the compactified dimension. For example, if RRR represents the radius of a circle, then it describes how "large" the curled-up dimension is.
3. **Alpha Prime (α'** or **alpha')**: This is a fundamental parameter in string theory related to the tension and properties of strings. It is a constant that helps determine the scale of the extra dimensions. Think of it as a measure of how "stretched" or "compressed" the strings are.
4. **Combined Expression**: The expression **α›/R** represents the effective size of the compactified dimension. If **R** is small, then α' is large, meaning the effective size of the dimension is significant in comparison to other scales.
5. **Geometric Picture**: Imagine you have a string vibrating in a three-dimensional space, but there's an extra dimension curled up very tightly, like a straw. The radius of this straw can be described by **R**. The term **α›/R** helps determine how this curled dimension affects the behavior of strings and particles.

Summary:

In simple terms, **"compactified dimension of radius α'/R"** means you have an extra dimension in string theory that is curled up into a small size, and the expression α'/R helps describe its effective size or properties based on the tension of the strings and the radius of the compactification. It plays a pivotal role in understanding how extra dimensions might influence physical phenomena in our universe.Top of Form

Glossary

Abelian group: A mathematical group of transformations where the outcome remains the same regardless of the order in which the transformations are applied.

Aberration: A small apparent displacement of fixed stars from their average positions on the celestial sphere, occurring with a period of one year.

Absolute temperature: Temperature measured on the Kelvin scale, where 0 K = -273.15°C.

The temperature measured on the Kelvin scale: zero kelvin = -273.15 degrees Celsius. Absolute temperature is directly related to (kinetic) energy via the equation $E = k^{bxT}$, where k^b is Boltzmann's constant. So, a temperature of K corresponds to zero energy, and room temperature, 300 K = 27-degrees C or (300 K = 27°C), corresponds to an energy of 0.025 eV.

Acceleration: The rate at which an object's velocity changes over time, typically measured in units like meters per second squared (m/s^2) or miles per hour per second. Acceleration can involve a change in speed, direction, or both. Forward acceleration occurs with a change in speed, while circular motion involves acceleration at right angles to the direction of movement.

Action at a distance: According to Newton, objects influence each other's motions by exerting forces across empty space, a concept referred to as action at a distance. This differs from Maxwell's theory of electromagnetism, which involves forces mediated by electromagnetic fields. The speed at which such forces propagate, including electromagnetic and gravitational effects, is approximately 300,000 km/s (186,000 miles per second), symbolized by ccc.

Air speed: The velocity of an aircraft relative to the surrounding air mass.

Alpha decay: A type of radioactive decay where an unstable nucleus spontaneously emits alpha particles.

Alpha particle: A particle consisting of two protons and two neutrons, equivalent to the nucleus of a helium atom.

Alpha ray: An alpha particle emitted during radioactive decay.

Alpha (α) particles: Particles first observed in radioactive α decay, later identified as helium nuclei consisting of two protons and two neutrons bound together.

Amplitude: See quantum-mechanical amplitude.

Angular momentum: The rotational equivalent of linear momentum, defined as mass x velocity x orbital radius. It is a vector quantity aligned with the axis of rotation. In quantum mechanics, orbital angular momentum is quantized in discrete multiples of $\hbar$ or (h-bar). This corresponds to only specific frequencies of rotation being allowed.

Antiparticles: Particles predicted by combining special relativity and quantum mechanics. For every particle, there exists an antiparticle with the opposite charge, magnetic moment, and internal quantum numbers (such as lepton number, baryon number, etc.), but the same mass, spin, and lifetime. Some neutral particles, like the photon, are their own antiparticles.

Asymptotic freedom: A phenomenon where the strength of the color force between quarks decreases as they get closer to each other. At extremely small separations, quarks behave as if they are free. This contrasts with the electromagnetic force, which increases in strength as charged particles come closer together.

Baryogenesis: The process responsible for generating the net baryon number in the universe, explaining why the universe is composed mostly of baryons (matter) and not antibaryons (antimatter).

Baryons: A class of strongly interacting particles with half-integer spin, such as protons, neutrons, and their heavier resonance states. Baryons are fermions, which means they follow the Pauli exclusion principle and have half-integer spins.

Beta particle: An electron or neutrino emitted during the spontaneous decay of certain radioactive nuclei.

Beta (β) particles: Particles first discovered in radioactive β decay, later identified as electrons.

B-factory: High-energy electron-positron collider experiments designed to study bottom quarks by tuning to the γ resonance.

Big-Bang Theory: The most widely accepted theory for the origin of the universe, stating that it began approximately 10 billion years ago from a singularity, a point of infinite energy density. Since then, the universe has been expanding, similar to how the surface of an inflating balloon grows, with every point moving away from every other point.

Black Body: An ideal object that absorbs all incoming radiation and is also a perfect emitter of radiation.

Black Dwarf: The remnant of a white dwarf star that has cooled down and no longer emits significant heat or light. This is the final stage of stellar evolution for stars like the Sun.

Black Hole: A region in space where gravity is so strong that nothing, not even light, can escape from it.

Boson: A particle that follows Bose-Einstein statistics, with integer spin values such as 0, $\hbar$, $2\hbar$, etc. Bosons are force carriers in particle physics.

Bosonic String Theory: A version of string theory that only includes bosons and lacks supersymmetry, meaning it cannot describe matter. It is mainly used as a theoretical model, requiring 26 space-time dimensions for internal consistency, and involves both open and closed strings.

Calculus: A branch of mathematics independently developed by Isaac Newton and Gottfried Wilhelm von Leibniz. It deals with variable quantities and their rates of change, described using functions. For example, the free fall of a body can be described in terms of its position as a function of time.

Cartesian Coordinate System: A system of coordinates that uses straight, perpendicular lines (axes) to locate points in space. The axes are divided into uniform units of length, such as centimeters. Points are represented by ordered sets of numbers (x, y, z), corresponding to their position along the axes.

Cabibbo Angle (α_c): A parameter that measures the probability of quark flavor changing quark (*mu*) (e.g., from *d* or *s* quarks) under the influence of the weak force.

CERN: The European Organization for Nuclear Research (formerly Conseil Européen pour la Recherche Nucléaire), located near Geneva, Switzerland. CERN is the world's largest laboratory for particle physics and is home to the Large Hadron Collider (LHC), one of the most significant high-energy physics projects in the world.

Charm: The fourth flavor (type) of quark, whose discovery in 1974 helped confirm the quark model and enhanced our understanding of quark dynamics. The "charm" property is conserved in strong interactions, just like other quantum properties.

Chirality: The "handedness" of a relativistic fermion, defined by how it transforms under Lorentz transformations. It distinguishes left-handed and right-handed particles.

Coherent: Light beams in which the waves vibrate in synchrony, maintaining a constant phase relationship.

Color: A property used to differentiate quarks of the same flavor. Quarks come in three "colors"—red, green, and blue—but these labels have no relation to visual color. In particle physics, "color" is the source of the strong force, which binds quarks together in baryons and mesons, similar to how electric charge creates electromagnetic forces.

Conservation Law: A principle in physics that states certain quantities, like energy or momentum, remain constant in isolated systems throughout a process.

Conservation of Energy: A fundamental principle stating that the total energy in a closed system remains constant. Energy cannot be created or destroyed, only transformed or transferred.

Cosmological Constant: A term added by Einstein to his equations of general relativity to account for a repulsive force (antigravity) that counteracts gravitational collapse. There is current evidence suggesting that a cosmological constant might be responsible for the accelerated expansion of the universe.

Cosmological Principle: The hypothesis that, on very large scales, the universe is both isotropic (the same in all directions) and homogeneous (the same at every point).

Coupling Constant: A number that defines the intrinsic strength of a fundamental force in particle interactions. For example, the electromagnetic coupling constant ($\alpha = e^2/\hbar c \approx 1/137$) quantifies the strength of the electromagnetic interaction.

Cross-section (σ): A measure of the likelihood that particles will interact when they collide. It represents the effective target area for interaction, usually measured in square centimeters. A typical unit is the barn (b), where $1\ b = 10^{-24}\ cm^2$. Neutrino cross-sections are often much smaller, around $10^{-39}\ cm^2$.Top of Form

Dark matter: A mysterious, invisible substance that cannot be directly observed but is inferred from its gravitational effects on visible matter, galaxies, and light.

Dark matter and energy: Unknown sources of energy density and matter needed to explain certain cosmological phenomena, such as the accelerated expansion of the universe. These components are essential in modern physics and cosmology but remain largely unexplained.

Deuteron: The nucleus of deuterium, an isotope of hydrogen. It consists of one proton and one neutron bound together.

Differential calculus: A branch of mathematics that studies how a function changes as its input changes. The derivative of a function represents the rate of change. For example, the derivative of a displacement function with respect to time gives the instantaneous speed, while the derivative of speed with respect to time gives acceleration.

Diffraction: A wave phenomenon where waves, such as light or sound, bend or spread as they pass through a narrow opening or encounter an obstacle. Diffraction distinguishes wave-like behavior and is most noticeable when the size of the opening is comparable to the wavelength.

Dimensions: Quantities in physics have associated dimensions that describe their physical nature. The fundamental dimensions are mass (M), length (L), and time (T). Other physical quantities, like momentum and energy, can be expressed in terms of these fundamental dimensions. Dimensionless quantities, which have no dimensions, are important because they are independent of measurement units.

Doppler effect: A change in the wavelength or frequency of a wave, such as sound or light, as perceived by an observer moving relative to the source. This effect causes shifts in pitch for sound and shifts in color (redshift or blueshift) for light.

Duality: A concept in superstring theory that shows equivalence between different string theories. Duality often reveals hidden relationships between seemingly distinct physical theories.

Eigenstate, eigenvalue: In quantum mechanics, an eigenvalue is a number λ that satisfies the equation $M\psi = \lambda\psi$, where M is a matrix representing a physical quantity (like position or energy) and ψ is the eigenstate. The eigenvalue represents the measured value of the quantity when the system is in that eigenstate.

Electric current: The rate at which electric charges move through a conductor or space.

Electron: A fundamental particle with a negative electric charge and no known substructure. It is a spin-$^{1/2}$ particle and interacts via electromagnetic, weak, and gravitational forces. Its mass is 0.511 *MeV/c²*, making it about 1,800 times lighter than a proton.

Elastic scattering: A type of particle interaction where the same particles emerge from the reaction as those that entered (e.g., $\pi^- + p \to \pi^- + p$). In contrast, in **inelastic scattering**, new particles may emerge, and energy is used to create these particles.

Energy: The capacity to perform work or the result of performing work.

Entropy: A measure of the disorder or randomness in a system. It quantifies the level of order or information within a physical system, typically expressed in units of energy per temperature.

Epicycles: Circular paths that, according to Ptolemy's geocentric model, planets followed around the Earth. This idea was later disproven by Copernicus with his heliocentric model.

Escape velocity: The minimum speed required for an object to leave Earth (or any celestial body) without falling back. For Earth, this speed is 11 kilometers per second (about 7 miles per second).

Ether: A hypothetical invisible substance once thought to fill all space and serve as the medium through which light travels. Einstein's theory of relativity rendered the ether concept obsolete.

Event horizon: The boundary surrounding a black hole beyond which nothing, not even light, can escape. It is defined by the Schwarzschild radius.

Fermilab: The Fermi National Accelerator Laboratory in Batavia, Illinois, USA. Fermilab houses the Tevatron, a proton-antiproton collider that was the most powerful particle accelerator until the Large Hadron Collider surpassed it.

Fermion: A particle with half-integer spin (e.g., $1/2\hbar$, $3/2\hbar$) that obeys the Pauli exclusion principle. Examples include electrons, protons, and neutrons.

Field: A region of space affected by a force due to the presence of a body, such as a gravitational, electric, or magnetic field. This distortion in space influences other bodies within the field.

Flavor: A term used to describe different types of quarks. There are six quark flavors: up, down, strange, charm, bottom, and top.

Galaxy: A vast collection of billions of stars, gas, dust, and dark matter, all bound together by gravity.

Gamma rays (Γ-rays): Electromagnetic radiation of extremely high energy, emitted by radioactive nuclei and certain cosmic processes. Gamma rays are a form of photon with much higher energy than visible light.

Geocentric: Earth-centered. The geocentric model, famously developed by Ptolemy, proposed that the Earth is the center of the Universe, with the Sun, planets, and stars revolving around it.

Gauge theory: A theory where the dynamics arise from a certain symmetry. The equations describing the system, particularly the Lagrangian, remain unchanged under specific symmetry transformations called "gauge transformations." Examples include quantum electrodynamics (QED) and quantum chromodynamics (QCD).

Generation: In particle physics, quarks and leptons are grouped into three sets, or generations. Each generation consists of two quarks and two leptons. The first generation includes (e, νe; u, d), the second (μ, νμ; c, s), and the third (τ, ντ; t, b).

Gluon: A massless gauge boson that mediates the strong force (color force) between quarks in quantum chromodynamics (QCD). Gluons carry color charge and can interact with each other, potentially forming particles called glue balls, although these have not yet been definitively observed.

Gravitational acceleration: The acceleration of an object due to the force of gravity at the Earth's surface, approximately 32 feet per second squared (9.8 m/s^2).

Graviton: A hypothetical massless spin-2 particle that mediates the force of gravity, much like how photons mediate the electromagnetic force. The graviton is predicted in quantum gravity theories but has not yet been experimentally detected.

Hadron: Any particle that experiences the strong nuclear force. Hadrons are either baryons (such as protons and neutrons) or mesons (such as pions).

Heisenberg's Uncertainty Principle: A fundamental concept in quantum mechanics stating that it is impossible to simultaneously know both the exact position and the exact momentum (velocity and direction) of a particle. The more precisely one is known, the less precisely the other can be determined.

Helicity: The projection of a particle's spin onto the direction of its motion. For a massless particle, helicity is the same as chirality.

Higgs boson: A fundamental particle, predicted by the Standard Model of particle physics, which was experimentally confirmed in 2012. It is associated with the Higgs field, and the interaction between particles and this field is what gives them mass. The Higgs boson is a scalar particle with no spin.

Higgs mechanism: The process by which particles acquire mass in certain gauge theories. In the Standard Model, the spontaneous breaking of electroweak symmetry by the Higgs field gives mass to the W± and Z bosons, responsible for mediating the weak nuclear force, while leaving the photon massless.

Hyperon: A type of baryon that contains one or more strange quarks. Examples include particles like the lambda (Λ) and sigma (Σ) baryons, which are more massive than protons and neutrons due to the inclusion of strange quarks.

Inertia: The property of an object that resists changes in its state of motion. It is related to mass—objects with greater mass have more inertia.

Integral calculus: The branch of calculus that deals with finding the total size, area, or value by adding up small quantities. It is essentially the reverse of differentiation and is used to calculate things like areas under curves and volumes of irregular shapes.

Interference: A phenomenon in which two or more waves overlap, resulting in a new wave pattern. Depending on how the waves align, the interference can be constructive (waves reinforce each other) or destructive (waves cancel each other out).

Isotopic spin (isospin): A quantum number introduced by Werner Heisenberg to explain the charge independence of the strong nuclear force. It treats protons and neutrons as different states of the same particle, called a nucleon. The concept of isospin is similar to spin, with nucleons having an isospin of 1/2, and the third component of isospin (I_3) determining whether the nucleon is a proton ($I_3 = +1/2$) or a neutron ($I_3 = -1/2$).

Kelvin: The SI unit of temperature that measures the absolute temperature, starting from absolute zero (0 K), the theoretical point at which all particle motion ceases.

K meson or kaon: A type of meson, specifically a spin-0 particle with a non-zero strangeness quantum number, meaning it contains a strange quark. Kaons play an important role in studying weak interactions.

Lagrangian: A mathematical function used to describe the dynamics of a physical system. It is the difference between the system's kinetic energy and potential energy and is central to Lagrangian mechanics, a fundamental approach in classical and quantum field theory.

Laser: A device that amplifies light by stimulating the emission of radiation. Lasers produce a narrow, intense beam of coherent light at a single wavelength.

Lepton: A class of elementary particles that do not experience the strong nuclear force. Examples include the electron, muon, tau, and their associated neutrinos. Leptons are spin-1/2 particles, making them fermions.

Mass: In the context of subatomic particles, mass is the property that determines an object's resistance to acceleration. For particles like the neutrino, the exact mass remains uncertain but is thought to be either zero or extremely small.

Mesons: Subatomic particles made up of one quark and one antiquark, classified as bosons due to their integral spin. Examples include the pion and kaon. Mesons mediate the strong force between nucleons in the nucleus.

M-Theory: A theoretical framework believed to unify all five previously known superstring theories. It suggests that fundamental particles are one-dimensional strings existing in 11 dimensions (10 spatial and 1 temporal). M-theory incorporates dualities like S-duality and T-duality to connect different types of string theories.

Neutron: A subatomic particle found in the nucleus of an atom, with no electric charge and a slightly greater mass than a proton.

Neutron star: A compact, incredibly dense object left behind after a massive star explodes in a supernova. Neutron stars are composed primarily of neutrons and are typically only about 10–20 kilometers in diameter, yet they can be more massive than the Sun.

Nucleons: Particles that make up the nucleus of an atom, namely protons and neutrons.

Pauli's exclusion principle: A principle in quantum mechanics that states no two identical fermions (such as electrons) can occupy the same quantum state within a given system.

Perturbation theory: A mathematical approach used to calculate approximate solutions to complex problems in quantum mechanics. It involves summing contributions from simpler systems, often represented by Feynman diagrams, to account for increasingly small effects.

Photon: The fundamental quantum of light and other forms of electromagnetic radiation. Photons are massless particles that carry energy and momentum, and they travel at the speed of light in a vacuum.

Photon (γ): In the context of quantum electrodynamics (QED), the photon is the gauge boson that mediates the electromagnetic force between charged particles. Virtual photons facilitate this force exchange even though they cannot be directly observed.

Planck units: A set of natural units that define fundamental physical constants, such as length, time, and mass, based on Planck's constant ($\hbar$), Newton's gravitational constant (G), and the speed of light (c). Planck units are essential in quantum gravity and the study of extremely small scales, like the Planck length.

Positron: The antiparticle of the electron, with the same mass and spin, but with positive electric charge. It was discovered by Carl Anderson in 1934 and plays a role in processes like particle-antiparticle annihilation.

Proton: A positively charged subatomic particle found in the nucleus of an atom. Protons, along with neutrons, form the atomic nucleus, and the number of protons in an atom determines its chemical element.

Pulsar: A highly magnetized, rotating neutron star that emits beams of electromagnetic radiation. Pulsars are detected by their regular pulses of radio waves and light, produced as the star rotates.

Quanta: Discrete packets or bundles of energy, which are fundamental in quantum theory. Electromagnetic radiation, such as light, consists of quanta, with photons being the quanta of light.

Quasar: An extremely luminous astronomical object, typically located in distant galaxies. Quasars are believed to be powered by supermassive black holes at the centers of galaxies and can outshine entire galaxies.

Quantum chromodynamics (QCD): The theory describing the strong interaction, which governs the behavior of quarks and gluons. It is a non-Abelian gauge theory with the symmetry group $SU(3)$, and the gluon is the mediator of the strong force between quarks.

Quantum electrodynamics (QED): The quantum field theory that describes how electrically charged particles interact via the electromagnetic force, mediated by photons. QED is a gauge theory with the symmetry group $U(1)$ and provides precise predictions for the behavior of electromagnetic interactions.

Quantum field theory (QFT): The theoretical framework for describing particles as excitations in underlying fields. It unifies quantum mechanics and special relativity and is used to explain the behavior of fundamental particles and forces.

Quantum-mechanical amplitude: A complex mathematical quantity in quantum mechanics, whose absolute square gives the probability of a specific event or process occurring. It is used in calculations involving the behavior of particles and fields.

Quantum theory: A fundamental theory in physics that describes the behavior of physical systems at atomic and subatomic scales. It introduces the concept of quantization, where certain physical quantities, such as energy and angular momentum, can only take on discrete values known as quanta.

Radioactive (α) decay: A type of radioactive decay in which an unstable atomic nucleus emits an alpha particle (two protons and two neutrons). This process allows the nucleus to decrease in size and move towards a more stable configuration on the binding energy curve.

Radioactivity: The natural process by which unstable atomic nuclei emit particles or radiation, transforming into more stable forms. This emission can occur in various forms, including alpha, beta, and gamma decay.

Red giant: A late stage in the life cycle of a star characterized by a significant increase in size and a decrease in surface temperature. The Sun is expected to become a red giant in its later years before eventually cooling down and becoming a black dwarf.

Reference frame: A coordinate system or viewpoint used to measure the position and motion of objects. It provides a standard from which to observe and describe the behavior of moving bodies.

Schwarzschild radius: The critical radius of a black hole, beyond which nothing, not even light, can escape the gravitational pull of the singularity at its center. It defines the event horizon, marking the boundary of the black hole.

Singularity: A point in space-time where gravitational forces cause matter to have an infinite density, leading to a breakdown of known physical laws. Singularities are often associated with black holes and the Big Bang.

Space-time: The four-dimensional continuum that merges the three dimensions of space with the dimension of time. In relativity, events are described within this framework, emphasizing that time and space are interwoven.

Spectrum: The range of electromagnetic radiation emitted by a source, organized according to wavelength or frequency. Different elements and compounds produce unique spectra, allowing scientists to identify their presence in various contexts.

Standard Model: A well-established theoretical framework that describes the fundamental particles and interactions in the universe, encompassing quantum chromodynamics (QCD) and the electroweak theory proposed by Glashow, Weinberg, and Salam.

String theory: A theoretical framework suggesting that the fundamental constituents of matter are not point-like particles but rather tiny, one-dimensional strings. These strings vibrate at specific frequencies, and their modes of vibration correspond to different particles.

Superstring theory: An extension of string theory that posits all elementary particles are manifestations of vibrating strings. Superstring theory aims to unify all fundamental forces in nature, potentially providing a comprehensive framework for understanding the universe.

Super-symmetry: A theoretical extension of Lorentz symmetry that links fermions (matter particles) and bosons (force-carrying particles). If super-symmetry exists in nature, each known particle will have a corresponding "super-partner" differing in spin by half a unit, which could help solve various problems in particle physics, such as the hierarchy problem.

Tachyons: Hypothetical particles that would always move faster than light. The concept of tachyons suggests that they could potentially travel backward in time, leading to intriguing implications for causality and the nature of reality.

Thermodynamics: The branch of physics focused on heat, energy, and the laws governing thermal effects. It examines how energy is transferred and transformed, providing a foundation for understanding various physical processes, including engines, refrigeration, and phase transitions.

Type I string theory: A variant of string theory that incorporates both bosons and fermions. It includes super-symmetry and has a gauge group known as SO(32). Consistency in this theory requires a ten-dimensional space-time framework.

Type II-A string theory: A version of string theory that features both open and closed strings, along with super-symmetry. Open strings in Type II-A have ends attached to higher-dimensional objects called D-branes. Unlike Type II-B, fermions in this theory are not chiral.

Type II-B string theory: Similar to Type II-A but distinguished by the presence of chiral fermions. This theory involves closed strings and has a gauge group of $E_8 \times E_8$. The left- and right-moving modes of the string require different numbers of space-time dimensions: ten and twenty-six, respectively. There are actually two distinct heterotic string theories.

Uniform motion: The motion of an object traveling at a constant speed in a straight line, with no acceleration or change in direction.

Vacuum: The lowest energy state of a quantum system, often referred to as the ground state. In this state, no real particles exist; however, due to Heisenberg's uncertainty principle, the vacuum is filled with virtual particles that momentarily appear and vanish.

Velocity: A vector quantity characterized by both magnitude (the speed of an object) and direction. It provides a complete description of an object's motion.

Virtual particle: A transient particle that exists for a brief moment as permitted by Heisenberg's uncertainty principle. Virtual particles play a crucial role in quantum field interactions, mediating forces between real particles.

Virtual processes: Quantum mechanical processes that do not conserve energy and momentum over very short timescales, as allowed by Heisenberg's uncertainty principle. These processes are inherently unobservable and illustrate the peculiarities of quantum mechanics.

Wave: A mechanism for transmitting energy through space, characterized by oscillations that propagate. Waves can be mechanical (requiring a medium, like sound waves) or electromagnetic (such as light), and they carry energy without the transport of matter.

Wavefunction (ψ): A fundamental mathematical function in quantum mechanics that describes the behavior and state of a quantum particle. The wavefunction is a solution to Schrödinger's wave equation and provides critical information about the system. The probability of locating the particle at a specific point in space is determined by taking the absolute square of the wavefunction, denoted as $|\psi|^2$.

Wavelength: The distance between consecutive points of similar phase in a wave, such as the distance between two crests or two troughs. It is a key parameter in wave mechanics, as it relates to the energy and frequency of the wave.

White dwarf: A small, dense stellar remnant left after a star like the Sun exhausts its nuclear fuel and sheds its outer layers. White dwarfs are composed mostly of electron-degenerate matter and are typically very hot but have low luminosity, eventually cooling over billions of years.

Wormhole: A hypothetical tunnel-like structure in space-time that connects two separate points in the universe, potentially allowing for shortcuts between distant regions. Wormholes are a fascinating concept in theoretical physics, but their existence and stability remain unproven.

References

The Fabric of the Cosmos—Space, Time, And the Texture of Reality—Brian Greene—Vintage eBooks

Warped Passages—Unraveling the Mysteries of the Universe's Hidden Dimensions—Lisa Randall—HarperCollins eBooks

pp. 58-70 Parallel Worlds—Michio Kaku—Anchor Books

In Search of the Multiverse—Parallel Worlds, Hidden Dimensions, and the Ultimate Quest for the Frontiers of Reality—John Gribbin—John Wiley & Sons, Inc.

Cosmos—Carl Sagan

In search of Schrödinger's' Cat—John Gribbin

Consciousness—A Very Short Introduction—Susan Blackmore

Philosophy of Religion—An Introduction—William L. Rowe

The Physics of Christianity—An Introduction—Frank J. Tilper

Einstein for Dummies—$E = MC^2$—Carlos L. Calle

Black Holes and Baby Universes—Stephen Hawking

Astronomy in Everyday Life - International Astronomical Union | IAU. (2020). https://www.iau.org/public/themes/astronomy_in_everyday_life/

ASTROPHYSICS definition | Cambridge English Dictionary. (2024). https://dictionary.cambridge.org/us/dictionary/english/astrophysics

ASTROPHYSICS Definition & Meaning | Dictionary.com. (2022). https://www.dictionary.com/browse/astrophysics

Astrophysics Definition & Meaning - Merriam-Webster. (2024). https://www.merriam-webster.com/dictionary/astrophysics

Benefits of Space Exploration on Daily Life | Florida Tech Ad Astra. (2018). https://news.fit.edu/archive/benefits-of-space-exploration/

Spectroscopy-Related Discoveries You Need To Know About. (2023). https://www.photonicsonline.com/doc/spectroscopy-related-discoveries-you-need-to-know-about-0001

James Webb Space Telescope findings that changed our ... (2023). https://www.space.com/james-webb-space-telescope-2023-discoveries

A "giant" rising in the desert: World's largest telescope comes ... (2024). https://www.space.com/the-universe/a-giant-rising-in-the-desert-worlds-largest-telescope-comes-together-photo

About - ELT | ESO. (1998). https://elt.eso.org/about/

Ai and image processing - Experienced Deep Sky Imaging. (2023). https://www.cloudynights.com/topic/865696-ai-and-image-processing/

AI in astronomical image processing - x-bit-astro-imaging. (2023). https://x-bit-astro-imaging.blogspot.com/2023/10/ai-in-astronomical-image-processing.html

Analysis: How AI is helping astronomers study the universe - PBS. (2023). https://www.pbs.org/newshour/science/analysis-how-ai-is-helping-astronomers-study-the-universe

Astronomers are enlisting AI to prepare for a data downpour. (2024). https://www.technologyreview.com/2024/05/20/1092636/astronomers-are-enlisting-ai-to-prepare-for-a-data-downpour/

Astronomers Use AI to More Clearly Observe Space - NCSA. (2024). https://www.ncsa.illinois.edu/astronomers-use-ai-to-more-clearly-observe-space/

CLASS furthers understanding of the Cosmic Microwave Background. (2024). https://www.jhunewsletter.com/article/2024/03/class-furthers-understanding-of-the-cosmic-microwave-background

Cosmic Microwave Background | Center for Astrophysics | Harvard ... (2023). https://www.cfa.harvard.edu/research/topic/cosmic-microwave-background

Cosmic Microwave Background - (Galaxies and the Universe). (n.d.). https://fiveable.me/key-terms/galaxies-universe/cosmic-microwave-background

Cosmic Microwave Background (CMB) - Nasa Lambda. (1996). https://lambda.gsfc.nasa.gov/education/graphic_history/microwaves.html

Cosmic Microwave Background (CMB) radiation - ESA. (2018). https://www.esa.int/Science_Exploration/Space_Science/Cosmic_Microwave_Background_CMB_radiation

Current advances in imaging spectroscopy and its state-of-the-art ... (2024). https://www.sciencedirect.com/science/article/pii/S095741742302674X

Current Trends in Analytical Spectroscopy: Technique and ... (2018). https://www.spectroscopyonline.com/view/current-trends-analytical-spectroscopy-technique-and-instrument-modifications-both-new-and-improved

ELT | ESO. (n.d.). https://elt.eso.org/

ELT (Extremely Large Telescope) - eoPortal. (2021). https://www.eoportal.org/other-space-activities/elt

How artificial intelligence is changing astronomy. (2022). https://www.astronomy.com/science/how-artificial-intelligence-is-changing-astronomy/

James Webb Space Telescope | Capabilities, First Images, Hubble ... (2021). https://www.rmg.co.uk/stories/topics/james-webb-space-telescope-vs-hubble-space-telescope

James Webb Space Telescope | Performance.gov. (2016). https://obamaadministration.archives.performance.gov/content/james-webb-space-telescope.html

James Webb Space Telescope - NASA Science. (2024). https://science.nasa.gov/mission/webb/

Latest Results from Cosmic Microwave Background Measurements. (2021). https://www.cfa.harvard.edu/news/latest-results-cosmic-microwave-background-measurements

Machine Learning | Center for Astrophysics | Harvard & Smithsonian. (2023). https://pweb.cfa.harvard.edu/research/topic/machine-learning

News and Multimedia - ELT | ESO. (2024). https://elt.eso.org/media/

Next-Generation Cosmic Observatory Hits South Pole Stumbling Block. (2024). https://www.scientificamerican.com/article/cosmic-microwave-background-observatory-hits-south-pole-stumbling-block/

Overview - NASA Science. (2023). https://science.nasa.gov/mission/webb/about-overview/

Reconstructing astronomical images with machine learning. (2023). https://researchoutreach.org/articles/reconstructing-astronomical-images-machine-learning/

science with the european extremely large telescope - ELT - ESO.org. (2021). https://www.eso.org/sci/facilities/eelt/science/

Spectroscopy | Center for Astrophysics | Harvard & Smithsonian. (2023). https://www.cfa.harvard.edu/research/topic/spectroscopy

Spectroscopy - ESO.org. (1999). https://www.eso.org/public/teles-instr/technology/spectroscopy/

Spectroscopy 101 – Introduction - Webb Space Telescope. (2022). https://webbtelescope.org/contents/articles/spectroscopy-101--introduction

Staying Updated with Spectroscopic Techniques: How Lead ... (2024). https://www.spectroscopyonline.com/view/staying-updated-with-spectroscopic-techniques-how-lead-investigators-adapt-to-a-changing-industry

Telescope Overview | Webb. (2000). https://webbtelescope.org/news/webb-science-writers-guide/telescope-overview

The CMB: how an accidental discovery became the key to ... (2015). https://theconversation.com/the-cmb-how-an-accidental-discovery-became-the-key-to-understanding-the-universe-45126

The European extremely large telescope ("elt") project - ESO.org. (2024). https://www.eso.org/sci/facilities/eelt/

The Extremely Large Telescope is astronomy's next big thing - ABC. (2024). https://www.abc.net.au/news/science/2024-09-27/extremely-large-telescope-chile-atacama-australian-astronomy/104305650

The James Webb Space Telescope's tech breakthroughs are ... (2023). https://www.space.com/james-webb-space-telescope-tech-simulation-spinoffs

Using Luminar AI for Astrophotography | Best Practices & Results. (2024). https://astrobackyard.com/luminar-ai-astrophotography/

Webb Innovations - NASA Science. (2022). https://science.nasa.gov/mission/webb/innovations/

What's New in Spectroscopy? - The Analytical Scientist. (2023). https://theanalyticalscientist.com/techniques-tools/whats-new-in-spectroscopy-9

World's largest mirror support installed | ELT Updates - ESO.org. (2024). https://www.eso.org/public/videos/eltu-m1-cell-progress-a/

Space milestones in 2023 that are "rewriting textbooks" | PBS News. (2024). https://www.pbs.org/newshour/science/5-space-milestones-in-2023-that-are-rewriting-textbooks

Space missions to look forward to in 2024 | PBS News. (2024). https://www.pbs.org/newshour/science/6-space-missions-to-look-forward-to-in-2024

Amazing Exoplanet Discoveries - Space.com. (2022). https://www.space.com/amazing-exoplanet-discoveries

James Webb Space Telescope findings that changed our ... (2023). https://www.space.com/james-webb-space-telescope-2023-discoveries

13 of the weirdest, strangest exoplanets in the Universe. (2024). https://www.skyatnightmagazine.com/space-science/weirdest-exoplanets-universe

13 record-breaking space discoveries of 2023. (2023). https://www.space.com/record-breaking-space-discoveries-2023

[1707.07983] Radial Velocities as an Exoplanet Discovery Method. (2017). https://arxiv.org/abs/1707.07983

[2411.03597] Future directions in cosmology - arXiv. (2024). https://arxiv.org/abs/2411.03597

A Shared Frontier? Collaboration and Competition in the Space ... (2022). https://hir.harvard.edu/a-shared-frontier-collaboration-and-competition-in-the-space-domain/

A unifying theory of dark energy and dark matter: Negative masses ... (n.d.). https://www.aanda.org/articles/aa/pdf/2018/12/aa32898-18.pdf

A Weird Form of Dark Energy Might Solve a Cosmic Conundrum. (2024). https://www.scientificamerican.com/article/could-early-dark-energy-resolve-the-mystery-of-cosmic-expansion/

Accelerating Expansion of the Universe : The Result of Dark Energy ... (n.d.). https://www.ebay.com/itm/355027602351

Advancements in Instrumentation and Technology for Astronomical ... (n.d.). https://www.frontiersin.org/research-topics/63347/advancements-in-instrumentation-and-technology-for-astronomical-observations-using-small-satellite-and-high-altitude-balloon-platforms

Advancements in space observation technology enhance views ... (2024). https://scopetrader.com/advancements-in-space-observation-technology-enhance-views-from-space/

Advances in Astronomy - Wiley Online Library. (2009). https://onlinelibrary.wiley.com/journal/5081

Advancing the Integration of Biosciences Data Sharing to Further ... (2020). https://www.sciencedirect.com/science/article/pii/S2211124720314303

An updated list of space missions: Current and upcoming voyages. (2024). https://www.astronomy.com/space-exploration/space-missions-a-list-of-current-and-upcoming-voyages/

Analysis: How AI is helping astronomers study the universe - PBS. (2023). https://www.pbs.org/newshour/science/analysis-how-ai-is-helping-astronomers-study-the-universe

Ancient rocks may bring dark matter to light | Virginia Tech News. (2024). https://news.vt.edu/articles/2024/10/science-ancient-rocks-dark-matter.html

Applications of artificial intelligence in astronomical big data. (n.d.). https://www.sciencedirect.com/science/article/abs/pii/B9780128190845000067

Artificial Intelligence: The Future of Astronomical Discoveries. (2024). https://www.linkedin.com/pulse/artificial-intelligence-future-astronomical-david-borish-80hhc

Astronomers accidentally discover "dark" primordial galaxy with no ... (2024). https://www.space.com/dark-primordial-galaxy-no-stars-green-bank-observatory

Astronomers Use AI to More Clearly Observe Space - NCSA. (2024). https://www.ncsa.illinois.edu/astronomers-use-ai-to-more-clearly-observe-space/

Astronomical adaptive optics: a review | PhotoniX | Full Text. (2024). https://photonix.springeropen.com/articles/10.1186/s43074-024-00118-7

Astronomy and Physics - Technology. (2017). https://www.jpl.nasa.gov/go/astronomy-and-physics/technology/

Astronomy News - ScienceDaily. (2024). https://www.sciencedaily.com/news/space_time/astronomy/

Black Holes - latest research news and features - Phys.org. (2024). https://phys.org/tags/black+holes/

Black Holes - NASA Science. (2024). https://science.nasa.gov/universe/black-holes/

Black holes Coverage - Space.com. (2024). https://www.space.com/the-universe/black-holes

Black Holes News - ScienceDaily. (2024). https://www.sciencedaily.com/news/space_time/black_holes/

Can We Find Life? - NASA Science. (2020). https://science.nasa.gov/exoplanets/can-we-find-life/

Centre for Theoretical Cosmology: Research - University of Cambridge. (2013). https://www.ctc.cam.ac.uk/research/

COLDSIM predictions of [C II] emission in primordial galaxies. (n.d.). https://www.aanda.org/articles/aa/full_html/2024/09/aa50332-24/aa50332-24.html

Collaborative Missions - ESA Science & Technology. (2019). https://sci.esa.int/web/collaborative-missions

Color-Shifting Stars: The Radial-Velocity Method. (2013). https://www.planetary.org/articles/color-shifting-stars-the-radial-velocity-method

Communicating the gravitational-wave discoveries of the LIGO-Virgo ... (2024). https://jcom.sissa.it/article/pubid/JCOM_2307_2024_N03/

Cosmic Microwave Background | Center for Astrophysics | Harvard ... (2023). https://www.cfa.harvard.edu/research/topic/cosmic-microwave-background

Cosmic Microwave Background Radiation | AMNH. (2012). https://www.amnh.org/learn-teach/curriculum-collections/cosmic-horizons-book/cosmic-microwave-background-radiation

Cosmology | Center for Astrophysics | Harvard & Smithsonian. (2024). https://www.cfa.harvard.edu/research/science-field/cosmology

Cosmology & the Universe - University of California Observatories. (2023). https://www.ucobservatories.org/enthusiasts/ask-an-astronomer/cosmology-the-universe/

Cosmology News | The Latest Updates from our Cosmos. (2024). https://skyandtelescope.org/astronomy-news/cosmology/

Cosmology News - ScienceDaily. (2024). https://www.sciencedaily.com/news/space_time/cosmology/

Dark Energy and Dark Matter | Center for Astrophysics - Harvard CfA. (2023). https://www.cfa.harvard.edu/research/topic/dark-energy-and-dark-matter

Dark energy, explained - UChicago News - The University of Chicago. (2024). https://news.uchicago.edu/explainer/dark-energy-explained

Dark matter universe - PNAS. (2015). https://www.pnas.org/doi/10.1073/pnas.1516944112

Data Governance in Space — Key Challenges and Opportunities. (2024). https://www.cdomagazine.tech/opinion-analysis/data-governance-in-space-key-challenges-and-opportunities

Data Sharing & Tools | NASA Center for Climate Simulation. (2024). https://www.nccs.nasa.gov/services/data-sharing-and-tools

Detector Technology | Center for Astrophysics | Harvard & Smithsonian. (2023). https://www.cfa.harvard.edu/research/topic/detector-technology

Discoveries Dashboard - NASA Science. (2024). https://science.nasa.gov/exoplanets/discoveries-dashboard/

Discovery Alert: A "Super-Earth" in the Habitable Zone - Exoplanets. (2024). https://exoplanets.nasa.gov/news/1774/discovery-alert-a-super-earth-in-the-habitable-zone/

DOE Explains…Cosmology | Department of Energy. (2024). https://www.energy.gov/science/doe-explainscosmology

DOE Explains…Dark Matter - Department of Energy. (2024). https://www.energy.gov/science/doe-explainsdark-matter

Entry Points to NASA Science Data. (n.d.). https://science.nasa.gov/learn/entry-points-to-nasa-data/

Europa Clipper | NASA Jet Propulsion Laboratory (JPL). (2024). https://www.jpl.nasa.gov/missions/europa-clipper/

Europa Clipper - NASA Science. (2024). https://science.nasa.gov/mission/europa-clipper/

Exoplanet Detection: Radial Velocity Method - NASA Science. (2022). https://science.nasa.gov/resource/exoplanet-detection-radial-velocity-method/

Exoplanet Exploration: Planets Beyond our Solar System. (2022). https://exoplanets.nasa.gov/alien-worlds/historic-timeline/

Exoplanets | Center for Astrophysics | Harvard & Smithsonian. (2023). https://www.cfa.harvard.edu/research/topic/exoplanets

Exoplanets - NASA Science. (n.d.). https://science.nasa.gov/exoplanets/

Exoplanets, extraterrestrial life and beyond - Oxford Academic. (2022). https://academic.oup.com/nsr/article/9/2/nwac008/6509500

Experimental and Observational Astrophysics and Cosmology. (2022). https://physics.stanford.edu/research/experimental-and-observational-astrophysics-and-cosmology

First gravitational-wave detection of a mass-gap object merging with … (2024). https://news.northwestern.edu/stories/2024/04/first-gravitational-wave-detection-of-mass-gap-object-merging-with-neutron-star/

Fossils of Primordial Galaxies Shed Light on Creation. (2012). https://reasons.org/explore/publications/nrtb-e-zine/fossils-of-primordial-galaxies-shed-light-on-creation

From Stars to Galaxies: Building the Pieces to Build Up the Universe. (n.d.). http://www.aspbooks.org/a/volumes/table_of_contents/?book_id=90

Future directions in cosmology - arXiv. (2024). https://arxiv.org/html/2411.03597v1

Future directions in cosmology - Inspire HEP. (2024). https://inspirehep.net/literature/2846195

Future gravitational wave observatories could see the earliest black ... (2024). https://phys.org/news/2024-09-future-gravitational-observatories-earliest-black.html

Habitable zone | Astrobiology, Exoplanets & Habitability - Britannica. (n.d.). https://www.britannica.com/science/habitable-zone

How astronomers search for life on exoplanets | The Planetary Society. (2023). https://www.planetary.org/articles/how-astronomers-search-for-life-on-exoplanets

How can astronomy improve life on earth? | Center for Astrophysics. (2024). https://www.cfa.harvard.edu/big-questions/how-can-astronomy-improve-life-earth

How do NASA and ESA work together? | The Planetary Society. (2024). https://www.planetary.org/articles/how-do-nasa-and-esa-work-together

How Do Scientists Know Dark Matter Exists? (2021). https://kids.frontiersin.org/articles/10.3389/frym.2021.576034

Hunting exoplanets using the transit method | by Martin Silvertant. (2019). https://medium.com/@msilvertant/hunting-exoplanets-using-the-transit-method-918e764e5576

Impact of Exoplanet Science on Society: Professional Contributions ... (2024). https://astrobiology.com/2024/10/impact-of-exoplanet-science-on-society-professional-contributions-citizen-science-engagement-and-public-perception.html

Important Milestones - Trottier Institute for Research on Exoplanets. (2022). https://exoplanetes.umontreal.ca/en/exoplanets-101/decouvertes-marquantes/

In Depth: Exoplanets - NASA Science. (2020). https://science.nasa.gov/exoplanets/facts/

International Space Exploration Coordination Group - NASA. (2024). https://www.nasa.gov/humans-in-space/international-space-exploration-coordination-group/

International Space Station Cooperation - NASA. (2023). https://www.nasa.gov/international-space-station/space-station-international-cooperation/

Introduction to Galaxy Formation and Evolution | Higher Education ... (2019). https://www.cambridge.org/highereducation/books/introduction-to-galaxy-formation-and-evolution/358E14A140A1234836437B9B3B0637DE

Introduction to Galaxy Formation and Evolution: From Primordial ... (n.d.). https://www.amazon.com/Introduction-Galaxy-Formation-Evolution-Present-Day/dp/1107134765

James Webb Space Telescope - NASA Science. (2023). https://science.nasa.gov/mission/webb/

Learn About the Universe With the James Webb Space Telescope. (2024). https://www.jpl.nasa.gov/edu/resources/teachable-moment/learn-about-the-universe-with-the-james-webb-space-telescope/

LIGO detected gravitational waves from neutron stars colliding - NPR. (2024). https://www.npr.org/2024/11/08/1211597243/neutron-star-gravitational-waves-detected-ligo

LIGO-Virgo-KAGRA (LVK) Collaboration Detected a Remarkable ... (2024). https://www.ligo.caltech.edu/news/ligo20240405

Mars 2020: Perseverance Rover - NASA Science. (2021). https://science.nasa.gov/mission/mars-2020-perseverance/

Massive Primordial Galaxies Found Swimming in Vast Ocean of ... (2017). https://public.nrao.edu/news/2017-alma-galaxies-dark-matter/

Mission Updates | News – NASA's Europa Clipper. (2022). https://europa.nasa.gov/news/mission-updates/

Model of ever-expanding universe confirmed by dark energy probe. (2024). https://www.science.org/content/article/model-ever-expanding-universe-confirmed-dark-energy-probe

NASA Europa Clipper (@EuropaClipper) / X. (n.d.). https://x.com/europaclipper?lang=en

NASA's Europa Clipper Mission Launches From Kennedy Space ... (2024). https://www.jpl.nasa.gov/videos/nasas-europa-clipper-mission-launches-from-kennedy-space-center-highlights/

NASA's Hubble Finds More Black Holes than Expected in the Early ... (2024). https://science.nasa.gov/missions/hubble/nasas-hubble-finds-more-black-holes-than-expected-in-the-early-universe/

NASA's James Webb Space Telescope Finds Most Distant Known ... (2024). https://webbtelescope.org/contents/early-highlights/nasas-james-webb-space-telescope-finds-most-distant-known-galaxy

New Detectable Gravitational Wave Source From Collapsing Stars ... (2024). https://www.simonsfoundation.org/2024/08/22/new-detectable-gravitational-wave-source-from-collapsing-stars-predicted-from-simulations/

New Directions in Philosophy of Cosmology. (2018). https://www.templeton.org/news/new-directions-philosophy-cosmology

New Insights from Cosmic Microwave Background - Editverse. (2024). https://editverse.com/cosmic-microwave-background/

New research challenges black holes as dark matter explanation. (2024). https://phys.org/news/2024-06-black-holes-dark-explanation.html

New technology is a "science multiplier" for astronomy - Phys.org. (2020). https://phys.org/news/2020-09-technology-science-astronomy.html

New Tools (Cosmology - American Institute of Physics. (1999). https://history.aip.org/exhibits/cosmology/tools/tools-new.htm

News | LIGO Lab | Caltech. (2021). https://www.ligo.caltech.edu/news

Observational Cosmology: Serjeant, Stephen - Amazon.com. (2010). https://www.amazon.com/Observational-Cosmology-Stephen-Serjeant/dp/0521157153

Origin of the Universe: How Did It Begin and How Will It End? (2024). https://www.apu.apus.edu/area-of-study/math-and-science/resources/origin-of-the-universe/

Our Expanding Universe: Delving into Dark Energy. (2017). https://www.energy.gov/science/articles/our-expanding-universe-delving-dark-energy

Our Mysterious Universe Still Evades Cosmological Understanding. (2023). https://www.aura-astronomy.org/blog/2023/03/06/our-mysterious-universe-still-evades-cosmological-understanding/

(PDF) AI-Assisted Astronomical Data Analysis Unveiling Patterns ... (2024). https://www.researchgate.net/publication/379600868_AI-Assisted_Astronomical_Data_Analysis_Unveiling_Patterns_and_Phenomena_in_the_Universe

(PDF) Future directions in cosmology - ResearchGate. (2024). https://www.researchgate.net/publication/385594785_Future_directions_in_cosmology

[PDF] Galaxy Formation and Evolution. - UMD Astronomy. (2024). https://www.astro.umd.edu/~richard/ASTRO620/MBW_Book_Galaxy.pdf

[PDF] Sharing space resources - Secure World Foundation. (n.d.). https://swfound.org/media/55858/al_crowdsourcing_article.pdf

[PDF] Understanding the Early Universe From the Cosmic Microwave ... (2024). https://ijsred.com/volume7/issue4/IJSRED-V7I4P99.pdf

Perseverance, NASA's newest Mars rover - The Planetary Society. (2019). https://www.planetary.org/space-missions/perseverance

Perseverance Rover Updates - NASA Science. (2024). https://science.nasa.gov/mission/mars-2020-perseverance/science-updates/

Philosophy of Cosmology. (2017). https://plato.stanford.edu/entries/cosmology/

Physicists discover first "black hole triple" - MIT News. (2024). https://news.mit.edu/2024/physicists-discover-first-black-hole-triple-1023

Primordial Galaxy Discovered, First of Its Kind. (2017). https://www.keckobservatory.org/primordial_galaxy_discovered_first_of_its_kind/

Primordial galaxy discovered is first of its kind. (2017). https://www.unimelb.edu.au/newsroom/news/2017/april/primordial-galaxy-discovered-is-first-of-its-kind

Program News and Announcements - Cosmic Origins (COR) - NASA. (2024). https://cor.gsfc.nasa.gov/news/2024/90_NASA_New_Class_of_Astrophysics_Missions.php

Radial Velocity Method - Las Cumbres Observatory. (2024). https://lco.global/spacebook/exoplanets/radial-velocity-method/

Radio astronomer answers biggest, trickiest questions about the ... (2024). https://www.skyatnightmagazine.com/space-science/questions-about-universe

Researchers study a million galaxies to find out how the universe ... (2023). https://www.sciencedaily.com/releases/2023/12/231222145443.htm

Revolutionizing astronomy: Technological advancements and future ... (2024). https://www.researchgate.net/publication/382426056_Revolutionizing_astronomy_Technological_advancements_and_future_horizons

Shining a Light on Dark Matter - NASA Science. (2021). https://science.nasa.gov/mission/hubble/science/science-highlights/shining-a-light-on-dark-matter/

Space exploration - International, Cooperation, Astronauts | Britannica. (2024). https://www.britannica.com/science/space-exploration/International-participation

Space Exploration Missions | The Planetary Society. (2019). https://www.planetary.org/space-missions

Space station missions highlight importance of international ... (2024). https://washingtondc.jhu.edu/news/space-station-missions-highlight-importance-of-international-collaboration/

Study: Early dark energy could resolve cosmology's two biggest ... (2024). https://news.mit.edu/2024/study-early-dark-energy-could-resolve-cosmologys-two-biggest-puzzles-0913

Telescope - Technology, Optics, Astronomy | Britannica. (2024). https://www.britannica.com/science/optical-telescope/Impact-of-technological-developments

The 10 most Earth-like exoplanets - Space.com. (2024). https://www.space.com/30172-six-most-earth-like-alien-planets.html

The Best Books on Cosmology - Five Books Expert Recommendations. (2022). https://fivebooks.com/best-books/cosmology-david-goldberg/

The Extravagant Universe | Princeton University Press. (n.d.). https://press.princeton.edu/books/paperback/9780691173184/the-extravagant-universe?srslti d=AfmBOorEpyPqiHbEytaOtNKOh4omt4dKKWGj-_KduSNzlYjmve1WnDbM

The Extravagant Universe: Exploding Stars, Dark Energy, and the ... (2014). https://www.amazon.com/Extravagant-Universe-Exploding-Accelerating-Princeton/dp/069111742X

The Habitable Zone - NASA Science. (2020). https://science.nasa.gov/exoplanets/habitable-zone/

The Little Book of Cosmology | Princeton University Press. (n.d.). https://press.princeton.edu/books/hardcover/9780691195780/the-little-book-of-cosmology?s rsltid=AfmBOorN88PilgzEQqhqhJuL229KpVv2JWwQ43z_mthdpjgffivylFUT

The most Earth-like exoplanets - The Planetary Society. (2023). https://www.planetary.org/video/the-most-earth-like-exoplanets

The past decade and the future of cosmology and astrophysics. (2019). https://www.technologyreview.com/2019/08/05/133873/a-transcendent-decade-the-past-decade-and-the-future-of-cosmology-and-astrophysics/

The Smallest Black Holes and Biggest Neutron Stars | astrobites. (2024). https://astrobites.org/2024/05/20/the-smallest-black-holes-and-biggest-neutron-stars/

The Technology of Cosmology - NASA ADS. (n.d.). https://adsabs.harvard.edu/full/2001ASPC..252...55L

Astronomy News Stories of 2023 - Sky & Telescope. (2023). https://skyandtelescope.org/astronomy-news/top-10-astronomy-news-stories-of-2023/

Space stories of 2023 - Astronomy Magazine. (2024). https://www.astronomy.com/science/top-10-space-stories-of-2023/

Transit Method - Las Cumbres Observatory. (n.d.). https://lco.global/spacebook/exoplanets/transit-method/

Transit Method - NASA Science. (2023). https://science.nasa.gov/mission/roman-space-telescope/transit-method/

Upcoming Planetary Events and Missions - the NSSDCA. (2024). https://nssdc.gsfc.nasa.gov/planetary/upcoming.html

Webb telescope's largest study of universe expansion confirms ... (2024). https://hub.jhu.edu/2024/12/09/webb-telescope-hubble-tension-universe-expansion/

What is Dark Energy? Inside our accelerating, expanding Universe. (2024). https://science.nasa.gov/universe/the-universe-is-expanding-faster-these-days-and-dark-energy-is-responsible-so-what-is-dark-energy/

What Is Dark Matter, and Where Is it Hiding? - Caltech Magazine. (2020). https://magazine.caltech.edu/post/where-is-dark-matter-hiding

What Is the Habitable Zone? | The Planetary Society. (2020). https://www.planetary.org/articles/what-is-the-habitable-zone

What is the Radial Velocity Method? - Universe Today. (2017). https://www.universetoday.com/138014/radial-velocity-method/

What makes a planet habitable - SEEC - NASA. (2024). https://seec.gsfc.nasa.gov/what_makes_a_planet_habitable.html

What's a transit? - NASA Science. (2020). https://science.nasa.gov/exoplanets/whats-a-transit/

Why We Search - NASA Science. (2020). https://science.nasa.gov/exoplanets/why-we-search/

Window to the Early Universe: The Cosmic Microwave Background. (2024). https://sedsvit.medium.com/window-to-the-early-universe-the-cosmic-microwave-background-42199022db97

2024 - Astrophysics Projects Division - NASA. (n.d.). https://apd440.gsfc.nasa.gov/archive.html

2024 Exoplanet Archive News. (2024). https://exoplanetarchive.ipac.caltech.edu/docs/exonews_archive.html

Analysis Of Habitability And Stellar Habitable Zones From Observed ... (2024). https://astrobiology.com/2024/08/analysis-of-habitability-and-stellar-habitable-zones-from-observed-exoplanets.html

Astronomers discover the fastest-feeding black hole in the early ... (2024). https://www.sciencedaily.com/releases/2024/11/241104112032.htm

Astronomers' theory of galaxy formation may be upended. (2024). https://thedaily.case.edu/james-webb-space-telescope-results-may-upend-astronomers-theory-of-galaxy-formation/

Black Holes News - ScienceDaily. (2024). https://www.sciencedaily.com/news/space_time/black_holes/

Brief review of recent advances in understanding dark matter and ... (n.d.). https://www.sciencedirect.com/science/article/pii/S1387647321000191

Cosmology News - ScienceDaily. (2024). https://www.sciencedaily.com/news/space_time/cosmology/

Dark Energy - latest research news and features - Phys.org. (2024). https://phys.org/tags/dark+energy/

Dark energy, explained - UChicago News - The University of Chicago. (2024). https://news.uchicago.edu/explainer/dark-energy-explained

Dark matter | CERN. (2024). https://home.cern/science/physics/dark-matter

Dark Matter - latest research news and features - Phys.org. (2024). https://phys.org/tags/dark+matter/

Dark Matter Hub - Interactions.org. (2024). https://interactions.org/hub/dark-matter-hub

Dark Matter News - ScienceDaily. (2024). https://www.sciencedaily.com/news/space_time/dark_matter/

Discoveries Dashboard - NASA Science. (2024). https://science.nasa.gov/exoplanets/discoveries-dashboard/

Discovery Alert: A "Super-Earth" in the Habitable Zone - NASA Science. (2024). https://science.nasa.gov/universe/exoplanets/discovery-alert-a-super-earth-in-the-habitable-zone/

Evidence mounts for dark energy from black holes | ScienceDaily. (2024). https://www.sciencedaily.com/releases/2024/10/241028131405.htm

Examining how stellar threats impact the habitable zone of exoplanets. (2024). https://phys.org/news/2024-11-stellar-threats-impact-habitable-zone.html

"Exoplanets" and the Search for Habitable Worlds – Teachable Moment. (2024). https://www.jpl.nasa.gov/edu/resources/teachable-moment/exoplanets-and-the-search-for-habitable-worlds/

Extrasolar Planets News - ScienceDaily. (2024). https://www.sciencedaily.com/news/space_time/extrasolar_planets/

First-ever Discovery of a "Black Hole Triple" Amazes Astronomers. (2024). https://www.newsweek.com/discovery-black-hole-triple-first-ever-discovery-space-1974447

Gravitational wave signatures of departures from classical black ... (2024). https://link.aps.org/doi/10.1103/PhysRevD.110.064029

International experiment sees new results in search for dark matter. (2024). https://www.psu.edu/news/eberly-college-science/story/international-experiment-sees-new-results-search-dark-matter

James Webb Space Telescope finds galaxies pointing toward a dark ... (2024). https://www.space.com/space-exploration/james-webb-space-telescope/james-webb-space-telescope-finds-galaxies-pointing-toward-a-dark-matter-alternative

LIGO-Virgo-KAGRA (LVK) Collaboration Detected a Remarkable ... (2024). https://www.ligo.caltech.edu/news/ligo20240405

LZ experiment sets new record in search for dark matter. (2024). https://news.umich.edu/lz-experiment-sets-new-record-in-search-for-dark-matter/

LZ Experiment Sets New Record in Search for Dark Matter | UC Davis. (2024). https://www.ucdavis.edu/curiosity/news/lz-experiment-sets-new-record-search-dark-matter

Monthly Roundup: Gravitational Wave Predictions and Comparisons. (2024). https://aasnova.org/2024/11/27/monthly-roundup-gravitational-wave-predictions-and-comparisons/

More evidence for black holes as the source of dark energy - EarthSky. (2024). https://earthsky.org/space/black-holes-as-the-source-dark-energy/

NASA's Hubble Finds More Black Holes than Expected in the Early ... (2024). https://hubblesite.org/contents/news-releases/2024/news-2024-032

NASA's Webb Finds Planet-Forming Disks Lived Longer in Early ... (2024a). https://science.nasa.gov/missions/webb/nasas-webb-finds-planet-forming-disks-lived-longer-in-early-universe/

NASA's Webb Finds Planet-Forming Disks Lived Longer in Early ... (2024b). https://webbtelescope.org/contents/news-releases/2024/news-2024-135

New Detectable Gravitational Wave Source From Collapsing Stars ... (2024). https://www.simonsfoundation.org/2024/08/22/new-detectable-gravitational-wave-source-from-collapsing-stars-predicted-from-simulations/

New evidence suggests that dark energy comes from black holes. (2024). https://www.earth.com/news/new-evidence-suggests-that-dark-energy-comes-from-black-holes/

New Findings Reveal Clearer Picture of Expanding Universe. (2024). https://news.utdallas.edu/science-technology/desi-findings-2024/

New map of the universe uses gravitational waves to reveal hidden ... (2024). https://phys.org/news/2024-12-universe-gravitational-reveal-hidden-black.html

New research challenges black holes as dark matter explanation. (2024). https://phys.org/news/2024-06-black-holes-dark-explanation.html

Physicists discover first "black hole triple" - MIT News. (2024). https://news.mit.edu/2024/physicists-discover-first-black-hole-triple-1023

Press Releases 2024 - ESA/Hubble. (2024). https://esahubble.org/news/archive/year/2024/

Press Releases 2024 - ESA/Webb. (2024). https://esawebb.org/news/archive/year/2024/

Press Releases 2024 - ESO.org. (1999). https://www.eso.org/public/news/archive/year/2024/

Shining a Light on Dark Matter - NASA Science. (2021). https://science.nasa.gov/mission/hubble/science/science-highlights/shining-a-light-on-dark-matter/

Team discovers ultra-massive galaxies in early Universe that ... (2024). https://news.ucsc.edu/2024/11/red-monsters.html

Webb Finds Early Galaxies Weren't Too Big for Their Britches After All. (2024). https://science.nasa.gov/missions/webb/webb-finds-early-galaxies-werent-too-big-for-their-britches-after-all/

Webb finds planet-forming discs lived longer in early Universe. (2024). https://esawebb.org/news/weic2430/

What is Dark Energy? Inside our accelerating, expanding Universe. (2024). https://science.nasa.gov/universe/the-universe-is-expanding-faster-these-days-and-dark-energy-is-responsible-so-what-is-dark-energy/

With Six New Worlds, 5500 Discovery Milestone Passed! (2024). https://science.nasa.gov/universe/exoplanets/discovery-alert-with-six-new-worlds-5500-discovery-milestone-passed/

Data Analytics Trends In 2024 - Exploding Topics. (2024). https://explodingtopics.com/blog/data-analytics-trends

data and analytics trends for 2024 from ThoughtSpot. (2024). https://www.thoughtspot.com/resources/ebook/data-analytics-trends-2024

AI and machine learning trends for 2024 | TechTarget. (2024). https://www.techtarget.com/searchenterpriseai/tip/9-top-AI-and-machine-learning-trends

2024 Trends in AI and Machine Learning | by Hunter Kempf - Medium. (2024). https://medium.com/@HunterKempf/2024-trends-in-ai-and-machine-learning-58f45a2007ca

Advancements in Artificial Intelligence and Machine Learning. (2024). https://online-engineering.case.edu/blog/advancements-in-artificial-intelligence-and-machine-learning

Artificial Intelligence - The most important AI trends in 2024 - IBM. (2024). https://www.ibm.com/think/insights/artificial-intelligence-trends

Big Data & Machine Learning (How Do They Relate?) - WEKA. (2021). https://www.weka.io/learn/glossary/ai-ml/big-data-machine-learning/

Big Data Analytics Trends in 2025 - XenonStack. (2024). https://www.xenonstack.com/blog/latest-trends-in-big-data-analytics

Big Data Trends 2024: Navigating the Future of Data Technology. (2024). https://innowise.com/blog/big-data-trends-2024/

Data science recent news | AI Business. (2024). https://aibusiness.com/data/data-science

Difference between Big Data and Machine Learning - GeeksforGeeks. (2023). https://www.geeksforgeeks.org/difference-between-big-data-and-machine-learning/

Emerging Data Analytics Trends to Watch in 2024 and Beyond. (2024). https://www.agilisium.com/blogs/emerging-data-analytics-trends-to-watch-in-2024-and-beyond

Exploring the Latest Vital Trends in Big Data Analytics - Ksolves. (2024). https://www.ksolves.com/blog/big-data/exploring-the-6-vital-trends-in-big-data-analytics

Five Key Trends in AI and Data Science for 2024. (2024). https://sloanreview.mit.edu/article/five-key-trends-in-ai-and-data-science-for-2024/

Gartner Identifies the Top Trends in Data and Analytics for 2024. (2024). https://www.gartner.com/en/newsroom/press-releases/2024-04-25-gartner-identifies-the-top-trends-in-data-and-analytics-for-2024

How Is Big Data Analytics Using Machine Learning? - Forbes. (2020). https://www.forbes.com/councils/forbestechcouncil/2020/10/20/how-is-big-data-analytics-using-machine-learning/

Machine learning discoveries honoured with 2024 Nobel Prize for ... (2024). https://www.embl.org/news/science-technology/machine-learning-discoveries-honoured-with-2024-nobel-prize-for-physics/

Machine Learning for Big Data | Udacity. (2020). https://www.udacity.com/blog/2020/08/machine-learning-for-big-data.html

Machine Learning Trends & Stats for 2024 - Encord. (2024). https://encord.com/blog/machine-learning-trends-statistics/

Machine Learning With Big Data | Coursera. (2021). https://www.coursera.org/learn/big-data-machine-learning

Our Impact on Big Data. (2024). http://ncats.nih.gov/research/our-impact/our-impact-big-data

(PDF) Machine Learning in Big Data - ResearchGate. (2024). https://www.researchgate.net/publication/327386452_Machine_Learning_in_Big_Data

[PDF] Machine learning on big data: Opportunities and challenges. (n.d.). https://www.sciencedirect.com/science/article/am/pii/S0925231217300577

Press release: The Nobel Prize in Physics 2024 - NobelPrize.org. (2024). https://www.nobelprize.org/prizes/physics/2024/press-release/

The discovery of tools key to machine learning wins the 2024 ... (2024). https://www.sciencenews.org/article/nobel-physics-2024-machine-learning

Breakthroughs in AI and Machine Learning for 2024 - Evonence. (2023). https://www.evonence.com/blog/top-5-breakthroughs-in-ai-and-machine-learning-for-2024

Machine Learning Trends for 2024 | Zuci Systems. (2022). https://www.zucisystems.com/blog/top-8-machine-learning-trends-for-2022/

Big Data Trends For 2025 | The Future of ... - GeeksforGeeks. (2024). https://www.geeksforgeeks.org/top-10-big-data-trends/

Data Science Trends in 2025 | Emerging & Future Trends. (2021). https://www.zucisystems.com/blog/top-10-data-science-trends-for-2022/

Machine Learning Trends To Impact Business in 2025. (2024). https://mobidev.biz/blog/future-machine-learning-trends-impact-business

Data Science Breakthroughs in 2021 - DSS Blog. (2021). https://roundtable.datascience.salon/top-data-science-breakthroughs-in-2021

Foundations and Trends in Machine Learning for 2024 - Kumo.ai. (2024). https://kumo.ai/learning-center/top-foundations-and-trends-in-machine-learning-what-you-need-to-know/

What is Big Data and Machine Learning - Javatpoint. (2021). https://www.javatpoint.com/what-is-big-data-and-machine-learning

Photo and Image Credits

All photos and images are from the author's archive. Every effort has been made to identify copyright holders; in case of oversight and on notification to the publisher, corrections will be made in the next edition.

[1] Book Cover Images Courtesy of NASA/JPL, 2007

[2] Image **Aristotle**—page 2—
Scientist Images—http:// http://www.123rf.com/stock-photo/scientist.html

[3] Image **Democritus**—
page 6—Scientist Images—http:// http://www.123rf.com/stock-photo/scientist.html

[4] Image **Democritus**—
page 6—Scientist Images—http://www.sciencekids.co.nz/pictures/scientists.html

[5] Image **Pythagoras**—
page 8—Scientist Images—http://www.sciencekids.co.nz/pictures/scientists.html

[6] Image **Archimedes**—
page 10—Scientist Images—http://www.sciencekids.co.nz/pictures/scientists.html

[7] Image **Hypatia**—page 14—
Scientist Images—http://www.123rf.com/stock-photo/scientist.html

[8] Image **Newton**—page 21—
Scientist Images—http://www.123rf.com/stock-photo/scientist.html

[9] Image **Huygens**—page 45—
Scientist Images—http://www.sciencekids.co.nz/pictures/scientists.html

[10] Image **Sir Isaac Newton**—page 46—Scientist Images—
http://www.sciencekids.co.nz/pictures/scientists.html

[11] Image **Young**—page 50—
Scientist Images—http://www.sciencekids.co.nz/pictures/scientists.html

[12] Image **Probability vs Determinism**—
Courtesy of the Niels Bohr Copenhagen Institute

[13] **Einstein around 1940**—Courtesy of California Institute of Technology and the Hebrew University of Jerusalem

[14] **Einstein around 1950**—Courtesy of California Institute of Technology and the Hebrew University of Jerusalem

[15] **Albert Einstein at age 4**—Courtesy of California Institute of Technology and the Hebrew University of Jerusalem

[16] **Einstein as a teenager**—Courtesy of California Institute of Technology and the Hebrew University of Jerusalem

[17] **Einstein at age 25**—Courtesy of California Institute of Technology and the Hebrew University of Jerusalem

[18] **Minkowski**—http://www.123rf.com/stock-photo/scientist.html

[19] **Schwarz**—Courtesy of California Institute of Technology

[20] **Witten**—Scientist Images—http://www.123rf.com/stock-photo/scientist.html

[21] **Multiverse**—Scientist Images—http://www.sciencekids.co.nz/pictures/scientists.html

Printed in the United States
by Baker & Taylor Publisher Services